Non-Newtonian Flow and Applied Rheology

Non-Newtonian Flow and Applied Rheology: Engineering Applications

Second Edition

R.P. Chhabra
Department of Chemical Engineering
Indian Institute of Technology
Kanpur
India

and

J.F. Richardson
Department of Chemical and Biological Process Engineering
University of Wales, Swansea
Swansea
Great Britain

ELSEVIER

AMSTERDAM • BOSTON • HEIDELBERG • LONDON • NEW YORK • OXFORD
PARIS • SAN DIEGO • SAN FRANCISCO • SINGAPORE• SYDNEY • TOKYO
Butterworth-Heinemann is an imprint of Elsevier

Butterworth-Heinemann is an imprint of Elsevier
Linacre House, Jordan Hill, Oxford OX2 8DP, UK
30 Corporate Drive, Suite 400, Burlington, MA 01803, USA

First published 1999

Second edition 2008

British Library Cataloguing in Publication Data
A catalogue record for this book is available from the British Library

Library of Congress Cataloging-in-Publication Data
A catalog record for this book is available from the Library of Congress

ISBN: 978-0-7506-8532-0

For information on all Butterworth–Heinemann publications
visit our website at books.elsevier.com

Typeset by Charon Tec Ltd., A Macmillan Company.
(www.macmillansolutions.com)

Printed and bound in Hungary
08 09 10 11 12   10 9 8 7 6 5 4 3 2

Contents

Preface to Second Edition

In presenting this new edition, we would like to thank the many individuals from all over the world who have pointed out errors, and, more importantly, have made suggestions for improvement in the text. Bearing in mind these ideas together with the points raised in independent reviews of the first edition, the entire text has been reviewed. Therefore where the need was recognized, the presentation has been improved by re-organizing, or by expanding the existing material, or by adding new text and/or illustrative examples throughout the whole book to facilitate comprehension of the material covered here. Apart from the overall general updating to include state-of-the-art information, the specific changes made to the first edition are summarized here. The discussion in *Chapter 1* has been sharpened to highlight the wide occurrence of, and the effects arising from, non-Newtonian flow behaviour in diverse industrial settings including food, pharmaceutical, personal care and household products, as well as in biomedical and biological process engineering applications. Included here is also a new section to emphasize the importance of the intimate link between the micro-structure and rheology of a fluid which is used widely to formulate new products, especially in food and personal-care product sectors to meet ever increasing expectations of the consumers. *Chapter 2* has been completely re-vamped to highlight the importance of proper rheological characterization of complex fluids, together with a discussion on the relative merits and demerits of various rheological techniques in current use. Conversely, online rheological measurements are increasingly used to monitor product quality continuously. This topic is addressed in a new section on online viscometry in this chapter. While very little new material has been added to *Chapter 3*, the existing coverage has been expanded significantly by providing detailed derivations of important relationships in order to improve clarity and several new illustrative examples have been introduced here. The role of non-Newtonian characteristics in continuous thermal treatment of foods is highlighted in a new section on the residence time distributions of the two phases in *Chapter 4*. In addition, this chapter provides a short introduction to the two-phase flow of gas and drag reducing polymer solutions in pipes, as practised in the oil industry. In *Chapter 5*, the existing discussion on the estimation of drag or the terminal falling velocity of non-spherical (isometric) particles settling in non-Newtonian media has been expanded to reflect the recent developments in this field. *Chapters 6 and 7*, which have been accepted in their existing form, have remain largely unchanged. Finally, the discussion on liquid mixing in *Chapter 8* has been strengthened in general by adding several new examples and by adding more detailed discussions on static mixers, novel impeller designs and the prediction of mixing times in particular. Lastly, several new exercises have been added at the end of the book for the benefit of the student.

With the rapid advances occurring in this vast interdisciplinary field, both the selection of material and its arrangement are becoming increasingly difficult, and must be to a great extent a matter of personal choice, but we hope that this new edition will continue to provide a sound basis for a study of the fundamentals of the subject and will also be of some value to practising professionals who must deal with such difficult materials on a day-to-day basis. In closing, we very much hope that our readers will continue to make suggestions for further improvements in this work.

R.P. Chhabra
J.F. Richardson

March 2008

Preface to First Edition

Non-Newtonian flow and rheology are subjects which are essentially interdisciplinary in their nature and which are also wide in their areas of application. Indeed non-Newtonian fluid behaviour is encountered in almost all the chemical and allied processing industries. The factors which determine the rheological characteristics of a material are highly complex, and their full understanding necessitates a contribution from physicists, chemists and applied mathematicians, amongst others, few of whom may have regarded the subject as central to their disciplines. Furthermore, the areas of application are also extremely broad and diverse, and require an important input from engineers with a wide range of backgrounds, though chemical and process engineers, by virtue of their role in the handling and processing of complex materials (such as foams, slurries, emulsions, polymer melts and solutions, etc.), have a dominant interest. Furthermore, the subject is of interest both to highly theoretical mathematicians and scientists and to practising engineers with very different cultural backgrounds.

Owing to this inter-disciplinary nature of the subject, communication across subject boundaries has been poor and continues to pose difficulties, and therefore, much of the literature, including books, is directed to a relatively narrow readership with the result that the engineer faced with the problem of processing such rheological complex fluids, or of designing a material with rheological properties appropriate to its end use, is not well served by the available literature. Nor does he have access to information presented in a form which is readily intelligible to the non-specialist. This book is intended to bridge this gap but, at the same time, is written in such a way as to provide an entrée to the specialist literature for the benefit of scientists and engineers with a wide range of backgrounds. Non-Newtonian flow and rheology is an area with many pitfalls for the unwary, and it is hoped that this book will not only forewarn readers but also equip them to avoid some of the hazards.

Coverage of topics is extensive and this book offers a unique selection of material. There are eight chapters in all.

The introductory material, *Chapter 1*, introduces the reader to the range of non-Newtonian characteristics displayed by materials encountered in every day life as well as in technology. A selection of simple fluid models which are used extensively in process design calculations is included here.

Chapter 2 deals with the characterization of materials and the measurement of their rheological properties using a range of commercially available instruments. The importance of adequate rheological characterization of a material under conditions as close as possible to that in the envisaged application cannot be overemphasized here. Stress is laid on the dangers of extrapolation beyond the range of variables covered in the experimental characterization. Dr. P.R. Williams (Reader, Department of Chemical Biological Process Engineering, Swansea, University of Wales, UK) who has contributed this chapter is in the forefront of the development of novel instrumentations in the field.

The flow of non-Newtonian fluids in circular and non-circular ducts encompassing both laminar and turbulent regimes is presented in *Chapter 3*. Issues relating to the transition from laminar to turbulent flow, minor losses in fittings and flow in pumps, as well as metering of flow, are also discussed in this chapter.

Chapter 4 deals with the highly complex but industrially important topic of multiphase systems – gas/non-Newtonian liquid and solid/non-Newtonian liquids – in pipes.

A thorough treatment of particulate systems ranging from the behaviour of particles and drops in non-Newtonian liquids to the flow in packed and fluidized beds is presented in *Chapter 5*.

The heating or cooling of process streams is frequently required. *Chapter 6* discusses the fundamentals of convective heat transfer to non-Newtonian fluids in circular and non-circular tubes under a range of boundary and flow conditions. Limited information on heat transfer from variously shaped objects – plates, cylinders and spheres – immersed in non-Newtonian fluids is also included here.

The basics of the boundary layer flow are introduced in *Chapter 7*. Heat and mass transfer in boundary layers and practical correlations for the estimation of transfer coefficients are included.

The final *Chapter 8* deals with the mixing of highly viscous and/or non-Newtonian substances, with particular emphasis on the estimation of power consumption and mixing time, and on equipment selection.

At each stage, considerable effort has been made to present the most reliable and generally accepted methods for calculations, as the contemporary literature is inundated with conflicting information. This applies especially in regard to the estimation of pressure gradients for turbulent flow in pipes. In addition, a list of specialist and/or advanced sources of information has been provided in each chapter as "Further reading".

In each chapter a number of worked examples have been presented, which, we believe, are essential to a proper understanding of the methods of treatment given in the text. It is desirable for both a student and a practising engineer to understand an appropriate illustrative example before tackling fresh practical problems themself. Engineering problems require a numerical answer and it is thus essential for the reader to become familiar with the various techniques so that the most appropriate answer can be obtained by systematic methods rather than by intuition. Further exercises which the reader may wish to tackle are given at the end of the book.

Incompressibility of the fluid has generally been assumed throughout the book, albeit this is not always stated explicitly. This is a satisfactory approximation for most non-Newtonian substances, notable exceptions being the cases of foams and froths. Likewise, the assumption of isotropy is also reasonable in most cases except perhaps for liquid crystals and for fibre-filled polymer matrices. Finally, although the slip effects are known to be important in some multiphase systems (suspensions, emulsions, etc.) and in narrow channels, the usual no-slip boundary condition is regarded as a good approximation in the type of engineering flow situations dealt with in this book.

In part, the writing of this book was inspired by the work of W.L. Wilkinson *Non-Newtonian Fluids*, published by Pergamon Press in 1960, and now long out-of-print, and it is hoped that readers will find it to be a welcome successor.

R.P. Chhabra
J.F. Richardson

Acknowledgements (First Edition)

The inspiration for this book originated in two works which have long been out-of-print and which have been of great value to those working and studying in the field of non-newtonian technology. They are W.L. Wilkinson's excellent introductory book, *Non-Newtonian Flow* (Pergamon Press, 1959), and J.M. Smith's chapter in the first two editions of Coulson and Richardson's *Chemical Engineering, Volume 3* (Pergamon Press, 1970 and 1978). The original intention was that R.P. Chhabra would join with the above two authors in the preparation of a successor but, unfortunately, neither of them had the necessary time available to devote to the task, and Raj Chhabra agreed to proceed on his own with my assistance. We would like to thank Bill Wilkinson and John Smith for their encouragement and support.

The chapter on Rheological Measurements has been prepared by Dr. P.R. Williams, Reader in the Department of Chemical and Biochemical Process Engineering at the University of Wales, Swansea – an expert in the field. Thanks are due also to Dr. D.G. Peacock, formerly of the School of Phamacy, University of London, for work on the compilation and processing of the Index.

J.F. Richardson
January 1999

Chapter 1
Non-Newtonian fluid behaviour

1.1 Introduction

One may classify fluids in two different ways; either according to their response to the externally applied pressure or according to the effects produced under the action of a shear stress. The first scheme of classification leads to the so called 'compressible' and 'incompressible' fluids, depending upon whether or not the volume of an element of fluid is dependent on its pressure. While compressibility influences the flow characteristics of gases, liquids can normally be regarded as incompressible and it is their response to shearing which is of greater importance. In this chapter, the flow characteristics of single-phase liquids, solutions and pseudo-homogeneous mixtures (such as slurries, emulsions, gas–liquid dispersions) which may be treated as a continuum if they are stable in the absence of turbulent eddies are considered depending upon their response to externally imposed shearing action.

1.2 Classification of fluid behaviour

1.2.1 Definition of a Newtonian fluid

Consider a thin layer of a fluid contained between two parallel planes a distance dy apart, as shown in Figure 1.1. Now, if under steady state conditions, the fluid is subjected to a shear by the application of a force F as shown, this will be balanced by an equal and opposite internal frictional force in the fluid. For an incompressible Newtonian fluid in laminar flow, the resulting shear stress is equal to the product of the shear rate and the viscosity of the fluid medium. In this simple case, the shear rate may be expressed as the velocity gradient in the direction perpendicular to that of the shear force, i.e.,

$$\frac{F}{A} = \tau_{yx} = \mu\left(-\frac{dV_x}{dy}\right) = \mu\dot{\gamma}_{yx} \tag{1.1}$$

Figure 1.1 *Schematic representation of unidirectional shearing flow*

Note that the first subscript on both τ and $\dot{\gamma}$ indicates the direction normal to that of shearing surface, while the second subscript refers to the direction of the force and the flow. By considering the equilibrium of a fluid layer, it can readily be seen that at any shear plane there are two equal and opposite shear stresses – a positive one on the slower moving fluid and a negative one on the faster moving fluid layer. The negative sign on the right hand side of equation (1.1) indicates that τ_{yx} is a measure of the resistance to motion. One can also view the situation from a different standpoint as: for an incompressible fluid of density ρ, equation (1.1) can be written as:

$$\tau_{yx} = -\frac{\mu}{\rho} \frac{d}{dy}(\rho V_x) \qquad (1.2)$$

The quantity 'ρV_x' is the linear momentum in the x-direction per unit volume of the fluid and hence τ_{yx} represents the momentum flux in the y-direction and the negative sign indicates that the momentum transfer occurs in the direction of decreasing velocity which is also in line with the Fourier's law of heat transfer and Fick's law of diffusive mass transfer.

The constant of proportionality, μ (or the ratio of the shear stress to the rate of shear) which is called the Newtonian viscosity is, by definition, independent of shear rate ($\dot{\gamma}_{yx}$) or shear stress (τ_{yx}) and depends only on the material and its temperature and pressure. The plot of shear stress (τ_{yx}) against shear rate ($\dot{\gamma}_{yx}$) for a Newtonian fluid, the so-called 'flow curve' or 'rheogram', is therefore a straight line of slope, μ, and passing through the origin; the single constant, μ, thus completely characterizes the flow behaviour of a Newtonian fluid at a fixed temperature and pressure. Gases, simple organic liquids, solutions of low molecular weight inorganic salts, molten metals and salts are all Newtonian fluids. The shear stress–shear rate data shown in Figure 1.2 demonstrate the Newtonian fluid behaviour of a cooking oil and a corn syrup; the values of the viscosity for some substances encountered in everyday life are given in Table 1.1.

Figure 1.1 and equation (1.1) represent the simplest case wherein the velocity vector which has only one component, in the x-direction and it varies only in the y-direction. Such a flow configuration is known as simple shear flow. For the more complex case of three-dimensional flow, it is necessary to set up the appropriate partial differential equations. For instance, the more general case of an incompressible Newtonian fluid may be expressed – for the x-plane (area oriented normal to the x-direction) – as follows (Bird *et al.*, 1987, 2002):

$$\tau_{xx} = -2\mu\frac{\partial V_x}{\partial x} + \frac{2}{3}\mu\left(\frac{\partial V_x}{\partial x} + \frac{\partial V_y}{\partial y} + \frac{\partial V_z}{\partial z}\right) \qquad (1.3)$$

$$\tau_{xy} = -\mu\left(\frac{\partial V_x}{\partial y} + \frac{\partial V_y}{\partial x}\right) \qquad (1.4)$$

$$\tau_{xz} = -\mu\left(\frac{\partial V_x}{\partial z} + \frac{\partial V_z}{\partial x}\right) \qquad (1.5)$$

Similar sets of equations can be drawn up for the forces acting on the y- and z-planes; in each case, there are two (in-plane) shearing components and a normal component.

Figure 1.2 *Typical shear stress–shear rate data for a cooking oil and a corn syrup*

Figure 1.3 shows the nine stress components schematically in an element of fluid. By considering the equilibrium of a fluid element, it can be shown that $\tau_{yx} = \tau_{xy}$; $\tau_{xz} = \tau_{zx}$ and $\tau_{yz} = \tau_{zy}$. The normal stresses can be visualized as being made up of two components: isotropic pressure and a contribution due to flow, i.e.,

$$P_{xx} = -p + \tau_{xx} \tag{1.6a}$$

$$P_{yy} = -p + \tau_{yy} \tag{1.6b}$$

$$P_{zz} = -p + \tau_{zz} \tag{1.6c}$$

where τ_{xx}, τ_{yy}, τ_{zz} contributions arising from flow are known as deviatoric normal stresses for Newtonian fluids and as extra stresses for non-Newtonian fluids. For an incompressible Newtonian fluid, the isotropic pressure is given by:

$$p = -\tfrac{1}{3}(P_{xx} + P_{yy} + P_{zz}) \tag{1.7}$$

Table 1.1   *Typical viscosity values at room temperature*

Substance	μ (mPa s)
Air	10^{-2}
Benzene	0.65
Water	1
Molten sodium chloride (1173 K)	1.01
Ethyl alcohol	1.20
Mercury (293 K)	1.55
Molten lead (673 K)	2.33
Ethylene glycol	20
Olive oil	100
Castor oil	600
100% Glycerine (293 K)	1500
Honey	10^4
Corn syrup	10^5
Bitumen	10^{11}
Molten glass	10^{15}

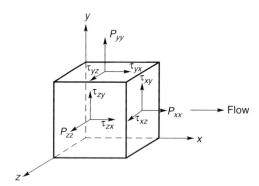

Figure 1.3   *Stress components in three-dimensional flow*

From equations (1.6) and (1.7) it follows that:

$$\tau_{xx} + \tau_{yy} + \tau_{zz} = 0 \tag{1.8}$$

For a Newtonian fluid in simple shearing motion, the deviatoric normal stress components are identically zero, i.e.,

$$\tau_{xx} = \tau_{yy} = \tau_{zz} = 0 \tag{1.9}$$

Thus, the complete definition of a Newtonian fluid is that it not only possesses a constant viscosity but it also satisfies the condition of equation (1.9), or simply that it satisfies the complete Navier–Stokes equations. Thus, for instance, the so-called constant viscosity Boger fluids (Boger, 1976; Prilutski *et al.*, 1983) which display constant shear viscosity but do not conform to equation (1.9) must be classed as non-Newtonian fluids. A cursory

Table 1.2   *Examples of substances exhibiting non-Newtonian fluid behaviour*

- Adhesives (wall paper paste, carpet adhesive, for instance)
- Ales (beer, liqueurs, etc.)
- Animal waste slurries from cattle farms
- Biological fluids (blood, synovial fluid, saliva, etc.)
- Bitumen
- Cement paste and slurries
- Chalk slurries
- Chocolates
- Coal slurries
- Cosmetics and personal care products (nail polish, lotions and creams, lipsticks, shampoos, shaving foams and creams, toothpaste, etc.)
- Dairy products and dairy waste streams (cheese, butter, yogurts, fresh cream, whey, for instance)
- Drilling muds
- Fire fighting foams
- Foodstuffs (fruit/vegetable purees and concentrates, sauces, salad dressings, mayonnaise, jams and marmalades, ice-cream, soups, cake mixes and cake toppings, egg white, bread mixes, snacks)
- Greases and lubricating oils
- Mine tailings and mineral suspensions
- Molten lava and magmas
- Paints, polishes and varnishes
- Paper pulp suspensions
- Peat and lignite slurries
- Polymer melts and solutions, reinforced plastics, rubber
- Printing colours and inks
- Pharmaceutical products (creams, foams, suspensions, for instance)
- Sewage sludge
- Wet beach sand
- Waxy crude oils

inspection of Table 1.2 reveals the widespread occurrence of non-Newtonian flow behaviour in materials encountered in everyday life as well as in diverse industrial settings.

1.2.2 Non-Newtonian fluid behaviour

A non-Newtonian fluid is one whose flow curve (shear stress versus shear rate) is non-linear or does not pass through the origin, i.e. where the apparent viscosity, shear stress divided by shear rate, is not constant at a given temperature and pressure but is dependent on flow conditions such as flow geometry, shear rate, etc. and sometimes even on the kinematic history of the fluid element under consideration. Such materials may be conveniently grouped into three general classes:

(1) fluids for which the rate of shear at any point is determined only by the value of the shear stress at that point at that instant; these fluids are variously known as 'time independent', 'purely viscous', 'inelastic' or 'generalized Newtonian fluids' (GNF);

(2) more complex fluids for which the relation between shear stress and shear rate depends, in addition, upon the duration of shearing and their kinematic history; they are called 'time-dependent fluids', and finally,

(3) substances exhibiting characteristics of both ideal fluids and elastic solids and showing partial elastic recovery, after deformation; these are categorized as 'visco-elastic fluids'.

This classification scheme is arbitrary in that most real materials often exhibit a combination of two or even all three types of non-Newtonian features. Generally, it is, however, possible to identify the dominant non-Newtonian characteristic and to take this as

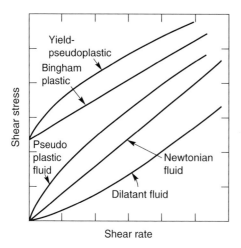

Figure 1.4   *Types of time-independent flow behaviour*

the basis for the subsequent process calculations. Also, as mentioned earlier, it is convenient to define an apparent viscosity of these materials as the ratio of shear stress to shear rate, though the latter ratio is a function of the shear stress or shear rate and/or of time. Each type of non-Newtonian fluid behaviour will now be dealt with in some detail.

1.3 Time-independent fluid behaviour

In simple shear, the flow behaviour of this class of materials may be described by a constitutive relation of the form,

$$\dot{\gamma}_{yx} = f(\tau_{yx}) \tag{1.10}$$

or its inverse form,

$$\tau_{yx} = f_1(\dot{\gamma}_{yx}) \tag{1.11}$$

This equation implies that the value of $\dot{\gamma}_{yx}$ at any point within the sheared fluid is determined only by the current value of shear stress at that point or vice versa. Depending upon the form of the function in equation (1.10) or (1.11), these fluids may be further subdivided into three types:

(a) shear-thinning or pseudoplastic
(b) viscoplastic
(c) shear-thickening or dilatant.

Qualitative flow curves on linear scales for these three types of fluid behaviour are shown in Figure 1.4; the linear relation typical of Newtonian fluids is also included.

1.3.1 Shear-thinning or pseudoplastic fluids

The most common type of time-independent non-Newtonian fluid behaviour observed is pseudoplasticity or shear-thinning, characterized by an apparent viscosity which decreases with increasing shear rate. Both at very low and at very high shear rates, most shear-thinning

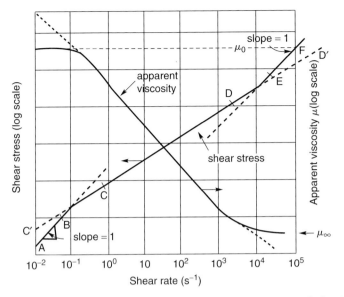

Figure 1.5 *Schematic representation of shear-thinning behaviour*

polymer solutions and melts exhibit Newtonian behaviour, i.e., shear stress–shear rate plots become straight lines, as shown schematically in Figure 1.5, and on a linear scale will pass through origin. The resulting values of the apparent viscosity at very low and high shear rates are known as the zero shear viscosity, μ_0, and the infinite shear viscosity, μ_∞, respectively. Thus, the apparent viscosity of a shear-thinning fluid decreases from μ_0 to μ_∞ with increasing shear rate. Data encompassing a sufficiently wide range of shear rates to illustrate this complete spectrum of pseudoplastic behaviour are difficult to obtain, and are scarce. A single instrument will not have both the sensitivity required in the low shear rate region and the robustness at high shear rates, so that several instruments are often required to achieve this objective. Figure 1.6 shows the apparent viscosity – shear rate behaviour of an aqueous polyacrylamide solution at 293 K over almost seven decades of shear rate. The apparent viscosity of this solution drops from 1400 mPa s to 4.2 mPa s, and so it would hardly be justifiable to assign a single average value of viscosity for such a fluid! The values of shear rates marking the onset of the upper and lower limiting viscosities are dependent upon several factors, such as the type and concentration of polymer, its molecular weight distribution and the nature of solvent, etc. Hence, it is difficult to suggest valid generalizations but many materials exhibit their limiting viscosities at shear rates below $10^{-2} \, \text{s}^{-1}$ and above $10^5 \, \text{s}^{-1}$, respectively. Generally, the range of shear rate over which the apparent viscosity is constant (in the zero-shear region) increases as molecular weight of the polymer falls, as its molecular weight distribution becomes narrower, and as polymer concentration (in solution) drops. Similarly, the rate of decrease of apparent viscosity with shear rate also varies from one material to another, as can be seen in Figure 1.7 for three aqueous solutions of chemically different polymers.

Mathematical models for shear-thinning fluid behaviour

Many mathematical expressions of varying complexity and form have been proposed in the literature to model shear-thinning characteristics; some of these are straightforward

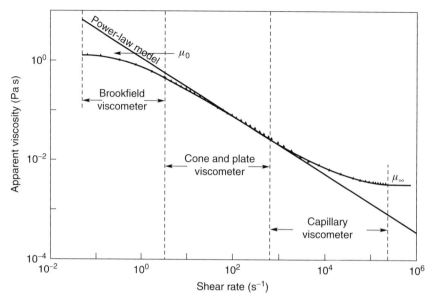

Figure 1.6   *Demonstration of zero shear and infinite shear viscosities for a shear-thinning polymer solution (Boger, 1977)*

Figure 1.7   *Representative shear stress and apparent viscosity plots for three pseudo-plastic polymer solutions*

attempts at curve fitting, giving empirical relationships for the shear stress (or apparent viscosity)–shear rate curves for example, while others have some theoretical basis in statistical mechanics – as an extension of the application of the kinetic theory to the liquid state or the theory of rate processes, etc. Only a selection of the more widely used viscosity models is given here; more complete descriptions of such models are available in many books (Bird *et al.*, 1987; Carreau *et al.*, 1997) and in a review paper (Bird, 1976).

(i) *The power-law or Ostwald de Waele model*
The relationship between shear stress and shear rate (plotted on double logarithmic coordinates) for a shear-thinning fluid can often be approximated by a straight line over a limited range of shear rate (or stress). For this part of the flow curve, an expression of the following form is applicable:

$$\tau_{yx} = m(\dot{\gamma}_{yx})^n \tag{1.12}$$

so the apparent viscosity for the so-called power-law (or Ostwald de Waele) fluid is thus given by:

$$\mu = \tau_{yx}/\dot{\gamma}_{yx} = m(\dot{\gamma}_{yx})^{n-1} \tag{1.13}$$

For $n < 1$, the fluid exhibits shear-thinning properties
  $n = 1$, the fluid shows Newtonian behaviour
  $n > 1$, the fluid shows shear-thickening behaviour

In these equations, m and n are two empirical curve-fitting parameters and are known as the fluid consistency coefficient and the flow behaviour index respectively. For a shear-thinning fluid, the index may have any value between 0 and 1. The smaller the value of n, the greater is the degree of shear-thinning. For a shear-thickening fluid, the index n will be greater than unity. When $n = 1$, equations (1.12) and (1.13) reduce to equation (1.1) which describes Newtonian fluid behaviour.

Although the power-law model offers the simplest representation of shear-thinning behaviour, it does have a number of shortcomings. Generally, it applies over only a limited range of shear rates and therefore the fitted values of m and n will depend on the range of shear rates considered. Furthermore, it does not predict the zero and infinite shear viscosities, as shown by dotted lines in Figure 1.5. Finally, it should be noted that the dimensions of the flow consistency coefficient, m, depend on the numerical value of n and therefore the m values must not be compared when the n values differ. On the other hand, the value of m can be viewed as the value of apparent viscosity at the shear rate of unity and will therefore depend on the time unit (e.g. second, minute or hour) employed. Despite these limitations, this is perhaps the most widely used model in the literature dealing with process engineering applications. Finally, while each non-Newtonian fluid is unique and its rheological behaviour must be evaluated directly, Table 1.3 provides a compilation of the power-law constants (m and n) for a variety of substances. It needs to be emphasized here that these values are provided here to give the readers a feel for their values.

(ii) *The Carreau viscosity equation*
When there are significant deviations from the power-law model at very high and very low shear rates as shown in Figure 1.6, it is necessary to use a model which takes account of the limiting values of viscosities μ_0 and μ_∞.

Based on the molecular network considerations, Carreau (1972) put forward the following viscosity model which incorporates both limiting viscosities μ_0 and μ_∞:

$$\frac{\mu - \mu_\infty}{\mu_0 - \mu_\infty} = \{1 + (\lambda\dot{\gamma}_{yx})^2\}^{(n-1)/2} \tag{1.14}$$

Table 1.3   *Typical values of power-law constants for a few systems*

System	Temperature (K)	n (−)	m (Pa s^n)
Agro- and food-related products			
Aerated poultry waste slurry (x is % volume of solids)	283–298	1.81–0.161 ln x	1.12×10^{-11} $(x)^{2.59}$
Ammonium alginate solution (3.37%)	297	0.5	13
Apple butter	–	0.15	200
Apple sauce	300	0.3–0.45	12–22
Apricot puree	300	0.3–0.4	5–20
Banana puree	293–315	0.33–0.5	4–10
Carrot puree	298	0.25	25
Chicken (minced)	296	0.10	900
Chocolate	303	0.5	0.7
Guava puree	296.5	0.5	40
Human blood	300	0.9	0.004
Mango pulp	300–340	0.3	3–10
Marshmallow cream	–	0.4	560
Mayonnaise	298	0.6	5–100
Papaya puree	300	0.5	10
Peach puree	300	0.38	1–5
Peanut butter	–	0.07	500
Pear puree	300	0.4–0.5	1–5
Plum puree	287	0.35	30–80
Tomato concentrate (5.8% solid)	305	0.6	0.22
Tomato ketch up	295	0.24	33
Tomato paste	–	0.5	15
Whipped desert toppings	–	0.12	400
Yoghurt	293	0.5–0.6	25
Polymer melts			
High density polyethylene (HDPE)	453–493	0.6	$3.75–6.2 \times 10^3$
High impact polystyrene	443–483	0.20	$3.5–7.5 \times 10^4$
Polystyrene	463–498	0.25	$1.5–4.5 \times 10^4$
Polypropylene	453–473	0.40	$4.5–7 \times 10^3$
Low density polyethylene (LDPE)	433–473	0.45	$4.3–9.4 \times 10^3$
Nylon	493–508	0.65	$1.8–2.6 \times 10^3$
Polymethylmethyacrylate (PMMA)	493–533	0.25	$2.5–9 \times 10^4$
Polycarbonate	553–593	0.65–0.8	$1–8.5 \times 10^3$

(Continued)

Table 1.3   (*Continued*)

System	Temperature (K)	$n\,(-)$	$M\,(\mathrm{Pa\,s}^n)$
Personal care products			
Nail polish	298	0.86	750
Mascara	298	0.24	200
Toothpaste	298	0.28	120
Sunscreen lotions	298	0.28	75
Ponds cold cream	298	0.45	25
Oil of Olay	298	0.22	25

Source: Modified after Steffe (1996) and Johnson (1999)

where $n(<1)$ and λ are two curve-fitting parameters. This model can describe shear-thinning behaviour over wide ranges of shear rates but only at the expense of the added complexity of four parameters. This model predicts Newtonian fluid behaviour $\mu = \mu_0$ when either $n = 1$ or $\lambda = 0$ or both.

(iii) *The Cross viscosity equation*
Another four parameter model which has gained wide acceptance is due to Cross (1965) which, in simple shear, is written as:

$$\frac{\mu - \mu_\infty}{\mu_0 - \mu_\infty} = \frac{1}{1 + k(\dot{\gamma}_{yx})^n} \tag{1.15}$$

In equation (1.15), $n\ (<1)$ and k are two fitting parameters whereas μ_0 and μ_∞ are the limiting values of the apparent viscosity at low and high shear rates, respectively. This model reduces to the Newtonian fluid behaviour as $k \to 0$. Similarly, when $\mu << \mu_0$ and $\mu >> \mu_\infty$, it reduces to the familiar power-law model, equation (1.13). Though initially Cross (1965) suggested that a constant value of $n = 2/3$ was adequate to approximate the viscosity data for many systems, it is now thought that treating the index, n, as an adjustable parameter offers considerable improvement over the use of the constant value of n (Barnes *et al.*, 1989).

(iv) *The Ellis fluid model*
When the deviations from the power-law model are significant only at low shear rates, it is perhaps more appropriate to use the Ellis model.

The three viscosity equations presented so far are examples of the form of equation (1.11). The three-constant Ellis model is an illustration of the inverse form, namely, equation (1.10). In simple shear, the apparent viscosity of an Ellis model fluid is given by:

$$\mu = \frac{\mu_0}{1 + (\tau_{yx}/\tau_{1/2})^{\alpha-1}} \tag{1.16}$$

In this equation, μ_0 is the zero shear viscosity and the remaining two constants $\alpha(>1)$ and $\tau_{1/2}$ are adjustable parameters. While the index α is a measure of the degree of shear-thinning behaviour (the greater the value of α, greater is the extent of shear-thinning),

$\tau_{1/2}$ represents the value of shear stress at which the apparent viscosity has dropped to half its zero shear value. Equation (1.16) predicts Newtonian fluid behaviour in the limit of $\tau_{1/2} \to \infty$. This form of equation has advantages in permitting easy calculation of velocity profiles from a known stress distribution, but renders the reverse operation tedious and cumbersome. It can easily be seen that in the intermediate range of shear stress (or shear rate), $(\tau_{yx}/\tau_{1/2})^{\alpha-1} \gg 1$, and equation (1.16) reduces to equation (1.13) with $n = (1/\alpha)$ and $m = \left\{ \mu_0 \tau_{1/2}^{\alpha-1} \right\}^{1/\alpha}$.

1.3.2 Viscoplastic fluid behaviour

This type of fluid behaviour is characterized by the existence of a yield stress (τ_0) which must be exceeded before the fluid will deform or flow. Conversely, such a material will deform elastically (or flow *en masse* like a rigid body) when the externally applied stress is smaller than the yield stress. Once the magnitude of the external stress has exceeded the value of the yield stress, the flow curve may be linear or non-linear but will not pass through origin (Figure 1.4). Hence, in the absence of surface tension effects, such a material will not level out under gravity to form an absolutely flat free surface. One can, however, explain this kind of fluid behaviour by postulating that the substance at rest consists of three-dimensional structures of sufficient rigidity to resist any external stress less than τ_0. For stress levels greater than τ_0, however, the structure breaks down and the substance behaves like a viscous material. In some cases, the build-up and breakdown of structure has been found to be reversible, i.e., the substance may regain its initial value of the yield stress.

A fluid with a linear flow curve for $|\tau_{yx}| > |\tau_0|$ is called a Bingham plastic fluid and is characterized by a constant plastic viscosity (the slope of the shear stress versus shear rate curve) and a yield stress. On the other hand, a substance possessing a yield stress as well as a non-linear flow curve on linear coordinates (for $|\tau_{yx}| > |\tau_0|$), is called a 'yield-pseudoplastic' material. Figure 1.8 illustrates viscoplastic behaviour as observed in a meat extract and in a polymer solution.

It is interesting to note that a viscoplastic material also displays an apparent viscosity which decreases with increasing shear rate. At very low shear rates, the apparent viscosity is effectively infinite at the instant immediately before the substance yields and begins to flow. It is thus possible to regard these materials as possessing a particular class of shear-thinning behaviour.

Strictly speaking, it is virtually impossible to ascertain whether any real material has a true yield stress or not, but nevertheless the concept of a yield stress has proved to be convenient in practice because some materials closely approximate to this type of flow behaviour, e.g. see Barnes and Walters (1985), Astarita (1990), Schurz (1990) and Evans (1992). Many workers in this field view the yield stress in terms of the transition from a solid-like (high viscosity) to a liquid-like (low viscosity) state which occurs abruptly over an extremely narrow range of shear rates or shear stress (Uhlherr *et al.*, 2005). It is not uncommon for the two values of viscosity to differ from each other by several orders of magnitude. The answer to the question whether a fluid has a yield stress or not seems to be related to the choice of a time scale of observation. Common examples of viscoplastic fluid behaviour include particulate suspensions, emulsions, foodstuffs, blood and drilling muds, etc. (Barnes, 1999).

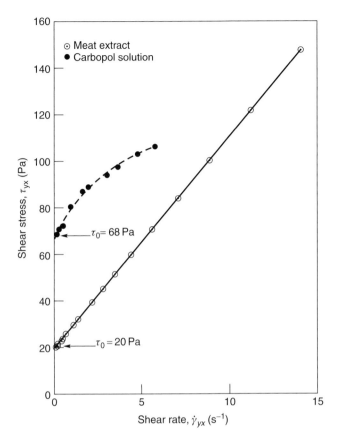

Figure 1.8   *Representative shear stress–shear rate data showing viscoplastic behaviour in a meat extract (Bingham plastic) and in an aqueous carbopol polymer solution (yield-pseudoplastic)*

Mathematical models for viscoplastic behaviour

Over the years, many empirical expressions have been proposed as a result of straightforward curve-fitting exercises. A model based on sound theory is yet to emerge. Three commonly used models for viscoplastic fluids are briefly described here.

(i) *The Bingham plastic model*
This is the simplest equation describing the flow behaviour of a fluid with a yield stress and, in steady one-dimensional shear, it is written as:

$$\tau_{yx} = \tau_0^B + \mu_B(\dot{\gamma}_{yx}) \quad \text{for } |\tau_{yx}| > |\tau_0^B|$$
$$\dot{\gamma}_{yx} = 0 \quad \text{for } |\tau_{yx}| < |\tau_0^B|$$

(1.17)

Often, the two model parameters τ_0^B and μ_B are treated as curve-fitting constants irrespective of whether or not the fluid possesses a true yield stress.

(ii) *The Herschel-Bulkley fluid model*

A simple generalization of the Bingham plastic model to embrace the non-linear flow curve (for $|\tau_{yx}| > |\tau_0^B|$) is the three constant Herschel–Bulkley fluid model. In one-dimensional steady shearing motion, it is written as:

$$
\begin{aligned}
\tau_{yx} &= \tau_0^H + m(\dot\gamma_{yx})^n & \text{for } |\tau_{yx}| &> |\tau_0^H| \\
\dot\gamma_{yx} &= 0 & \text{for } |\tau_{yx}| &< |\tau_0^H|
\end{aligned}
\tag{1.18}
$$

Note that here too, the dimensions of m depend upon the value of n. The physical meaning of m and n (<1) in equation (1.18) is similar to that in equation (1.12). With the use of the third parameter, this model provides a somewhat better fit to some experimental data.

(iii) *The Casson fluid model*

Many foodstuffs and biological materials, especially blood, are well described by this two constant model as:

$$
\begin{aligned}
(|\tau_{yx}|)^{1/2} &= (|\tau_0^C|)^{1/2} + (\mu_c|\dot\gamma_{yx}|)^{1/2} & \text{for } |\tau_{yx}| &> |\tau_0^C| \\
\dot\gamma_{yx} &= 0 & \text{for } |\tau_{yx}| &< |\tau_0^C|
\end{aligned}
\tag{1.19}
$$

This model has often been used for describing the steady shear stress–shear rate behaviour of blood, yoghurt, tomato pureé, molten chocolate, etc. The flow behaviour of some particulate suspensions also closely approximates to this type of behaviour.

The comparative performance of these three as well as several other models for visco-plastic behaviour has been thoroughly evaluated in an extensive review paper by Bird *et al.* (1983) and a through discussion on the existence, measurement and implications of yield stress has been provided by Barnes (1999).

1.3.3 Shear-thickening or dilatant fluid behaviour

Dilatant fluids are similar to pseudoplastic systems in that they show no yield stress but their apparent viscosity increases with increasing shear rate; thus these fluids are also called shear-thickening. This type of fluid behaviour was originally observed in concentrated suspensions and a possible explanation for their dilatant behaviour is as follows: At rest, the voidage is minimum and the liquid present is sufficient to fill the void space. At low shear rates, the liquid lubricates the motion of each particle past others and the resulting stresses are consequently small. At high shear rates, on the other hand, the material expands or dilates slightly (as also observed in the transport of sand dunes) so that there is no longer sufficient liquid to fill the increased void space and prevent direct solid–solid contacts which result in increased friction and higher shear stresses (as shown schematically in Figure 1.9). This mechanism causes the apparent viscosity to rise rapidly with increasing rate of shear.

The term dilatant has also been used for all other fluids which exhibit increasing apparent viscosity with increasing rate of shear. Many of these, such as starch pastes, are not true suspensions and show no dilation on shearing. The above explanation therefore is not applicable but nevertheless such materials are still commonly referred to as dilatant fluids.

Of the time-independent fluids, this sub-class has received very little attention; consequently very few reliable data are available. Until recently, dilatant fluid behaviour

Figure 1.9 *Schematic representation of shear-thickening behaviour*

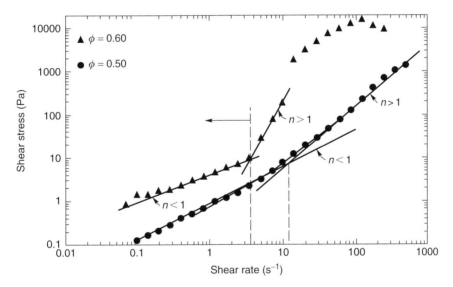

Figure 1.10 *Shear stress–shear rate behaviour of polyvinylchloride (PVC) in dioc-tylphthalate (DOP) dispersions at 298 K showing regions of shear-thinning and shear-thickening (Boersma* et al., *1990)*

was considered to be much less widespread in the chemical and processing industries. However, with the recent growing interest in the handling and processing of systems with high solids loadings, it is no longer so, as is evidenced by the number of recent review articles on this subject (Barnes *et al.*, 1987; Barnes, 1989; Boersma *et al.*, 1990; Goddard and Bashir, 1990). Typical examples of materials exhibiting dilatant behaviour include concentrated suspensions of china clay, titanium dioxide (Metzner and Whitlock, 1958) and of corn flour in water (Griskey *et al.*, 1985). Figure 1.10 shows the dilatant behaviour of dispersions of polyvinylchloride in dioctylphthalate (Boersma *et al.*, 1990).

The limited information reported so far suggests that the apparent viscosity–shear rate data often result in linear plots on double logarithmic coordinates over a limited shear rate range and the flow behaviour may be represented by the power-law model, equation (1.13), with the flow behaviour index, n, greater than unity, i.e.,

$$\mu = m(\dot{\gamma}_{yx})^{n-1} \tag{1.13}$$

One can readily see that for $n > 1$, equation (1.13) predicts increasing viscosity with increasing shear rate. The dilatant behaviour may be observed in moderately concentrated suspensions at high shear rates, and yet, the same suspension may exhibit pseudoplastic behaviour at lower shear rates, as shown in Figure 1.10; it is not yet possible to ascertain whether these materials also display limiting apparent viscosities.

Example 1.1

The following shear stress–shear rate data were obtained for an aqueous polymer solution at 291 K.

$\dot{\gamma}_{yx}\,(\mathrm{s}^{-1})$	$\tau_{yx}(\mathrm{Pa})$	$\dot{\gamma}_{yx}\,(\mathrm{s}^{-1})$	$\tau_{yx}(\mathrm{Pa})$
0.14	0.12	4.43	3.08
0.176	0.14	5.57	3.79
0.222	0.17	7.02	4.68
0.28	0.21	8.83	5.41
0.352	0.28	11.12	6.53
0.443	0.35	14	8.11
0.557	0.446	17.62	9.46
0.702	0.563	22.2	11.50
0.883	0.69	27.9	13.5
1.11	0.85	35.2	16.22
1.4	1.08	44.3	18.92
1.76	1.31	55.7	22.10
2.22	1.63	70.2	26.13
2.8	2.01	88.3	30
3.52	2.53	111.2	34.8

(a) Plot the flow curve on log–log coordinates
(b) Can the power-law model fit this data over the entire range? What are the values of m and n?
(c) Can the Ellis fluid model (equation (1.16)) fit this data better than the power-law model? Evaluate the values of μ_0, $\tau_{1/2}$ and α?

Solution

Figure 1.11 shows the flow curve for this polymer solution.

The plot is not linear on log–log coordinates and therefore the power-law model cannot fit the data over the whole range; however, it is possible to divide the plot in two parts, each of which can be represented by power-law model as:

$$\tau = 0.75 \, \dot{\gamma}^{0.96} \quad \text{for} \quad \dot{\gamma} < {\sim}5 \, s^{-1}$$

and

$$\tau = 1.08 \, \dot{\gamma}^{0.76} \quad \text{for} \quad 5 \le \dot{\gamma} \le 100 \, s^{-1}$$

Note that at $\dot{\gamma} = 5 \, s^{-1}$, both equations yield nearly equal values of the shear stress and the boundary between the two zones has been taken simply as the intersection of the two equations. Note that if the extrapolation is based on the second equation, it will over-estimate the value of shear stress at shear rates below $5 \, s^{-1}$ whereas the first equation will also over-predict shear stress at shear rates above $5 \, s^{-1}$.

The fitting of the data to the Ellis model, equation (1.16), however, needs a non-linear regression approach to minimize the sum of squares. The best values are found to be:

$$\mu_0 = 0.79 \, Pa \, s, \quad \tau_{1/2} = 21.55 \, Pa \quad \text{and} \quad \alpha = 2.03$$

The predications of this model are also plotted in Figure 1.11 where a satisfactory fit can be seen to exist. □

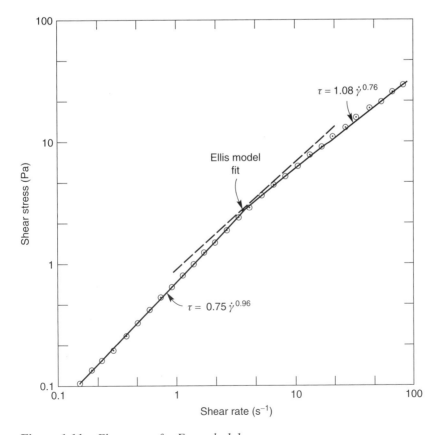

Figure 1.11   *Flow curve for Example 1.1*

1.4 Time-dependent fluid behaviour

The flow behaviour of many industrially important materials cannot be described by a simple rheological equation like (1.12) or (1.13). In practice, apparent viscosities may depend not only on the rate of shear but also on the time for which the fluid has been subjected to shearing. For instance, when materials such as bentonite–water suspensions, red mud suspensions (waste stream from aluminium industry), cement paste, crude oils and certain foodstuffs are sheared at a constant rate following a long period of rest, their apparent viscosities gradually become less as the 'internal' structure of the material is progressively broken down. As the number of structural 'linkages' capable of being broken down decreases, the rate of change of apparent viscosity with time drops progressively to zero. Conversely, as the structure breaks down, the rate at which linkages can re-form increases, so that eventually a state of dynamic equilibrium is reached when the rates of build-up and of breakdown are balanced.

Time-dependent fluid behaviour may be further sub-divided into two categories: thixotropy and rheopexy or negative thixotropy.

1.4.1 Thixotropy

A material is said to exhibit thixotropy if, when it is sheared at a constant rate, its apparent viscosity (or the corresponding shear stress) decreases with the time of shearing, as can be seen in Figure 1.12 for a red mud suspension (Nguyen and Uhlherr, 1983). If the flow curve is measured in a single experiment in which the shear rate is steadily increased at a constant rate from zero to some maximum value and then decreased at the same rate to zero again, a hysteresis loop of the form shown in Figure 1.13 is obtained; the height, shape and enclosed area of the hysteresis loop depend on the duration of shearing, the rate of increase/decrease of shear rate and the past kinematic history of the sample. Broadly speaking, the larger the enclosed area, stronger is the time-dependent behaviour of the materials. Thus, no hysteresis loop is observed for time-independent fluids, that is, the enclosed area of the loop is zero. Figure 1.14 shows the hysteresis loop

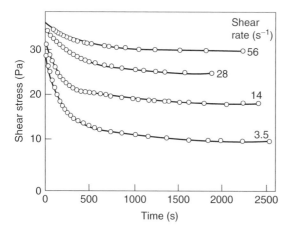

Figure 1.12   *Representative data showing thixotropy in a 59% (by weight) red mud suspension*

for a cement paste. Furthermore, in some cases, the breakdown of structure by shearing is reversible in the sense that upon removal of shearing and following a period of rest, it is possible for the fluid to regain (build-up of structure) the initial value of viscosity, as can be seen in Figure 1.15 for a proprietary lotion. In this case, the viscosity is seen to drop from ~80 to 10 Pa s in about 5–10 s when it is sheared at $\dot{\gamma} = 100\,\text{s}^{-1}$ and after the removal of shear, the viscosity builds up to almost its initial value in about 50–60 s.

The term 'false body' has been introduced to describe the thixotropic behaviour of viscoplastic materials. Although the thixotropy is associated with the build-up of structure at rest and breakdown of structure under shear, viscoplastic materials do not lose

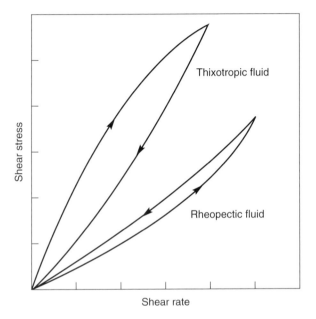

Figure 1.13   *Schematic shear stress–shear rate behaviour for time-dependent fluid behaviour*

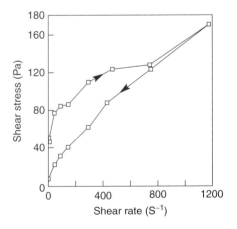

Figure 1.14   *Thixotropic behaviour of a cement paste (replotted from Struble and Ji, 2001)*

Figure 1.15   *Breakdown and re-formation of structure in a proprietary body lotion (replotted from Schramm, 1994)*

their solid-like properties completely and can still exhibit a yield stress, though this is usually less than the original value of the virgin sample which is regained (if at all) only after a long recovery period.

Other examples of materials exhibiting thixotropic behaviour include concentrated suspensions, laponite and bentonite clay suspensions, emulsions, drilling fluids, waxy crude oils, protein solutions and foodstuffs, etc. (Barnes, 1997).

1.4.2 Rheopexy or negative thixotropy

The relatively few fluids for which the apparent viscosity (or the corresponding shear stress) increases with time of shearing are said to display rheopexy or negative thixotropy. Again, hysteresis effects are observed in the flow curve, but in this case it is inverted, as compared with a thixotropic material, as can be seen in Figure 1.13.

In a rheopectic fluid, the structure builds up by shear and breaks down when the material is at rest. For instance, Freundlich and Juliusburger (1935), using a 42% aqueous gypsum paste, found that, after shaking, this material re-solidified in 40 min if at rest, but in only 20 s if the container was gently rolled in the palms of hands. This indicates that gentle shearing motion (rolling) facilitates structure build-up but more intense motion destroys it. Thus, there is a critical amount of shear beyond which re-formation of structure is not induced but breakdown occurs. It is not uncommon for the same dispersion to display both thixotropy as well as rheopexy depending upon the shear rate and/or the concentration of solids. Figure 1.16 shows the gradual onset of rheopexy for a saturated polyester at 60°C (Steg and Katz, 1965). Similar behaviour is reported to occur with suspensions of ammonium oleate, colloidal suspensions of vanadium pentoxide at moderate shear rates (Tanner, 2000), coal–water slurries (Keller and Keller Jr, 1990) and protein solutions (Pradipasena and Rha, 1977).

Figure 1.16   *Onset of rheopexy in a saturated polyester (Steg and Katz, 1965)*

Owing to the wide occurrence of thixotropic behaviour in numerous industrial settings (Mewis, 1979; Barnes, 1997; Mujumdar *et al.*, 2002), significant efforts have been directed at the development of constitutive relations for thixotropic behaviour (Mujumdar *et al.*, 2002). Broadly, most of the currently available models are based on three distinct approaches, namely, continuum, micro-structural and structural kinetics. Within the framework of the continuum approach, existing constitutive equations (such as the Bingham, Herschel–Bulkley and Reiner–Rivlin models) are modified by making the viscosity function, yield stress, etc. to be time dependent. It is, however, not easy to relate the model parameters to the underlying physical processes responsible for the structural changes in the material under shear and for the subsequent build-up of structure at rest in the case of reversible thixotropy. The models based on the consideration of micro-structure necessitate a detailed knowledge of inter-particle forces which are seldom available for real systems encountered in engineering applications thereby severely limiting the utility of this class of models. Finally, the models based on the structural kinetic considerations rely on the value of a scalar parameter, ξ, which varies from zero (corresponding to the completely broken-down structure) to unity (denoting the other extreme of the complete build-up of structure). This family of models consists of two equations: the first equation connects the shear stress to shear rate for a fixed values of ξ. The second equation describes the variation of ξ with time, akin to a reversible chemical reaction. More detailed discussions about the merits and demerits of these approaches are available in the literature (Mujumdar *et al.*, 2002; Dullaert and Mewis, 2005, 2006). One such model which has been reasonably successful in describing thixotropic behaviour of scores of fluids is due to Houska (1981) which is really a generalization of the Herschel–Bulkley model. In simple one-dimensional shearing flow, it is written as:

$$\tau_{yx} = (\tau_{y0} + \tau_{y1}) + (m_0 + \xi m_1)\dot{\gamma}_{yx}^n \tag{1.20}$$

$$\frac{d\xi}{dt} = a(1 - \xi) - b\xi\dot{\gamma}_{yx}^\varepsilon \tag{1.21}$$

where τ_{y0} and m_0 are the so-called permanent yield stress and consistency coefficient, respectively whereas τ_{y1} and m_1 are the corresponding time-dependent contributions which are postulated to be linearly dependent on the 'instantaneous' structure, i.e., the value of ξ. Thus, equation (1.20) applies for a constant value of ξ. The second equation (1.21) denotes a dynamic equilibrium between the rates of build-up (first term) and of breakdown of structure (second term) thereby giving the rate of change of ξ. Note that the rate of breakdown depends on both the value of ξ and the value of shear rate whereas the rate of build-up depends only on the number of structural units available at any instant of time. Equations (1.20) and (1.21) contain eight parameters, namely, τ_{y0}, τ_{y1}, m_0, m_1, n, a, b and ε, out of which the first five are material parameters and the last three are kinetic parameters. While some experimental protocols have been developed to evaluate the model parameters for this (and other similar) models for thixotropic fluids (Cawkwell and Charles, 1989; Dullaert and Mewis, 2005, 2006), these are not only cumbersome and often entail transient experiments, but these might vary from one substance to another.

It is thus not possible to put forward simple mathematical equations of general valid-ity to describe time-dependent fluid behaviour, and it is usually necessary to make meas-urements over the range of conditions of interest. The conventional shear stress–shear rate curves are of limited utility unless they relate to the particular history of interest in the application. For example, when the material enters a pipe slowly and with a mini-mum of shearing, as from a storage tank directly into the pipe, the shear stress–shear rate–time curve should be based on tests performed on samples which have been stored under identical conditions and have not been subjected to shearing by transference to another vessel for example. At the other extreme, when the material undergoes vigorous agitation and shearing, as in passage through a pump, the shear stress–shear rate curve should be obtained using highly sheared pre-mixed material. Assuming then that reli-able flow property data are available, the zero shear and infinite shear flow curves can be used to form the bounds for the design of a flow system. For a fixed pressure drop, the zero shear limit (maximum apparent viscosity) will provide a lower bound and the infi-nite shear conditions (minimum apparent viscosity) will provide the upper bound on the flowrate. Conversely, for a fixed flowrate, the zero and infinite shear data can be used to establish the maximum and minimum pressure drops or pumping power.

For many industries (notably foodstuffs) the way in which the rheology of the materi-als affects their processing is much less significant than the effects that the process has on their rheology. Implicit here is the recognition of the importance of the time-dependent properties of materials which can be profoundly influenced by mechanical working on the one hand or by an aging process during a prolonged shelf life on the other.

The above brief discussion of time-dependent fluid behaviour provides an introduc-tion to the topic, but Mewis (1979) and Barnes (1997) have given detailed accounts of recent developments in the field. Govier and Aziz (1982), moreover, have focused on the practical aspects of the flow of time-dependent fluids in pipes.

Example 1.2

The following steady shear data has been reported for a polymer solution at 298 K.

$\dot{\gamma}(s^{-1})$	0.205	0.280	0.377	0.510	0.692	0.944	1.29	1.582	1.77	2.17	2.43
τ(Pa)	49.69	51.94	54.46	56.44	59.02	61.93	65.48	68.07	70.0	71.6	73.85

Evaluate the model parameters for the following viscosity equations (i) Power law (ii) Bingham plastic model (iii) Casson model, and (iv) Herschel–Bulkley model.

Solution

The given data is plotted on linear coordinates in Figure 1.17 and a non-linear regression method was used to fit the above-noted viscosity models. The values of the model parameters together with the values of the sum of the residuals ($\delta = \Sigma(\tau_{pred} - \tau_{meas})^2$) are summarized in Table 1.4.

The respective fits are also included in Figure 1.17. An inspection of Figure 1.17 together with the corresponding values of the residuals clearly shows the inadequacy of the Bingham plastic in this case. For the remaining three models, the choice is not immediately obvious. Further examination of Figure 1.17 suggests the possibility of a yield stress. The vane method (see Section 2.5) gave a value of $\tau_0 = 38.2$ Pa which is significantly different from the values of $\tau_0^B = 50.45$ Pa, $\tau_0^C = 42.58$ Pa and $\tau_0^H = 32$ Pa listed in Table 1.4. Figure 1.18 shows the log–log plot of $(\tau - 38.2)$ versus shear rate. Since the slope of this line is less than 1 and therefore, this polymer solution also shows shear-thinning behaviour beyond $\tau > \tau_0$. The Herschel–Bulkley model was re-fitted with $\tau_0^H = 38.2$ Pa, and the new values of the constants are: $m = 24.31$ Pa s^n and $n = 0.44$, with the sum of residuals, $\delta = 1.54$. This fit is also included in Figure 1.18 (log-coordinates) and in Figure 1.19 (linear coordinates) where the degree of fit is seen to be satisfactory.

It is therefore always desirable to measure the value of yield stress by independent means as far as possible rather than treating it as a fitting parameter. $\square$

Figure 1.17   *Shear stress–shear rate data for the polymer solution of Example 1.2*

Table 1.4   *Summary of calculations*

Model	Values of parameters	Value of δ (Pa^2)
Power law	$m = 63.34\,\text{Pa}\,\text{s}^n$	3.21
	$n = 0.16$	
Bingham plastic	$\mu_B = 10.41\,\text{Pa}\,\text{s}$	26.17
	$\tau_0^B = 50.45\,\text{Pa}$	
Casson equation	$\mu_C = 1.83\,\text{Pa}\,\text{s}$	4.67
	$\tau_0^C = 42.58\,\text{Pa}$	
Herschel–Bulkley equation	$m = 30.79\,\text{Pa}\,\text{s}^n$	0.93
	$n = 0.34$	
	$\tau_0^H = 32\,\text{Pa}$	

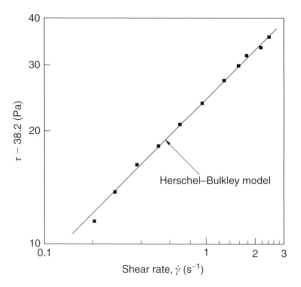

Figure 1.18   *Plot of (τ-38.2) Pa against shear rate on log–log coordinates for the polymer solution of Example 1.2*

1.5 Visco-elastic fluid behaviour

In the classical theory of elasticity, the stress in a sheared body is directly proportional to the strain. For tension, Hooke's law applies and the coefficient of proportionality is known as Young's modulus, G:

$$\tau_{yx} = -G\frac{\mathrm{d}x}{\mathrm{d}y} = G(\gamma_{yx}) \tag{1.22}$$

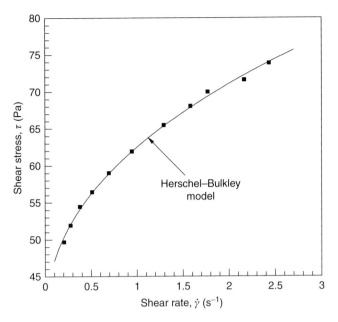

Figure 1.19   *Flow curve together with the prediction of Herschel–Bulkley model with the vane value of yield stress, $\tau_0 = 38.2\,Pa$*

where dx is the shear displacement of two elements separated by a distance dy. When a perfect solid is deformed elastically, it regains its original form on removal of the stress. However, if the applied stress exceeds the characteristic yield stress of the material, complete recovery will not occur and 'creep' will take place – that is, the 'solid' will have flowed. Table 1.5 lists representative values of the Young's modulus (G) for a selection of substances including metals, engineering plastics, polymer and colloidal solutions, foodstuffs, etc.

At the other extreme, in a Newtonian fluid, the shearing stress is proportional to the rate of shear, equation (1.1). Many materials show both elastic and viscous effects under appropriate circumstances. In the absence of the time-dependent behaviour mentioned in the preceding section, the material is said to be visco-elastic. Perfectly elastic deformation and perfectly viscous flow are, in effect, limiting cases of visco-elastic behaviour. For some materials, it is only these limiting conditions that are observed in practice. The elasticity of water and the viscosity of ice may generally pass unnoticed! The response of a material depends not only on its structure but also on the conditions (kinematic) to which it has been subjected; thus the distinction between 'solid' and 'fluid' and between 'elastic' and 'viscous' is to some extent arbitrary and subjective. In other words, it is not uncommon for a material to behave like a viscous fluid in one situation and like an elastic solid in another situation.

Many materials of practical interest (such as polymer melts, polymer and soap solutions, synovial fluid) exhibit visco-elastic behaviour; they have some ability to store and recover shear energy, as shown schematically in Figure 1.20. Perhaps the most easily observed experiment is the 'soup bowl' effect. If a liquid in a dish is made to rotate by means of gentle stirring with a spoon, on removing the energy source (i.e. the spoon), the inertial circulation will die out as a result of the action of viscous forces. If the liquid is

Table 1.5   *Representative (approximate) values of Young's modulus*

Material	Value of G
Glass	70 GPa
Aluminium, copper and alloys	100 GPa
Steel	200 GPa
High modulus oriented fibres	>300 GPa
Concrete	10–20 GPa
Stones	40–60 GPa
Wood	1–10 GPa
Ice	10 GPa
Engineering plastics	5–20 GPa
Leather	1–100 MPa
Rubber	0.1–5 MPa
Polymer and colloidal solutions	1–100 Pa
Dry spaghetti	3 GPa
Carrots	20–40 MPa
Pears	10–30 MPa
Potatoes	6–14 MPa
Peach	2–20 MPa
Raw apples	6–14 MPa
Gelatin Gel	0.2 MPa
Banana	0.8–3 MPa

Source: Steffe (1996), Malkin and Isayev (2006)

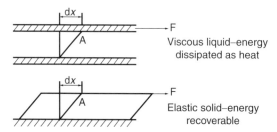

Viscous liquid–energy dissipated as heat

Elastic solid–energy recoverable

Figure 1.20   *Qualitative differences between a viscous fluid and an elastic solid*

visco-elastic (as some of the proprietary soups are), the liquid will be seen to slow to a stop and then to unwind a little. This type of behaviour is closely linked to the tendency for a gel structure to form within the fluid; such an element of rigidity makes simple shear less likely to occur – the shearing forces tending to act as couples to produce rotation of the fluid elements as well as pure slip. Such incipient rotation produces a stress perpendicular to the direction of shear. Numerous other unusual phenomena often ascribed to fluid visco-elasticity include die swell, rod climbing (Weissenberg effect), tubeless

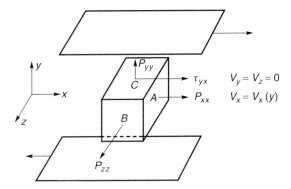

Figure 1.21 *Components of stress in one-dimensional steady shearing motion of a visco-elastic fluid*

syphon, bouncing of spheres, and the development of secondary flows at low Reynolds numbers. Most of these have been illustrated photographically in an excellent book (Boger and Walters, 1992). A detailed treatment of visco-elastic fluid behaviour is beyond the scope of this book and interested readers are referred to several excellent books available on this subject (e.g. see Schowalter, 1978; Bird *et al.*, 1987; Carreau *et al.*, 1997; Tanner and Walters, 1998; Larson, 1998; Morrison, 2001). Here we shall describe the 'primary' and 'secondary' normal stress differences observed in steady shearing flows which are used both to classify a material as visco-elastic or viscoinelastic as well as to quantify the importance of visco-elastic effects in an envisaged application.

Normal stresses in steady shear flows

Let us consider the one-dimensional shearing motion of a fluid; the stresses developed by the shearing of an infinitesimal element of fluid between two planes are shown in Figure 1.21. By nature of the steady shear flow, the components of velocity in the y- and z-directions are zero (i.e., $V_y = 0$; $V_z = 0$) while that in the x-direction is a function of y only, i.e., $V_x(y)$. Note that in addition to the shear stress, τ_{yx}, there are three normal stresses denoted by P_{xx}, P_{yy} and P_{zz} within the sheared fluid which are given by equation (1.6). Weissenberg (1947) was the first to observe that the shearing motion of a visco-elastic fluid gives rise to unequal normal stresses. Since the pressure in a non-Newtonian fluid cannot be defined by equation (1.7) the differences, $P_{xx} - P_{yy} = N_1$ and $P_{yy} - P_{zz} = N_2$, are more readily measured than the individual stresses, and it is therefore customary to express N_1 and N_2 together with τ_{yx} as functions of the shear rate $\dot{\gamma}_{yx}$ to describe the rheological behaviour of a visco-elastic material in a simple shear flow. Sometimes the first and second normal stress differences N_1 and N_2 are expressed in terms of two coefficients, ψ_1, and ψ_2 defined as follows:

$$\psi_1 = \frac{N_1}{\dot{\gamma}_{yx}^2} \tag{1.23}$$

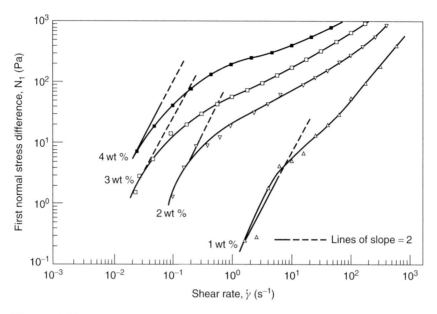

Figure 1.22   *Representative first normal stress difference data for polystyrene-in-toluene solutions at 298 K (Kulicke and Wallbaum, 1985)*

and

$$\psi_2 = \frac{N_2}{\dot{\gamma}_{yx}^2} \tag{1.24}$$

A typical dependence of the first normal stress difference on shear rate is shown in Figure 1.22 for a series of polystyrene-in-toluene solutions. Usually, the rate of decrease of ψ_1 with shear rate is greater than that of the apparent viscosity. At very low shear rates, the first normal stress difference, N_1, is expected to be proportional to the square of shear rate – that is, ψ_1 tends to a constant value ψ_0; this limiting behaviour is seen to be approached by some of the experimental data shown in Figure 1.22. It is common that the first normal stress difference N_1 is higher than the shear stress τ at the same value of shear rate. The ratio of N_1 to τ is often taken as a measure of how elastic a liquid is; specifically $(N_1/2\tau)$ is used and is called the recoverable shear. Recoverable shears greater than 0.5 are not uncommon in polymer solutions and melts. They indicate a highly elastic behaviour of the fluid. There is, however, no evidence of ψ_1 approaching a limiting value at high shear rates. It is fair to mention here that the first normal stress difference has been investigated much less extensively than the shear stress.

Even less attention has been given to the study and measurement of the second normal stress difference. The most important points to note about N_2 are that it is an order of magnitude smaller than N_1, and that it is negative. Until recently, it was thought that $N_2 = 0$; this so-called Weissenberg hypothesis is no longer believed to be correct. Some data in the literature even seem to suggest that N_2 may change sign. Typical forms of the dependence of N_2 on shear rate are shown in Figure 1.23 for the same solutions as used in Figure 1.22.

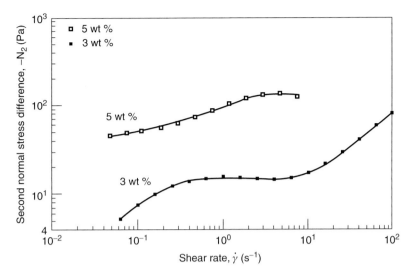

Figure 1.23   *Representative second normal stress difference data for polystyrene-in-toluene solutions at 298 K (Kulicke and Wallbaum, 1985)*

The two normal stress differences defined in this way are characteristic of a material, and as such are used to categorize a fluid either as purely viscous ($N_1 \sim 0$) or as visco-elastic, and the magnitude of N_1 in comparison with τ_{yx}, is often used as a measure of visco-elasticity.

Aside from the simple shearing motion, the response of visco-elastic materials in a variety of other well-defined flow configurations including the cessation/initiation of flow, creep, small amplitude sinusoidal shearing, extension, etc. also lies in between that of a perfectly viscous fluid and a perfectly elastic solid, as will be seen in Chapter 2. Conversely, these tests may be used to infer a variety of rheological information about a material. Detailed discussions of the subject are available in a number of books, e.g. see Walters (1975), Whorlow (1992) and Macosko (1994).

Elongational flow

Flows which result in fluids being subjected to stretching in one or more directions occur in many processes, fibre spinning and polymer film blowing being only two of the most common examples. Again, when two bubbles coalesce, a very similar stretching of the liquid film between them takes place until rupture occurs. Another important example of the occurrence of extensional effects is the flow of polymer solutions in porous media, as encountered in the enhanced oil recovery process, in which the fluid is stretched as the extent and shape of the flow passages change. There are three main forms of elongational flow: uniaxial, biaxial and planar, as shown schematically in Figure 1.24.

Fibre spinning is an example of uniaxial elongation (but the stretch rate varies from point to point along the length of the fibre). Tubular film blowing which involves extruding of polymer through a slit die and pulling the emerging sheet forward and sideways is an example of biaxial extension; here, the stretch rates in the two directions can normally be specified and controlled. Another example is the manufacture of plastic tubes

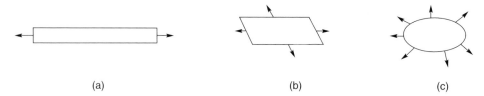

(a)   (b)   (c)

Figure 1.24   *Schematic representation of uniaxial (a), biaxial (b) and planar (c) extension*

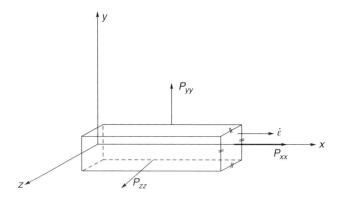

Figure 1.25   *Uniaxial extensional flow*

which may be made either by extrusion or by injection moulding, followed by heating and subjection to high pressure air for blowing to the desired size. Due to symmetry, the blowing step is an example of biaxial extension with equal rates of stretching in two directions. Irrespective of the type of extension, the sum of the volumetric rates of extension in the three directions must always be zero for an incompressible fluid.

Naturally, the mode of extension affects the way in which the fluid resists deformation and, although this resistance can be referred to loosely as being quantified in terms of an elongational or extensional viscosity (which further depends upon the type of elongational flow, i.e., uniaxial, biaxial or planar), this parameter is, in general, not necessarily constant. For the sake of simplicity, consideration may be given to the behaviour of an incompressible fluid element which is being elongated at a constant rate $\dot{\varepsilon}$ in the x-direction, as shown in Figure 1.25. For an incompressible fluid, the volume of the element must remain constant and therefore it must contract in both the y- and z-directions at the rate of $(\dot{\varepsilon}/2)$, if the system is symmetrical in those directions. The normal stresses P_{yy} and P_{zz} will then be equal. Under these conditions, the three components of the velocity vector V are given by:

$$V_x = \dot{\varepsilon}x, \quad V_y = -\frac{\dot{\varepsilon}}{2}y \text{ and } V_z = -\frac{\dot{\varepsilon}}{2}z \tag{1.25}$$

and clearly, the rate of elongation in the x-direction is given by:

$$\dot{\varepsilon} = \frac{\partial V_x}{\partial x} \tag{1.26}$$

In uniaxial extension, the elongational viscosity μ_E is then defined as:

$$\mu_E = \frac{P_{xx} - P_{yy}}{\dot{\varepsilon}} = \frac{\tau_{xx} - \tau_{yy}}{\dot{\varepsilon}} \tag{1.27}$$

or P_{yy} and τ_{yy} can be replaced by P_{zz} and τ_{zz}, respectively.

The earliest determinations of elongational viscosity were made for the simplest case of uniaxial extension, the stretching of a fibre or filament of liquid. Trouton (1906) and many later investigators found that, at low strain (or elongation) rates, the elongational viscosity μ_E was three times the shear viscosity μ (Barnes *et al.*, 1989). The ratio μ_E/μ is referred to as the Trouton ratio, T_r and thus:

$$T_r = \frac{\mu_E}{\mu} \tag{1.28}$$

The value of 3 for Trouton ratio for an incompressible Newtonian fluid applies to all values of shear and elongation rates. By analogy, one may define the Trouton ratio for a non-Newtonian fluid:

$$T_r = \frac{\mu_E(\dot{\varepsilon})}{\mu(\dot{\gamma})} \tag{1.29}$$

The definition of the Trouton ratio given by equation (1.29) is somewhat ambiguous, since it depends on both $\dot{\varepsilon}$ and $\dot{\gamma}$, and some convention must therefore be adopted to relate the strain rates in extension and shear. To remove this ambiguity and at the same time to provide a convenient estimate of behaviour in extension, Jones *et al.* (1987) proposed the following definition of the Trouton ratio:

$$T_r = \frac{\mu_E(\dot{\varepsilon})}{\mu(\sqrt{3}\dot{\varepsilon})} \tag{1.30}$$

i.e., in the denominator, the shear viscosity is evaluated at $\dot{\gamma} = \sqrt{3}\dot{\varepsilon}$. They also suggested that for inelastic isotropic fluids, the Trouton ratio is equal to 3 for all values of $\dot{\varepsilon}$ and $\dot{\gamma}$, and any departure from the value of 3 can be ascribed unambiguously to visco-elasticity. In other words, equation (1.30) implies that for an inelastic shear-thinning fluid, the extensional viscosity must also decrease with increasing rate of extension (so-called 'tension-thinning'). Obviously, a shear-thinning visco-elastic fluid (for which the Trouton ratio will be greater than 3, often reaching upto 10^3) will thus have an extensional viscosity which increases with the rate of extension; this property is also called 'strain-hardening'. Many materials including polymer melts and solutions thus exhibit shear-thinning in simple shear and strain-hardening in uniaxial extension. Except in the limit of vanishingly small rates of deformation, there does not appear to be any simple relationship between the elongational viscosity and the other rheological properties of the fluid and, to date, its determination rests entirely on experiments which themselves are often constrained by the difficulty of establishing and maintaining an elongational flow field for long enough for the steady state to be reached (Gupta and Sridhar, 1988; James and Walters, 1994). Measurements made on the same fluid using different methods seldom show quantitative agreement, especially for low to medium viscosity fluids (Tirtaatmadja and Sridhar, 1993). The Trouton ratios for biaxial and planar

Figure 1.26   *Schematic representation of the Maxwell model*

extensions at low strain rates have values of 6 and 9, respectively for all inelastic fluids and for Newtonian fluids under all conditions.

Mathematical models for visco-elastic behaviour

Though the results of experiments in steady and transient shear or in an elongational flow field may be used to calculate viscous and elastic properties for a fluid, in general the mathematical equations need to be quite complex in order to describe a real fluid adequately. Certainly, the most striking feature connected with the deformation of a visco-elastic substance is its simultaneous display of 'fluid-like' and 'solid-like' characteristics. It is thus not at all surprising that early attempts at the quantitative description of visco-elastic behaviour hinged on the notion of a linear combination of elastic and viscous properties by using mechanical analogues involving springs (elastic component) and dashpots (viscous action). The Maxwell model represents the cornerstone of the so-called linear visco-elastic models; though it is crude, nevertheless it does capture the salient features of visco-elastic behaviour, at least qualitatively.

A mechanical analogue of this model is obtained by series combinations of a spring and a dashpot (a vessel whose outlet contains a flow constriction over which the pressure drop is proportional to flow rate), as shown schematically in Figure 1.26. If the individual strain rates of the spring and the dashpot, respectively, are $\dot{\gamma}_1$ and $\dot{\gamma}_2$, then the total strain rate $\dot{\gamma}$ is given by the sum of these two components:

$$\dot{\gamma} = \dot{\gamma}_1 + \dot{\gamma}_2 = \frac{d\gamma_1}{dt} + \frac{d\gamma_2}{dt} \tag{1.31}$$

Combining equation (1.31) with the Hooke's law of elasticity, equation (1.22) and Newton's law of viscosity, one can obtain:

$$\tau + \lambda\dot{\tau} = \mu\dot{\gamma} \tag{1.32}$$

where $\dot{\tau}$ is the time derivative of τ; μ is the viscosity of the dashpot fluid and $\lambda(=\mu/G)$ is the relaxation time, which is a characteristic of the fluid. In order to illustrate the behaviour of a Maxwell fluid, the simplest experiment one can perform is to apply instantaneously a small strain (strain jump) which is maintained at a constant level. For an ideal elastic substance (spring), the stress will follow the strain and reach a constant value in accordance with the Hooke's law of elasticity, i.e., $\tau = G\gamma$. In the Maxwell model, the spring will therefore respond instantly in such a strain-jump experiment.

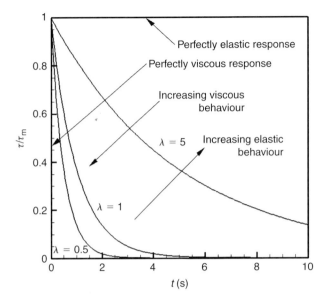

Figure 1.27 *Stress decay behaviour of a Maxwell fluid in a strain-jump experiment*

On the other hand, the piston will move slowly through the dashpot and therefore, the stress in the spring will gradually reduce with time. For a constant value of γ, the rate of change of strain is zero, i.e., $\dot{\gamma} = 0$. Equation (1.32) thus reduces to:

$$\tau + \lambda\dot{\tau} = 0 \tag{1.33}$$

which can be re-written as:

$$\frac{d\tau}{dt} = -\frac{\tau}{\lambda} \tag{1.34}$$

Integration with respect to time together with the initial condition at $t = 0$, $\tau = \tau_m$ (where τ_m is the maximum value of stress) leads to:

$$\tau = \tau_m \exp(-t/\lambda) \tag{1.35}$$

This equation describes the decay of stress with time when a rapid strain is applied (and thereafter maintained at this value) to a Maxwell fluid (Figure 1.27). This is similar to a first-order chemical reaction in which the concentration of the reactant is depleted exponentially in accordance with equation (1.35).

In early stages ($t \to 0$) of the strain-jump experiment, the exponential term $\exp(-t/\lambda)$ tends to one and the response is simply that of the spring, i.e., of an elastic fluid. When $t = \lambda$, the stress has dropped from its maximum value τ_m to τ_m/e. Thus, the value of λ dictates the rate of stress relaxation. Conversely, such simple tests can be used to evaluate the material constants like shear modulus (G) and relaxation time (λ). For instance, as noted above, for small value of t, the response of a Maxwell fluid in a strain-jump test is of the spring and therefore the initial value of stress (τ_m) divided by the strain gives the value of shear modulus (stress divided by strain) of the spring. Similarly, the value of the relaxation time, λ, can be extracted from a stress–time curve.

Figure 1.28   *Schematic representation of the Voigt model*

An important feature of the Maxwell model is its predominantly fluid-like response. A more solid-like behaviour is obtained by considering the so-called Voigt model which is represented by the parallel arrangement of a spring and a dashpot, as shown in Figure 1.28.

In this case, the strain in the two components is equal and the equation describing the stress–strain behaviour of this system is obtained by adding the individual stresses in the two elements as:

$$\tau = G\gamma + \mu\dot{\gamma} \tag{1.36}$$

If the stress is constant at τ_0 and the initial strain is zero, upon the removal of the stress, the strain decays exponentially with a time constant, $\lambda(=\mu/G)$.

The more solid-like response of this model is clear from the fact that it does not exhibit unlimited non-recoverable viscous flow and it will come to rest when the spring has taken up the load.

One of the main virtues of such linear models is that they can be conveniently superimposed by introducing a spectrum of relaxation times, as exhibited in practice, by polymeric systems or by including higher derivatives. Alternatively, using the idea of superposition, one can assume the stress to be due to the summation of a number of small partial stresses, each pertaining to a partial strain, and each stress relaxing according to some relaxation law. This approach yields the so-called 'integral' models. In addition, many other ideas have been employed to develop elementary models for visco-elastic behaviour including the dumb-bell, bead-spring representations, network and kinetic theories. Invariably, all such attempts entail varying degrees of idealization and empiricism; their most notable limitation is the restriction to small strain and strain rates.

The next generation of visco-elastic fluid models has attempted to relax the restriction of small deformation and deformation rates, thereby leading to the so-called non-linear models. Excellent critical appraisals of the developments in the field, together with the merits and de-merits of a selection of models, as well as some guidelines for selecting an appropriate equation for an envisaged process application, are available in the literature (e.g. see references Bird *et al.*, 1987; Larson, 1988, 1999; Barnes *et al.*, 1989; Macosko, 1994; Bird and Wiest, 1995; Tanner, 2000; Han, 2007).

1.6 Dimensional considerations for visco-elastic fluids

It has been a common practice to describe visco-elastic fluid behaviour in steady shear in terms of a shear stress (τ_{yx}) and the first normal stress difference (N_1); both of which are functions of shear rate. Generally, a fluid relaxation or characteristic time, λ_f, (or a spectrum) is defined to quantify the visco-elastic behaviour. There are several ways of defining a characteristic time by combining shear stress and the first normal stress difference, e.g. the so-called Maxwellian relaxation time is given by:

$$\lambda_f = \frac{N_1}{2\tau_{yx}\dot{\gamma}_{yx}} \tag{1.37}$$

Since, in the limit of $\dot{\gamma}_{yx} \to 0$, both $\psi_1 (= N_1/\dot{\gamma}_{yx}^2)$ and $\mu \ (= \tau_{yx}/\dot{\gamma}_{yx})$ approach constant values, λ_f also approaches a constant value. Though equation (1.37) defines a fluid characteristic time as a function of shear rate, its practical utility is severely limited by the fact that in most applications, the kinematics (or shear rate) is not known *a priori*. Many authors (Leider and Bird, 1972; Grimm, 1978) have obviated this difficulty by introducing the following alternative definition of λ_f :

$$\lambda_f = \left(\frac{m_1}{2m}\right)^{1/(p_1-n)} \tag{1.38}$$

This definition is based on the assumption that both N_1 and $\dot{\gamma}_{yx}$ can be approximated as power-law functions of shear rate in the range of conditions of interest, that is,

$$N_1 = m_1(\dot{\gamma}_{yx})^{p_1} \tag{1.39}$$

and

$$\tau_{yx} = m(\dot{\gamma}_{yx})^n \tag{1.12}$$

By re-defining λ_f in this manner, it is not necessary to extend the rheological measurements to the zero-shear region. Note that in the limit of $\dot{\gamma}_{yx} \to 0$, $p_1 \to 2$, $n \to 1$ and thus equation (1.38) coincides with equation (1.37).

For Newtonian fluids, the state of flow can be described by two dimensionless groups, usually the Reynolds number, *Re* (inertial forces/viscous forces) and Froude number, *Fr* (inertial/gravity forces). For a visco-elastic fluid, at least one additional group involving elastic forces is required.

The Reynolds number represents the ratio of inertial to viscous forces, and it might be reasonable to expect that such a ratio would provide a useful parameter. Unfortunately, attempts to achieve meaningful correlations have not been very successful, perhaps being defeated most frequently by the complexity of natural situations and real materials. One simple parameter that may prove of value is the ratio of a characteristic time of deformation to a natural time constant for the fluid. The precise definition of these times is a matter for argument, but it is evident that for processes that involve very slow deformation of the fluid elements it is possible for the elastic forces to be released by the normal processes of relaxation as they build up. As examples of the flow of rigid (apparently infinitely viscous) materials over long periods of time, even the thickening of the lower parts of medieval glass windows is insignificant compared with the plastic flow and deformation that lead to the folded strata of geological structures. In operations that are carried out rapidly, the extent of viscous flow will be minimal and the deformation will be followed by recovery when the stress is removed. To get some idea of the possible regions in which such an analysis can provide guidance, consider the flow of a 1% aqueous polyacrylamide solution. Typically, this solution might have a relaxation time of the order of 10 ms. If the fluid were flowing through a packed bed, it would be subjected to alternating acceleration and deceleration as it flowed through the interstices of the bed. With a particle size of 25 mm, say, and a superficial velocity of 0.25 m/s, the deformation or process time will be of the order of $25 \times 10^{-3}/0.25 = 0.1$ s which is greater than the fluid relaxation time. Thus, the fluid elements can adjust to the changing flow area and one would therefore not expect the elastic properties to influence the flow significantly. However, in a bed composed of 250 μm sand grains, the characteristic process time would be 1 ms and therefore, the fluid element now will not be able to relax the elastic stresses and one can expect significant visco-elastic effects in this case. Similarly, in a free jet discharge with a velocity of 30 m/s through a nozzle of 3 mm diameter, the deformation or process time of $(3 \times 10^{-3})/30 = 0.1$ ms which is 100 times smaller than the fluid response time and the fluid elements will not have sufficient time to re-adjust; it might therefore be reasonable to expect some evidence of elastic behaviour near the point of discharge from the nozzle. A parameter which might be expected to be important is the Deborah number, De, is defined as (Metzner *et al.*, 1966):

$$De = \frac{\text{Fluid response time}}{\text{Process characteristic time}} \tag{1.40}$$

In the examples above, for the packed bed, $De = 0.1$, 10 and for the free jet, $De = 100$. The greater the value of De, the more likely the elasticity to be of practical significance. The same material might well exhibit strongly elastic response under certain conditions of deformation and essentially an inelastic response under other conditions, as seen above in the packed bed flow example. The extent of visco-elastic phenomenon is thus determined by a combination of material properties and flow kinematics; the limits of viscous and elastic behaviour are given as $De \rightarrow 0$ and $De \rightarrow \infty$, respectively.

Unfortunately, this group depends on the assignment of a single characteristic time to the fluid (a relaxation time?). While this is better than no description at all, it appears to be inadequate for many visco-elastic materials which show different relaxation behaviour under differing conditions.

From the preceding discussion in this chapter, it is abundantly clear that each non-Newtonian fluid is unique in its characteristics, and the only real information about the rheology of a material comes from the experimental points or flow curves that are obtained using some form of rheometers, as discussed in Chapter 2. Provided that there are sufficient experimental points, interpolation can usually be satisfactory; extrapolation should, however, be avoided as it can frequently lead to erroneous results. Certainly, the fitting of an empirical viscosity model to limited data should not be used as a justification for extrapolating the results beyond the experimental range of shear rates or shear stresses. Similarly, it is usually possible to fit a number of different equations (e.g. the power-law and the Bingham plastic model) to a given set of data equally well, and the choice is largely based on convenience or individual preference. Frequently, it is not possible to decide whether a true yield stress exists or not. Therefore, some workers prefer to refer to an 'apparent yield stress' which is an operational parameter and its evaluation involves extrapolation of data to zero shear rate, often the value depending upon the range of data being used to evaluate it. For instance, it is therefore likely that the values of the apparent yield stress fitted in the Bingham, Herschel–Bulkley and Casson models may be quite different for the same fluid (see example 1.2). Thus, extreme caution must be exercised in analysing, interpreting and using rheological data.

1.7 Influence of micro-structure on rheological behaviour

By now it must be abundantly clear that non-Newtonian flow characteristics in most cases are observed in the so-called 'structured' fluids. Undoubtedly, there is a direct and strong link between the type and extent of non-Newtonian flow behaviour on one hand and the response of the structure to externally applied shearing forces on the other. In principle, it is therefore possible to synthesize a material of desired rheological characteristics by manipulating its structure using a range of additives. Conversely, rheological measurements can be used to gain useful insights about the micro-structure of such a system. Though this link is exploited extensively on a routine basis in formulating materials into a convenient form in food, personal care, cleaning aids and pharmaceutical sectors, its significance can be illustrated by considering the following examples of products we use every day: a decorative paint, chocolate, mayonnaise and toothpaste. Let us look in detail the function of a paint. The function of a paint is twofold: to provide a polymeric film to protect the surface and to achieve a decorative finish via the paint as a carrier for colouring agents (pigments). Ideally, in the storage conditions (i.e, in a can) the pigment particles should remain in suspension and this is accomplished by producing a weak-gel (by inducing a yield stress or a high viscosity at low stress levels). This characteristic also leads to another desirable effect, namely, the non-drip behaviour of the paint after its application. On the other hand, during the application of the paint by a brush, it must easily 'thin' (shear-thinning) to be readily laid on the surface. Also, we would like to achieve as uniform a thickness of the coating as possible (controlled by surface tension forces) and of course, without it running (caused by gravity) prior to drying. Thus, after its application on to a surface, the paint should start gelling again but at a slow rate such that adequate leveling may occur. In other words, an ideal paint should be a soft solid (in the can) that becomes a viscous liquid during its application (by a brush or a roller) and finally, it must re-solidify at a controlled rate, slow enough for adequate leveling to occur, but quickly enough to minimize the tendency for running of a vertical surface (or leading to uneven film thickness on a horizontal surface) due to the effect of gravity, i.e., under its own weight. All these attributes are imparted by manipulating the inter-particle forces responsible for the structuring of these systems. Similarly, one can consider the role of micro-structure in chocolates, which are like soft solids when in a wrapper, and transform into a sweet and creamy viscous liquid in the mouth by the shearing action of the tongue. A molten chocolate contains sugar, cocoa and powdered milk particles in a Newtonian fat melt, typically cocoa butter. Depending upon the prevailing levels of shearing stress, the dispersed particles can organize themselves in a number of forms including platelets, fibres or coils which can form agglomerates or form edge–face or face–face like assemblies (see Figure 1.29). At low stress levels, these three-dimensional structures offer a very high resistance which is frequently interpreted in terms of a yield stress. With a gradual increase in the applied stress, chocolates can exhibit shear-thinning and shear-thickening characteristics under appropriate conditions (Figure 1.29). Thus, the challenge for chocolatiers and food scientists is to understand the link between the micro-structure and the bulk rheology which will not only lead to a consistent product quality, but will also help in meeting the ever changing demands of the consumers in a systematic manner rather than relying on a trial and error approach. Our next example, mayonnaise is an emulsion of a vegetable oil in vinegar or in lemon juice which is stabilized by lecithin, a natural surfactant present in egg yolk. Mayonnaise more or less holds its shape (like a soft solid) against gravity, but it flows readily on the application of

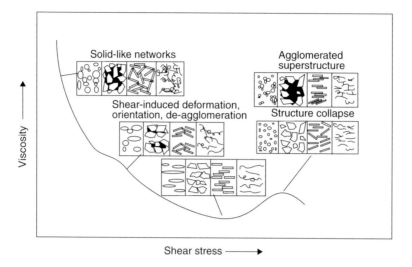

Figure 1.29   *Typical relationship between the viscosity and structure of a chocolate (replotted from Windhab, 2006)*

small stresses (like during pouring and spreading); therefore it must be a viscoplastic substance with a small yield stress. The magnitude of the yield stress is primarily governed by the size of oil droplets ($10\,\mu$m diameter is typical for a good mayonnaise) and by the surface tension of the droplets. The final example of toothpaste is equally illustrative of the strong structure–rheology link. The toothpaste flows out of the tube only when it is squeezed gently and it stops flowing immediately after it has been transferred on to the brush. This prevents it from sinking into the bristles under its own weight and therefore, some of the initial yield stress is regained. Finally, it shows almost water-like consistency under high shear rates achieved during brushing. Thus, the toothpaste has a yield stress and it exhibits shear-thinning upon flowing. Therefore, in addition to the functional ingredients like the gently abrasive particles for cleaning action (silica powder), fluoride ions for preventing tooth-decay and cavity formation, flavouring and foaming agents, good toothpaste also contains a suitable polymer like carboxymethyl cellulose for controlling its rheology. In essence, toothpaste is a thick suspension! Such examples abound in food, pharmaceutical, house-hold and personal-care product sectors where the desired rheological response of a product is obtained by tailoring its micro-structure by adjusting the composition and/or physico-chemical environment of the system. Thus, there is no question that an adequate understanding of the structure-rheology link is of great importance in numerous industrial settings.

Before examining the role of micro-structures on the macroscopic flow behaviour of these systems, it is appropriate to review two key assumptions implicit in the concept of shear or elongational viscosity or any other rheological material functions like yield stress, first normal stress difference, etc. Firstly, in this approach, the so-called continuum description of such materials is assumed to be valid, i.e., the micro-structural details are deemed unimportant in evaluating the gross-flow characteristics, albeit no real fluids are truly 'structureless continua'. Thus, the use of viscosity as a space- and time-averaged fluid property poses no difficulty for low molecular weight substances with molecular dimensions of the order of 1–$10\,$nm. Similarly, concentrated polymer solutions or melts, colloidal systems, foams, worm-like micellar systems, etc. all possess 'micro-structures' of a size approaching 1–2 μm which can be regarded to be a continuous medium, except perhaps during their flow in very fine and/or twisted flow channels (such as in porous rocks). In other words, such two-phase systems can be treated as pseudo-homogeneous (or single phase) fluids and assigned average values of the viscosity and the other material functions if the distribution of particles in the liquid phase is not influenced by the channel dimensions. The second issue concerns the homogeneity of a substance so that the use of a space-averaged fluid property is justified. Both these conditions, namely, continuum description and homogeneity of substances are assumed to be valid throughout the discussion in this book.

As noted earlier, though all fluids are structured to varying extents, what sets the two-phase or macromolecular rheologically complex systems apart from simple fluids (like hydrocarbons) is the fact that the 'structures' present in the rheologically complex systems are not only transient in nature, but can easily be perturbed by the application of relatively low stresses. For instance, for an organic fluid like cyclohexane, its structure remains unperturbed by the application of stresses almost up to about 1 MPa. On the other hand, the corresponding value for a polymer of moderate molecular weight is only about 100 Pa and for a colloidal system (of radius 100 nm), it is only of the order of 200 mPa. It is this ease of being able to change the micro-structure by applied stresses which translates into a range of useful rheological features experienced in day-to-day activities such as pouring of liquids, spreading and rubbing of lotions, etc.

Figure 1.30 shows qualitatively the different types of micro-structures encountered in rheologically complex systems in a rest state (relevant to storage conditions) and how these get modified under the action of shear or in a flowing state. Most systems contain irregularly shaped particles with size distribution, droplets in case of emulsions, or branched and/or highly entangled long chain polymeric molecules in solutions and melts, or loosely formed aggregates of particles in suspensions, etc. At rest, these structures are oriented randomly corresponding to their minimum energy state. At low levels of externally applied stress, the system resists any deformation and strives to retain its structure thereby offering a very high resistance by exhibiting either a very large value of the apparent viscosity or a yield stress, as seen in the case of chocolate (Figure 1.29). As the level of shearing stress is gradually increased, the structural units respond to the applied stress either by aligning themselves in the direction of flow (Figure 1.30), or by deforming to orient along the streamlines, or by way of decomposition of the aggregates into primary particles. Similarly, coiled and entangled polymeric molecules may become disentangled and finally, may fully straighten out (Figure 1.31). All these micro-structural changes facilitate bulk flow in the system. This can be seen as the lowering of the effective viscosity with increasing rate of shear, as seen in viscoplastic and shear-thinning substances. Of course, at very high values of the applied stress, no more changes are possible in shape and/or in their orientation and some fluids (especially polymeric solutions) once again display nearly a constant viscosity, the so-called infinite shear viscosity (see Figure 1.6).

Aside from the aforementioned types of micro-structural changes induced by shearing motion, many other possibilities exist and contribute to micro-structural changes depending upon the relative magnitudes of inter-particle forces. For instance, in suspensions of fine (sub-micron range) particles, the van der Waals attraction forces between particles can cause the particles to stick to each other on touching. This effect is fairly significant in colloidal suspensions (colloidal particles of ∼1 μm) and it leads to coagulation. Similarly, a strong repulsive force stems from repulsion between like-charges on the surface of neighbouring particles. If these repulsive forces are sufficiently large and

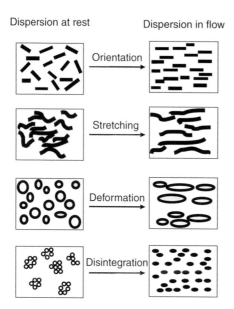

Figure 1.30   *Schematics of 'structures' in rheologically complex systems under rest and under shearing conditions (from Brummer, 2006)*

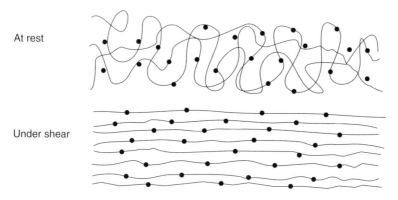

Figure 1.31   *Qualitative role of shearing in uncoiling and stretching of an entwined macromolecule*

long-range, then the effect of the attractive van der Waals forces is countered thereby yielding a stable colloidal dispersion. In practice, this is achieved by using block copolymers, which have one end soluble in the solvent and the other end insoluble and capable of being absorbed on to the particle surface. In this case, the protruding polymer rings and strands prevent the particles coming in contact with each other thereby minimizing the tendency for flocculation. This effect is called *steric* repulsion. A similar effect is also achieved by using absorbed polyelectrolytes, i.e., by using polymers with charges distributed along their chain. While the nature and intensity of the resulting forces may vary from one case to another, the overall effect is a net repulsive force. Another attractive force may also manifest when the suspending medium itself is a polymer solution. In this case, when the two particles approach each other, the gap between them may be too small for a large polymer molecule or a micelle to be present here. This can lead to a polymer-depleted region which in turn leads to a local gradient in osmotic pressure.

The net result is that the two neighbouring particles attract each other. This combined with van der Waals attractive forces can also lead to flocculation. Finally, from an application standpoint, flocculation is encountered far more frequently than the completely dispersed suspensions of primary particles. In these situations, the attractive and repulsive forces typically range from a few $k_B T$ to ~ 10–$20\ k_B T$ where k_B is the Boltzmann constant and T is the absolute temperature. Naturally, no flocculation occurs when these forces are small, because the Brownian motion prevents particle–particle contact. Conversely, one can infer that the resulting flocs are not very strong and the shearing levels encountered in typical pipe flows are sufficient to break (at least partially) such flocs, albeit these might re-form under quiescent conditions. The rheology, especially viscosity, of a flocculated suspension is strongly influenced by the shape and size of a floc. Naturally, intense shearing will reduce the flocs to primary particles and low to moderate shearing might break it into smaller flocs. The influence of shearing on the size of a floc is approximated as (Barnes, 1992):

$$\frac{R_s}{R_0} = \left[1 + \left(b \dot{\gamma} \right)^c \right]^{-1} \tag{1.41}$$

where R_0 and R_s are the effective initial and reduced sizes of a floc; b and c are two constants which vary from one system to another. On the other hand, the form (or shape) of a floc is governed by the interaction forces and the flow history. This process of breakdown of structure under shear and rebuild up under quiescent conditions manifests itself as thixotropy, as seen in Figure 1.15.

In view of the aforementioned link between rheology and structure, it is possible to control the rheological behaviour of a colloidal suspension by altering the balance of attractive and repulsive forces. Two commonly used approaches to achieve this goal are via the addition of an electrolyte to a suspension of charged particles and by altering the pH of the solution. In the first case, similar charges on particle surface lead to a strong repulsion between neighbouring particles, which in turn, results in a pseudo-crystalline state with particles occupying regular lattice positions. Obviously, this arrangement restricts the mobility (except for Brownian effects) of particles even in the presence of shear thereby resulting in a very high value of viscosity. Now, with the addition of a salt to the aqueous phase, the charges on the particle are shielded (thus lowering the repulsion) and this improves the mobility of the system (lower viscosity), at least at low shear rates. With the further addition of salt, at some stage, there is no net charge (neutral) on the particles and this corresponds to minimum viscosity. Further addition reduces the repulsion and the van der Waals attractive forces take over leading to flocculation which is accompanied by an increase in viscosity. The second idea of modulating solution pH relies on the fact that the ubiquituous hydroxyl groups present in several mineral particles (quartz, chalk, titanium, etc.) become charged. This provides a convenient scheme to control the charge on a particle; low pH leads to a large positive charge and high pH to a large negative charge, the point of zero charge is known as the isoelectric point. As the surface charge changes from being positive to negative, the viscosity increases appreciably. At the isoelectric point, the system flocculates which is accompanied not only by an enormous rise in viscosity, but also by showing shear-thinning behaviour. Finally, a somewhat similar effect can also be achieved by adding a surfactant molecule to a flocculated system. Essentially, the particle surface is covered or shielded thereby reducing the attractive forces which, in turn, leads to some deflocculation and hence lowering of the viscosity.

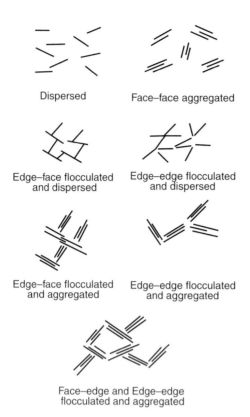

Figure 1.32 *Possible forms of agglomerates in kaolin suspensions (redrawn from Schramm, 2005)*

The rheological behviour of aqueous kaolin suspensions can be similarly modified by adjusting the solution pH or by adding a surfactant. The kaolin consists of plate-like particles and depending upon the surface charges, it can form aggregates, which are either edge–face ('house of cards') type or face–face type in quiescent conditions. These two possibilities result in completely different rheological behaviour, as under some conditions groups of platelets form such aggregates (Figure 1.32). As expected the 'house of cards' configuration exhibits a yield stress due to the formation of three-dimensional networks whereas when this configuration is changed to face–face by adding a surfactant (hexadecyl trimethyl ammonium bromide, HDTAB in this case), not only the yield stress disappears altogether, but the suspension also shows a much lower viscosity (Figure 1.33). The surfactant adsorbs on the face of platelets thereby disrupting the structure. Thus, suitable tuning of the properties of other similar clays – natural and synthetic (like laponite) also allows the control of rheological characteristics in these systems.

Polymeric systems form a class of industrially significant materials which also exhibit complex rheological characteristics. Owing to their wide occurrence in nature (gum-arabic, cellulose, guar gum, alginates, pectins, gelatin, starches, etc.) and their wide ranging applications in technology, the rheology of polymeric systems – melts and solutions with and without fillers – has been explored extensively. Synthetic polymers represent a significant industry, whose products impinge on our everyday life ranging from house-hold items like containers, utensils, cabinets of TV and PC to highly specialized

Figure 1.33   *Effect of surfactant (HDTAB) on the steady shear behaviour of a 2% (by volume) sodium kaolinite suspension in water at pH of 4 (replotted from Goodwin and Hughes, 2000)*

applications like plastic lenses for optical use, artificial organs, and fibres to transmit data in electronic forms. One only needs to walk down the aisles of supermarkets and departmental stores to gauge their impact in modern-day living. The special properties of these systems like visco-elasticity, shear-thinning, strain hardening, time-dependence, etc. stem from the ability of macromolecules to interact, to entangle and to knot up with their neighbours, and many possible configurations at a rest state which a macromolecule can acquire. It is not at all difficult to perturb their configuration by the application of relatively small forces. A macromolecule responds to such a force via straightening of chains (Figure 1.31), disentangling of loose networks, uncoiling and stretching, etc. in the presence of an imposed flow or force. In spite of their relatively large size, Brownian effects tend to randomize the flow units present in such polymeric systems. As seen above, the rheological behaviour of two-phase systems with fluid or solid particles is influenced strongly by pair interactions, the polymeric systems also exhibit a rich spectrum of interesting and complex characteristics once the polymer molecules begin to interact, entangle, knot up with each other. The significant factors governing the rheological behaviour of a polymeric system are the molecular weight (or molecular weight distribution), the structure of the molecule and their possible configurations, and the chemical composition. In the case of polymeric solutions, polymeric blends and filled systems, additional factors including polymer–solvent interactions (for polymer solutions), compatibility of the constituents in case of blends and physico-chemical properties of fillers, volume fraction, etc. also exert varying levels of influence in determining their rheology. It is therefore virtually impossible to develop universally applicable frameworks, as each system comes with a degree of its own peculiarity. However, some general observations can be made which shed some light on, or at least illustrate the role of micro-structure in determining the bulk rheological behaviour of these systems.

For a wide range of polymeric systems, the zero shear viscosity increases linearly with the molecular weight ($\mu_0 \propto M$) below a critical value of the molecular weight, M_c, which denotes the onset of entanglement. Above this value, the zero shear viscosity scales with molecular weight rather strongly, typically as $\propto M^{3.5}$. While the transition between the two regimes is not sharp, experimental data can often be represented by two straight lines (on log–log coordinates) with slopes of 1 and 3.5, respectively, and these lines when

$M < M_c$

$M > M_c$

No entanglement   Entangled system

Figure 1.34   *Schematic representation of entangled and disentangled molecules*

extrapolated intersect at $M = M_c$. The two regimes are shown schematically in Figure 1.34. Conversely, the low molecular weight regime relates to the state of no or little entanglement. The linear dependence of the zero shear viscosity on the molecular weight for short-chain polymers is predicted reasonably well by the so-called Rouse model. Table 1.6 lists typical values of the critical molecular weight for a selection of polymers. It is seen that the critical value of the molecular weight lies between 10 000 and 40 000 for most polymers, and it generally increases with the increasing stiffness of the chain. Such entanglements can be visualized as temporary cross-links which reduce the mobility of the system. Conversely, the resistance to flow or viscosity increases significantly with the increasing degree of entanglement. This phenomenon is analogous to the fact that a greater force is required to pull a strand from a ball of tangled yarn than when the ball is not tangled. As the molecular weight is progressively increased, both the probability and density of entanglement increase due to the increase in the chain length (Graessley, 1974).

In qualitative terms, below the critical molecular weight (M_c), the polymer chains are too short to interact with each other to become entangled and it behaves more or less as a non-interacting system. In this regime, doubling the molecular weight only increases the number of non-interacting (and perhaps slightly longer) entities and therefore the viscosity increases proportionally by a factor of two. On the other hand, in the regime characterized by $M > M_c$, polymeric chains become entangled and the bulk flow becomes extremely difficult, i.e., the viscosity increases significantly; doubling the molecular weight increases the viscosity by more than an order of magnitude ($\sim 2^{3.5} = 11.3$). Furthermore, most real polymers often display a distribution of molecular weights rather than a uniform molecular weight. For a narrow molecular weight distribution, the onset of shear-thinning is rather abrupt and occurs at a higher shear rate as opposed to that for a polymer with polydispersity or a wider MWD. At this juncture, it is useful to ask if there is a minimum chain length, or the number of monomer units for the onset of entanglement at $M = M_c$. Experimental studies with several polymers show that indeed there is a critical chain length Z_c which corresponds to the onset of entanglement. Furthermore, in contrast to the 10-fold variation seen in the value of M_c (Table 1.6), the critical chain length Z_c ranges from about 300 to 700 main chain atoms (Berry and Fox, 1968). In essence, the idea of entanglement is equivalent to an inter-molecular interaction which is not easy to characterize.

Chain architecture, on the other hand, is a description of the physical arrangement of the building blocks (monomers) of a polymer. While the linear (homopolymers) chains in which the monomers are arranged sequentially are easy to analyse and study, most systems of industrial significance are of branched-type, including long-chain branching, short-chain branching, radial branching (star-like), dense branching upon branches, known as hyperbranching and comb-like structure (see Figure 1.35). Dendrimers, a new class of polymers, consists of branches only and it resembles fractals (Figure 1.35).

Table 1.6   *Typical values of M_c for polymers*

Polymer	M_c
Linear polyethylene	4 000
Polyisobutylene	17 000
Polyvinyl acetate	29 200
Polystyrene	38 000
Polydimethyl siloxane	35 200
Polymethyl methacrylate	10 400
Caprolactum	
– Linear	19 200
– Tetra branched	22 000
– Octa branched	31 100

Source: Gupta (2000)

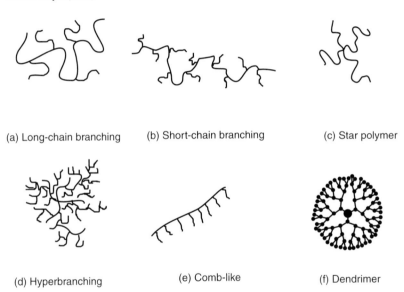

(a) Long-chain branching   (b) Short-chain branching   (c) Star polymer

(d) Hyperbranching   (e) Comb-like   (f) Dendrimer

Figure 1.35   *Representative architectures of branched polymer molecules*

Qualitatively speaking, branching influences the viscosity – shear rate relationship much the same way as does the degree of polydispersity. The effect of branching, (especially long-chain) is also modulated by shear rate. Since polymers relax by motion along their backbones (Figure 1.36), long-chain branching impedes relaxation thereby adding to the viscosity of a polymer at low shear rates. Also, once a branched polymer has been proc-essed for some time, it attains some chain configurations which require a very long time to randomize thereby introducing a dependence on the shear history. On the other hand at high shear rates, the viscosity of a branched system can be lower than that of a linear system of the same molecular weight. This is often explained by recognizing that both linear and branched chains disentangle under the action of shear and flow past each other as whole units and the compact form of branched molecules reduces the interactions

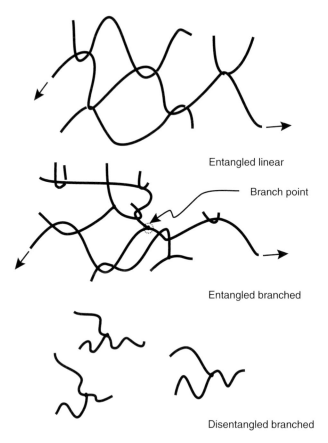

Entangled linear

Branch point

Entangled branched

Disentangled branched

Figure 1.36   *Schematics of relaxation in linear and branched polymers*

(and hence viscosity). Branched-chain polymers are preferred in processes with a strong elongational character such as in film blowing where the long-chain branching increases strain-hardening tendency (high values of Trouton ratio) which, in turn, stabilizes the blown film. In short-chain branched systems, the molecules do not extend far enough to have significant interactions with the neighbouring molecules. Therefore, their viscosity is less than that of a linear polymer of the same molecular weight. Similarly, since the hyperbranched polymers are very compact in form, not only these exhibit lower levels of viscosity than linear polymers of equal molecular weight, but also do not show shear-thinning behaviour. The molecular weight of the branches influences the value of viscosity much more than the number of chains. It is these differences in branching which cause the zero shear viscosity of a high-density polyethylene (HDPE) to be higher than that of the linear low-density polyethylene (LLDPE), but lower than that of a low-density polyethylene (Saini and Shenoy, 1983).

In addition, the rheology of polymers is also influenced by the nature of chains whether these are rod-like (rigid, semi-rigid) or coils and springs. For example, xanthan gum (MW $\sim 2 \times 10^6$) is naturally a rod-like molecule with length-to-diameter ratio of the order of 600 and the chain length of the order of ~ 0.5–$1\,\mu$m. Similarly, in the case of a polyelectrolyte in a weak electrolyte solution, the repulsive forces arising from the

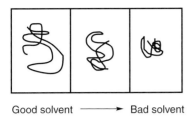

Good solvent ⟶ Bad solvent

Figure 1.37 *Influence of solvent quality on polymer coil dimensions*

charge along the polymer chain causes the convoluted chain to become straight thereby imparting some features of rod-like systems. At low shear rates, these long molecules are oriented randomly, and the resulting increase in their shear and elongational viscosities is strongly dependent on the value of (L/d). As the shear rate is progressively increased, the molecules become aligned along the direction of flow and the role of (L/d) gradually diminishes. This effect, in turn, manifests in the form of shear-thinning in these systems (Scheraga, 1955).

Another way of visualizing polymers is in terms of coils and springs. This description arises from the fact that polymers are made up of freely joined segments and their many configurations are possible. For instance at low concentration, when the linear coils are isolated and almost non-interacting with each other, the shape is nearly spherical which deforms into ellipsoids under the action of shear which are less of a bluff body (i.e., slightly aligned in the direction of shear). At high shear rates, coils can open up and straighten out into strings which are aligned fully in the direction of flow (Figure 1.31). As the number density of coils increases (by concentration in solution or by molecular weight in melts), inter-chain interactions in the form of entanglement and intertwining increase. When subject to an external force, these 'structures' continuously slide over each other during which it can knot up and disengage from individual entanglements. The visco-elasticity in polymeric melts and concentrated polymer solutions originates from the tension in the chains between disentanglement while the entangled chains slide over each other. However, their elastic behaviour is on a short-time scale governed by the life span of these transient networks, but as the stretched chains slide over each other, they relax and display the usual viscous behaviour.

In the case of polymer solutions, the solvent properties also play an important role. Figure 1.37 illustrates the effect of organic solvents from 'good' to 'bad' and the effect of this on viscosities is obvious in terms of the size of the 'flow unit'. Similarly, if the polymer has a charge distributed along the chains, the amount of electrolyte in the aqueous solvent influences the overall shape of the chain. Thus, without the addition of any electrolyte, the charge on the polymer chain is unshielded and the strong repulsive force causes the chain to stretch out. This effect is suppressed by the addition of a salt due to the shielding effect (Figure 1.38), thereby the chain tends to go towards its natural random configuration. This 'rod-to-sphere' change in the shape of the polymer chain lowers the viscosity of the solution.

By a judicious combination of molecular proportions and electrolyte concentrations, mixtures of many surfactant molecules organize themselves into worm-like or rod-like threads (also known as living polymers), as shown in Figure 1.39. At moderate concentrations, these 'rods' or 'worms' behave like a polymer molecule and become

Low electrolyte level ⟶ High electrolyte level

Figure 1.38   *Effect of electrolyte concentration on coil dimensions in a poly electrolyte system (from Barnes, 2000)*

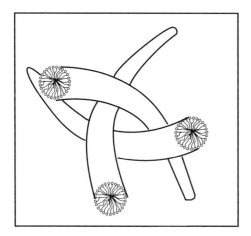

Figure 1.39   *Schematics of a worm-like micelle in a living polymer (from Barnes, 2000)*

entangled with their neighbours. This leads to visco-elastic behaviour. This class of polymers often finds applications as thickening agents for products like shower gels, shampoos, etc. (Barnes, 2000). However, these so-called living polymers differ from their synthetic counter parts, in so far that the worm-like micelles can be broken into two halves in the presence of a strong extensional field, albeit the broken micelles can re-form under quiescent conditions. On the other hand, these systems are much more susceptible to chemical breakdown, especially by the presence of metal ions or electrolytes. At low shear rates when the system is highly entangled, the viscosity is high which progressively decreases with the gradual alignment of the structures with the direction of flow. Interestingly, this change from very high to very low viscosity occurs fairly quickly and over a very narrow window of shear stress which is also consistent with our everyday experience with shower gels and shampoos. Similar structuring also occurs in the so-called associative polymers. This class of polymers has 'hydrophobic' ends attached to their 'hydrophilic' chains, akin to surfactant molecules. Therefore, these 'ends' interact either with similar groups on the same chain and/or on the adjacent chains. This leads to the formation of networks which get dislodged by Brownian effects, but are re-formed sometime later involving the end groups from another network. Even though the polymer has a low molecular weight, but these gel-like networks induce high viscosity and visco-elasticity in these systems. As the shear rate (or shear

Cetyl hydrophobe

Hydroxyethyl cellulose backbone

Surfactant molecules

Figure 1.40   *Schematics of the formation of networks in an aqueous surfactant solution mixed with an associative polymeric thickener (hydroxyethyl cellulose with cetyl branches (from Clarke, 1993)*

stress) is progressively increased, these gel-like networks are destroyed and this, in turn, results in shear-thinning behaviour. This class of polymers finds applications as thickeners in aqueous coating fluids, drilling muds, laundry detergents, etc. Depending upon the type of application, the interactions between the hydrophobic ends can be minimized by blocking or shielding the sites by surfactants and this leads to the lowering of viscosity. Conversely, the formation of gel-like networks is promoted by the addition of an electrolyte if thickening (high viscosity) is a requirement. Figure 1.40 shows the formation of networks in an aqueous surfactant solution mixed with an associative thickening polymer (hydroxyethyl cellulose) with cetyl (16-carbon alkyl) branches. This strategy is used to impart rheological features which cause many shampoos, mousses, lotions and conditioners to flow easily from the bottle into our hands with very little tendency for dripping. It is this ease of manipulation of the viscosity and possibly visco-elasticity which leads to their wide ranging applications for inducing desirable rheological features in drilling muds, laundry detergents, aqueous coating suspensions and in many other everyday use products.

In addition to the molecular weight (or distribution) and chain architecture, the rheological properties are extremely sensitive to chemical composition. In polymer industry, it is a common technique for producing new rheological properties by blending two or more polymers, or by adding fillers (like carbon black, glass and metallic fibres). For instance, it is now well established that a filled polymer has greater stiffness and larger resistance to flow (high viscosity) at high temperatures than the pure polymer melt. In principle, the extent of modifications of viscosity–shear rate curves for filled systems depends upon the size and shape of fillers, but their concentration is probably the main factor influencing the viscosity of filled melts. Broadly speaking, while the addition of fillers can increase the viscosity by up to an order of magnitude at low shear rates, there is a very little increase in their viscosity at high shear rates. This can lead to some problems in moulding operations. For instance, the flow of a filled melt into a cold mould can be quite slow and in this low shear rate step, the viscosity could be significantly higher than that of the virgin polymer melt, thereby necessitating higher pressures and/ or longer processing times (Shenoy, 1999).

In spite of the fact that most polymers are incompatible to varying extents, blending of polymers to achieve improved rheological characteristics (as compared to that of the constituents) is frequently practiced in industry. Similar to the mixing of two immiscible liquids, a polymer blend consists of phase-separate domains due to the lack of their miscibility. Depending upon the system and the method of preparation, the size of such domains can range from a few hundred nanometres to several micrometres. It is not uncommon for such a blend to exhibit viscosity larger than that of its constituents. For instance, for the nylon-branched polyolefin system, nylon–lean blends exhibit viscosity which is not only higher than that of nylon and of polyolefin on its own, but such blends also show shear-thinning behaviour. This is in stark contrast to the behaviour of its constituents. On its own, nylon does not shear-thin and polyolefin shear-thins only at high shear rates. In this case, stiff polyolefin domains behave like fillers and this effect is regarded to be responsible for the enhancement of viscosity. Another way to obtain improved mechanical properties is to create a block copolymer. Such a molecule has long sequences of one type of polymer connected to long sequences of another polymer. This class of polymers also finds extensive application in diverse settings due to their tough and elastic behaviour at room temperature, but ease of flow at high temperatures.

Finally, surfactant systems also constitute an important class of industrial materials which exhibit complex rheological behaviour. The surfactant molecules on their own are rather small, but the presence of hydrophobic (tails) and hydrophilic (heads) ends or groups facilitate the formation of large structures (up to 1 μm in size). When a surfactant (soap, detergent) is added to water, the molecules are dispersed monomerically at low concentrations. However, beyond a certain concentration (CMC – critical micelle concentration), micelles comprising several molecules begin to form. The 'structures' so formed are temporary and loose networks which are in a state of flux, i.e., these are continuously being broken down and re-forming. Their presence gives rise to rheological features, similar to that seen in polymeric and/or in multiphase systems. The shape of micelles is influenced by several variables, including the shape of the surfactant molecule, temperature and electrolytic nature of water. Commonly encountered shapes of micelles include spherical, rod-like, lamellar sheets and lamellar droplets, as shown schematically in Figure 1.41.

This section is concluded by re-iterating that undoubtedly there is a strong link between the rheological behaviour of a substance and its structure. The ultimate goal of

Spherical micelle   Lamellar-phase droplets   Lamellar-phase sheet

Figure 1.41   *Schematics of shapes of micelles in surfactant solutions (from Barnes, 2000)*

course is to be able to predict *a priori* the 'type of structure' needed to impart desirable rheological properties to a product for its satisfactory functional performance as well as to manufacture it in a convenient form. Over the years, based on theoretical considerations aided by experimental observations, considerable progress has been made in developing this link to the extent that it can be used to meet the ever increasing quality standards and to develop new products, especially in food and in house-hold products sectors. However, as seen above, some of these developments are strongly system dependent and as such cannot be extrapolated to other products. The aforementioned discussion provides only a general introduction to this vast subject (drawn mainly from Barnes (2000) and Morrison (2001)), for each product comes with a degree of its own complexity and peculiarities (in terms of requirements and specifications) which need to be incorporated into the structure–rheology link. More detailed discussions on this topic are available in several excellent books (Larson, 1998; Gupta, 2000; Morrison, 2001; Dinger, 2002; Graessley, 2004; Han, 2007).

This chapter is concluded by providing a list of materials displaying a spectrum of non-Newtonian flow characteristics in diverse applications to reinforce idea yet again of the ubiquitous nature of such flow behaviour (Table 1.7).

Similarly, since much has been written about the importance of the measurement of rheological data in the same range of shear or deformation rates as those likely to be

Table 1.7   *Non-Newtonian characteristics of some common materials*

Practical fluid	Characteristics	Consequence of non-Newtonian behaviour
Toothpaste	Bingham Plastic	Stays on brush and behaves more liquid-like while brushing
Drilling muds	Bingham Plastic	Good lubrication properties and ability to convey debris
Non-drip paints	Thixotropic	Thick in the tin, thin on the brush
Wallpaper paste	Pseudoplastic and Visco-elastic	Good spreadability and adhesive properties
Egg white	Visco-elastic	Easy air dispersion (whipping)
Molten polymers	Visco-elastic	Thread-forming properties
'Bouncing Putty'	Visco-elastic	Will flow if stretched slowly, but will bounce (or shatter) if hit sharply
Wet cement aggregates	Dilatant and thixotropic	Permit tamping operations in which small impulses produce almost complete settlement
Printing inks	Pseudoplastic	Spread easily in high speed machines yet do not run excessively at low speeds
Waxy crude oils	Viscoplastic and thixotropic	Flows readily in a pipe, but difficult to restart the flow

Table 1.8   *Shear rates typical of some familiar materials and processes*

Situation	Typical range of shear rates (s^{-1})	Application
Sedimentation of fine powders in a suspending liquid	10^{-6}–10^{-4}	Medicines, paints, printing inks
Levelling due to surface tension	10^{-2}–10^{-1}	Paints, printing inks
Draining under gravity	10^{-1}–10^{1}	Painting and coating, toilet bleaches
Extruders	10^{0}–10^{2}	Polymers
Chewing and swallowing	10^{1}–10^{2}	Foods
Dip coating	10^{1}–10^{2}	Paints, confectionary
Pouring	10^{1}–10^{2}	Pharmaceutical formulations
Mixing and stirring	10^{1}–10^{3}	Manufacturing liquids
Pipe flow	10^{0}–10^{3}	Pumping, blood flow
Spraying and brushing	10^{3}–10^{4}	Spray drying, painting, fuel atomization, spraying of aerosols
Spreading and coating	10^{3}–10^{4}	Application of nail polish and lipsticks
Rubbing	10^{4}–10^{5}	Application of creams and lotions to the skin
Milling pigments in fluid bases	10^{3}–10^{5}	Paints, printing inks
High speed coating	10^{5}–10^{6}	Paper coating
Lubrication	10^{3}–10^{7}	Gasoline engines

Source: Barnes *et al.* (1989)

encountered in the envisaged application, Table 1.8 gives typical orders of magnitudes for various processing operations in which non-Newtonian fluid behaviour is likely to be significant.

Further reading

Barnes, H.A., Hutton, J.F. and Walters, K., *An Introduction to Rheology*, Elsevier, Amsterdam (1989).
Bird, R.B., Armstrong, R.C. and Hassager, O., *Dynamics of Polymeric Liquids 2nd edn*, Vol. 1, Wiley, New York (1987).
Carreau, P.J., Dekee, D. and Chhabra, R.P., *Rheology of Polymeric Systems: Principles and Applications*, Hanser, Munich (1997).
Laba, D., ed., *Rheological Properties of Cosmetics and Toileteries*, Marcel-Dekker, New York (1993).
Larson, R.G., *The Structure and Rheology of Complex Fluids*, Oxford University Press, New York (1998).
Macosko, C.W., *Rheology: Principles, Measurements and Applications*, VCH, Munich (1994).

Steffe, J.F., *Rheological Methods in Food Process Engineering*, Freeman Press, East Lansing, MI (1996).
Tanner, R.I. and Walters, K., *Rheology: An Historical Perspective*, Elsevier, Amsterdam (1998).

References

Astarita, G., *J. Rheol.* 34 (1990) 275; *ibid* **36** (1992) 1317.
Barnes, H.A., *J. Rheol.* **33** (1989) 329.
Barnes, H.A. (1992). Keynote address. In: '*The 1992 IChem E Research Event*', pp.24–29, IChem E, Rugby, U. K. (1992).
Barnes, H.A., *J. Non-Newt. Fluid Mech.* **70** (1997) 1.
Barnes, H.A., *J. Non-Newt. Fluid Mech.* **81** (1999) 133.
Barnes, H.A., *A Handbook of Elementary Rheology*, Institute of Non-Newtonian Fluid Mechanics, University of Wales, Aberystwyth, U.K. (2000).
Barnes, H.A., Edwards, M.F. and Woodcock, L.V., *Chem. Eng. Sci.* **42** (1987) 591.
Barnes, H.A., Hutton, J.F. and Walters, K., *An Introduction to Rheology*, Elsevier, Amsterdam (1989).
Barnes, H.A. and Walters, K., *Rheol. Acta* **24** (1985) 323.
Berry, G.C. and Fox, T.G., *Adv. Polym. Sci.* **5** (1968) 261.
Bird, R.B., *Annu. Rev. Fluid Mech.* **8** (1976) 13.
Bird, R.B. and Wiest, J.M., *Annu. Rev. Fluid Mech.* **27** (1995) 169.
Bird, R.B., Stewart, W.E. and Lightfoot, E.N., *Transport Phenomena*, 2nd edition, Wiley, New York (2002).
Bird, R.B., Dai, G.C. and Yarusso, B.J., *Rev. Chem. Eng.* **1** (1983) 1.
Bird, R.B., Armstrong, R.C. and Hassager, O., *Dynamics of Polymeric Liquids. Fluid Dynamics*, 2nd edition, Vol. 1, Wiley, New York (1987).
Boersma, W.H., Levan, J. and Stein, H.N., *AIChE J* **36** (1990) 321.
Boger, D.V., *J. Non-Newt. Fluid Mech.* **3** (1976) 87.
Boger, D.V., *Nature* **265** (1977) 126.
Boger, D.V. and Walters, K., *Rheological Phenomena in Focus*, Elsevier, Amsterdam (1992).
Brummer, R., *Rheology Essentials of Cosmetics and Food Emulsions*, Springer, New York (2006).
Carreau, P.J., *Trans. Soc. Rheol.* **16** (1972) 99.
Carreau, P.J., Dekee, D. and Chhabra, R.P., *Rheology of Polymeric Systems: Principles and Applications*, Hanser, Munich (1997).
Cawkwell, M.G. and Charles, M.E., *J. Pipelines*, **7** (1989) 251.
Clarke, M.T., *Rheological Properties of Cosmetics and Toiletries* (edited by Laba, D.), Marcel Dekker, New York (1993).
Cross, M.M., *J. Colloid Sci.* **20** (1965) 417.
Dinger, D.R., *Rheology for Ceramists*, Ceramic Consulting Services, Clemsen, SC (2002).
Dullaert, K. and Mewis, J., *J. Rheol.* **49** (2005) 1213.
Dullaert, K. and Mewis, J., *J. Non-Newt. Fluid Mech.* **139** (2006) 21.
Evans, I.D., *J. Rheol.* **36** (1992) 1313.
Freundlich, H. and Juliusburger, F., *Trans. Faraday Soc.* **31** (1935) 920.
Goddard, J.D. and Bashir, Y., *Recent Developments in Structured Continua II (Chapter 2)*, Longman, London (1990).
Goodwin, J.W and Hughes, R.W., *Rheology for Chemists*, Royal Society of Chemistry, London (2000).
Govier, G.W. and Aziz, K., *The Flow of Complex Mixtures in Pipes*, Krieger, Malabar, FL (1982).
Graessley, W.W., *Adv. Polym. Sci.* **16** (1974) 1.
Graessley, W.W., *Polymer Liquids and Networks: Structure and Properties*, Garland Science, New York (2004).
Grimm, R.J., *AIChE J* **24** (1978) 427.
Griskey, R.G., Nechrebecki, D.G., Notheis, P.J. and Balmer, R.T., *J. Rheol.* **29** (1985) 349.
Gupta, R.K., *Polymer and Composite Rheology*, 2nd edition, Marcel Dekker, New York (2000).
Gupta, R.K. and Sridhar, T., *Rheological Measurements* (edited by Collyer, A.A. and Clegg, D.W.), Elsevier, 211, Amsterdam, p. 211 (1988).
Han, C.D., *Rheology and Processing of Polymeric Materials*, Vol. 1, Oxford University Press, New York (2007).
Houska, M., PhD thesis, Czech Technical University, Prague (1981).
James, D.F. and Walters, K., *in Techniques in Rheological Measurements* (edited by Collyer, A.A.), Elsevier, Amsterdam (1994).

Johnson, A.T., *Biological Process Engineering*, Wiley, New York (1999).

Jones, D.M., Walters, K. and Williams, P.R., *Rheol. Acta* **26** (1987) 20.

Keller, D.S. and Keller, D.V., Jr, *J. Rheol.* **34** (1990) 1267.

Kulicke, W.-M. and Wallbaum, U., *Chem. Eng. Sci.* **40** (1985) 961.

Larson, R.G., *Constitutive Equations for Polymer Melts and Solutions*, Butterworth-Heinemann, Stoneham, MA (1988).

Larson, R.G., *The Structure and Rheology of Complex Fluids*, Oxford University Press, New York (1998).

Leider, P.J. and Bird, R.B., *Ind. Eng. Chem. Fundam.* **13** (1972) 342.

Macosko, C.W., *Rheology: Principles, Measurements and Applications*, VCH, Munich (1994).

Mujumdar, A., Beris, A.N. and Metzner, A.B., *J. Non-Newt. Fluid Mech.*, **102** (2002) 157.

Malkin, A.Ya. and Isayev, A.I., *Rheology: Concepts, Methods and Applications*, ChemTec Publishing, Toronto (2006).

Metzner, A.B. and Whitlock, M., *Trans. Soc. Rheol.* **2** (1958) 239.

Metzner, A.B., White, J.L. and Denn, M.M., *AIChE J* **12** (1966) 863.

Mewis, J., *J. Non-Newt. Fluid Mech.* **6** (1979) 1.

Morrison, F.A., *Understanding Rheology*, Oxford University Press, New York (2001).

Nguyen, Q.D. and Uhlherr, P.H.T., *Proc. 3rd Nat. Conf. on Rheol*, Melbourne, Australia (1983) 63.

Pradipasena, P. and Rha, C., *J. Texture Studies* **8** (1977) 311.

Prilutski, G., Gupta, R.K., Sridhar, T. and Ryan, M.E., *J. Non-Newt. Fluid Mech.* **12** (1983) 233.

Saini, D.R. and Shenoy, A.V., *Eur. Polym. J.* **19** (1983) 811.

Schramm, G., *A Practical Approach to Rheology and Rheometry*, Haake, Karlsruhe, Germany (1994).

Schramm, L.L., *Emulsions, Foams and Suspensions*, Wiley-VCH, New York (2005).

Scheraga, H.A., *J. Chem. Phys.* **23** (1955) 1526.

Schowalter, W.R., *Mechanics of non-Newtonian Fluids*, Pergamon, Oxford (1978).

Schurz, J., *Rheol. Acta.* **29** (1990) 170.

Shenoy, A.V., *Rheology of Filled Polymer Systems*, Springer, New York (1999).

Steg, I. and Katz, D., *J. Appl. Polym. Sci.* **9** (1965) 3177.

Struble, L.J. and Ji, X., *Handbook of Analytical Techniques in Concrete Science and Technology Chapter 9* (edited by Ramachandran, V.S. and Beaudoin, J.J.), William Andrew Inc, New York (2001).

Tanner, R.I., *Engineering Rheology*, 2nd edition, Oxford University Press, Oxford (2000).

Tanner, R.I. and Walters, K., *Rheology: An Historical Perspective*, Elsevier, Amsterdam (1998).

Tirtaatmadja, V. and Sridhar, T., *J. Rheol.* **37** (1993) 1081.

Trouton, F.T., *Proc. Roy. Soc.* A77 (1906) 426.

Uhlherr, P.H.T., Guo, J., Zhang, X-M., Zhou, J.Z.-Q. and Tiu, C., *J. Non-Newt. Fluid Mech.* **125** (2005) 101.

Walters, K., *Rheometry*, Chapman and Hall, London (1975).

Weissenberg, K., *Nature* **159** (1947) 310.

Windhab, E.J., *Physics Today* **82** (2006) 82.

Whorlow, R.W., *Rheological Techniques*, 2nd edition, Ellis Horwood, London (1992).

Nomenclature

		Dimensions in $\mathbf{M, L, T, \theta}$
A	area (m^2)	$\mathbf{L}^2$
b	constant, equation (1.41) (s)	$\mathbf{T}$
c	constant, equation (1.41) ($-$)	$\mathbf{M^0 L^0 T^0}$
De	Deborah number (–)	$\mathbf{M^0 L^0 T^0}$
F	force (N)	$\mathbf{MLT}^{-2}$
G	Young's modulus (Pa)	$\mathbf{ML^{-1}T^{-2}}$
k_B	Boltzmann constant (J/K)	$\mathbf{ML^2 T^{-2} \theta^{-1}}$
M	molecular weight	
m	power-law consistency coefficient (Pa s^n)	$\mathbf{ML^{-1}T^{n-2}}$
m_1	power-law consistency coefficient for first normal stress difference (Pa·s^{P_1})	$\mathbf{ML^{-1}T^{P_1-2}}$
N_1	first normal stress difference (Pa)	$\mathbf{ML^{-1}T^{-2}}$
N_2	second normal stress difference (Pa)	$\mathbf{ML^{-1}T^{-2}}$

n	power-law index (–)	$M^0L^0T^0$
p	pressure (Pa)	$ML^{-1}T^{-2}$
P	total normal stress (Pa)	$ML^{-1}T^{-2}$
p_1	power-law index for first normal stress difference (–)	$M^0L^0T^0$
R	radius of floc (m)	L
t	time (s)	T
T	absolute temperature (k)	θ
T_r	Trouton ratio (–)	$M^0L^0T^0$
V	velocity (m/s)	LT^{-1}
Z_c	critical chain length (m)	L
x, y, z	coordinate system (m)	L

Greek letters

α	fluid parameter in Ellis fluid model (–)	$M^0L^0T^0$
γ	strain (–)	$M^0L^0T^0$
$\dot{\gamma}$	shear rate (s^{-1})	$M^0L^0T^{-1}$
$\dot{\varepsilon}$	rate of extension (s^{-1})	$M^0L^0T^{-1}$
λ	fluid parameter in Carreau viscosity equation or characteristic time in Maxwell model (s)	T
μ	Newtonian or apparent viscosity (Pa s)	$ML^{-1}T^{-1}$
ρ	fluid density (kg/m^3)	ML^{-3}
τ	component of stress tensor (Pa)	$ML^{-1}T^{-2}$
ϕ	volume fraction (–)	$M^0L^0T^0$
ψ_1	first normal stress difference coefficient (Pa s^2)	ML^{-1}
ψ_2	second normal stress difference coefficient (Pa s^2)	ML^{-1}

Subscripts/superscripts

B	pertaining to Bingham fluid model
C	relating to Casson fluid model
E	extensional
H	relating to Herschel–Bulkley fluid model
m	maximum value
0	zero shear or initial value at $t = 0$
xx, yy, zz	normal stress components
xy, yz, zx, yx, zy, xz	shear stress components
∞	infinite shear

Chapter 2

Rheometry for non-Newtonian fluids

2.1 Introduction

The rheological characterization of non-Newtonian fluids is widely acknowledged to be far from straightforward. In some non-Newtonian systems, such as concentrated suspensions, rheological measurements may be complicated by non-linear, dispersive, dissipative and thixotropic mechanical properties; and the rheometrical challenges posed by these features may be compounded by an apparent yield stress.

For non-Newtonian fluids, even the apparently simple determination of a shear rate versus shear stress relationship is problematical as the shear rate can only be determined directly if it is constant (or nearly so) throughout the measuring system employed. While very narrow shearing gap coaxial cylinder and cone-and-plate measuring geometries provide good approximations to this requirement, such systems are often of limited utility in the characterization of non-Newtonian multiphase products such as suspensions, foams and emulsions whose particulate/aggregate constituents preclude the use of narrow gaps. As most measuring geometries do not approximate to constant shear rate, various measurement strategies have been devised to overcome this limitation. The basic features of these rheometrical approaches, and of the main instrument types for their implementation, are considered below.

2.2 Capillary viscometers

Capillary viscometers are the most commonly used instruments for the measurement of viscosity due, in part, to their relative simplicity, low cost and (in the case of long capillaries) accuracy. However, when pressure drives a fluid through a pipe, the velocity is a maximum at the centre: the velocity gradient or shear rate $\dot{\gamma}$ are a maximum at the wall and zero at the centre of the pipe. The flow is therefore non-homogeneous and capillary viscometers are restricted to measuring *steady* shear functions, i.e., steady shear stress–shear rate behaviour for time independent fluids. Due to their inherent similarity to many process flows, which typically involve pipes, capillary viscometers are widely employed in process engineering applications and are often converted or adapted (with relative ease) to produce slit or annular flows.

2.2.1 Analysis of data and treatment of results

The fully developed, steady flow of an incompressible fluid in a tube of radius, R, is shown in Figure 2.1a. The flow is caused by the pressure difference imposed across the two ends of the pipe. Since there is no angular velocity and the fluid is flowing at a

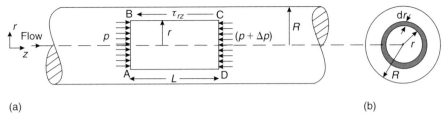

(a)   (b)

Figure 2.1   *Schematics of flow in a tube*

steady state, the linear momentum balance (in the direction of flow, z) on a fluid element *ABCD* of radius r and length l, may be written as:

$$p(\pi r^2) - (p + \Delta p)(\pi r^2) = \tau_{rz}(2\pi r L)$$

i.e.,

$$\tau_{rz} = \left(-\frac{\Delta p}{L}\right)\left(\frac{r}{2}\right) \tag{2.1}$$

This shows the linear variation of the shear stress across the tube cross-section, increasing from zero at the axis of the tube to a maximum value at the wall of the tube. It should be emphasized here that equation (2.1) is applicable to both laminar and turbulent flow of any incompressible fluid in steady, fully developed conditions, since it is based simply on a force balance and no assumption has been made regarding either the type of fluid or the flow pattern. Equation (2.1) thus provides a convenient basis for determining the shear stress at the wall of the tube τ_w as:

$$\tau_w = \left(\frac{R}{2}\right)\left(\frac{-\Delta p}{L}\right) \tag{2.2}$$

The shear stress may then be evaluated in terms of the shear rate at the wall, $\dot\gamma_w$ or $(-dV_z/dr)_w$ to yield steady shear stress–shear rate data for a fluid. This relationship may be obtained, however, because the z-component of the velocity is a function only of the radial coordinate, i.e., $V_z(r)$. The volumetric flow rate through the annulus formed by two concentric fluid elements at radial positions of r and $(r + dr)$, as shown in Figure 2.1b.

The volumetric flow rate is given by:

$$dQ = 2\pi r V_z(r)dr \tag{2.3}$$

For the sake of simplicity, $V_z(r)$ will now be written as V_z. The total volumetric flow rate is obtained by integrating equation (2.3) over the cross-section of the tube as:

$$Q = \int_0^R 2\pi r V_z \, dr \tag{2.4}$$

Carrying out the integration in equation (2.4) by parts:

$$Q = 2\pi \left\{ \left[\left(\frac{r^2}{2}V_z\right)\right]_0^R + \int_0^R \frac{r^2}{2}\left(\frac{-dV_z}{dr}\right)dr \right\} \tag{2.5}$$

Assuming the no-slip condition at the wall of the tube, i.e. $V_z = 0$ when $r = R$, the first term on the right hand side of equation (2.5) is identically zero at both limits of integration. Thus, equation (2.5) simplifies to:

$$Q = \pi \int_0^R r^2 \left(\frac{-dV_z}{dr} \right) dr \tag{2.6}$$

It is then necessary to make assumptions regarding the nature of the flow and of the characteristics of the fluid. For the laminar flow of time-independent fluids, the shear rate $(-dV_z/dr)$ is determined only by the value of the shear stress, i.e. the corresponding value of τ_{rz}. Thus, without any loss of generality, it is convenient to write this functional relationship as:

$$-\frac{dV_z}{dr} = f(\tau_{rz}) \tag{2.7}$$

where f is an unspecified function. Also, combining equations (2.1) and (2.2):

$$\frac{\tau_{rz}}{\tau_w} = \frac{r}{R} \tag{2.8}$$

On differentiation (for constant values of R and τ_w),

$$dr = \left(\frac{R}{\tau_w} \right) d\tau_{rz} \tag{2.9}$$

Now substituting equations (2.7–2.9) into equation (2.6), the volumetric flow rate Q is now given as:

$$Q = \frac{\pi R^3}{\tau_w^3} \int_0^{\tau_w} \tau_{rz}^2 f(\tau_{rz}) d\tau_{rz} \tag{2.10}$$

Equation (2.10) can be used in two ways:

(i) By integration directly for a specific fluid model, i.e., a known form of $f(\tau_{rz})$. For instance, for a power-law model fluid,

$$\tau_{rz} = m \left(-\frac{dV_z}{dr} \right)^n \tag{2.11}$$

and therefore,

$$-\frac{dV_z}{dr} = f(\tau_{rz}) = \left(\frac{\tau_{rz}}{m} \right)^{1/n} \tag{2.12}$$

This approach is used in Section 3.2.4, or

(ii) to establish general shear stress–shear rate characteristics of a time-independent fluid to develop rheological data for a fluid, as shown here.

It is useful to rearrange equation (2.10) as:

$$\left(\frac{Q}{\pi R^3} \right) \tau_w^3 = \int_0^{\tau_w} \tau_{rz}^2 f(\tau_{rz}) d\tau_{rz} \tag{2.13}$$

Since the right hand side of equation (2.13) embodies a definite integral, the final result depends only on the value of the wall shear stress, τ_w and not on the nature of the continuous function $f(\tau_{rz})$. Therefore, it is necessary to evaluate only the wall shear stress τ_w (equation 2.2) and the corresponding shear rate at the wall $(-dV_z/dr)$ at $r = R$, or simply $f(\tau_w)$.

By applying the Leibnitz rule enabling the differential of a definite integral of the form $(d/ds')\left\{\int_0^{s'} s^2 f(s)\,ds\right\}$ to be written as $(s')^2\, f(s')$, where s is a dummy variable of integration (τ_{rz} here) and s' is naturally identified as τ_w. Applying this rule to differentiate equation (2.13) with respect to τ_w:

$$\frac{d}{d\tau_w}\left\{\left(\frac{Q}{\pi R^3}\right)\tau_w^3\right\} = \frac{d}{d\tau_w}\left\{\int_0^{\tau_w} \tau_{rz}^2 f(\tau_{rz})\,d\tau_{rz}\right\}$$

which upon simplification yields,

$$\left(3\tau_w^2\right)\left(\frac{Q}{\pi R^3}\right) + \tau_w^3\,\frac{d}{d\tau_w}\left(\frac{Q}{\pi R^3}\right) = \tau_w^2 f(\tau_w) \tag{2.14}$$

Equation (2.14) can be rearranged as:

$$f(\tau_w) = 3\left(\frac{Q}{\pi R^3}\right) + \tau_w\,\frac{d}{d\tau_w}\left(\frac{Q}{\pi R^3}\right) \tag{2.15}$$

Introducing a factor of 4 on the right hand side and using the identity $d\ln x = dx/x$, equation (2.15) may be written to give:

$$f(\tau_w) = \left(-\frac{dV_z}{dr}\right)_w = \frac{4Q}{\pi R^3}\left[\frac{3}{4} + \frac{1}{4}\,\frac{d\ln(4Q/\pi R^3)}{d\ln\tau_w}\right] \tag{2.16}$$

or, in terms of the average velocity over the cross-section, $V = Q/\pi R^2$, and pipe diameter D:

$$-\frac{dV_z}{dr}\bigg|_w = \left(\frac{8V}{D}\right)\left[\frac{3}{4} + \frac{1}{4}\,\frac{d\ln(8V/D)}{d\ln(\tau_w)}\right] \tag{2.17}$$

The Hagen–Poiseuille equation, which applies to the laminar, fully developed and steady flow of incompressible Newtonian fluids gives the mean velocity, V, as:

$$V = \left(-\frac{\Delta p}{L}\right)\left(\frac{R^2}{8\mu}\right) \tag{2.18}$$

Rearranging in terms of the wall shear stress,

$$\left(-\frac{\Delta p}{L}\right)\left(\frac{R}{2}\right) = \tau_w = \mu\left(\frac{8V}{D}\right) \tag{2.19}$$

Thus, $(8V/D)$ is seen to be the true shear rate at the wall for a Newtonian fluid, but equation (2.17) shows that a correction factor must be added for non-Newtonian fluids. However, $(8V/D)$ is often used as the nominal (or apparent) shear rate at the wall for non-Newtonian fluids also.

Equation (2.17) may be written in terms of the slope, n', of the log–log plot of τ_w against $(8V/D)$. Thus,

$$n' = \frac{d \ln \tau_w}{d \ln(8V/D)} \tag{2.20}$$

Therefore, the true shear rate at the wall for a non-Newtonian fluid is obtained by combining equations (2.17) and (2.20) as:

$$\dot{\gamma}_w = \left(-\frac{dV_z}{dr} \right)_w = \left(\frac{8V}{D} \right) \left(\frac{3n' + 1}{4n'} \right) \tag{2.21}$$

where the slope n' may vary with the nominal shear rate.

For shear-thinning fluids $(n' < 1)$, the apparent shear rate at the wall is obviously lower than the true shear rate, with the converse applying near the centre of the tube (Laun, 1983). Thus, at some radial location, say x^*R, the true shear rate of a fluid of apparent viscosity μ equals that of a Newtonian fluid of the same viscosity. The value of shear stress at this radius $x^*\tau_w$ is independent of the fluid properties and thus the true viscosity at this radius is equal to the apparent viscosity at the wall and the viscosity calculated using equation (2.19) is the true viscosity at $\tau = x^*\tau_w$. Laun (1983) found this 'single-point' method for correcting viscosity to be as good as the Weissenberg–Rabinowitsch method. Macosko (1994) has shown that the value of x^* is weakly dependent

(a)   (b)

Figure 2.2 *Schematics of a capillary viscometer: (a) constant pressure device and (b) constant flow rate device* on the value of n', but the use of $x^* = 0.83$ leads to errors less than 2% in the range $0.2 < n < 1.3$, which rises to about 8% for $n = 1$.

It is thus, in principle, possible to obtain shear stress–shear rate data from a series of volumetric flow rate–pressure drop measurements in a tube of known length and diameter using equations (2.2) and (2.21) provided that the simplifying assumptions in the above analysis are met in the experimental methods.

In practice, a capillary viscometer may operate in a constant flow rate mode (Figure 2.2b), or in a constant pressure drop mode (Figure 2.2a), the latter being common for low to medium consistency fluids while the constant flow rate device is more appropriate for highly viscous polymer melts, resins and thick suspensions. In addition, pipe viscometers consisting of a straight section of pipe are also used in industrial settings for quick on-site measurements. Irrespective of their mode of operation, the basic measurement of this method is the volumetric flow rate as a function of the pressure drop (or vice versa) in tubes of known length and diameter. Example 2.1 illustrates the utility of this approach.

Example 2.1

The following capillary viscometer data on a high pressure polyethylene melt at 190°C have been reported in the literature (Metzger, A.P. and Brodkey, R.S., *J. Appl. Polymer Sci.* **7** (1963) 399). Obtain the true shear stress–shear rate data for this polymer. Assume the end effects to be negligible.

$\left(\dfrac{8V}{D}\right)(\text{s}^{-1})$	τ_{w} (kPa)
10	22.4
20	31
50	43.5
100	57.7
200	75
400	97.3
600	111
1000	135.2
2000	164

Solution

From equation (2.21):

$$f(\tau_{\text{w}}) = \dot{\gamma}_{\text{w}} = \left\{ \frac{3n' + 1}{4n'} \cdot \frac{8V}{D} \right\}$$

and using equation (2.2),

$$\tau_{\text{w}} = \frac{R}{2}\left(\frac{-\Delta p}{L}\right)$$

where τ_w is the true shear stress at the wall irrespective of the type of fluid behaviour, whereas $(8V/D)$ is the corresponding shear rate at the wall only for Newtonian fluids. The factor $(3n' + 1)/4n'$ corrects the nominal wall shear rate $(8V/D)$ for the non-Newtonian fluid behaviour.

where $n' = \dfrac{d \log \tau_w}{d \log(8V/D)}$

Thus, the given data is first plotted on log–log coordinates as shown in Figure 2.3 and the value of n' is evaluated at each point (value of $8V/D$), as summarized in Table 2.1.

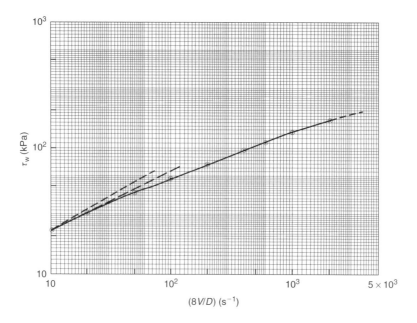

Figure 2.3   *Rheological data of Example 2.1*

Table 2.1   *Summary of calculations*

τ_w (kPa)	$\left(\dfrac{8V}{D}\right)(s^{-1})$	$n' = \dfrac{d \log \tau_w}{d \log(8V/D)}$	$\dot\gamma_w = \left(\dfrac{3n'+1}{4n'}\right)\left(\dfrac{8V}{D}\right)(s^{-1})$
22.4	10	0.50	12.5
31	20	0.47	25.6
43.5	50	0.43	66.6
57.7	100	0.42	135
75	200	0.40	275
97.3	400	0.36	578
111	600	0.34	891
135.2	1000	0.31	1556
164	2000	0.31	3113

As can be seen, the value of the correction factor $(3n' + 1)/(4n')$ varies from 25% to 55.6%. Thus, the values of $(\tau_w, \dot{\gamma}_w)$ represent the true shear stress–shear rate data for this polymer melt which displays shear-thinning behaviour as can be seen from the values of $n' < 1$. □

2.2.2 Sources of errors

There are several possible sources of errors which may falsify the data obtained using a capillary device (van Wazer *et al.*, 1963; Macosko, 1994). These include kinetic energy losses (associated with the acceleration of the fluid from the barrel velocity to that in the capillary tube), entrance and exit effects, departure from laminar flow conditions, viscous heating effects (relevant in high shear devices and/or with highly viscous materials), wall-slip effects, variable head in the barrel, etc., although some of these can be minimized by suitable design of the viscometer. For instance, since the cross-sectional area of the barrel is usually much greater than that of the capillary, one can neglect the effect of variable head and the fluid friction in the barrel. Similarly, it is possible to operate the viscometer at flow rates where changes in the kinetic energy of the emerging stream from the capillary are negligible compared with the other terms in the mechanical energy balance for the system. The two most common (and also probably most severe) sources of errors are the so-called end effects and wall-slip effects, and these are addressed here in detail.

End effects

The 'end effect' is really a consequence of the fact that near the entry to the capillary tube, the flow is not fully developed and the pressure drop is increased as a result. The contribution due to the exit effect is usually negligible. The magnitude of the entrance effect is dependent on the type of non-Newtonian fluid behaviour. It is usually safe to neglect it as long as the length-to-diameter ratio (L/D) of the capillary is of the order of 100–120, at least for purely viscous fluids (Dervisoglu and Kokini, 1986; Macosko, 1994). For visco-elastic substances, even greater values of (L/D) are required, but as of now, no reliable estimates are available.

There may be situations when it is not possible to use such large values of (L/D) and therefore, the resulting shear stress values must be corrected for end effects. Bagley (1957) suggests that $Q - (-\Delta p)$ data be obtained for several (at least three) tubes of the same diameter but of different lengths. These data are then plotted in terms of $(-\Delta p)$ against Q as a function of tube length L for a constant value of diameter D, as shown schematically in Figure 2.4. These data may be interpolated either by fitting a polynomial (usually second order) or by using a graphical method to obtain a series of $(-\Delta p) - L$ points for a range of constant values of Q. Alternatively, these data may be re-plotted on linear coordinates as $(-\Delta p)$ versus L for constant values of the apparent shear rate at the wall $(8V/D)$ (as shown schematically in Figure 2.5). Usually a linear dependence of $(-\Delta p)$ on L is found and one can thus easily extrapolate to $L = 0$ to obtain the entrance correction, $(-\Delta p_e)$, for each value of $(8V/D)$, and this procedure must now be repeated for other values of the capillary diameter, D. Finally, the corrected pressure drop $(-\Delta p_c)$ corresponding to the fully developed condition is obtained as:

$$(-\Delta p_c) = (-\Delta p) - (-\Delta p_e) \tag{2.22}$$

The corrected value of the pressure drop can now be used to estimate the value of the wall shear stress by simply replacing $-\Delta p$ by $-\Delta p_c$ in equation (2.2). This procedure is illustrated in Example 2.2.

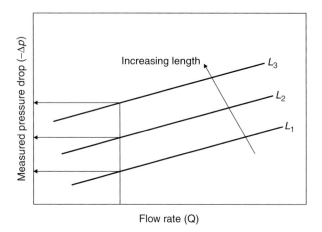

Figure 2.4   *Dependence of pressure drop on flow rate and length for the same diameter tubes*

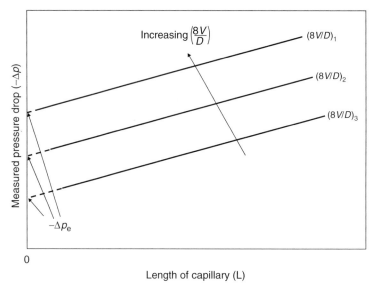

Figure 2.5   *Dependence of pressure drop on the length of capillary for constant values of the nominal shear rate at wall*

   Undoubtedly, this method is employed widely to correct the experimental data for end effects (e.g. see Kurath and Larson, 1990; Lam *et al.*, 2007), it is obviously cumbersome and it may always not be possible to carry out tests using capillaries of different diameters and lengths. Also, there are situations when the $(-\Delta p)$ versus L plots for constant values of $(8V/D)$, referred to above, show deviations from linearity. Possible reasons for such a non-linear dependence include wall slip, time-dependent fluid behaviour, viscous heating, pressure-dependent viscosity or capillaries too short for the flow to have become fully developed. While pressure-dependent viscosity causes a positive deviation from a straight line when (L/D) is large, wall slip, thixotropy and viscous heating cause negative deviations (Laun and Hirsch, 1989; Kim and Dealy, 2001).

Example 2.2

The following experimental data has been obtained for a polymer solution (4% carboxymethyl cellulose in water) using an extrusion (constant flow rate) rheometer. The diameters of the barrel and of the capillary, respectively, are 23.3 and 0.408 mm.

Velocity of ram V_{ram} (mm/s)	$(-\Delta p)$(kPa)		
	$L = 49.5$ mm	$L = 74.5$ mm	$L = 99.5$ mm
0.09	448.16	655.0	882.5
0.12	489.53	717.1	965.3
0.15	537.80	779.1	1048
0.21	606.74	882.5	1165
0.24	620.53	910.1	1186

Obtain the shear stress–shear rate for this polymer solution. Neglect the friction in the barrel. (Data courtesy BHR Group, Cranfield, UK.)

Solution

Since the data is available for capillaries of the same diameter and of three different lengths, it is possible to establish the magnitude of end effects in this case.

The nominal shear rate at the wall of the capillary is calculated as:

$$\dot{\gamma}_{wn} = \frac{8V}{D}$$

The average velocity V in the capillary is obtained by using the continuity equation:

$$\frac{\pi}{4} \times D_b^2 \times V_{ram} = \frac{\pi}{4} \times D^2 \times V$$

where D_b is the diameter of the barrel.

For instance, for the first data point, the average velocity in the capillary is calculated as:

$$V = \left(\frac{23.3 \times 10^{-3}}{0.408 \times 10^{-3}} \right)^2 \times 0.09 \times 10^{-3} = 0.294 \text{ m/s}$$

and therefore the nominal shear rate at the wall is given as:

$$\dot{\gamma}_{wn} = \frac{8V}{D} = \frac{8 \times 0.294}{0.408 \times 10^{-3}} = 5755 \text{ s}^{-1}$$

The shear stress at the wall is given by equation (2.2):

$$\tau_w = \left(\frac{-\Delta p_c}{L} \right) \left(\frac{D}{4} \right)$$

where the corrected pressure drop $(-\Delta p_c)$ is obtained as:

$$(-\Delta p_c) = (-\Delta p) - (-\Delta p_e)$$

The end effect correction $(-\Delta p_e)$ is evaluated by plotting the measured values of $(-\Delta p)$ against the length of the capillary for a constant value of $(8V/D)$, as shown in Figure 2.6. These

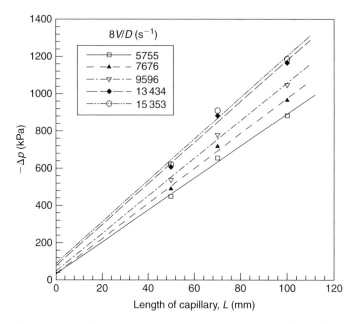

Figure 2.6   *Measured pressure drop as a function of length of capillary*

Table 2.2   *Summary of calculations*

L (mm)	V_{ram} (mm/s)	V (m/s)	$(8V/D)$ (s^{-1})	$(-\Delta p)$ (kPa)	$(-\Delta p_e)$ (kPa)	τ_w (Pa)	τ_{avg} (Pa)	$\dot{\gamma}_w$ (s^{-1})
49.5	0.09	0.294	5 755	448.16	14.6	893.4	886	9 548
	0.12	0.391	7 676	489.53	15.1	977.6	970	12 731
	0.15	0.489	9 596	537.80	28.1	1 050	1 041	15 914
	0.21	0.685	13 434	606.74	52.7	1 142	1 139	22 280
	0.24	0.783	15 353	620.53	63.1	1 149	1 153	25 463
74.5	0.09	0.294	5 755	655.00	14.6	876	–	–
	0.12	0.391	7 676	717.10	15.1	961	–	–
	0.15	0.489	9 596	779.10	28.1	1 028	–	–
	0.21	0.685	13 434	882.50	52.7	1 136	–	–
	0.24	0.783	15 353	910.10	63.1	1 160	–	–
99.5	0.09	0.294	5 755	882.50	14.6	889	–	–
	0.12	0.391	7 676	965.30	15.1	950	–	–
	0.15	0.489	9 596	1 048	28.1	1 020	–	–
	0.21	0.685	13 434	1 165	52.7	1 112	–	–
	0.24	0.783	15 353	1 186	63.1	1 123	–	–

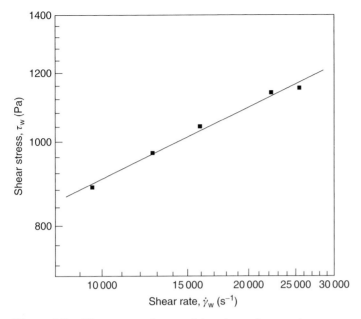

Figure 2.7 *Flow curve after applying the end correction*

linear plots can readily be extrapolated to $L = 0$ and the intercept on the y-axis is the value of $(-\Delta p_e)$ for the fixed value of $(8V/D)$. The resulting values of $(-\Delta p_e)$ are summarized in Table 2.2 and can now be used to evaluate the corrected wall shear stress by substituting $(-\Delta p_c)$ in equation (2.2). For instance, for $(8V/D) = 5755\,\mathrm{s}^{-1}$, $(-\Delta p_e) = 14.6\,\mathrm{kPa}$, and therefore,

$$\tau_w = \frac{(448.16 - 14.6) \times 10^3 \times 0.408 \times 10^{-3}}{4 \times 49.5 \times 10^{-3}} = 893.4\,\mathrm{Pa}$$

Table 2.2 presents a summary of these calculations. Note that after the end correction has been applied, the shear stress values exhibit a small degree of scatter and a mean value has been used for each value of $(8V/D)$. Thus, for instance, for $\dot{\gamma}_{wn} = 5755\,\mathrm{s}^{-1}$, the average wall shear stress, τ_w, is given as:

$$\tau_w = \frac{1}{3}(893.4 + 876 + 889) = 886\,\mathrm{kPa}$$

One can now plot $\log \tau_w$ vs $(8V/D)$, and the slope of this line, $n' = 0.275$. Finally, the true shear rate at the wall is given by equation (2.21):

$$\dot{\gamma}_w = \left(\frac{8V}{D}\right)\left(\frac{3n' + 1}{4n'}\right)$$

For the first data point, the value of the wall shear rate is calculated as:

$$\dot{\gamma}_w = 5755 \times \left(\frac{3 \times 0.275 + 1}{4 \times 0.275}\right) = 9548\,\mathrm{s}^{-1}$$

These values are also included in Table 2.2. Figure 2.7 shows the linear plot of corrected data for this polymer solution. ☐

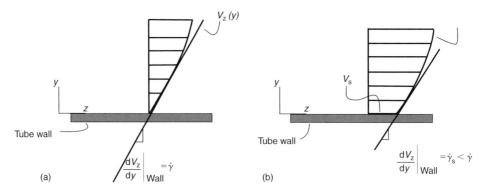

Figure 2.8   *Schematic representation of flow without slip (a) and with slip (b) at the wall*

Wall-slip effects

Slip may occur, e.g., in the case of multiphase mixtures including concentrated suspensions, foams, emulsions, etc. and with large polymer molecules confined in narrow passages where the layer of fluid near the wall is depleted, and this thin, dilute layer has a much lower viscosity, resulting in an apparent slippage of the bulk fluid along the wall. Whether true slip occurs or not, the concept of an apparent slip at the wall has proved to be of value in explaining anomalous results for the flow of heterogeneous mixtures and polymeric systems in small diameter tubes such as that encountered in food processing applications (Steffe, 1996), in thin gaps (such as in concentric cylinders, parallel plates, cone-and-plate viscometers, e.g. see Barnes, 1995), in the extrusion of pasty materials and flow of macromolecular polymeric melts and solutions in narrow passages, etc. The phenomenon of slip is generally associated with the flow of structured materials in narrow gaps as in porous media, micro-fluidic devices, extrusion dies used in food and polymer processing applications. Excellent reviews of slip phenomena are available in the literature (Agarwal *et al.*, 1994; Barnes, 1995; Neto *et al.*, 2005).

The presence of the depleted layer near the wall is responsible for a higher flow rate than expected at a given value of the wall shear stress, or conversely, at a fixed shear rate, a lower than expected shear stress is encountered. Figure 2.8 shows the phenomenon of slip schematically. It can easily be seen that for a constant value of wall shear stress, the apparent shear rate (velocity gradient) at the wall is reduced when there is slip and owing to the occurrence of a positive velocity at the wall, the observed flow rate is larger than expected.

While the exact reasons or mechanisms for slip are not at all obvious (Neto *et al.*, 2005), serious errors in interpreting experimental measurements may occur due to wall slip with highly structured fluids if appropriate allowance is not made. In the context of capillary viscometry, wall slip is likely to give rise to a situation when the data for wall shear stress (τ_w) and apparent wall shear rate ($8V/D$) obtained with tubes of different diameters appear to be inconsistent even when the results have been corrected for all other known effects. It is customary (Mooney, 1931) to quantify the effect of slip by assuming that a slip velocity, V_s, exists at the wall. This implies that at the wall of the tube ($r = R$), the fluid has a velocity $V_z = V_s$ instead of $V_z = 0$ (as was used in the derivation of equation (2.15)).

Referring back to equation (2.5):

$$Q = 2\pi \left\{ \left(\frac{r^2}{2} V_z \right) \bigg|_0^R + \int_0^R \frac{r^2}{2} \left(\frac{-dV_z}{dr} \right) dr \right\}$$

When the fluid is flowing at a slip velocity V_s at the walls ($r = R$), the above equation yields:

$$Q = \pi R^2 V_s + \pi \int_0^R r^2 \left(\frac{-dV_z}{dr} \right) dr \tag{2.23}$$

On the right hand side of equation (2.23), the two terms can be identified as the contributions to the total flow rate stemming from slip (Q_s) and that occurring when there is no-slip (Q_{ns}). The Q_{ns} would still be expected to be given by equation (2.10) which is clearly only a function of the wall shear stress, τ_w. Furthermore, the slip velocity is generally assumed to be a function of the wall shear stress only, albeit there is some evidence that it might be a function of tube diameter also (Jastrzebski, 1967). Equation (2.23) can therefore be rearranged as:

$$\frac{Q}{\pi R^3 \tau_w} = \frac{V_s}{R \tau_w} + \frac{Q_{ns}}{\pi R^3 \tau_w} \tag{2.24}$$

Re-writing equation (2.24) in terms of the apparent wall shear rate:

$$\left(\frac{8V}{D} \right) \frac{1}{\tau_w} = \frac{8V_s}{D \tau_w} + \frac{32 Q_{ns}}{\pi D^3 \tau_w} \tag{2.25}$$

Noting that the second term on the right hand side of equation (2.25) is constant for a fixed value of the wall shear stress, it is seen that the slip velocity V_s can be evaluated from the slope of ($8V/D$) versus ($1/D$) plots on linear scales for a range of constant values of the wall shear stress (τ_w) and thereby evaluating the slip velocity as a function of the wall shear stress. Finally, the apparent shear rate at the wall is simply given as:

$$\dot{\gamma}_{wn} = \frac{8(V - V_s)}{D} \tag{2.26}$$

In summary, thus this method of correcting the data for wall slip proceeds as follows: flow rate (Q)–pressure drop ($-\Delta p$) data is needed for different diameter tubes (at least three). From the plots of the wall shear stress (τ_w) versus the apparent shear rate ($8V/D$) for different values of D, interpolation is used to generate a series of ($8V/D$)-D data points for constant values of the wall shear stress. Next, in accordance with equation (2.25), these data are plotted on linear coordinates in terms of ($8V/D$) against ($1/D$) for a series of constant values of τ_w. If the dependence in this plot can be approximated by a straight line thereby confirming the validity of equation (2.25), the value of the slip velocity, V_s, is evaluated simply from the slope of ($8V/D$) versus ($1/D$) plots referred to above. This, in turn, can be used to calculate the corrected apparent wall shear rate using equation (2.26). However, as mentioned earlier, sometimes the slip velocity shows additional dependence on the tube diameter (Jastrzebski, 1967). This situation is dealt with by postulating the slip velocity V_s to vary as D^ϕ and equation (2.25) may now be modified as:

$$\left(\frac{8V}{D}\right)\frac{1}{\tau_w} = \frac{8V_s}{\tau_w D^{\phi+1}} + \frac{32Q_{ns}}{\pi D^3 \tau_w} \qquad (2.27)$$

Therefore, the value of ϕ is varied (typically $\phi \approx 1$ works well) until the plots of $(8V/D)$ versus $(1/D)^{\phi+1}$ yield linear dependence for a range of constant values of the wall shear stress. Once again, one can evaluate the value of V_s from the slope of these lines. The application of this procedure is illustrated in Example 2.3.

Example 2.3

For a 0.5% partially hydrolyzed polyacrylamide-in-water solution, the following data is available for four capillaries of the same length ($L = 374\,\text{mm}$) but of different diameters. Obtain the true shear stress–shear rate for this polymer solution. End effects may be assumed to be negligible, as the minimum value of (L/D) is 340 (Based on Y. Cohen and A.B. Metzner, *J. Rheol.*, 29 (1985) 67).

D = 0.191 mm		D = 0.266 mm		D = 0.626 mm		D = 1.10 mm	
$Q \times 10^{11}$ (m³/s)	$(-\Delta p)$ (kPa)	$Q \times 10^{11}$ (m³/s)	$(-\Delta p)$ (kPa)	$Q \times 10^{10}$ (m³/s)	$(-\Delta p)$ (kPa)	$Q \times 10^{9}$ (m³/s)	$(-\Delta p)$ (kPa)
3.66	45.65	9.98	35.61	–	–	–	–
7.66	56.48	18.5	42.03	15.9	16.14	12.1	11.25
10.3	63.58	30.9	51.06	34.3	22.33	17.3	13.08
19.6	86.13	42.2	58.86	53.4	28.07	23.2	15.08
24.8	97.31	56.3	68.10	65.2	31.24	31.5	17.68
33.3	113.6	75.5	79.64	78.4	34.48	47.3	21.99
39.7	124.2	88	86.54	107	40.32	62	25.23
45.3	132.2	103	94.37	122	42.69	77.8	27.92
51.8	140.1	112	98.69	155	46.47	90.4	29.47
60.1	148.1	140	110.6	169	47.41	104	30.53

Solution

One can readily convert the raw data to the corresponding wall shear stress and nominal wall shear rate values. For $D = 0.191\,\text{mm}$ capillary, and for the first set of $Q - (-\Delta p)$ values:

$$\tau_w = \left(\frac{-\Delta p}{L}\right)\frac{D}{4} = \frac{45.65 \times 10^3}{374 \times 10^{-3}} \times \frac{0.191 \times 10^{-3}}{4} = 5.83\,\text{Pa}$$

$$\dot\gamma_{wn} = \left(\frac{8V}{D}\right)$$

and the mean velocity, $V = Q/(\pi D^2/4)$, and hence

$$\dot{\gamma}_{wn} = \left(\frac{32Q}{\pi D^3}\right) = \frac{32 \times 3.66 \times 10^{-11}}{3.14 \times (0.191 \times 10^{-3})^3} = 53.5\,\text{s}^{-1}$$

A summary of $\tau_w - \dot{\gamma}_{wn}$ values for all others experimental data is given in Table 2.3.

Figure 2.9 shows the plot of wall shear stress versus nominal wall shear rate, and clearly data for different diameter capillaries do not superimpose thereby suggesting the possible occurrence

Table 2.3   *Values of wall shear stress–nominal shear rate*

D = 0.191 mm		D = 0.266 mm		D = 0.626 mm		D = 1.10 mm	
τ_w(Pa)	(8V/D) (s^{-1})	τ_w(Pa)	(8V/D) (s^{-1})	τ_w(Pa)	(8V/D) (s^{-1})	τ_w (Pa)	(8V/D) (s^{-1})
5.83	53.5	6.33	54.0	–	–	–	–
7.21	111.1	7.47	100.1	6.75	66.02	8.27	92.6
8.12	150.6	9.08	167.2	9.34	142.4	9.62	132.4
11.0	286.5	10.46	228.4	11.74	221.7	11.10	177.5
12.42	362.5	12.10	304.7	13.07	270.7	13.00	241.1
14.50	486.8	14.16	408.6	14.42	325.5	16.17	362
15.86	580.3	15.39	476.3	16.87	444.3	18.55	474.5
16.88	662.2	16.78	557.4	17.86	506.6	20.53	595.4
17.89	757.2	17.55	606.1	19.44	643.6	21.67	691.8
18.90	878.6	19.66	757.7	19.83	701.7	22.45	796

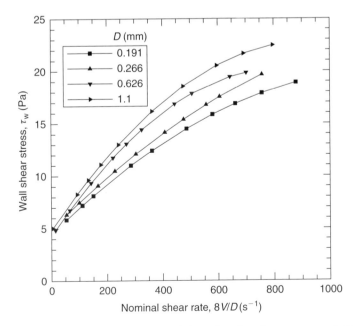

Figure 2.9   *Raw experimental data showing the presence of slip*

of wall slip in this case. Therefore, a correction for wall slip must be applied to this set of data as outlined in the preceding section.

As a first step, data shown in Figure 2.9 is interpolated to produce a series of data points in terms of $(8V/D)$ versus diameter for a range of constant values of the wall shear stress, as summarized in Table 2.4.

Figure 2.10 shows the variation of the nominal wall shear rate $(8V/D)$ with the reciprocal of diameter (D) on linear coordinates. The dependence is seen to be linear, except for a little scatter at $\tau_w = 25$ Pa. This confirms the assumption of the slip velocity being independent of the capillary diameter, in accordance with equation (2.25), or conversely, $\phi = 0$ in equation (2.27). According to equation (2.25), the slope of each line in Figure 2.10 is equal to $8V_s$, and the resulting values of the slip velocity, V_s, are summarized in Table 2.5.

Table 2.4 *Interpolated results from Figure 2.9*

τ_w(Pa)	Value of $(8V/D)$ (s^{-1})			
	$D = 0.191$ mm	$D = 0.266$ mm	$D = 0.626$ mm	$D = 1.10$ mm
5	45.37	14.70	31.75	24.85
10	223.6	203.3	145.5	128.7
15	529.0	456.6	362.3	316.7
20	961.7	774.5	682.2	588.8
25	1521	1158	1105	945

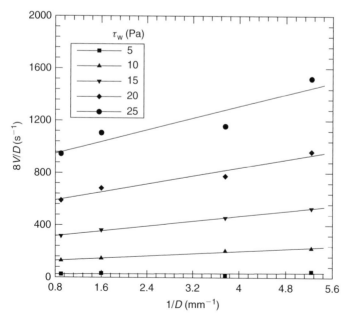

Figure 2.10 *Interpolated data plotted in accordance with equation (2.25)*

Table 2.5 *Values of slip velocity (V_s)*

τ_w(Pa)	Slope, $8V_s$ (mm/s)	V_s(mm/s)
5	2.6	0.325
10	22.6	2.83
15	47.7	5.96
20	77.9	9.74
25	113.2	14.15

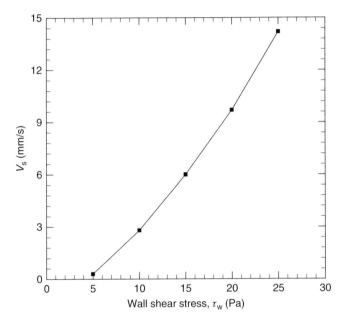

Figure 2.11 *Slip velocity as a function of wall shear stress*

In order to facilitate interpolation of these data for the intermediate values of the wall shear stress, the values of the slip velocity shown in Table 2.5 are plotted in Figure 2.11 and this dependence is adequately described by the following power-law-type relation:

$$V_s = 1 \times 10^{-5}\tau_w^2 + 3 \times 10^{-4}\tau_w - 0.0015$$

where the slip velocity is in mm/s and τ_w is in Pa. This relationship can now be used to calculate the value of V_s corresponding to each value of τ_w in Table 2.3. This, in turn, allows the estimation of the corrected nominal wall shear rate as:

$$\dot{\gamma}_{wn} = \frac{8(V - V_s)}{D}$$

For instance, for $D = 0.191$ mm and $\tau_w = 5.83$ Pa, $V_s = 0.59$ mm/s which yields a value of $\dot{\gamma}_{wn} = 28.88 \, \text{s}^{-1}$.

Table 2.6   *Summary of calculations (SI units)*

D = 0.191 mm			D = 0.266 mm			D = 0.626 mm			D = 1.10 mm		
τ_w	(8V/D)	$\dfrac{8(V - V_s)}{D}$	τ_w	(8V/D)	$\dfrac{8(V - V_s)}{D}$	τ_w	(8V/D)	$\dfrac{8(V - V_s)}{D}$	τ_w	(8V/D)	$\dfrac{8(V - V_s)}{D}$
5.83	53.5	28.9	6.33	54.0	29.9	–	–	–	–	–	–
7.21	111.1	61.53	7.47	100.1	61.0	6.75	66.02	53.47	8.27	92.6	80.5
8.12	150.6	83.8	9.08	167.2	105.6	9.34	142.4	114.6	9.62	132.4	115.6
11.0	286.5	160.5	10.46	228.4	146.1	11.74	221.7	178.2	11.1	177.5	155.3
12.42	362.5	204.6	12.1	304.7	196.5	13.07	270.7	217.9	13.0	241.1	211.3
14.5	486.8	279.3	14.16	408.6	265.6	14.42	325.5	262.8	16.17	362	318.6
15.86	580.3	338.6	15.39	476.2	311.3	16.87	444.3	362.4	18.55	474.5	419.9
16.88	662.2	393.5	16.78	557.4	366.5	17.86	506.6	416.5	20.53	595.4	530.9
17.89	757.2	461.3	17.55	606.1	400.3	19.44	643.6	539.9	21.67	691.8	621.3
18.9	878.6	554.1	19.66	757.7	509.1	19.83	701.7	594.5	22.45	796	721.1

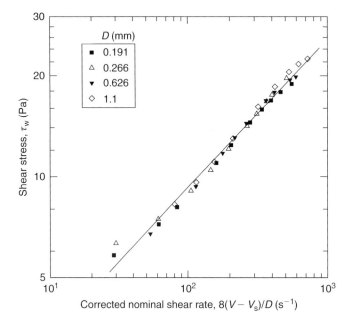

Figure 2.12   *Flow curve after correcting for slip effects*

A summary of these calculations is presented in Table 2.6, and the shear stress–shear rate are plotted in Figure 2.12 where results obtained with different capillaries collapse onto one curve after the wall-slip correction has been applied.

It is worthwhile to point out here that the wall-slip correction is quite significant in small diameter capillaries and/or at large values of the wall shear stress. Finally, note that the corresponding values of the true shear rate at the wall still need to be calculated as $\{8(V - V_s)/D\}\{(3n' + 1)/4n')\}$. From figure 2.12, the slope of the line is 0.44, i.e., $n' = 0.44$ and thus the value of the Rabinowitsch–Mooney factor $\{(3 \times 0.44 + 1)/(4 \times 0.44)\} = 1.318$. Therefore, the true wall shear stress is 1.318 times the value of $8(V - V_s)/D$ listed in Table 2.6. □

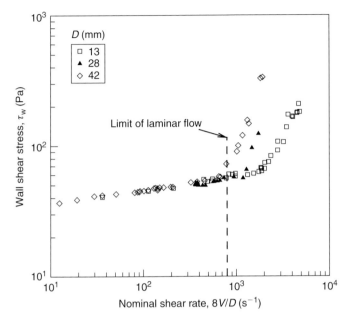

Figure 2.13 *Typical data showing the end of laminar region for a 15% kaolin suspension (Source: Data courtesy Dr V Fester, Cape Peninsula University of Technology, Capetown)*

Another possible source of error in capillary measurements arises from data points (especially at high values of the wall shear stress and/or of the apparent wall shear rate) being outside the range of laminar flow conditions. This effect usually shows up as a sudden change in the slope of log τ_w − log($8V/D$) plots, as shown in Figure 2.13 for the flow of a 15% aqueous kaolin suspension in several tubes of different diameters. Clearly, the data points deviating from the laminar line must be excluded from further consideration. It is also appropriate to add here that while in principle equations (2.2) and (2.15) can also be used for visco-elastic fluids, but in practice considerable confusion exists regarding the end effects, and from die swell at the exit end of the capillary.

In summary, capillary (or tube) viscometry offers a simple and robust method of measuring steady shear stress–shear rate data for time-independent and visco-elastic fluids over wide ranges of shear rate (~1 to $10^5 \, \text{s}^{-1}$). Some designs are available for high temperature and pressure applications also. On the other hand, some of the disadvantages include large volume of fluid, tedious experimental protocol, cleaning problems, etc.

2.3 Rotational viscometers

Due to their relative importance as tools for the rheological characterization of non-Newtonian fluid behaviour, we concentrate on this class of rheometers by considering the two main types, namely: the controlled shear rate instruments (also known as controlled *rate* devices) and controlled *stress* instruments. Both types are usually supplied with the same range of measuring geometries, principally the concentric cylinder, cone-and-plate and parallel-plate systems. The relative merits, potential drawbacks, working

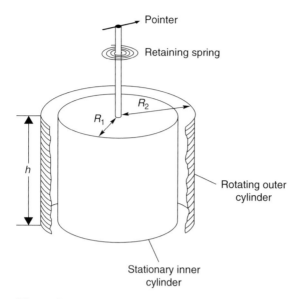

Figure 2.14   *Partial section of a concentric cylinder viscometer*

equations and other formulae associated with these designs have been described in great detail elsewhere (e.g. see Walters, 1975; Whorlow, 1992; Macosko, 1994) and so only their most basic aspects are covered here.

2.3.1 The concentric cylinder geometry

It is appropriate to begin by considering this geometry as it was the basis of the first practical rotational rheometer. Ideally the sample is contained in a *narrow* gap between two concentric cylinders (as shown in Figure 2.14). Typically the outer 'cup' rotates and the torque T on the inner cylinder, which is usually suspended from a torsion wire or bar, is measured.

Working equations relating the measured torque to the requisite shear stress, and angular velocity (of the cup) to the requisite shear rate, are widely available along with their derivations (see the list of references above). It is noteworthy that the working formulae quoted in many instances ignore the curvature of the surfaces of the measuring geometry. The determination of the shear stress and shear rate within the shearing gap is thus valid only for *very* narrow gaps wherein κ, the ratio of inner to outer cylinder radii, is >0.99.

Several designs have been described which overcome end effects due to the shear flow at the bottom of the concentric cylinder geometry. These include the recessed bottom system which usually entails trapping a bubble of air (or a low viscosity liquid such as mercury) beneath the inner cylinder of the geometry. Alternatively the 'Mooney–Ewart' design, which features a conical bottom may, with suitable choice of cone angle, cause the shear rate in the bottom to match that in the narrow gap between the sides of the cylinders, see Figure 2.15.

In this example, the sample temperature is controlled by circulation of liquid through the outer cylinder housing (flow marked in and out in Figure 2.15) and h denotes the

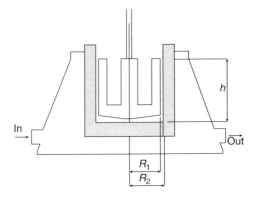

Figure 2.15   *The Mooney–Ewart geometry*

sample height within the shearing gap. For $k > 0.99$, the shear rate may be calculated from:

$$\dot{\gamma} = \frac{2\Omega}{1 - k^2} \approx \frac{R_2 \Omega}{R_2 - R_1} \tag{2.28}$$

where R_2 and R_1 are the outer and inner cylinder radii respectively, and Ω is the angular velocity. It is worthwhile to emphasize here that equation (2.28) is only an approximation valid for a close gap arrangement. Strictly speaking, the shear rate for non-Newtonian fluids depends upon the viscosity model itself. Thus, for the commonly used power-law fluid model, the shear rate is a function of the power-law index. For instance, one such approximate expression valid in the range $0.5 \le k \le 1$ is:

$$\dot{\gamma}_{R_1} = \frac{\Omega}{-\ln k} \left[1 - \frac{1}{n} \ln k + \left(\frac{1}{n} \ln k \right)^{2/3} - \left(\frac{1}{n} \ln k \right)^{4/45} + \cdots \right] \tag{2.29}$$

where n is the flow behaviour index and is given by the slope of a plot of $\ln T$ versus $\ln \Omega$ as $(d \ln T/d \ln \Omega)$. Usually, it is sufficient to retain the first three to four terms of the series appearing in equation (2.29).

For $k > 0.5$ and if the value of $(d \ln T/d \ln \Omega)$ is constant over the range of interest $(\tau_{R_1}$ to $\tau_{R_2})$, one can use the following expressions for evaluating the shear rates at $r = R_1$ and $r = R_2$ respectively:

$$\dot{\gamma}_{R_1} = \frac{2\Omega}{n \left(1 - k^{2/n} \right)} \tag{2.30}$$

$$\dot{\gamma}_{R_2} = \frac{-2\Omega}{n \left(1 - k^{-2/n} \right)} \tag{2.31}$$

Furthermore, many commercial instruments employ $k > 0.9$ and it is not uncommon to calculate the shear rate by assuming the fluid to be Newtonian. It is therefore useful to ascertain the extent of uncertainty in using this approximation. The ratio of the shear rates at $r = R_1$, for a power-law fluid ($\dot\gamma_{PL}$) and for a Newtonian fluid ($\dot\gamma_N$) is given as (Macosko, 1994):

$$\frac{\dot\gamma_{PL}}{\dot\gamma_N} = \frac{1 - k^2}{n\left(1 - k^{2/n}\right)} \tag{2.32}$$

Evidently for a Newtonian fluid, $n = 1$, this ratio is unity. For typical shear-thinning substances encountered in industrial practice, the flow behaviour index ranges from ~0.2 to 1. Over this range and for $k > 0.99$, the error in using equation (2.28) is at most 3%. However, it rises to 10% for $k = 0.98$ and $n = 0.2$.

In this geometry, the shear stress is evaluated from torque data. The torque T is acting on the fluid adhering to the outer surface of the rotating cup is calculated by multiplying the shearing force acting on the surface area $(\tau_{R_1})(2\pi R_1 h)$ by R_1 and therefore the torque is given as: $T = (2\pi R_1 h \tau_{R_1})R_1$, or,

$$\tau_{R_1} = \frac{T}{2\pi R_1^2 h}. \tag{2.33}$$

One can similarly evaluate the shear stress acting at $r = R_2$ as:

$$\tau_{R_2} = \frac{T}{2\pi R_2^2 h}.$$

Thus, the shear stress varies as $(1/r^2)$ from τ_{R_1} at $r = R_1$ to τ_{R_2} at $r = R_2$. For $k > 0.99$, $R_1 \simeq R_2$ and therefore, the two values are very close, and the shear stress is given as:

$$\tau = \frac{T}{2\pi R_1^2 h} \tag{2.34}$$

To minimize end effects, the lower end of the inner cylinder is a truncated cone. The shear rate in this region is equal to that between the cylinders if the cone angle, α, is related to the cylinder radii by:

$$\alpha = \tan^{-1}\frac{R_2 - R_1}{R_2} \tag{2.35}$$

The main sources of error in the concentric cylinder type measuring geometry arise from end effects (see above), wall slip, inertia and secondary flows, viscous heating effects and eccentricities due to misalignment of the geometry (Macosko, 1994).

Secondary flows are of particular concern in the controlled stress instruments which usually employ a rotating *inner* cylinder, in which case inertial forces cause a small axisymmetric cellular secondary motion ('Taylor' vortices). The dissipation of energy by these vortices leads to overestimation of the torque. The stability criterion for a Newtonian fluid in a narrow gap is:

$$\text{Ta} = \rho^2\Omega^2(R_2 - R_1)^3 R_1/\mu^2 < 3400 \tag{2.36}$$

where Ta is the 'Taylor' number.

In the case of non-Newtonian polymer solutions (and narrow gaps), the stability limit increases. In situations where the outer cylinder is rotating, stable Couette flow may be maintained until the onset of turbulence at a Reynolds number, Re, of ca. 50 000 where $Re = \rho\Omega R_2(R_2 - R_1)/\mu$ (Van Wazer *et al.*, 1963).

2.3.2 The wide gap rotational viscometer: determination of the flow curve for a non-Newtonian fluid

An important restriction on the use of the concentric cylinder measuring geometry for the determination of the shear rate versus shear stress relationship for a non-Newtonian fluid is the requirement, noted above, for a narrow shearing gap between the cylinders. As indicated in the introduction to this chapter, direct measurements of shear rates can only be made if the shear rate is constant (or very nearly so) throughout the shearing gap but many coaxial measuring systems do not fulfil this requirement. In addition, many (if not most) non-Newtonian fluid systems, particularly those of industrial or commercial interest such as pastes, suspensions or foods, may contain relatively large particles or aggregates of particles. Thus the requisite shearing gap size to ensure that adequate bulk measurements are made, i.e. a gap size approximately $10-100$ times the size of the largest 'particle' size (Van Wazer *et al.*, 1963), may conflict with the gap size required to ensure near constant shear rate, within the gap.

Procedures for extracting valid shear stress versus shear rate data from measurements involving wide gap coaxial cylinder systems (the Brookfield viscometer being an extreme example of wide gap devices) are therefore of considerable interest in making quantitative measurements of the flow properties of non-Newtonian products. Most of these data treatment procedures necessarily involve some assumption regarding the functional form of the flow curve of the material. One example is that made in the derivation of data from the Brookfield-type instrument, which assumes that the speed of rotation of the cylinder or spindle is proportional to the shear rate experienced by the fluid. This assumption implies that the flow curve is adequately described by a simple power law (which for many shear-thinning non-Newtonian fluids may be acceptable), but this assumption is widely taken to *exclude* all fluids which display an apparent yield stress and/or non-power-law-type behaviour.

The starting point lies in considering the basic equation for the coaxial rotational viscometer, which has been solved for various sets of boundary conditions (Krieger and Maron, 1952):

$$\Omega = \frac{1}{2} \int_{R_b}^{R_c} \frac{f(\tau)}{\tau} \, d\tau \tag{2.37}$$

where Ω is the angular velocity of the spindle with respect to the cup, τ is the shear stress in the fluid at any point in the system, $f(\tau) = \dot{\gamma}$ is the rate of shear at the same point and the subscripts b and c refer to the bob and the cup, respectively.

The particular solution to equation (2.37) for a finite cylindrical bob rotating in an infinite cup can provide valuable quantitative rheological data for systems whose particulate constituents, and practical limitations on the size of the measuring geometry in terms of cylinder radius, preclude the use of conventional narrow gap geometries. The infinite cup boundary condition may be closely approximated by using a narrow cylindrical *spindle* (such as are supplied with instruments of the Brookfield type) in place of the more commonly used bob.

Assuming the infinite cup boundary condition, τ_c (shear stress on the cup) in equation (2.37) becomes equal to zero and the expression may be differentiated with respect to τ_b giving:

$$\dot{\gamma}_b = -f(\tau_b) = 2d\Omega/d\ln\tau_b = 2\Omega \frac{d(\ln\Omega)}{d(\ln\tau_b)} \tag{2.38}$$

and thus the rate of shear may be obtained by evaluating (graphically) either of the derivatives on the right hand side of equation (2.38).

The derivation of equation (2.38) assumes that a cup of infinite radius is filled with *fluid*. Implicitly this would exclude all systems which display a yield stress, as such systems would not behave as a fluid for values of stress below the yield value.

As many non-Newtonian systems are sufficiently 'structured' to display an apparent yield stress, this requirement would appear to severely restrict the application of what would otherwise appear to be a very useful technique. However, on closer inspection, it has been shown that for a fluid which displays a yield stress, a more general derivation than that reported by Krieger and Maron (1952) may be obtained, and that the restriction of infinite outer boundary (i.e. cup) radius may in fact be eliminated (Jacobsen, 1974).

In a system which displays yield stress behaviour, the integral in the general expression for the rate of shear need not be evaluated from the bob all the way to the cup. This is due to the fact that, for such a system, no shearing takes place where τ is less than the yield value, τ_0. Thus the integral need only be evaluated from the bob to the critical radius, R_{crit}, the radius at which $\tau = \tau_0$. This gives

$$\Omega = (1/2) \int_{R_b}^{R_{crit}} f(\tau) d\tau / \tau \tag{2.39}$$

where the 'critical' radius, R_{crit}, is given as:

$$R_{crit} = R_b (\tau_b / \tau_0)^{1/2} \tag{2.40}$$

This derivation relies on the fact that the condition of differentiability is not that one limit of the integral be zero (as is the case in the infinite cup solution) but that one limit be *constant*. Thus, for systems which may be described in terms of a constant value of yield stress, equation (2.39) may be differentiated, giving:

$$\dot{\gamma}_b = -f(\tau_b) = 2d\Omega/d \ln \tau_b = 2\Omega \frac{d(\ln \Omega)}{d(\ln \tau_b)} \tag{2.41}$$

i.e. exactly the same result is obtained as that derived for the case of the infinite cup, equation (2.38).

In practice, shear stress data are plotted against Ω and the slopes ($d\Omega/d\tau_b$) are taken at each point. Given that the graphical solution may be somewhat tedious, and that a rapid evaluation of the general form of the flow curve is often all that is required (e.g. in a product 'quality control' context), the form of the Ω versus τ_b plots is sometimes taken as giving the general form of the corresponding $\dot{\gamma}$ versus τ_b curve (although, of course, the curves will differ quantitatively). In the absence of an apparent yield stress (over the experimental time scale), the general character of the $\dot{\gamma}$ versus τ_b curves may sometimes be correctly inferred by this procedure; the situation, however, is quite different when the system exhibits an apparent yield stress and this situation poses a trap for the unwary.

An examination of equation (2.41) shows that for any fluid with a finite yield point, the Ω versus τ_b curve approaches the τ_b axis at zero slope, due to the requirement for such a system that the shear rate must become zero at finite τ_b. This may lead to apparent shear-thinning characteristics being ascribed to systems, *irrespective of the actual form of their flow curves above the yield point,* i.e. whether Bingham plastic, shear-thickening (with a yield stress) or shear-thinning (with a yield stress).

An instrument called the 'rotating disc indexer' is also widely used in quality control applications and involves a rotating disc in a 'sea' of fluid. Williams (1979) has described a numerical method for obtaining true $\mu - \dot{\gamma}$ data with this device.

In summary, the concentric cylinder geometry is well suited for low to medium viscosity systems ($<\sim 100\,\mathrm{Pa\,s}$) and high shear rates. The gravity-induced separation in multiphase samples is not as problematic as in the case of cone-and-plate geometry (Section 2.3.3).

Example 2.4

The following steady shear data for a salad dressing has been obtained at 295 K using a concentric cylinder viscometer ($R_1 = 20.04\,\mathrm{mm}$; $R_2 = 73\,\mathrm{mm}$; $h = 60\,\mathrm{mm}$). Obtain the true shear stress data for this fluid. (Data taken from Steffe, 1996.)

Table 2.7   *Torque–speed measurements*

Ω (rad/s)	$T \times 10^4$ (Nm)	$\dot{\gamma}_b(\mathrm{s}^{-1})$	$\tau_b(\mathrm{Pa})$
0.146	6.09	0.8	4.025
0.512	9.98	2.796	6.595
1.036	12.64	5.657	8.353
2.087	16.23	11.396	10.725
4.163	20.33	22.732	13.435
6.276	24.30	34.271	16.058
8.359	27.08	45.645	17.895
10.49	29.70	57.282	19.627
12.59	31.49	68.749	20.81
14.68	33.35	80.162	22.039
16.77	35.09	91.57	23.189

Solution

Since $R_1/R_2 = 20.04/73 = 0.275$ ($<< 0.99$), one cannot use the close gap approximation.
The shear stress at the surface of the rotating bob, τ_b, is given by equation 2.34 as:

$$\tau_b = \frac{T}{2\pi h R_1^2}$$

For the first data point,

$$\tau_b = \frac{6.09 \times 10^{-4}}{2 \times 3.14 \times 60 \times 10^{-3} \times (20.04 \times 10^{-3})^2} = 4.025\,\mathrm{Pa}$$

The resulting values of shear stress are also included in Table 2.7.
The calculation of the corresponding shear rate $\dot{\gamma}_b$ using equation (2.38) requires a knowledge of the slope, $\mathrm{d}\ln(\Omega)/\mathrm{d}\ln(\tau_b)$. The given data is plotted in terms of $\ln(\Omega)$ versus $\ln(\tau_b)$ in Figure 2.16. The dependence is seen to be linear and the slope is 2.73, and therefore the shear rate is calculated as:

$$\dot{\gamma}_b = 2\Omega\,\frac{\mathrm{d}\ln\Omega}{\mathrm{d}\ln\tau_b} = 2 \times 0.146 \times 2.73 = 0.8\,\mathrm{s}^{-1}$$

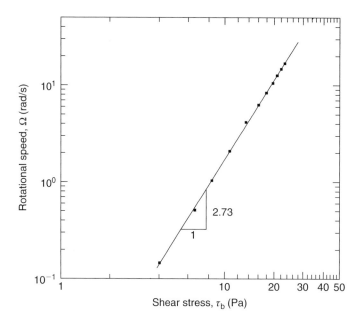

Figure 2.16   *Evaluation of the value of d ln(Ω)/d ln(τ$_b$) for Example 2.4*

Figure 2.17   *Shear stress–shear rate curve for the salad dressing in Example 2.4*

The resulting values of the shear rate at the surface of the bob are also included in Table 2.7.
Finally, Figure 2.17 shows the rheogram for this material and it appears that this substance has an apparent yield stress of about ~3–4 Pa. □

2.3.3 The cone-and-plate geometry

In the cone-and-plate geometry, the test sample is contained between an upper rotating cone and a stationary flat plate (see Figure 2.18a). In the example shown, the cone is 40 mm in diameter, with a cone angle of 1°59′ relative to the plate, and a truncation (or the gap) of 51 μm.

The small cone angle ($<4°$) ensures that the shear rate is constant throughout the shearing gap, this being of particular advantage when investigating time-dependent systems because all elements of the sample experience the same shear history, but the small angle can lead to serious errors arising from eccentricities and misalignment.

The small gap size dictates the practical constraints for the geometry: a gap-to-maximum particle (or aggregate) size ratio of >100 is desirable to ensure the adequate measurement of bulk material properties. This geometry is, therefore, limited to systems containing small particles or aggregates, and the strain sensitivity is fixed. Normal stress differences may be determined from pressure and thrust measurements on the plate.

By considering the torque acting on an element of fluid bounded by $r = r$ and $r = r + \mathrm{d}r$ (Figure 2.19a),

$$\mathrm{d}T = (2\pi r\, \mathrm{d}r)(\tau)r \tag{2.42}$$

which can be integrated to obtain the total torque T as:

$$T = \int_0^T \mathrm{d}T = \int_0^R 2\pi r^2 \tau\, \mathrm{d}r \tag{2.43}$$

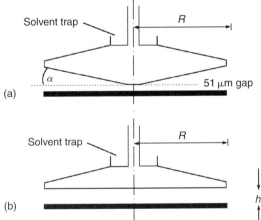

(a)

(b)

Figure 2.18   *(a) Cone-and-plate and (b) parallel-plate (lower) geometries*

(a)   Stationary plate   (b)   Stationary plate

Figure 2.19   *Schematics for the calculation of (a) shear stress and (b) shear rate*

For a constant value of τ,

$$T = \frac{2\pi R^3 \tau}{3} \qquad (2.44)$$

or the shear stress is given as:

$$\tau = \frac{3T}{2\pi R^3} \qquad (2.45)$$

The corresponding expression for shear rate is obtained by considering the angular velocity gradient, as shown in Figure 2.19b. The fluid particle adhering to the rotating cone has a velocity of $r\Omega$ and that adhering to the stationary plate is at rest. Therefore, the velocity gradient or shear rate is estimated as:

$$\text{Shear rate} = \frac{r\Omega - 0}{r \tan \alpha} = \frac{\Omega}{\tan \alpha} \qquad (2.46)$$

Since shear rate does not depend upon the value of r, the fluid everywhere experiences the same level of shearing. For small values of α, it is justified to use the approximation $\tan \alpha \approx \alpha$ in equation (2.46).

The influence of geometry misalignment and other factors, such as flow instabilities arising from fluid elasticity, have been extensively studied in the case of this geometry (Macosko, 1994). Unlike the concentric cylinder geometry, where fluid inertia causes a depression around the inner cylinder rather than the well-known 'rod-climbing' effect due to visco-elastic normal stresses, in the cone-and-plate geometry the effect of inertia is to draw the plates together, rather than push them apart (Walters, 1975).

Many experimentalists employ a 'sea' of liquid around the cone (often referred to as a 'drowned edge'), partly in an attempt to satisfy the requirement that the velocity field be maintained to the edge of the geometry.

The main advantages of a cone–plate rheometer include a homogeneous shear field (for cone angles up to about 4°) thereby making the analysis of data much easier than that for the coaxial cylinders and parallel plate devices and small sample volumes required ($<\sim 5\,\text{cm}^3$). On the other hand, this geometry is not at all suitable for suspensions due to the possibility of particle jamming. For highly volatile systems, the test sample may dry out at the edge of the cone and this tendency can be minimized by applying a thin layer of a silicone oil of comparable viscosity as that of the fluid being tested. In summary, the cone-and-plate geometry is suitable for low to medium viscosity materials, especially when it is necessary to subject the whole sample to a uniform shear rate such as in the characterization of shear sensitive foodstuffs, personal care products, etc. This geometry is also frequently used for measuring the first normal stress difference (see Section 2.6).

Example 2.5

A 25 mm radius cone–plate system ($\alpha = 1°18'45''$) is used to obtain the following steady shear data for a food product at 295 K. Obtain shear stress–shear rate data for this substance.

Table 2.8   *Raw data and shear rate–shear stress values*

$\Omega \times 10^3$ (rad/s)	$T \times 10^4$ (Nm)	$\dot{\gamma}(s^{-1})$	τ(Pa)
2	1.34	0.1	4.1
5	1.65	0.25	5.04
13	2.16	0.65	6.60
25	2.59	1.25	7.92
40	2.98	2.00	9.11
63	3.42	3.15	10.45
100	3.97	4.99	12.13
159	4.58	7.95	14.00
252	5.28	12.60	16.14
399	6.24	19.94	19.10
632	7.33	31.60	22.41

Solution

For a cone-and-plate geometry with $\alpha < 4°$, the shear rate is given by equation (2.46) and the corresponding shear stress is given by equation (2.45). Therefore for the first data point:

$$\text{Shear stress, } \tau = \frac{3T}{2\pi R^3} = \frac{3 \times 1.34 \times 10^{-4}}{2 \times 3.14 \times (25 \times 10^{-3})^3} = 4.1\,\text{Pa}$$

The value of α in radians is ~0.02 and for small values of α, $\tan\alpha \approx \alpha$:

$$\text{Shear rate, } \dot{\gamma} = \frac{\Omega}{\tan\alpha} = \frac{2 \times 10^{-3}}{\tan(0.02)} = \frac{2 \times 10^{-3}}{0.02} = 0.1\,\text{s}^{-1}$$

Figure 2.20   *Shear stress–shear rate curve for the food product in Example 2.5*

The resulting values of $\tau - \dot{\gamma}$ for the remaining data are included in Table 2.8. Figure 2.20 shows the rheogram for this material. □

2.3.4 The parallel plate geometry

In this measuring geometry, the sample is contained between an upper rotating or oscillating flat stainless steel plate and a lower stationary plate (see Figure 2.18b). The upper plate in the example shown is 40 mm in diameter. In contrast to the cone-and-plate geometry, the shear strain is proportional to the gap height, h, and may be varied to adjust the sensitivity of shear rate, a feature which readily facilitates testing for wall (slip) effects (Yoshimura and Prud'homme, 1988).

The large gap sizes available can be used to overcome the limitations encountered using the cone-and-plate geometry, such as its sensitivity to eccentricities and misalignment. However, it should be borne in mind that, as in the case of the wide gap Couette devices, shear rate is not constant in a parallel plate device. Usually the strain reported is that measured at the outer rim, which provides a maximum value of the spatially varying strain within the gap.

Loading and unloading of samples can often prove easier than in the cone-and-plate or concentric cylinder geometries, particularly in the case of highly viscous liquids or 'soft solids' such as foods, gels, etc. The parallel plate geometry is particularly useful for obtaining apparent viscosity and normal stress data at high shear rates, the latter being increased either by increasing Ω or by decreasing the shearing gap size. An additional benefit of the latter approach is that errors due to secondary flows, edge effects and shear heating may all be reduced.

The form factors for the parallel plate geometry, in terms of the shear stress and the shear rate at $r = R$ are given below:

Shear stress:

$$\tau = \frac{T}{2\pi R^3}\left[3 + \frac{d \ln T}{d \ln \Omega}\right] \tag{2.47}$$

Shear rate:

$$\dot{\gamma} = \frac{\Omega R}{h} \tag{2.48}$$

where h is the plate separation (m), Ω is the angular velocity (rad/s) and R is the plate radius (m). Thus to evaluate the shear stress for an unknown fluid, sufficient number of torque (T) – rotational velocity (Ω) data points must be available to determine the value of the slope (d ln T/d ln Ω) with sufficient accuracy. In practice, this step does not pose any problem, since most materials do exhibit power-law behaviour. For a shear-thinning substance, the value of (d ln T/d ln Ω) < 1 and therefore even an uncertainty of 10% in its value will lead to an error of only 3% in the value of shear stress. For Newtonian fluids, the slope will be equal to one and therefore, the shear stress is given by:

$$\tau = \frac{2T}{\pi R^3} \tag{2.49}$$

This value is also known as the apparent or nominal shear stress for non-Newtonian fluids. A full derivation of the working equations may be found elsewhere (e.g. Macosko, 1994).

Example 2.6

The following rotational speed–torque data has been obtained for a 3% aqueous Methocel solution at 297 K using a parallel plate device ($h = 0.7$ mm; $R = 25$ mm). Obtain shear stress–shear rate data for this polymer solution.

Table 2.9 *Torque–rotational speed data*

$\Omega \times 10^4$ (rad/s)	$T \times 10^5$ (N m)	τ(Pa)	$\dot{\gamma}$(s^{-1})
3.56	1.16	0.454	0.0127
5.54	2.11	0.826	0.0198
8.88	3.34	1.308	0.0317
14.1	4.42	1.731	0.0504
22.3	8.07	3.16	0.0796
35.3	12.6	4.93	0.126
56	20.3	7.95	0.200
88.7	29.8	11.67	0.317
140	45.4	17.78	0.500
223	66.9	26.20	0.796
352	93.4	36.57	1.26
558	129	50.51	1.99
884	173	67.74	3.16
1400	227	88.88	5.00
2220	293	114.7	7.93
3510	373	146	12.54

Solution

For a parallel plate device, equations (2.47) and (2.48) can be used to convert the given raw data into shear stress–shear rate coordinates corresponding to the rim, i.e., at $r = R$. For the first pair of data points, the shear rate is calculated as:

$$\dot{\gamma} = \left(\frac{R}{h}\right)\Omega = \frac{25 \times 10^{-3}}{0.7 \times 10^{-3}} \times 3.56 \times 10^{-4} = 0.0127\,\text{s}^{-1}$$

For calculating the corresponding value of the shear stress, the value of (d ln T/d ln Ω) is calculated first by plotting the experimental data in the form of ln T versus ln Ω (Figure 2.21).

The slope of ln T versus ln Ω line is found to be 0.842, i.e. (d ln T/d ln Ω) = 0.842. Thus, in turn, can be used to evaluate the shear stress via equation (2.47) as:

$$\tau = \frac{T}{2\pi R^3}\left[3 + \frac{\text{d ln } T}{\text{d ln } \Omega}\right] = \frac{1.16 \times 10^{-5}}{2 \times 3.14 \times \left(25 \times 10^{-3}\right)^3}\left[3 + 0.842\right] = 0.454\,\text{Pa}$$

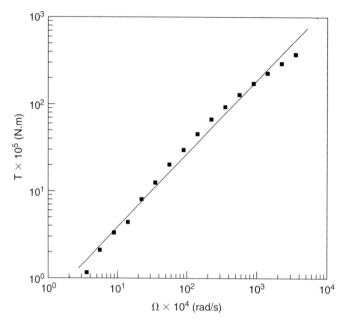

Figure 2.21   *Evaluation of the value of (d ln T/d ln Ω)*

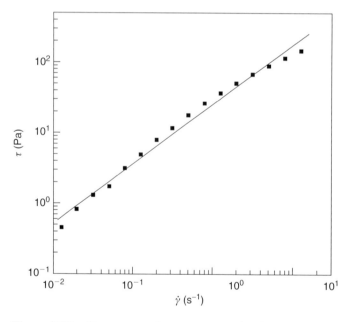

Figure 2.22   *Shear stress–shear rate curve for the polymer solution in Example 2.6*

Similarly, the values of τ and $\dot{\gamma}$ can be calculated for the remaining data points and these values are summarized in Table 2.9 and Figure 2.22 shows the shear-thinning flow behaviour of this product. □

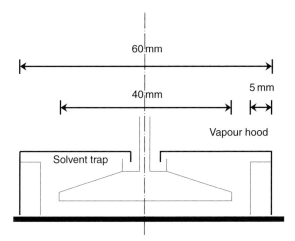

60 mm

40 mm

5 mm

Vapour hood

Solvent trap

Figure 2.23 *Vapour hood employed with a parallel plate geometry*

2.3.5 Moisture loss prevention: the vapour hood

When dealing with high concentration samples of low volume, even low moisture loss can have a critical effect on measured rheological properties (Barnes *et al.*, 1989). During prolonged experiments, moisture loss may be minimized by employing a vapour hood incorporating a solvent trap, as shown in Figure 2.23.

As noted above, edge effects can be encountered with each of the geometries considered here. They become of particular importance when dealing with samples which form a surface 'skin' in contact with the atmosphere, due mainly to evaporation. Conditions at the outer edge of the parallel-plate and cone-and-plate geometries strongly influence the measured torque value. Stresses in this region act on a larger area and are operating at the greatest radius. To ensure homogeneous bulk sample conditions, the evaporation process at the sample surface may be minimized by employing a vapour hood, as shown in Figure 2.23 for the parallel plate system.

2.4 The controlled-stress rheometer

Since the mid-1980s and the advent of reliable 'second generation' controlled-stress rheometers, the controlled-stress technique has become widely established. The facility which most of this type of instrument offers, i.e. of performing three different types of test (steady shear, oscillation and creep), makes them particularly cost effective.

The instrument referred to here for illustration is a TA Instruments CSL 100 controlled-stress rheometer (TA Instruments, UK). The rheometer (typically operated under the control of a microcomputer) and ancillary equipment required for its operation consist of the following main components (see Figure 2.24).

An electronically controlled induction motor incorporates an air bearing, which supports and centres a rotating hollow spindle. The spindle incorporates a threaded draw rod, onto which the components of the required measuring geometry is secured and the air bearing prevents any contact between fixed and moving parts. A digital encoder consisting of a light source and a photocell is arranged either side of a transparent disc

Figure 2.24   *Schematic of a TA Instruments CSL 100/CSL² controlled-stress rheometer*

attached to the spindle. Fine lines (similar to diffraction grating lines) are photographically etched around the disc edge and, through the use of a stationary diffraction grating
between the light source and the disc, diffraction patterns are set up as the disc moves
under an applied torque. These are directly related to the angular displacement of the
measuring system.

The non-rotating lower platen of the measuring assembly is fixed to a height-
adjustable pneumatic ram which may be raised to provide the desired gap setting, with
micrometer-fine adjustment. A temperature control unit is incorporated within the lower
plate. This is usually of the peltier type, using a thermoelectric effect enabling it to function as a heat pump with no moving parts. Control of the magnitude and direction of
the electrical current allows the desired temperature adjustment within the lower platen
(control to 0.1°C), and thus within the sample, for cone-and-plate and parallel-plate
geometries. For the concentric cylinder geometry a temperature-controlled recirculating
water bath is generally used.

Due in part to its ability to produce extremely low shear rates, the controlled-stress
technique has been found to be highly suited to the determination of apparent yield stress,
and in this respect the controlled-stress instrument is widely claimed to be more successful than its controlled-shear rate counterparts. This is usually attributed to the fact that, for
suitably low stresses, the structure of the material may be preserved under the conditions
of test. Indeed, the introduction of the 'second generation' of controlled-stress instruments
can be said to have provoked considerable interest in, and debate surrounding, the field of
yield stress determination, with some early advocates of the controlled-stress technique
advancing the controversial notion of the 'yield stress myth' (Barnes and Walters, 1985).

Apart from the range of instruments described in the preceding sections, several of
the inexpensive viscometers used for quality control in industry give rise to complicated flow and stress fields (which may be neither known nor uniform), but they have
the great advantage that their operation is simple. In the case of Newtonian fluids, the

use of such methods does not pose any problem, since the instruments can be calibrated against a standard Newtonian liquid of known viscosity. However, for non-Newtonian fluids, the analysis and interpretation of results obtained by using such devices is not simple and straightforward. Such devices can be broadly classified into two types. The first have what might loosely be called 'flow constrictions', as exemplified by the Ford cup arrangement, in which the time taken for a fixed volume of liquid to drain through the constriction is measured. Such a device can cope with different ranges of viscosities by changing the size of the constriction. This robust and convenient instrument is used widely in the petroleum and oil industries. The second class of instruments involves the flow around an obstruction as in the falling ball and rolling ball methods (van Wazer *et al.*, 1963) where the time taken for the sphere to settle or roll through a known distance is measured. Although such 'shop-floor' viscometers can perhaps be used for qualitative comparative purposes for purely inelastic fluids, great care needs to be exercised when attempting to characterise visco-elastic and time-dependent non-Newtonian materials even qualitatively (Barnes *et al.*, 1989; Chhabra, 2006).

2.5 Yield stress measurements

Notwithstanding the continuing debate over the very existence of a 'true' yield stress, the concept of an *apparent* yield stress has been found to be an extremely useful empiricism in many areas of science and engineering (Hartnett and Hu, 1989) (see also Chapter 1). A comprehensive review (Barnes, 1999) has critically assessed the various issues raised in the definition, measurement and application of apparent yield stress behaviour.

Any operational definition of apparent yield stress should take into account both the inevitable rheometrical limitations in its determination, and the characteristic time of the process to which it pertains. Such an *operational* definition has been proposed for a true yield stress in the context of the classical stress relaxation experiment (Spaans and Williams, 1995).

Notwithstanding the inherent advantages of the controlled-stress technique in yield studies, it should be borne in mind that an interpretation of the results of creep-compliance measurements in terms of a 'real' yield stress (i.e. a stress below which the sample exhibits Hookean elastic behaviour) is subject to the usual experimental limitations of machine resolution (i.e. of angular displacement) and the role of time scale in the sample's response to applied stress.

Numerous workers have described the role of wall-slip effects on measurements made with conventional smooth-walled geometries (Barnes, 1995). Slip can occur in suspensions at high (*ca.* 60%) solids volume fraction, and can involve fluctuating torque in a rotational viscometer under steady rotation (Cheng and Richmond, 1978; Cheng, 1986). Given that conventional smooth-walled rotational devices tend to slip when in contact with many systems which display an apparent yield stress, several workers have adopted the vane measuring geometry (Nguyen and Boger, 1985).

Typically, the vane geometry consists of a small number (usually four) of thin blades arranged around a cylindrical shaft of small diameter (shown schematically in Figure 2.25 whereas Figure 2.26 shows pictures of a series of vanes of different dimensions). When the material in which the vane is immersed undergoes yielding, it does so within the body of the material, not at a solid boundary, thereby overcoming wall slip. An additional advantage of the vane is that on its insertion into a material there is minimal disturbance of the sample structure compared with that experienced within the narrow shearing

Figure 2.25 *Schematics of a vane used for measuring yield stress in viscoplastic systems*

Figure 2.26 *Typical vanes used for yield stress measurement in silica suspensions and carbopol solutions (Photograph Courtesy Dr P.H.T. Uhlherr, Monash University, Clayton, Australia)*

gaps of conventional measuring geometries. This feature is important for mechanically weak structured systems, such as gels and colloidal dispersions. In practice, the depth of the suspension and the diameter of the container should be at least twice as large as the height and diameter of the vane to minimize the end and wall effects.

Vane rheometry provides a direct measurement of the shear stress at which flow is evident under the conditions of test (i.e. within the time scale of the measurement). In a constant rate (CR) experiment, the material is sheared at a low but constant rate according to the speed of rotation of the viscometer's spindle, and the corresponding torque response with time is recorded, as shown schematically in Figure 2.27. With constant stress (CS) experiments, it is the inferred *deformation* that is recorded as a function of time, under the application of a series of controlled and constant shear stresses (Figure 2.28).

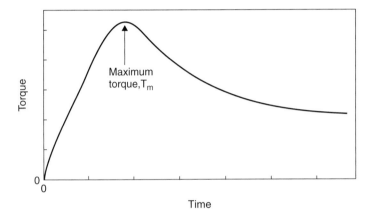

Figure 2.27   *Evolution of torque with time in a viscoplastic system*

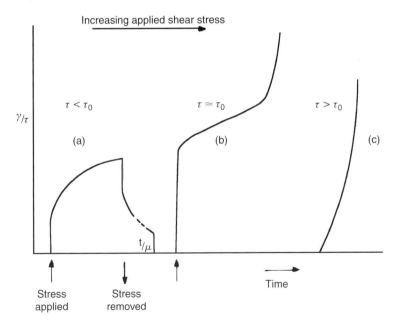

Figure 2.28   *Schematics of a creep-type test for a yield stress material*

The vane technique has its origins as a method for *in situ* measurements of the shear strength of soils and an important assumption in the method is that the yielding surface that results from the vane's rotation is cylindrical, and of the same diameter as the vane (corrections can be made at a later stage if this is proven otherwise but they are difficult to assess for opaque materials). This assumption dictates that the material between the blades acts as a solid cylinder of dimensions equal to those of the vane, and the issue of the yield surface of visco-elastic and plastic fluids in a vane viscometer has been addressed in several studies (e.g. see Yan and James, 1997; Keentok *et al.*, 1985; Barnes and Nguyen, 2001).

The maximum torsional moment (T_m) coincides with the material yielding along this cylindrical surface; thus, a torque balance at the point of yielding provides the following equation:

Total torque = torque from the vane cylindrical shearing surface
+ torque from both vane end shearing surfaces

If τ_c is the shear stress on the cylindrical shearing surface, τ_e for each of the vane end surfaces, the torque balance can be written as:

$$T_m = (\pi DH)\left(\frac{D}{2}\right)\tau_c + 2\int_0^{D/2} 2\pi\tau_e r^2 \, dr \qquad (2.50)$$

where D is the diameter of the cylinder prescribed by the rotating vane; H is the vane height and r is the radial distance at the vane ends.

Since the shear stress, τ_e, acting on the end surfaces of the vane varies from being zero at the centre ($r = 0$) to the wall shear stress value τ_c at $r = D/2$, it is necessary to make some assumption about this relationship. It is customary to postulate either a linear dependence of τ_e on r or a power-law-type dependence as follows:

(a) for a uniformly increasing stress distribution,

$$\tau_e = \left(\frac{2r}{D}\right)\tau_0 \qquad (2.51)$$

(b) for a power-law stress distribution,

$$\tau_e = \left(\frac{2r}{D}\right)^s \tau_0 \qquad (2.52)$$

The constant, *s*, is referred to as the power-law exponent and τ_0 is the yield stress.

At the point of yielding, the value for stress at the cylindrical shearing surface is equal to the yield stress and therefore, $\tau_c = \tau_0$. Therefore, equation (2.50) can be solved for the two cases given by equations (2.51) and (2.52):

(a) for a uniformly increasing stress distribution,

$$\tau_0 = \frac{2T_m}{\pi D^3}\left[\frac{H}{D} + \frac{1}{4}\right]^{-1} \qquad (2.53)$$

(b) for a power-law stress distribution,

$$\tau_0 = \frac{2T_m}{\pi D^3}\left[\frac{H}{D} + \frac{1}{(s + 3)}\right]^{-1} \qquad (2.54)$$

Science and Engineering Research Council
Separation Processes Sub-Committee

Programme Manager: Professor J F Richardson
Department of Chemical Engineering
University College
SWANSEA SA2 8PP

Telephone: 0792 295194

4 - 9 . 08

Dear Pasky,

We were delighted to
see you both. Thank you
for coming and for your gift.

You may like to have
a copy of the N-N book with Raj
All the best Jack

Please excuse out-of-date paper

For $H \gg D$, the contributions of stress from the vane ends becomes negligible. A height-to-diameter ratio, H/D, in excess of 3 is desirable, and it should be as large as is practical.

For a very thin vane ($D \rightarrow 0$), one can argue that the shear stress is uniform everywhere, i.e. $s = 0$, and therefore, equation (2.23) simplifies to:

$$\tau_0 = \frac{2T_\mathrm{m}}{\pi D^3}\left[\frac{H}{D} + \frac{1}{3}\right]^{-1} \tag{2.55}$$

Some of the early studies are based on this approximation. Thus, while the construction of the vane with a very small shearing diameter would seem the ideal way of achieving a high $H{:}D$ ratio, but caution needs to be taken with small diameter vanes as it is likely that torque contributions from the resistance to rotation at the exposed shaft surfaces, above the vane, may then become important.

The vane has been used in conjunction with controlled-stress rheometers to determine apparent yield stresses in cohesive clay suspensions (James *et al.*, 1987); a similar technique has been reported for time-independent materials (Yoshimura *et al.*, 1987). It is important in this type of test that the material attains an equilibrium micro-structural state prior to test. The time, t_e, required to achieve this may be assessed using information obtained from separate, small strain experiments (James *et al.*, 1987).

In a typical controlled-stress vane measurement, the sample is allowed to equilibrate mechanically for a period $t > t_\mathrm{e}$ following insertion of the vane. Subsequently a (constant) stress is applied and maintained for a period, then suddenly removed (see Figure 2.28).

The resulting time-dependent deformation response is recorded (in terms of the resulting angular displacement of the vane) both under applied stress, and following its removal. The material is then allowed to recover for a period $t > t_\mathrm{e}$ and the process is repeated, at a slightly higher value of applied stress. By repeating this sequence, using gradually increased levels of stress, the elastic and viscous components of the deformation of a material may be studied from a state of virtually undisturbed structure, to the onset of an apparent yield behaviour (see Figure 2.28c).

The latter procedure is an adaptation of that usually employed to generate creep-compliance data: a 'creep' yield stress is determined by extrapolating creep rate data to zero rate, with the corresponding value of applied stress being taken as the yield stress (Lohnes *et al.*, 1972). In addition to its use in yield stress measurement, the vane has found application as the basis of a viscometer for systems where severe cases of wall slip may be anticipated (Barnes and Carnali, 1990).

Example 2.7

The following torque–time data is available for a 67.6% (by wt) red mud suspension using a four-bladed vane of diameter $D = 26.15\,\mathrm{mm}$ and of height $H = 50.2\,\mathrm{mm}$.

t(s)	8.3	16.6	25	33.3	41.7	50	66.7	83.3	100	116.7	133.3	150	200
T(mN m)	5.55	8.33	10	10.73	10.9	10.67	10.33	10	9.83	9.50	9.33	9.15	8.55

Estimate the yield stress of this suspension.

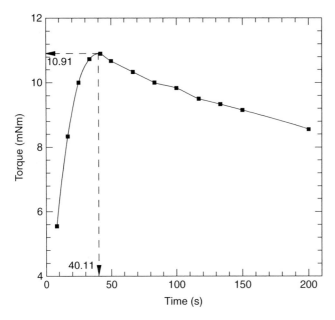

Figure 2.29   *Variation of torque with time*

Solution

Figure 2.29 shows the evolution of torque with time and the maximum value of the torque $T_m = \sim 10.91 \times 10^{-3}$ Nm at about $t \sim 40$ s and therefore, it corresponds to the yield point.

Substituting values of H, D and T_m in equation (2.53):

$$\tau_0 = \frac{2 \times 10.91 \times 10^{-3}}{3.14 \times \left(26.15 \times 10^{-3}\right)^3} \left[\frac{50.2}{26.15} + \frac{1}{4}\right]^{-1} = 179.2 \text{ Pa}$$

On the other hand if the assumption of constant shear stress on the ends surfaces is made, i.e. $s = 0$, equation (2.55) gives,

$$\tau_0 = 172.3 \text{ Pa}$$

Similarly, the resulting value is 183 Pa if $s = 2$ is used. The effect of the value of s will further diminish if a vane with $H/D > 3$ is used. □

2.6 Normal stress measurements

Whorlow (1992) notes that, of the many methods which have been proposed for the measurement of various combinations of the first and second normal stress differences, N_1 and N_2, respectively, few can give reliable estimates of N_2. Combined pressure gradient and total force measurements in the cone-and-plate geometry, or combined cone-and-plate and plate–plate force measurements, appear to give reliable values (Walters, 1975); and satisfactory results may also be obtained from techniques based on the measurement of the elevation of the surface of a liquid as it flows down an inclined open duct (Kuo and Tanner, 1974).

During rotational flow of liquids which display normal stresses, the tension along the circular streamlines is always greater than that in other directions so that streamlines tend to contract unless prevented from doing so by an appropriate pressure distribution. Determination of the pressure distribution (say, over the area of the plate in a cone-and-plate system) therefore provides a means of determining the total force exerted on the plate. Alternatively, and in practice more generally, the total force is measured directly. As the pressure distribution and total force measured in a (parallel) plate–plate system depend in a different manner on the normal stress differences, both cone-and-plate systems and plate–plate systems have been used to obtain values for both N_1 and N_2. Numerous modifications to instruments (such as the Weissenberg Rheogoniometer) to permit more accurate normal force measurements have been described, some involving the use of piezoelectric crystals as very stiff load cells (Higman, 1973).

In the work reported by Jackson and Kaye (1966), the spacing, h, between a cone and a plate was varied and the normal force measured as a function of gap size. The same method was used by Marsh and Pearson (1968) who showed that,

$$N_2(\dot{\gamma}_R) = \frac{h + R\theta}{h} \left(N_1(\dot{\gamma}_R) - \frac{2F}{\pi R^2} + \frac{h}{\pi R^2} \frac{dF}{dh} \right) \tag{2.56}$$

where the shear rate at the rim, $\dot{\gamma}_R$, is given by $\Omega R/(h + R\theta)$; θ is the cone angle. As the value of zero gap spacing is used to find N_1, equation (2.56) can only be used for non-zero values of h, and the geometry (being intermediate between that of the conventional cone-and-plate and parallel-plate systems) may not give as good estimates of N_2 as those obtained by other methods (Walters, 1975).

The direct determination of N_1 by measuring the total force on the cone in a cone-and-plate system is limited to low shear rates. Binding and Walters (1976) have described a *torsion balance rheometer* which can provide normal force and viscosity data on low viscosity fluids at very high shear rates. This instrument is particularly useful on systems which display an apparent yield stress (Binding *et al.*, 1976). Encouraging agreement has been reported between the results of measurements made in a rheogoniometer, a torsion balance rheometer and a 'Stressmeter' (Lodge *et al.*, 1987), the latter instrument (which can provide data at extremely high shear rates) exploiting the 'hole pressure effect' identified by Broadbent *et al.* (1968).

For a detailed and systematic treatment of the various approaches to normal stress measurements, the reader is referred to the text by Walters (1975), Whorlow (1992) and Macosko (1994).

2.7 Oscillatory shear measurements

At the outset, it is useful to consider the response of a Newtonian fluid and of a Hookean solid to a strain which varies sinusoidally with time as:

$$\gamma = \gamma_m \sin \omega t \tag{2.57}$$

where γ_m is the amplitude and ω is the frequency. For a Hookean elastic solid, the stress is related linearly to strain, the constant of proportionality being Young's modulus, i.e.,

$$\tau = G\gamma \tag{2.58}$$

Owing to the linear relationship shown in equation (2.58) both τ and γ are in phase or the phase lag is zero in this case.

On the other hand, for a Newtonian fluid, the shear stress is related to the rate of shear, i.e.,

$$\tau = \mu\dot{\gamma} \tag{2.59}$$

Differentiating equation (2.57) with respect to time to obtain the expression for $\dot{\gamma}$:

$$\dot{\gamma} = \frac{d\gamma}{dt} = \gamma_m \omega \cos \omega t = \gamma_m \omega \sin\left(\omega t + \frac{\pi}{2}\right) \tag{2.60}$$

and therefore,

$$\tau = \mu\gamma_m \omega \sin\left(\omega t + \frac{\pi}{2}\right) \tag{2.61}$$

Obviously in this case, the resulting shear stress and the applied strain have a phase difference of $(\pi/2)$, i.e. the applied strain lags behind the stress (Figure 2.30). Thus, the measurement of the phase angle δ which can obviously vary between 0 (purely viscous) and $\pi/2$ (purely elastic) provides a convenient method to quantify the level of visco-elasticity of a substance. Suffice it to add here that small values of δ correspond to predominantly viscous behaviour whereas large values of δ imply strongly visco-elastic nature of the fluid in oscillatory shear motion.

Of the techniques used to characterize the linear visco-elastic behaviour displayed by many non-Newtonian fluids, the oscillatory shear technique which involves either an applied stress or shear rate which varies harmonically with time, is perhaps the most convenient and widely used.

The definition of linear visco-elasticity may be expressed in the following form: the ratio of the applied stress to strain for any shear history is a function of time alone, and independent of stress magnitude: each stress applied to a material produces a strain which is independent of that produced by any other stresses. However, the total strain experienced by a material is equal to the sum of all the changes induced in the material by the applied stress throughout its history.

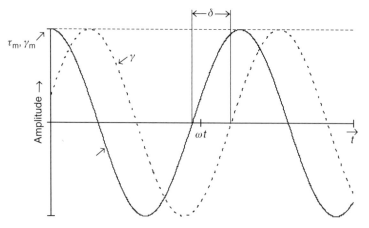

Figure 2.30   *Oscillatory shear strain (---) out of phase with stress (–) by a phase angle δ*

The foregoing is an expression of the Boltzmann *superposition principle* (Bird *et al.*, 1987) which may also be expressed in the following terms:

$$\tau(t) = \int_{-\infty}^{t} G(t - t')\dot{\gamma}(t')dt' \tag{2.62}$$

where $G(t)$ is the stress relaxation modulus and the integration is performed over all past times t' up to the current time. In visco-elastic liquids, the function $G(s)$, where $s = t - t'$, approaches zero as s approaches infinity, giving rise to the following alternative expression, in terms of strain history rather than rate of strain:

$$\tau(t) = \int_{-\infty}^{t} m(t - t')\gamma(t, t')dt' \tag{2.63}$$

where the 'memory function', $m(t)$, is given by $-dG(t)/dt$.

These 'constitutive' equations, involving $G(t)$, describe the response of linear visco-elastic materials to various time-dependent patterns of stress and strain in simple shear (Ferry, 1980). In many cases it is more convenient, in terms of experimental technique, to consider the complex shear modulus, G^*, which is measured using *oscillatory* shear.

Within the region of linear visco-elastic behaviour, an imposed stress of angular frequency, ω, results in a harmonic strain of amplitude proportional to the stress amplitude, and with phase lag δ relative to the stress, which is independent of the applied stress amplitude (Ferry, 1980).

Figure 2.30 represents a controlled harmonic shear stress applied to a linear visco-elastic system, which results in a harmonic strain response waveform involving a phase lag, δ, on the applied stress. This harmonic shear stress, and the resultant strain, may be conveniently expressed as complex quantities, where $\gamma^* = \gamma_m e^{i\omega t}$, $\tau^* = \tau_m e^{i(\omega t + \delta)}$, and γ_m and τ_m are the peak strain and peak stress amplitudes, respectively.

The complex shear modulus, G^*, is then defined as:

$$G^* = \frac{\tau^*}{\gamma^*} = \frac{\tau_m}{\gamma_m} \frac{e^{i(\omega t + \delta)}}{e^{i\omega t}} = \frac{\tau_m}{\gamma_m} e^{i\delta} \tag{2.64}$$

where $G^* = G' + iG''$.

Hence,

$$G' = \frac{\tau_m}{\gamma_m} \cos \delta \tag{2.65}$$

and

$$G'' = \frac{\tau_m}{\gamma_m} \sin \delta \tag{2.66}$$

where G' is the dynamic rigidity, defined as the stress in phase with the strain divided by the strain in a sinusoidal deformation: it is a measure of energy stored and recovered per cycle of deformation, i.e., the extent of elastic behaviour. G'' is the loss modulus, defined as the stress in quadrature (90° out of phase) with the strain, divided by the strain: it is a measure of energy dissipated per cycle, i.e., the extent of viscous behaviour. By definition:

$$\delta = \tan^{-1} \frac{G''}{G'} \quad \text{or} \quad \tan \delta = \frac{G''}{G'}$$

Phase relationships may also be expressed in terms of the complex viscosity, μ^*:

$$\mu^* = \mu' + i\mu'' \tag{2.67}$$

where the dynamic viscosity, μ', is defined as the stress in phase with the rate of strain divided by the rate of strain in a sinusoidal deformation, and μ'' as the stress in quadrature to the rate of strain in a sinusoidal deformation. These are related to G^* as follows:

$$G^* = \omega\mu'' + i\omega\mu' \tag{2.68}$$

hence

$$\mu'' = \frac{G'}{\omega} \quad \text{and} \quad \mu' = \frac{G''}{\omega} \tag{2.69}$$

Controlled strain (or, more properly, controlled displacement) oscillatory shear instruments, exemplified by the Weissenberg Rheogoniometer (Macsporran and Spiers, 1982) readily facilitate tests in which independent measurements of both changing length scales and time scales of applied deformation can be performed. In this way it is, in principle, possible to separate effects due to strain and strain *rates* as the frequency of oscillation may be held constant while the maximum (cyclic) shear strain amplitude is varied. Alternatively, the frequency of deformation can be varied at constant maximum shear strain amplitude.

Rheometers capable of performing oscillatory shear are widely available as commercial instruments, in addition to more specialized devices (Te Nijenhuis and van Donselaar, 1985). An example of the latter is an oscillating plate rheometer (Eggers and Richmann, 1995) which requires a very small liquid sample volume (<0.3 ml) and has a (potentially) great frequency range (2 Hz to 1 kHz). This latter feature is unusual as it spans the gap between specialized devices and commercially available oscillatory shear instruments, whose frequency range is typically 10^{-3} Hz to *ca.* 10^2 Hz.

Usually, the deformation of a sample undergoing oscillatory shear is monitored by measuring the sinusoidally varying motion of a transducer-controlled driving surface in contact with the sample. However, in turning to the subsequent calculation of shear strain amplitude in dynamic measurements, it must be recognized that conversion of experimentally determined forces and displacements to the corresponding stresses and strains experienced by a sample can involve consideration of the role of sample inertia.

Sample inertial forces will be small if the sample density, ρ, is small compared with (G'/h^2f^2) or (G''/h^2f^2) where h is the shearing gap thickness and f is the frequency of oscillation (in Hz). Under these conditions, the states of stress or strain may be considered uniform throughout the shearing gap and to experience no periodic spatial variation. This represents the 'gap loading' condition (Ferry, 1980).

Under 'surface loading' conditions, shear waves propagate within the medium, decaying in amplitude from the driving surface to the opposite side of the shearing gap. Thus, under 'surface loading' condition, sample inertia effects are dominant. However, it should be borne in mind that the gap loading condition is not one in which wave propagation is unimportant, and for high precision measurements the shear wavelength, λ should be at least 40 times as large as the shearing gap size, h (a ratio of λ/h of *ca.* 10 may be tolerable (Schrag, 1977)).

Fluid inertia effects have been found to be very small for the cone-and-plate geometries typically supplied with these instruments. While inertial corrections are found to be unimportant for the parallel plate geometries, for shearing gaps of the order of 2 mm or less (except possibly for very 'thin' fluids), they must be taken into account in the concentric cylinder geometry (especially for high density, mobile fluids). Evaluation methods are available for the complex viscosity, μ^* in the case of cylindrical and plane Couette flow, taking into account fluid inertia (Aschoff and Schummer, 1993).

2.8 Extensional flow measurements

Extensional flows may be generated within a flow field that experiences a sudden change in geometry, such as contractions or orifice plates. The need for appropriate rheological data pertaining to such flows represents a major obstacle to the development of improved process simulation, monitoring and control for non-Newtonian fluids, and requires, not only a knowledge of fluid shear viscosity, but also an *appropriate* measure of its *extensional* viscosity, μ_E. The problematical effects of extensional viscosity in process engineering become evident when one considers that the value of the extensional viscosity, μ_E, may be several *orders of magnitude* higher than the corresponding shear viscosity in some non-Newtonian liquids. In other words, the value of the Trouton ratio (defined in equation (1.30)) may be of the order of $10^3 - 10^4$ for strongly elastic substances.

In marked contrast to measurements of shear rheological properties, such as apparent viscosity in steady shear, or of complex viscosity in small amplitude oscillatory shear, extensional viscosity measurements are far from straightforward. This is particularly so in the case of mobile elastic liquids whose rheology can mitigate against the generation of well-defined extensional flow fields.

Techniques for measuring the extensional properties of fluids can be divided (broadly) into those of the 'flow-through' and 'stagnation-point' types (Hermansky and Boger, 1995). The former usually involve 'spinnable' fluids, a feature exploited in instruments such as the Carri-Med (now TA Instruments) elongational viscosity rheometer (Ferguson and Hudson, 1990).

In spin-line experiments, fluid is delivered through a nozzle and subsequently stretched by an applied force. Procedures are available for obtaining useful semi-quantitative estimates of μ_E from spin-line experiments, and the technique is able to distinguish between fluids that are tension-thinning and tension-thickening (Jones *et al.*, 1987). Despite being limited to low rates of strain, and generally suitable for highly viscous or elastic fluids, perhaps the most successful device of this type, is the filament-stretching technique (Sridhar *et al.*, 1991) which is described below in Section 2.8.2.

When a fluid is not 'spinnable', the various orifice flow techniques, which involve pressure drop measurements across a contraction (Binding, 1988, 1993), can provide a means of estimating the extensional-viscosity behaviour of *shear-thinning* polymer solutions.

For low viscosity fluids such as dilute polymer solutions, the various stagnation-point devices can also prove useful. This category is the commercially available Rheometrics RFX opposing jet device, a development of earlier instruments (Cathey and Fuller, 1988), which has the potential to produce a wide range of strain rates, and which has been used to study fluids with a viscosity approaching that of water (Hermansky and Boger, 1995).

2.8.1 Lubricated planar stagnation die flows

Lubricated dies have been proposed as a means of generating extensional flows in which the lubricant protects the sample of interest from the shear effects of the die walls, and serves to transmit forces to the walls where they can be measured. A theoretical analysis of steady, two-dimensional flow through a lubricated planar stagnation die has shown that pure planar extension through the die is prevented by the interaction of the die shape with the normal viscous stresses at the free interface between the lubricant and the test fluid (Secor *et al.*, 1987). Interestingly, two regions of the flow (near the inlet and outlet of the die) are nearly extensional, and are capable of being represented by an approximate expression which, in addition to being independent of the constitutive relation of the fluid, provides a simple relationship between wall pressure measurements and extensional viscosity.

In practice, the technique is beset by experimental difficulties, such as the need to measure extremely small pressures. In addition, the difficulty of maintaining adequately lubricated flows requires the provision of large sample volumes. The application of converging flows for determining the extensional flow behaviour of polymer melts has been reviewed (Rides and Chakravorty, 1997).

2.8.2 Filament-stretching techniques

In this method, the sample is held between two discs, the lower of which is attached to a shaft whose movement is controlled by a computer capable of generating an exponentially varying voltage, the shaft velocity being proportional to the applied voltage. The upper disc is attached to a load measuring device and an optical system is used to measure filament diameter (Sridhar *et al.*, 1991).

A reverse-flow near the plates causes a delay in the development of the uniform cylindrical column, and a difference between the local and imposed extension rates at early times (Shipman *et al.*, 1991). Problems may also be encountered due to adhesion of the fluid to the plates (Spiegelberg and McKinley, 1996) but this may be compensated for (Tirtaatmadja and Sridhar, 1993). Notwithstanding these difficulties, the technique can provide meaningful extensional viscosity data for polymer solutions, as the deformation is uniform and independently imposed, and the total strain is measurable.

Detailed accounts of the available experimental methods for measuring the elongational viscosity of polymeric melts and solutions have been provided by James and Walters (1994), Gupta and Sridhar (1998) and McKinley and Sridhar (2002).

2.9 On-line viscometry

The rheological characteristics, particularly viscosity and yield stress, of end products and/or of intermediate process streams are routinely used as indicators of quality in many industrial settings including food, personal care products, pharmaceutical products, paints and polymers. Until recently, it was deemed adequate to carry out off-line periodic quality checks together with statistical analysis to maintain the product specifications within the acceptable tolerance limits. However, with the increasing competitiveness coupled with the ever-increasing quality expectations, this approach is grossly inadequate. Furthermore, the development and use of on-line and in-line sensors is not only necessary for rapid and continuous monitoring of product quality, but is also a prerequisite for process automation and control. This would further reduce the manufacturing of products failing to meet the desired quality standards. Lastly, it is not always possible to link an off-line measurement to a key processing variable for an appropriate correction to be identified and implemented. For instance, a milk concentrate (containing

~50 wt% solids) might flow readily into a sampling device from a process line, but frequently it might undergo rapid gelation thereafter even in a controlled temperature environment; and therefore, the off-line rheological measurements would hardly be relevant to the actual processing conditions. Similarly, chocolate displays markedly thixotropic behaviour and hence an off-line measurement on a sample which has been subjected to an ill-defined shear history during sampling is unlikely to yield useful information about the *in situ* quality of the product. Finally, sampling valves themselves can sometimes damage the structure (for instance, droplet size in an emulsion) thereby making the off-line measurement of little utility. In view of all these issues, much attention has been given to the development of on-line or in-line measurements of viscosity for a range of rheologically complex substances over the past 20 years or so (Cheng *et al.*, 1984; Dealy, 1990; Steffe, 1996; Roberts, 2001). The terms *on-line* and *in-line* are used here to distinguish the location of the sensor being in a bye-pass line or in the actual process line itself, but both refer to continuous monitoring of quality parameters.

Early attempts (which are still used in numerous situations) relied on single-point measurement, which is quite satisfactory for a Newtonian fluid, but its relevance diminishes for a non-Newtonian substance, as shown schematically in Figure 2.31. This approach can obviously lead to quite an erroneous information about the overall shear-thinning nature of the products as shown in Figure 2.31. Therefore, it is essential that the sensor should be capable of giving a series of points covering the relevant range of shear rate. In addition, such a sensor should be robust and stable, able to operate at high temperatures and pressures, immune to plant noise, compatible with the product or the process stream (hygiene-risk free in food, pharmaceutical settings, for instance) and quick (fast response time) to follow the changes in the characteristics. Most of the developments in this field follow the well-established technologies of the capillary and rotational viscometry. In its simplest form, an on-line capillary viscometer is a bye-pass line into which a measured quantity of the stream is diverted continuously using a positive displacement pump, such as a gear pump, and the corresponding pressure drop is measured (using pressure transducers) as a function of the flow rate. This information can readily be converted into the wall shear stress–shear rate data, as detailed in Section 2.2.1

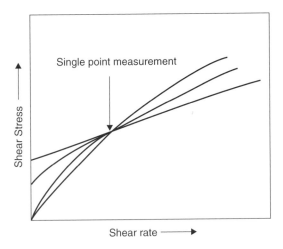

Figure 2.31   *Limitations of a single-point shear stress (or viscosity) measurement for non-Newtonian substances*

(e.g. see Perona and Sordo, 2006). Depending upon the specific application, the diverted stream may be discarded or recycled back into the process line. In principle, this is a cheap and simple on-line viscometer, but caution is necessary for materials exhibiting wall slip. Also, in some instances, the response time may be too long to monitor the quick changes in the micro-structure (and hence in viscosity) of a product.

In recent years, there has been a growing interest in the use of non-invasive methods involving magnetic resonance imaging, MRI (McCarthy *et al.*, 1992; Li *et al.*, 1994; Arola *et al.*, 1997; Gibbs *et al.*, 1997; Goloshevsky *et al.*, 2005), ultrasound Doppler velocimetry (Roberts, 2001; Greenwood and Bamberger, 2002; Ouriev *et al.*, 2003; Brunn *et al.*, 2004; Pfund *et al.*, 2006) and LDA (laser doppler anemometer) techniques. The MRI technique employs two magnetic field gradients, one in the direction of flow and the other in the direction normal to the flow. In the ultrasonic method, an impulse of known frequency directed into the flow is then reflected from a moving particle or droplet. The technique relies on the presence of reflective particles in the stream. The methods just described yield detailed velocity profiles across the cross-section of a capillary and hence the local shear rate can be estimated by the differentiation of the velocity distribution, which, in turn, can be linked to the corresponding shear stress from a knowledge of the pressure gradient together with the application of equation (2.1). Hence, one can generate a series of shear stress–shear rate data points over the relevant range of shear rate. At this juncture, it is probably appropriate to say that MRI is a useful research tool, but the cost and portability severely limit its routine application, at least at present, as an on-line method. Sensors based on ultrasonics have been used to monitor the flow of chocolates and starch solutions, but this technique is not suitable for bubbly fluids due to the low signal-to-noise ratio. Similarly, while this method can probably provide reliable information about the yield stress of material in the form of a plug formation, but it does not work well close to the wall and therefore, wall-slip measurements are not possible. Similarly, though LDA devices have been used extensively for flow visualization in a variety of settings, their cost and the subsequent somewhat intricate data processing have impeded their use as routine sensors in process engineering applications.

Within the class of rotational devices, the most widely used configuration is the concentric cylinders or bob-in-cup type. These devices are usually mounted in a bye-pass loop, as shown in Figure 2.32. A potential problem with these devices is their poor sensitivity and long response time. The reservoir must be sealed to avoid contamination, but the additional friction on the inner rotating cylinder (bob) due to this seal may adversely influence the sensitivity, especially for low viscosity substances. While many in-house designs operating at a specific shear rate relevant to the application are available (e.g. see Roberts, 2001), perhaps the most commonly employed on-line rotational viscometers are those manufactured by Brookfield, Inc. for a range of process applications.

Other modifications of this principle include the imposing of an axial flow on the rotation imposed by the bob thereby resulting in the so-called helical flow. This idea was exploited by Akroyd and Nguyen (2003) for on-line viscometry of aqueous clay, fly-ash and mine-tailing slurries. Another variation of the rotational viscometry is the use of an impeller to keep the suspension homogeneous, and to use the approach of Metzner and Otto (1957), as detailed in Chapter 8 and in Chhabra (2003), to infer the mean value of the viscosity based on an average shear rate (proportional to the rotational speed of the impeller). Needless to say, most of these devices require calibration using Newtonian fluids of known viscosity as close as possible to that of the material to be monitored. In other words, such devices are capable of yielding results which are adequate

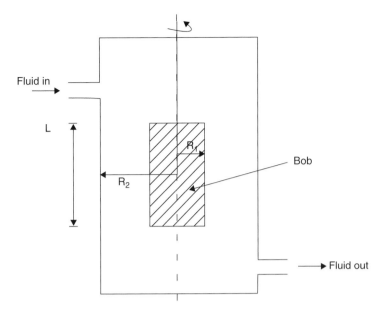

Figure 2.32 *Schematic of an on-line cub-and-bob viscometer*

for comparative purposes only, but in many situations that is sufficient for quality control purpose.

In food process engineering applications, vibration viscometers are also used routinely to monitor viscosity and sometimes even elasticity. These probes, in the form of a cylinder or a sphere, operate typically at 10^2–10^6 Hz, but they undergo damping due to the motion of the fluid past them. These devices are robust and are easy to install in the main process line or in a mixing vessel (holding tank) or a fermenter without requiring a bye-pass loop and the associated accessories. The so-called Nametre vibrating sphere (Figure 2.33) is used widely in the process industries. The power required to maintain a constant amplitude of vibration is related to the product of the density and the viscosity of the fluid. Since viscosity is more sensitive than density to composition, this device can operate over the viscosity range of ~10 mPa s to ~100 Pa s. Another variant of this principle is the flow through an oscillating tube (Roberts, 2001). While these high frequency devices are quite sensitive to the changes in composition of a product, their performance deteriorates rapidly for highly aerated (bubbly) fluids. Also, owing to the complex flow field produced in these devices, it is virtually impossible to link such on-line measurements to off-line rheograms obtained in well-defined configurations.

In summary, the on-line and in-line sensors based on the principle of a capillary viscometer can yield true rheological data which may be linked to the off-line measurements, provided the usual precautions regarding end effects, wall-slip effects, etc. are taken. On the other hand, this link is not as straightforward for the sensors based on rotational systems (bob-in-cup, impeller), or operating on a vibrational principle. The latter category also necessitates calibration with Newtonian fluids, but these measurements are adequate for comparative quality control purposes in many instances.

Before closing this chapter, it is useful to emphasize that the choice of a viscometer (rheometer) is strongly dependent on the nature of the material to be tested and the purpose of the measurements. For instance, for homogeneous (and pseudo-homogeneous)

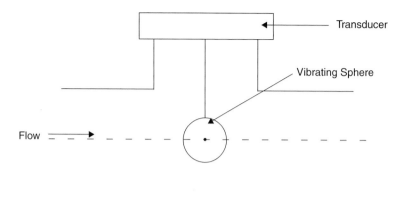

Figure 2.33   *Schematic of a vibrating sphere viscometer*

substances like polymer melts and their solutions, stable suspensions of fine particles and stable micro-emulsions, the choice of a device is really governed by the range of shear rates over which data are needed and on what rheological aspect (like other than shear stress, N_1, N_2, G' or G'', etc.). This choice is far from obvious for heterogeneous systems like fibrous suspensions, slurries (settling or non-settling, particle size, concentration, etc.), foams and emulsions. The single most complicating feature in all these cases is their inherent unstable nature and general tendency to exhibit wall slippage. Broadly speaking, it is recommended that the gap or capillary diameter should be 50 times the particle/floc, droplet size to avoid jamming and segregation arising from the migration of particles in shearing flows. Further difficulties may arise in the case of thixotropic materials or mixtures in highly volatile solvents, slurries with abrasive particles, etc. Therefore, an appropriate choice of a measuring instrument and geometry depends strongly on the nature of the material and the range of conditions over which it needs to be characterized. Some practical guidelines in this regard are available in the literature (Heywood, 1985; ESDU 95012; Coussot, 2005).

Further reading

Coussot, P., *Rheometry of Pastes, Suspensions and Granular Materials*, Wiley, New York (2005).

James, D.F. and Walters, K., A critical appraisal of available methods for the measurement of extensional properties of mobile systems. *Techniques in Rheological Measurements* (edited by Collyer, A.A.), Elsevier, London (1993).

Macosko, C.W. and Souza Mendes, P.R., Fluid mechanics measurements in non-Newtonian fluids. *Fluid Mechanics Measurements* (edited by Goldstern, R.J.), 2nd edition, Taylor and Francis, New York (1996).

Walters, K., *Rheometry*, Chapman and Hall, London (1975).

Whorlow, R.W., *Rheological Techniques*, 2nd edition, Ellis Horwood, London (1992).

References

Agarwal, U.S., Dutta, A. and Mashelkar, R.A., *Chem. Eng. Sci.* **49** (1994) 1693.

Akroyd, T.J. and Nguyen, Q.D., *Exp. Therm Fluid Sci* **27** (2003) 507.

Arola, D.F., Barrall, G.A., Powell, R.L., McCarthy, K.L. and McCarthy, M.J., *Chem. Eng. Sci.* **52** (1997) 2049.

Aschoff, D. and Schummer, P., *Rheol. Acta* **37** (1993) 1237.

Bagley, E.B., *J. Appl. Phys.* **28** (1957) 624.

Barnes, H.A., *J. Non-Newt. Fluid Mech.* **56** (1995) 221.

Barnes, H.A., *J. Non-Newt. Fluid Mech.* **81** (1999) 133.

Barnes, H.A. and Carnali, J.O., *J. Rheol.* **34** (1990) 841.

Barnes, H.A. and Walters, K., *Rheol. Acta* **24** (1985) 323.

Barnes, H.A., Hutton, J.F. and Walters, K., *An Introduction to Rheology*, Elsevier, Amsterdam (1989).

Barnes, H.A. and Nguyen, Q.D., *J. Non-Newt. Fluid Mech.* **98** (2001) 1.

Binding, D.M., *J. Non-Newt. Fluid Mech.* **27** (1988) 173.

Binding, D.M., *Techniques in Rheological Measurement* (edited by Collyer, A.A.), Chapman & Hall, London (1993).

Binding, D.M. and Walters, K., *J. Non-Newt. Fluid Mech.* **1** (1976) 259, and 277.

Binding, D.M., Hutton, J.F. and Walters, K., *Rheol. Acta* **15** (1976) 540.

Bird, R.B., Hassager, O., Armstrong, R.C. and Curtiss, C.F., *Dynamics of Polymeric Liquids, Kinetic Theory*, 2nd edition, Vol. 2, John Wiley & Sons, New York (1987).

Broadbent, J.M., Kaye, A., Lodge, A.S. and Vale, D.G., *Nature* **217** (1968) 55.

Brunn, P.O., Wunderlich, T. and Muller, M., *Flow Meas. Instrum.* **15** (2004) 139.

Cathey, C.A. and Fuller, G.G., *J. Non-Newt. Fluid Mech.* **30** (1988) 303.

Cheng, D.C-H., *Rheol. Acta* **25** (1986) 542.

Cheng, D.C-H. and Richmond, R.A., *Rheol. Acta* **17** (1978) 446.

Cheng, D.C.-H., Hunt, J.A. and Madhvi, P., *Status Report on Process Control Viscometers: Current Applications and Future Needs*, Warren Springs Laboratory, Stevenage, UK (1984).

Chhabra, R.P., *Adv. Heat Tran.* **37** (2003) 77.

Chhabra, R.P., *Bubbles, Drops and Particles in Non-Newtonian Fluids*, 2nd edition, CRC Press, Boca Raton, Florida (2006).

Coussot, P., *Rheometry of Pastes, Suspensions and Granular Materials*, Wiley, New York (2005).

Dealy, J.M., *Rheol. Acta* **29** (1990) 519.

Dervisoglu, M. and Kokini, J.L., *J. Food Sci.* **51** (1986) 541.

Eggers, F. and Richmann, K.-H., *Rheol. Acta* **34** (1995) 483.

Ferguson, J. and Hudson, N., *J. Non-Newt. Fluid Mech.* **35** (1990) 197.

Ferry, J.D., *Visco-elastic Properties of Polymers*, 3rd Edition, Wiley, New York (1980).

Gibbs, S.J., Haycock, D.E., Frith, W.J., Ablett, S. and Hall, L.D., *J. Magn. Reson* **125** (1997) 43.

Goloshevsky, A.G., Walton, J.H., Shutov, M.V., deRopp, J.S., Collins, S.D. and McCarthy, M.J., *Meas. Sci. Technol.* **16** (2005) 513.

Greenwood, M.S. and Bamberger, J.A., *Ultrasonics* **39** (2002) 623.

Gupta, R.K. and Sridhar, T., *Rheological Measurements* (edited by Collyer, A.A. and Clegg, D.W.), 2nd edition, Chapman and Hall, London, p. 516 (1998).

Hartnett, J.P. and Hu, R.Y.Z., *J. Rheol.* **33** (1989) 671.

Hermansky, C.G. and Boger, D.V., *J. Non-Newt. Fluid Mech.* **56** (1995) 1.

Higman, R.W., *Rheol. Acta* **12** (1973) 533.

Heywood, N.I., Selecting a viscometer. *Chem. Eng.* **415** (1985) 16.

Jackson, R. and Kaye, A., *Brit. J. Appl. Phys.* **17** (1966) 1355.

Jacobsen, R.T., *J. Coll. Interf. Sci.* **48** (1974) 437.

James, A.E., Williams, D.J.A. and Williams, P.R., *Rheol. Acta* **26** (1987) 437.

James, D.F. and Walters, K., *Techniques in Rheological Measurements* (edited by Collyer, A.A.), Elsevier, Amsterdam, p. 33 (1994).

Jastrzebski, Z.D., *Ind. Eng. Chem. Fund.* **6** (1967) 445.

Jones, T.E.R., Davies, J.M. and Thomas, A., *Rheol. Acta* **26** (1987) 1419.

Joseph, D.D., Narain, A. and Riccius, O., *J. Fluid Mech.* **171** (1986) 289.

Keentok, M., Milthorpe, J.F. and O'Donovan, E., *J. Non-Newt. Fluid Mech.* **17** (1985) 23.

Kim, S. and Dealy, J.M., *J. Rheol.* **45** (2001) 1413.

Krieger, I.M. and Maron, S., *J. Appl. Phys.* **23** (1952) 147.

Kuo, Y. and Tanner, R.I., *Rheol. Acta* **13** (1974) 443.

Kurath, S.F. and Larson, W.S., *Tappi J.* **119** (1990) 235.

Lam, C.Y., Wang, Z.Y., Chen, X. and Joshi, S.C., *Powder Technol.* **177** (2007) 162.

Laun, H.M., *Rheol. Acta* **22** (1983) 171.

Laun, H.M. and Hirsch, G., *Rheol. Acta* **28** (1989) 267.

Li, T.Q., Powell, R.L., Odberg, L., McCarthy, M.J. and McCarthy, K.L., *Tappi J.* **123** (1994) 456.

Lodge, A.S., Al-Hadithi, T.S.R. and Walters, K., *Rheol. Acta* **26** (1987) 516.

Lohnes, R.A., Millan, A. and Hanby, R.L., *J. Soil Mech. Found. Div.* **98** (1972) 143.

Macosko, C.W., *Rheology: Principles, Measurements and Applications*, Wiley-VCH, New York (1994).

Macsporran, W.C. and Spiers, R.P., *Rheol. Acta* **21** (1982) 184.

Marsh, B.D. and Pearson, J.R.A., *Rheol. Acta* **7** (1968) 326.

McCarthy, M.J., Maneval, J.E. and Powell, R.L., *Adv. Food Eng.* **1** (1992) 87.

McKinley, G.H. and Sridhar, T., *Annu. Rev. Fluid Mech.* **34** (2002) 375.

Metzner, A.B. and Otto, R.E., *AIChE J.* **3** (1957) 3.

Mooney, M., *Trans. Soc. Rheol.* **2** (1931) 210.

Neto, C., Evans, D.R., Bonnaccurso, E., Butt, H.-J. and Craig, V.S.J., *Rep. Prog. Phys.* **68** (2005) 2859.

Nguyen, Q.D. and Boger, D.V., *J. Rheol.* **29** (1985) 335.

Ouriev, B., Windhab, E., Braun, P., Zeng, Y. and Birkhofer, B., *Rev. Sci. Instr.* **74** (2003) 5255.

Perona, P. and Sordo, S., *Proc. 4th Int. Sym. Food Rheology & Structure (ISFRS)*, Zurich, p. 203 (2006).

Pfund, D.M., Greenwood, M.S., Bamberger, J.A. and Pappas, R.A., *Ultrasonics* **44** (2006) e477.

Rides, M. and Chakravorty, S., *National Physical Laboratory Report CMMT (A)* **80**, Aug 1997. ISSN 1361–4061.

Roberts, I., *Instrumentation and Sensors for the Food Industry* (edited by Kress-Rogers, E. and Brimelow, C. J.B.), 2nd edition, CRC Press, Boca Raton (2001), Chapter 14.

Schrag, J.L., *Trans. Soc. Rheol.* **21** (1977) 399.

Secor, R.B., Macosko, C.W. and Scriven, L.E., *J. Non-Newt. Fluid Mech.* **23** (1987) 355.

Shipman, R.W.G., Denn, M.M. and Keunings, R., *J. Non-Newt. Fluid Mech.* **40** (1991) 281.

Spaans, R.D. and Williams, M.C., *J. Rheol.* **39** (1995) 241.

Spiegelberg, S.H. and McKinley, G.H., *J. Non-Newt. Fluid Mech.* **67** (1996) 49.

Sridhar, T., Tirtaatmadja, V., Nguyen, D.A. and Gupta, R.K., *J. Non-Newt. Fluid Mech.* **40** (1991) 271.

Steffe, J.F., *Rheological Methods in Food Process Engineering*, Freeman Press, East Lansing, MI (1996).

Te Nijenhuis, K. and van Donselaar, R., *Rheol. Acta* **24** (1985) 47.

Tirtaatmadja, V. and Sridhar, T., *J. Rheol.* **37** (1993) 1081.

Van Wazer, J.R., Lyons, J.W., Lim, K.Y. and Colwell, R.E., *Viscosity and Flow Measurement*, Wiley, New York (1963).

Walters, K., *Rheometry*, Chapman and Hall, London (1975).

Whorlow, R.W., *Rheological Techniques*, 2nd edition, Ellis Horwood, London (1992).

Williams, R.W., *Rheol. Acta* **18** (1979) 345.

Yan, J. and James, A.E., *J. Non-Newt. Fluid Mech.* **70** (1997) 237.

Yoshimura, A.S., Princen, H.M. and Kiss, A.D., *J. Rheol.* **31** (1987) 699.

Yoshimura, A.S. and Prud'homme, R.K., *J. Rheol.* **32** (1988) 53.

Nomenclature

		Dimensions in $\mathbf{M, L, T}$
D	capillary or tube diameter (m)	$\mathbf{L}$
F	axial force (N)	$\mathbf{MLT}^{-2}$
G'	dynamic rigidity or storage modulus (Pa)	$\mathbf{ML}^{-1}\mathbf{T}^{-2}$
G''	loss modulus (Pa)	$\mathbf{ML}^{-1}\mathbf{T}^{-2}$
$G*$	complex shear modulus (Pa)	$\mathbf{ML}^{-1}\mathbf{T}^{-2}$
h	gap in parallel plate system (m)	$\mathbf{L}$
k	ratio R_1/R_2 (−)	$\mathbf{M}^0\mathbf{L}^0\mathbf{T}^0$
L	length of capillary (m)	$\mathbf{L}$
m	power law consistency coefficient (Pa s^n)	$\mathbf{ML}^{-1}\mathbf{T}^{-2-n}$
N_1	first normal stress difference (Pa)	$\mathbf{ML}^{-1}\mathbf{T}^{-2}$
N_2	second normal stress difference (Pa)	$\mathbf{ML}^{-1}\mathbf{T}^{-2}$
n	power law flow behaviour index (−)	$\mathbf{M}^0\mathbf{L}^0\mathbf{T}^0$
n'	apparent power law index, equation (2.20) (−)	$\mathbf{M}^0\mathbf{L}^0\mathbf{T}^0$
$-\Delta p$	pressure drop (Pa)	$\mathbf{ML}^{-1}\mathbf{T}^{-2}$
$-\Delta p_c$	corrected pressure drop (Pa)	$\mathbf{ML}^{-1}\mathbf{T}^{-2}$
$-\Delta p_e$	pressure drop due to entrance effects (Pa)	$\mathbf{ML}^{-1}\mathbf{T}^{-2}$

$\dfrac{-\Delta p}{L}$	pressure gradient (Pa/m)	$\mathbf{ML^{-2}T^{-2}}$
Q	volumetric flow rate (m³/s)	$\mathbf{L^3T^{-1}}$
R	radius of capillary or of plate (m)	$\mathbf{L}$
R_1	inner radius in coaxial cylinder viscometer (m)	$\mathbf{L}$
R_2	outer radius in coaxial cylinder viscometer (m)	$\mathbf{L}$
r	radial coordinate (m)	$\mathbf{L}$
Re	Reynolds number ($-$)	$\mathbf{M^0L^0T^0}$
T	torque (N m)	$\mathbf{ML^2T^{-2}}$
Ta	Taylor number ($-$)	$\mathbf{M^0L^0T^0}$
T_r	Trouton ratio ($-$)	$\mathbf{M^0L^0T^0}$
V	average velocity in capillary (m/s)	$\mathbf{LT^{-1}}$
V_s	slip velocity (m/s)	$\mathbf{LT^{-1}}$
V_z	point velocity (m/s)	$\mathbf{LT^{-1}}$

Greek letter

γ	shear strain	$\mathbf{M^0L^0T^0}$
$\dot{\gamma}$	shear rate (s⁻¹)	$\mathbf{T^{-1}}$
δ	phase shift in oscillatory test ($-$)	$\mathbf{M^0L^0T^0}$
$\dot{\varepsilon}$	rate of extension (s⁻¹)	$\mathbf{T^{-1}}$
θ	cone angle of cone-and-plate viscometer	$\mathbf{M^0L^0T^0}$
λ	relaxation time (s)	$\mathbf{T}$
μ	viscosity (Pa·s)	$\mathbf{ML^{-1}T^{-1}}$
ρ	fluid density (kg/m³)	$\mathbf{ML^{-3}}$
τ	shear stress (Pa)	$\mathbf{ML^{-1}T^{-2}}$
ω	frequency (Hz)	$\mathbf{T^{-1}}$
Ω	angular velocity (rad/s)	$\mathbf{T^{-1}}$

Subscripts

Crit	location at which $\tau = \tau_o$
E	extensional
m	maximum value
N	Newtonian
P	plastic
PL	for power-law fluid
R	at $r = R$
rz	point value
w	at wall
0	yield point

Superscripts

*	complex
′	in phase component
″	out of phase component

Chapter 3

Flow in pipes and in conduits of non-circular cross-sections

3.1 Introduction

In the chemical and process industries, it is often required to pump fluids over long distances from storage to various processing units and/or from one plant site to another. There may be a substantial frictional pressure loss in both the pipe line and in the individual units themselves. It is thus often necessary to consider the problems of calculating the power requirements for pumping through a given pipe network, the selection of optimum pipe diameter, measurement and control of flow rate, etc. A knowledge of these factors also facilitates the optimal design and layout of flow networks which may represent a significant part of the total plant cost. Aside from circular pipes, one also encounters conduits of other cross-sections and may be concerned with axial flow in an annulus (as in a double-pipe heat exchanger), rectangular, triangular and elliptic conduits as employed in nuclear reactors and for extrusion through dies. Furthermore, the velocity profile established in a given flow situation strongly influences the heat and mass transfer processes. For instance, the analysis and interpretation of data obtained in a standard falling-film absorber used for the determination of diffusion coefficients relies on the knowledge of flow kinematics. This chapter deals with engineering relationships describing flow in a variety of geometries. The treatment here is, however, restricted to the so-called purely viscous or time-independent type of fluids for which the viscosity model describing the flow curve is already known. However, subsequently a generalized treatment for the laminar flow of time-independent fluids in circular tubes is presented. Notwithstanding the existence of time-dependent and visco-elastic fluid behaviour, experience has shown that the shear rate dependence of the viscosity is the most significant factor in most engineering applications which invariably operate under steady state conditions. Visco-elastic behaviour does not significantly influence laminar flow through circular tubes. Visco-elastic effects begin to manifest themselves for flow in non-circular conduits in the form of secondary flows and/or in pipe fittings due to sudden changes in the cross-sectional area available for flow thereby leading to acceleration/deceleration of a fluid element. Even in these circumstances, it is often possible to develop predictive expressions purely in terms of steady-shear viscous properties.

Many of the formulae to be developed here will relate the frictional pressure drop $(-\Delta p)$ to the volumetric flow rate (Q). The major application of such relationships is in the mechanical energy balance which is written to calculate the total head loss in a pipe network which in turn allows the estimation of the required power to be delivered by a pump. The total loss term, representing the conversion of the mechanical energy into thermal energy as a result of fluid friction, consists of two components: one due to the frictional pressure drop associated with fully developed flow through the conduit and the other due to the frictional losses associated with entrance effects, pipe fittings, valves, flow metering devices, etc. Because of their generally high viscosities, laminar flow is more commonly encountered in practice for non-Newtonian fluids, as opposed to Newtonian fluids.

3.2 Laminar flow in circular tubes

Consider the laminar, steady, incompressible fully developed flow of a time-independent fluid in a circular tube of radius, R, as shown in Figure 3.1. Since there is no angular velocity, the force balance in the z-direction on a fluid element situated at distance, r, can be written as:

$$p(\pi r^2) - (p + \Delta p)\pi r^2 = \tau_{rz} 2\pi r L \tag{3.1}$$

i.e.

$$\tau_{rz} = \left(\frac{-\Delta p}{L}\right)\frac{r}{2} \tag{3.2}$$

This shows the familiar linear shear stress distribution across the pipe cross-section, the shear stress being zero at the axis of the tube, as shown in Figure 3.2. Note that equation (3.2) is applicable to both laminar and turbulent flow of any fluid since it is based on a simple force balance and no assumption has been made so far concerning the type of flow or fluid behaviour.

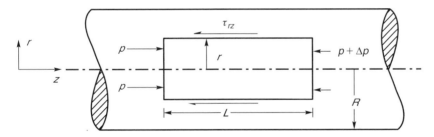

Figure 3.1 *Flow through a horizontal pipe*

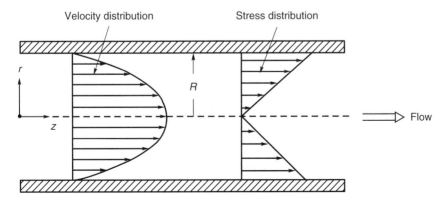

Figure 3.2 *Schematic representation of shear stress and velocity distribution in fully developed laminar flow in a pipe*

3.2.1 Power-law fluids

For a power-law fluid in a pipe, the shear stress is related to the shear rate by (Coulson and Richardson, 1999):

$$\tau_{rz} = m\left(-\frac{dV_z}{dr}\right)^n \tag{3.3}$$

where V_z is the velocity in the axial direction at radius r. Now combining equations (3.2) and (3.3) and the separation of variables leads to:

$$dV_z = -\left\{\left(\frac{-\Delta p}{L}\right)\frac{1}{2m}\right\}^{1/n} r^{1/n}\ dr \tag{3.4}$$

Equation (3.4) can now be integrated with respect to r to obtain the following expression for the velocity distribution:

$$V_z = -\left(\frac{n}{n+1}\right)\left\{\left(\frac{-\Delta p}{L}\right)\frac{1}{2m}\right\}^{1/n} r^{(n+1)/n} + \text{constant} \tag{3.5}$$

At the walls of the pipe (i.e. when $r = R$), the velocity V_z must be zero in order to satisfy the no-slip condition. Substituting the value $V_z = 0$, when $r = R$ in equation (3.5):

$$\text{constant} = \left(\frac{n}{n+1}\right)\left\{\left(\frac{-\Delta p}{L}\right)\frac{1}{2m}\right\}^{1/n} R^{(n+1)/n} \tag{3.6}$$

and therefore, the velocity distribution in the pipe is given by:

$$V_z = \left(\frac{n}{n+1}\right)\left(\frac{-\Delta p}{mL}\cdot\frac{R}{2}\right)^{1/n} R\left\{1 - \left(\frac{r}{R}\right)^{(n+1)/n}\right\} \tag{3.7}$$

The velocity profile may be expressed in terms of the average velocity, V, which is given by:

$$V = \frac{Q}{\pi R^2} = \frac{1}{\pi R^2}\int_0^R 2\pi r V_z\ dr \tag{3.8}$$

where Q is the volumetric flow rate of the liquid. On substitution for V_z from equation (3.7), and integration yields,

$$V = \frac{2\pi}{\pi R^2}\left(\frac{n}{n+1}\right)\left(\frac{-\Delta p}{L}\cdot\frac{R}{2m}\right)^{1/n} R\int_0^R r\left\{1 - \left(\frac{r}{R}\right)^{(n+1)/n}\right\} dr$$

$$= 2\left(\frac{n}{n+1}\right)\left(\frac{-\Delta p}{L}\cdot\frac{R}{2m}\right)^{1/n} R\int_0^R \frac{r}{R}\left\{1 - \left(\frac{r}{R}\right)^{(n+1)/n}\right\} d\left(\frac{r}{R}\right)$$

$$= 2\left(\frac{n}{n+1}\right)\left(\frac{-\Delta p}{L}\cdot\frac{R}{2m}\right)^{1/n} R\left[\frac{1}{2} - \frac{n}{3n+1}\right]$$

and therefore:

$$V = \left(\frac{n}{3n+1}\right)\left(-\frac{\Delta p \cdot R}{2mL}\right)^{1/n} R \tag{3.9}$$

Equation (3.7) can now be rewritten in terms of the average velocity as:

$$\frac{V_z}{V} = \left(\frac{3n+1}{n+1}\right)\left\{1 - \left(\frac{r}{R}\right)^{(n+1)/n}\right\} \tag{3.10}$$

The velocity profiles calculated from equation (3.10) are shown in Figure 3.3, for various values of n. Compared with the parabolic distribution for a Newtonian fluid ($n = 1$), the profile is flatter for a shear-thinning fluid ($n < 1$) and sharper for a shear-thickening ($n > 1$) fluid. The velocity is seen to be a maximum when $r = 0$, i.e. at the pipe axis. Thus the maximum velocity V_{zmax}, at the pipe axis, is given by equation (3.10) when $r = 0$ and:

$$V_{zmax} = \left(\frac{3n+1}{n+1}\right)V$$

Hence:

$$\frac{V_{zmax}}{V} = \left(\frac{3n+1}{n+1}\right) \tag{3.11}$$

Thus, it can be readily seen from equation (3.11) that the value of the centre-line velocity drops from 2.33V to 1.18V as the value of the power-law index n decreases from 2 to 0.1. Of course, as expected, this ratio is 2 for a Newtonian fluid.

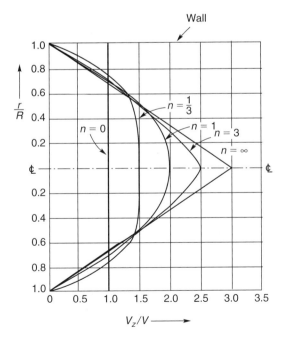

Figure 3.3 *Velocity distribution for power-law fluids in laminar flow in pipes*

Rewriting equation (3.9) in terms of the volumetric flow rate and the pressure gradient:

$$Q = \pi R^2 V = \pi \left(\frac{n}{3n+1}\right)\left(\frac{-\Delta p}{2mL}\right)^{1/n} R^{(3n+1)/n} \tag{3.12}$$

For a given power-law fluid and fixed pipe radius, $-\Delta p \propto Q^n$, i.e. for a shear-thinning fluid ($n < 1$), the pressure gradient is less sensitive than for a Newtonian fluid to changes in flow rate. The flow rate, on the other hand, shows a rather stronger dependence on the radius of pipe. For instance, for $n = 1$, $Q \propto R^4$ whereas for $n = 0.5$, $Q \propto R^5$.

It is useful to rewrite equation (3.12) in dimensionless form by introducing a friction factor defined as $f = (\tau_w/(1/2)\rho V^2)$ where $\tau_w = (-\Delta p/L)(D/4)$ and defining a suitable Reynolds number which will yield the same relationship as that for Newtonian fluids, that is, $f = (16/Re)$.

Thus, writing the wall shear stress in terms of the friction factor, f:

$$\tau_w = \frac{1}{2}\rho V^2 f \tag{3.13}$$

Now rewriting the wall shear stress in terms of the pressure gradient:

$$\frac{-\Delta p}{L} = \frac{2f\rho V^2}{D} \tag{3.14}$$

Substituting for $(-\Delta p/L)$ from equation (3.14) in equation (3.9) and writing it in terms of pipe diameter:

$$V = \left(\frac{n}{3n+1}\right)\left(\frac{D}{4m}\frac{2f\rho V^2}{D}\right)^{1/n}\left(\frac{D}{2}\right) \tag{3.15}$$

which, in turn, can be rearranged to give:

$$f = \frac{16}{Re_{PL}} \tag{3.16}$$

where the new Reynolds number Re_{PL} is defined by:

$$Re_{PL} = \frac{\rho V^{2-n}D^n}{8^{n-1}m\left(\frac{3n+1}{4n}\right)^n} \tag{3.17}$$

It will be seen later (equation (3.57)) that this definition of the Reynolds number coincides with that of Metzner and Reed (1955), and hence hereafter it will be written as Re_{MR}, that is, $Re_{PL} = Re_{MR}$.

Example 3.1

A polymer solution (density = 1075 kg/m³) is being pumped at a rate of 2500 kg/h through a 25 mm inside diameter pipe. The flow is known to be laminar and the power-law constants for the solution are $m = 3\,Pa\,s^n$ and $n = 0.5$. Estimate the pressure drop over a 10 m length of straight pipe and the centre-line velocity for these conditions. How does the value of pressure drop change if a pipe of 37 mm diameter is used?

Solution

$$\text{Volumetric flow rate, } Q = \frac{2500}{3600} \times \frac{1}{1075} = 6.46 \times 10^{-4} \text{ m}^3/\text{s}$$

$$\text{Pipe radius, } R = \frac{25}{2} \times 10^{-3} = 0.0125 \text{ m}$$

Substituting in equation (3.12) and solving for pressure drop gives:

$$-\Delta p = 110 \text{ kPa}$$

$$\text{Average velocity in pipe, } V = \frac{Q}{\pi R^2} = \frac{6.46 \times 10^{-4}}{\pi (0.0125)^2}$$
$$= 1.32 \text{ m/s}$$

The centre-line velocity is obtained by using equation (3.11),

$$V_z|_{r=0} = V_{zmax} = V\left(\frac{3n+1}{n+1}\right) = 1.32 \times \frac{3 \times 0.5 + 1}{0.5 + 1}$$
$$= 2.2 \text{ m/s}$$

For the pipe diameter of 37 mm, for the same value of the flow rate, equation (3.12) suggests that

$$-\Delta p \propto R^{-(3n+1)}$$

Hence, the pressure drop for the new pipe diameter can be estimated:

$$-\Delta p_{new} = -\Delta p_{old}\left(\frac{R_{new}}{R_{old}}\right)^{-(3n+1)}$$

$$= 110\left(\frac{37/2}{25/2}\right)^{-2.5} = 41.3 \text{ kPa}$$

Note that if the fluids were Newtonian ($n = 1$) of a viscosity of 3 Pa s, the new value of the pressure drop would be 23 kPa, about half of that for the polymer solution. □

3.2.2 Bingham plastic and yield-pseudoplastic fluids

A fluid with a yield stress will flow only if the applied stress (proportional to pressure gradient) exceeds the yield stress. There will be a solid plug-like core flowing in the middle of the pipe where $|\tau_{rz}|$ is less than the yield stress, as shown schematically in Figure 3.4. The radius of the plug, R_p, will depend upon the magnitude of the yield stress and on the wall shear stress (i.e., the value of $-\Delta p/L$). From equation (3.2),

$$\frac{\tau_0^B}{\tau_w} = \frac{R_p}{R} \tag{3.18}$$

where τ_w is the shear stress at the wall of the pipe.

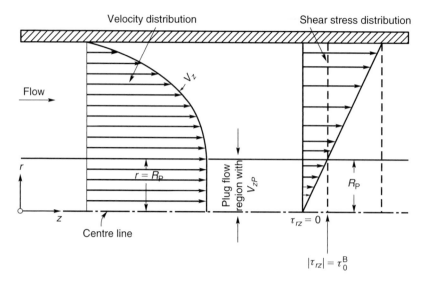

Figure 3.4   *Schematic velocity distribution for laminar flow of a Bingham plastic fluid in a pipe*

In the annular area $R_p < r < R$, the velocity will gradually decrease from the constant plug velocity to zero at the pipe wall. The expression for this velocity distribution will now be derived.

For the region $R_p < r < R$, the value of shear stress will be greater than the yield stress of the fluid, and the Bingham fluid model for pipe flow is given by (equation (1.17)):

$$\tau_{rz} = \tau_0^B + \mu_B \left(-\frac{dV_z}{dr} \right) \tag{3.19}$$

Now combining equations (3.2) and (3.19) followed by an integration with respect to r yields the following expression for the velocity distribution

$$V_z = -\frac{1}{\mu_B} \left\{ \left(\frac{-\Delta p}{L} \right) \frac{r^2}{2} - r\tau_0^B \right\} + \text{constant} \tag{3.20}$$

At the walls of the pipe (i.e. when $r = R$), the velocity V_z must be zero to satisfy the condition of no-slip. Substituting the value $V_z = 0$, when $r = R$ in equation (3.20):

$$\text{constant} = \frac{1}{\mu_B} \left\{ \left(\frac{-\Delta p}{L} \right) \frac{R^2}{2} - R\tau_0^B \right\} \tag{3.21}$$

and, therefore, the velocity distribution in the fluid-like region is given by:

$$V_z = \left(\frac{-\Delta p}{L} \right) \frac{R^2}{4\mu_B} \left(1 - \frac{r^2}{R^2} \right) - \frac{\tau_0^B}{\mu_B} R \left(1 - \frac{r}{R} \right) \tag{3.22}$$

Clearly, equation (3.22) is applicable only when $|\tau_{rz}| > |\tau_0^B|$ and $r \geq R_p$. The corresponding velocity, V_{zp}, in the plug region ($0 \leq r \leq R_p$) is obtained by substituting $r = R_p$ in equation (3.22) to give:

$$V_{zp} = \left(\frac{-\Delta p}{L}\right)\frac{R^2}{4\mu_B}\left(1 - \frac{R_p}{R}\right)^2 \qquad 0 \leq r \leq R_p \tag{3.23}$$

The corresponding expression for the volumetric flow rate, Q, is obtained by evaluating the integral

$$Q = \int_0^R 2\pi r V_z \, dr = \int_0^{R_p} 2\pi r V_{zp} \, dr + \int_{R_p}^R 2\pi r V_z \, dr \tag{3.24}$$

Substituting from equations (3.22) and (3.23):

$$Q = \int_0^{R_p} \left(\frac{-\Delta p}{L}\right)\left(\frac{R^2}{4\mu_B}\right)\left(1 - \frac{R_p}{R}\right)^2 (2\pi r)dr$$

$$+ \int_{R_p}^R \left[\left(\frac{-\Delta p}{L}\right)\left(\frac{R^2}{4\mu_B}\right)\left(1 - \frac{r^2}{R^2}\right) - \frac{\tau_0^B R}{\mu_B}\left(1 - \frac{r}{R}\right)\right](2\pi r)dr \tag{3.25}$$

Evaluating the definite integrals appearing in equation (3.25):

$$Q = 2\pi\left(\frac{-\Delta p}{L}\right)\left(\frac{R^2}{4\mu_B}\right)\left(1 - \frac{R_p}{R}\right)^2 \left\{\frac{r^2}{2}\right\}\bigg|_0^{R_p} + 2\pi\left(\frac{-\Delta p}{L}\right)\left(\frac{R^2}{4\mu_B}\right)\left\{\frac{r^2}{2} - \frac{r^4}{4R^2}\right\}\bigg|_{R_p}^R$$

$$- 2\pi\frac{\tau_0^B R}{\mu_B}\left\{\frac{r^2}{2} - \frac{r^3}{3R}\right\}\bigg|_{R_p}^R \tag{3.26}$$

Now denoting the ratios $\left(\tau_0^B/\tau_w\right) = \left(R_p/R\right) = \phi$, equation (3.26) reduces to:

$$Q = \frac{\pi R^4}{8\mu_B}\left(\frac{-\Delta p}{L}\right)\left[2\phi^2(1 - \phi)^2 + 2(1 - \phi^2) - (1 - \phi^4) - 8\phi\left(\frac{1 - \phi^2}{2} - \frac{1 - \phi^3}{3}\right)\right] \tag{3.27}$$

Further simplification leads to the following relationship for the volumetric flow rate, Q:

$$Q = \frac{\pi R^4}{8\mu_B}\left(\frac{-\Delta p}{L}\right)\left(1 - \frac{4}{3}\phi + \frac{1}{3}\phi^4\right) \tag{3.28}$$

where $\quad \phi = \tau_0^B/\tau_w$.

It is useful to rewrite equation (3.28) in a dimensionless form by introducing the following definitions:

Bingham Reynolds number, $Re_B = \dfrac{\rho V D}{\mu_B}$ (3.29)

Friction factor, $f = \dfrac{\tau_w}{\left(\dfrac{1}{2}\right)\rho V^2} = \dfrac{D(-\Delta p/L)}{2\rho V^2}$ (3.30)

Hedstrom number, $He = \dfrac{\rho D^2 \tau_0^B}{\mu_B^2}$ (3.31)

Rewriting equation (3.28) in terms of the mean velocity, $V = Q/\pi R^2$,

$\phi = \left(\dfrac{\tau_0^B}{\tau_w}\right) = \left(\dfrac{\tau_0^B}{(1/2)f\rho V^2}\right)$ and pipe diameter, D:

$$V = \dfrac{(D/2)^2}{8\mu_B}\left(\dfrac{-\Delta p}{L}\right)\left\{1 - \dfrac{8}{3}\dfrac{\tau_0^B}{f\rho V^2} + \dfrac{1}{3}\left(\dfrac{2\tau_0^B}{f\rho V^2}\right)^4\right\}$$ (3.32)

Dividing both sides by ρV^2 and slight rearrangement leads to:

$$\dfrac{\mu_B}{\rho V D} = \dfrac{D}{32\rho V^2}\left(\dfrac{-\Delta p}{L}\right)\left\{1 - \dfrac{8}{3}\dfrac{\tau_0^B}{f\rho V^2} + \dfrac{16}{3}\left(\dfrac{\tau_0^B}{f\rho V^2}\right)^4\right\}$$ (3.33)

i.e.,

$$\dfrac{1}{Re_B} = \dfrac{f}{16}\left\{1 - \dfrac{8}{3}\dfrac{\tau_0^B}{f\rho V^2} + \dfrac{16}{3}\left(\dfrac{\tau_0^B}{f\rho V^2}\right)^4\right\}$$ (3.34)

or,

$$\dfrac{16}{Re_B} = f - \dfrac{8}{3}\dfrac{\tau_0^B}{\rho V^2} + \dfrac{16}{3}\left(\dfrac{\tau_0^B}{\rho V^2}\right)^4\dfrac{1}{f^3}$$ (3.35)

This can now be rearranged to obtain the final form as:

$$f = \dfrac{16}{Re_B}\left(1 + \dfrac{1}{6}\dfrac{He}{Re_B} - \dfrac{1}{3}\dfrac{He^4}{f^3 Re_B^7}\right)$$ (3.36)

It should be noted that equation (3.36) is implicit in pressure gradient (because τ_w, and hence ϕ, is a function of pressure gradient) and therefore, for a specified flow rate, an

Figure 3.5   *Friction factor as a function of the Reynolds number and Hedström numbers, as predicted by equation (3.36)*

iterative method is needed to evaluate the pressure drop. The predictions of equation (3.36) are shown in Figure 3.5. It is clearly seen that the lower is the value of the Hedström number (Hedstrom, 1952), smaller is the deviation from the Newtonian line, i.e., $f = 16/Re_B$. It is also appropriate to add here that many authors prefer to use another dimensionless parameter, namely, Bingham number, Bi, which is defined as:

$$\mathrm{Bi} = \frac{\tau_0^B D}{\mu_B V} \qquad (3.37)$$

It is easy to show that $\mathrm{He} = Re_B\,\mathrm{Bi}$. Equation (3.36) can be recast in terms of the Bingham number as:

$$f = \frac{16}{Re_B}\left[1 + \frac{Bi}{6} - \frac{1}{3}\frac{Bi^4}{(fRe_B)^3}\right] \qquad (3.38)$$

In this case, one can expect the product $(f \cdot Re_B)$ to be a function of the Bingham number and this relationship is shown in Figure 3.6. Once again, it is seen that the value of $(f \cdot Re_B)$ deviates very little from its Newtonian value of 16 up to about $\mathrm{Bi} \sim 0.1$ thereby suggesting little effect of the yield stress of the fluid on the pressure gradient. Figure 3.6 is convenient to calculate the unknown pressure gradient for a given flow rate, pipe diameter and rheological characteristics (μ_B, τ_0^B); the reverse calculation of an unknown flow rate for a given pressure gradient requires an iterative solution, however.

This analysis can readily be extended to the laminar flow of Herschel–Bulkley model fluids (equation (1.18)), and the resulting final expressions for the point velocity, the

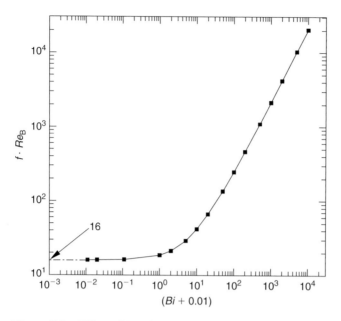

Figure 3.6   *Effect of Bingham number on the value of $(f \cdot Re_B)$ as predicted by equation (3.38)*

plug velocity and volumetric flow rate are given as follows (Skelland, 1967; Govier and Aziz, 1982; Bird *et al.*, 1983, 1987):

$$V_z = \frac{nR}{(n+1)}\left(\frac{\tau_w}{m}\right)^{1/n}\left\{(1-\phi)^{(n+1)/n} - \left(\frac{r}{R} - \phi\right)^{(n+1)/n}\right\} \tag{3.39}$$

The plug velocity, V_{zp}, is given as:

$$V_{zp} = \frac{nR}{(n+1)}\left(\frac{\tau_w}{m}\right)^{1/n}(1-\phi)^{(n+1)/n} \tag{3.40}$$

and the volumetric flow rate, Q, as:

$$Q = \pi R^3 n\left(\frac{\tau_w}{m}\right)^{1/n}(1-\phi)^{(n+1)/n}\left\{\frac{(1-\phi)^2}{3n+1} + \frac{2\phi(1-\phi)}{2n+1} + \frac{\phi^2}{n+1}\right\} \tag{3.41}$$

where ϕ is now the ratio (τ_0^H/τ_w). These expressions are also implicit in pressure gradient.

Example 3.2

The rheological properties of a china clay suspension can be approximated by either a power-law or a Bingham plastic model over the shear rate range 10 to $100\,\mathrm{s}^{-1}$. If the yield stress is 15 Pa and the plastic viscosity is 150 mPa s, what will be the approximate values of the power-law consistency coefficient and flow behaviour index?

Estimate the pressure drop when this suspension is flowing under laminar conditions in a pipe of 40 mm diameter and 200 m long, when the centre-line velocity is 0.6 m/s, according to the Bingham plastic model. Calculate the centre-line velocity for this pressure drop for the power-law model.

Solution

Using the Bingham plastic model,

$$\tau_{rz} = \tau_0^B + \mu_B(-dV_z/dr)$$

when $-dV_z/dr = 10\,s^{-1}, \tau_{rz} = 15 + 150 \times 10^{-3} \times 10 = 16.5\,Pa$

$-dV_z/dr = 100\,s^{-1}, \tau_{rz} = 15 + 150 \times 10^{-3} \times 100 = 30\,Pa$

Now using the power-law model,

$$\tau_{rz} = m\left(-\frac{dV_z}{dr}\right)^n$$

Substituting the values of τ_{rz} and $(-dV_z/dr)$:

$16.5 = m(10)^n$ and $30 = m(100)^n$

Solving for m and n gives:

$n = 0.26, \quad m = 9.08\,Pa\,s^n$

For a Bingham plastic fluid, equation (3.23) gives:

$$V_{z\,max} = V_{zp} = 0.6 = \left(\frac{-\Delta p}{L}\right)\left(\frac{(20 \times 10^{-3})^2}{4 \times 0.15}\right)\left(1 - \frac{R_p}{R}\right)^2$$

Substitution for (R_p/R) from equation (3.18) and writing the wall shear stress in terms of pressure gradient gives:

$$0.6 = \left(\frac{-\Delta p}{L}\right)\left(\frac{(20 \times 10^{-3})^2}{4 \times 0.15}\right)\left[1 - \frac{15}{\dfrac{20 \times 10^{-3}}{2}\left(\dfrac{-\Delta p}{L}\right)}\right]^2$$

A trial and error procedure leads to

$$\frac{-\Delta p}{L} = 3200\,Pa/m$$

and therefore the total pressure drop over the pipe length, $-\Delta p = 3200 \times 200 = 640\,kPa$.

The centre-line velocity according to the power-law model is given by equation (3.7) with $r = 0$, i.e.

$$V_{z\,max} = \left(\frac{n}{n+1}\right)\left(-\frac{\Delta p}{mL}\frac{R}{2}\right)^{1/n} \cdot R$$

$$= \left(\frac{0.26}{0.26+1}\right)\left(\frac{3200 \times 20 \times 10^{-3}}{9.08 \times 2}\right)^{1/0.26} \times 20 \times 10^{-3}$$

$$= 0.52\,m/s$$

Finally, the value of the shear stress at the wall is calculated as:

$$\tau_w = \left(\frac{-\Delta p}{L}\right)\frac{D}{4} = \frac{3200 \times 40 \times 10^{-3}}{4} = 32\,\text{Pa}$$

Now using equation (3.18):

$$\frac{R_p}{R} = \frac{\tau_0^B}{\tau_w} = \frac{15}{32} = 0.47$$

Therefore, one can easily see that the plug-like motion occurs across a substantial portion of the cross-section as $R_p = 0.47\,R$. □

Example 3.3

The rheological behaviour of a clay slurry (density = 1500 kg/m³) can be approximated by a Herschel–Bulkley model over a limited range of shear rates relevant to its laminar flow in a pipeline (40 mm diameter, 500 m long). The values of the model parameters are: $\tau_0^H = 17\,\text{Pa}$; $m = 0.83\,\text{Pa s}^n$ and $n = 0.5$. Estimate the pressure drop when this slurry is flowing under laminar conditions with a mean velocity of 0.5 m/s. What is the plug velocity in the middle of the pipe under these conditions? Also, calculate the size of the plug.

Solution

For Herschel–Bulkley model fluids, equation (3.41) relates the volumetric flow rate (and hence the mean velocity) to the model parameters and the shear stress at the wall. Since this relationship is implicit in wall shear stress, τ_w, a trial and error procedure is needed to solve for the unknown pressure gradient. Noting that $\phi = (\tau_0^H/\tau_w)$ and $V = Q/\pi R^2$, equation (3.41) can be slightly rearranged as follows:

$$V = \frac{Q}{\pi R^2} = nR\left(\frac{\tau_0^H}{m\phi}\right)^{1/n}(1-\phi)^{(n+1)/n}\left\{\frac{(1-\phi)^2}{3n+1} + \frac{2\phi(1-\phi)}{2n+1} + \frac{\phi^2}{n+1}\right\}$$

Now substituting the following values as: $m = 0.83\,\text{Pa s}^n$, $n = 0.5$, $\tau_0^H = 17\,\text{Pa}$, $R = (40/2) \times 10^{-3} = 0.02\,\text{m}$ and V = 0.5 m/s as,

$$0.5 = 0.5 \times 0.02\left(\frac{17}{0.83\phi}\right)^{1/0.5}(1-\phi)^{(0.5+1)/0.5}\left\{\frac{(1-\phi)^2}{3\times0.5+1} + \frac{2\phi(1-\phi)}{2\times0.5+1} + \frac{\phi^2}{0.5+1}\right\}$$

Further simplification yields,

$$1 = 0.02\left(\frac{17}{0.83\phi}\right)^2(1-\phi)^3\left\{\frac{(1-\phi)^2}{2.5} + \phi(1-\phi) + \frac{\phi^2}{1.5}\right\}$$

Solving for ϕ by a trial and error procedure gives $\phi = 0.58$. (The other two roots are complex and hence are unrealistic.)

Therefore, the value of τ_w is now obtained as:

$$\phi = 0.58 = \frac{\tau_0^H}{\tau_w} \quad \text{or} \quad \tau_w = \frac{\tau_0^H}{0.58} = \frac{17}{0.58} = 29.31\,\text{Pa}$$

and finally, using $\tau_w = (D/4)(-\Delta p/L)$ yields the value of pressure drop $(-\Delta p)$:

$$(-\Delta p) = 29.31 \times \frac{4}{40 \times 10^{-3}} \times 500 = 1.46 \times 10^6 \, \text{Pa}$$

The plug velocity, V_{zp}, is now calculated using equation (3.40):

$$V_{zp} = \frac{nR}{n+1}\left(\frac{\tau_w}{m}\right)^{1/n}[1 - \phi]^{(n+1)/n}$$

Substituting values,

$$V_{zp} = \frac{0.5 \times \left(\frac{40}{2} \times 10^{-3}\right)}{0.5 + 1}\left(\frac{29.31}{0.83}\right)^{1/0.5}(1 - 0.58)^{(0.5+1)/0.5} = 0.62 \, \text{m/s}$$

Finally, recognizing that the plug radius is also related to the value of ϕ as:

$$\phi = \frac{R_p}{R} = 0.58, \text{ i.e., } R_p = 0.58 \times 20 \times 10^{-3} = 11.6 \times 10^{-3} \, \text{m} \quad \text{or} \quad 11.6 \, \text{mm}$$

In other words, the plug-like motion occurs in the region $0 \leqslant r \leqslant 11.6 \, \text{mm}$. □

Before concluding this section, it is appropriate to mention here that one can establish similar $Q-\Delta p$ relations for the other commonly used viscosity models in an identical manner. A summary of such relations can be found in standard textbooks (Skelland, 1967; Govier and Aziz, 1982).

3.2.3 Average kinetic energy of fluid

In order to obtain the kinetic energy correction factor, α, for insertion in the mechanical energy balance, it is necessary to evaluate the average kinetic energy per unit mass in terms of the average velocity of flow. The calculation procedure is exactly similar to that used for Newtonian fluids (e.g., see Coulson and Richardson, 1999).

Average kinetic energy/unit mass:

$$= \frac{\int \frac{1}{2} V_z^2 \, d\dot{m}}{\int d\dot{m}} = \frac{\int_0^R \frac{1}{2} V_z^2 \, 2\pi r V_z \rho \, dr}{\int_0^R 2\pi r V_z \rho \, dr} \tag{3.42}$$

$$= \frac{V^2}{2\alpha} \tag{3.43}$$

where α is a kinetic energy correction factor to take account of the non-uniform velocity over the cross-section. For power-law fluids, substitution for V_z from equation (3.10) into equation (3.42) and integration gives:

$$\alpha = \frac{(2n + 1)(5n + 3)}{3(3n + 1)^2} \tag{3.44}$$

The corresponding expression for a Bingham plastic is cumbersome. However, Metzner (1956) gives a simple expression for α which is accurate to within 2.5%:

$$\alpha = \frac{1}{2 - \phi} \tag{3.45}$$

Again, both equations (3.44) and (3.45) reduce to $\alpha = 1/2$ for Newtonian fluid behaviour. Note that as the degree of shear-thinning increases, i.e., the value of n decreases, the kinetic energy correction factor approaches unity at $n = 0$ as would be expected, as all the fluid is flowing at the same velocity (Figure 3.3). For shear-thickening fluids, on the other hand, it attains a limiting value of 0.37 for the infinite degree of shear-thickening behaviour $(n = \infty)$.

3.2.4 Generalized approach for laminar flow of time-independent fluids

Approach used in Section 3.2 for power-law, Bingham plastic and Herschel–Bulkley models fluids can be extended to other fluid models. Even if the relationship between shear stress and shear rate is not known exactly, it is possible to use the following approach to the problem. It depends upon the fact that the shear stress distribution over the pipe cross-section is not a function of the fluid rheology and is given simply by equation (3.2), which can be rewritten in terms of the wall shear stress, i.e.,

$$\frac{\tau_{rz}}{\tau_w} = \frac{r}{R} \tag{3.2}$$

As was shown in Section 2.2.1, the volumetric flow rate is given by equation (2.10), i.e.,

$$Q = \frac{\pi R^3}{\tau_w^3} \int_0^{\tau_w} \tau_{rz}^2 \, f(\tau_{rz}) \, d\tau_{rz} \tag{3.46}$$

The form of the function $f(\tau_{rz})$ will therefore depend on the viscosity model chosen to describe the rheology of the fluid. Equation (3.46) can be used in two ways:

(i) to determine general non-Newtonian characteristics of a time-independent fluid, as demonstrated in Section 2.2.1, or
(ii) to be integrated directly for a specific fluid model to obtain volumetric flow rate–pressure drop relationship. This is demonstrated here for the flow of a power-law fluid, for which the shear rate is given by equation (3.3):

$$-\frac{dV_z}{dr} = f(\tau_{rz}) = \left(\frac{\tau_{rz}}{m}\right)^{1/n} \tag{3.47}$$

Substituting equation (3.47) into equation (3.46) followed by integration and rearrangement gives:

$$V = \frac{Q}{\pi R^2} = \left(\frac{n}{3n + 1}\right)\left(\frac{\tau_w}{m}\right)^{1/n} R \tag{3.48}$$

Substituting for $\tau_w = (R/2)(-\Delta p/L)$ in equation (3.48) gives:

$$V = \left(\frac{n}{3n+1}\right)\left(\frac{(-\Delta p)R}{2mL}\right)^{1/n} R \tag{3.49}$$

which is identical to equation (3.9).

A similar analysis for a Bingham plastic fluid will lead to the same expression for Q as equation (3.28) and for a Herschel–Bulkley fluid to equation (3.41). Thus, these are alternative methods of obtaining the flow rate–pressure gradient relation for any specific model to describe the fluid rheology. The approach given above provides a quicker method of obtaining the relation between pressure gradient and flow rate, but has the disadvantage that it does not provide a means of obtaining the corresponding velocity profile.

Example 3.4

The shear-dependent viscosity of a commercial grade of polypropylene at 403 K can satisfactorily be described using the three constant Ellis fluid model (equation 1.16), with the values of the constants: $\mu_0 = 1.25 \times 10^4\,\mathrm{Pa\,s}$, $\tau_{1/2} = 6900\,\mathrm{Pa}$ and $\alpha = 2.80$. Estimate the pressure drop required to maintain a volumetric flow rate of $4\,\mathrm{cm^3/s}$ through a 50 mm diameter and 20 m long pipe. Assume the flow to be laminar.

Solution

Since we need the $Q-(-\Delta p)$ relation to solve this problem, such a relationship will be first derived using the generalized approach outlined in Section 3.2.4. For laminar flow in circular pipes, the Ellis fluid model is given as:

$$\tau_{rz} = \frac{\mu_0(-\mathrm{d}V_z/\mathrm{d}r)}{1 + (\tau_{rz}/\tau_{1/2})^{\alpha-1}} \tag{1.16}$$

or

$$-\frac{\mathrm{d}V_z}{\mathrm{d}r} = f(\tau_{rz}) = \frac{1}{\mu_0}\left[\tau_{rz} + \frac{\tau_{rz}^\alpha}{\tau_{1/2}^{\alpha-1}}\right] \tag{3.50}$$

Substituting equation (3.50) into equation (3.46) followed by integration and rearrangement gives:

$$Q = \frac{\pi R^3 \tau_w}{4\mu_0}\left[1 + \left(\frac{4}{\alpha+3}\right)\left(\frac{\tau_w}{\tau_{1/2}}\right)^{\alpha-1}\right]$$

Note that in the limit $\tau_{1/2} \to \infty$, i.e. for Newtonian fluid behaviour, this equation reduces to the Hagen–Poiseuille equation.

Now substituting the numerical values:

$$4 \times 10^{-6} = \frac{3.14 \times (0.025)^3 \tau_w}{4 \times 1.25 \times 10^4}\left[1 + \left(\frac{4}{2.8+3}\right)\left(\frac{\tau_w}{6900}\right)^{2.8-1}\right]$$

or

$$4074.37 = \tau_w(1 + 8.4 \times 10^{-8}\tau_w^{1.8})$$

A trial and error procedure gives $\tau_w = 3412\,\mathrm{Pa}$

$$\therefore \ -\Delta p = \frac{4\tau_w L}{D} = \frac{4 \times 3412 \times 20}{0.05} = 5.46 \times 10^6\,\mathrm{Pa}$$

i.e. the pressure drop across the pipe will be 5.46 MPa. □

3.2.5 Generalized Reynolds number for the flow of time-independent fluids

It is useful to define an appropriate Reynolds number which will result in a unique friction factor–Reynolds number curve for all time-independent fluids in laminar flow in circular pipes. Metzner and Reed (1955) outlined a generalized approach obviating this difficulty. The starting point is equation (3.46):

$$Q = \frac{\pi R^3}{\tau_w^3} \int_0^{\tau_w} \tau_{rz}^2 f(\tau_{rz}) d\tau_{rz} \tag{3.46}$$

Equation (3.46) embodies a definite integral, the value of which depends only on the values of the integral function at the limits, and not on the nature of the continuous function that is integrated. For this reason it is necessary to evaluate only the wall shear stress τ_w and the associated velocity gradient at the wall $(-dV_z/dr)$ at $r = R$ or $f(\tau_w)$. This is accomplished by the use of the Leibnitz rule, as was shown in Section 2.2.1, to obtain the following expression for the shear rate at the wall of the pipe:

$$f(\tau_w) = \left(-\frac{dV_z}{dr} \right)_{wall} = \frac{4Q}{\pi R^3} \left\{ \frac{3}{4} + \frac{1}{4} \cdot \frac{\dfrac{d(4Q/\pi R^3)}{(4Q/\pi R^3)}}{\dfrac{d\tau_w}{\tau_w}} \right\} \tag{3.51}$$

or in terms of average velocity V and pipe diameter D,

$$\left(-\frac{dV_z}{dr} \right)_{wall} = \frac{8V}{D} \left\{ \frac{3}{4} + \frac{1}{4} \frac{d \log(8V/D)}{d \log \tau_w} \right\} \tag{3.52}$$

Here, $(8V/D)$ is the wall shear rate for a Newtonian fluid and is referred to as the nominal shear rate for a non-Newtonian fluid which is identical to equation (2.17). Alternatively, writing it in terms of the slope of $\log \tau_w - \log(8 V/D)$ plots,

$$\dot{\gamma}_w = \left(-\frac{dV_z}{dr} \right)_{wall} = \left(\frac{8V}{D} \right) \left(\frac{3n' + 1}{4n'} \right) \tag{3.53}$$

where $n' = (d \log \tau_w / d \log(8V/D))$ which is not necessarily constant at all shear rates. Equation (3.53) is identical to equation (2.21) in Chapter 2.

Thus, the index n' is the slope of the log–log plots of the wall shear stress τ_w versus $(8V/D)$ in the laminar region (the limiting condition for laminar flow is discussed in Section 3.3). Plots of τ_w versus $(8V/D)$ thus describe the flow behaviour of time-independent non-Newtonian fluids and may be used directly for scale-up or process design calculations.

Over the range of shear rates over which n' is approximately constant, one may write a power-law type equation for this segment as:

$$\tau_w = \frac{D}{4} \left(\frac{-\Delta p}{L} \right) = m' \left(\frac{8V}{D} \right)^{n'} \tag{3.54}$$

Substituting for τ_w in terms of the friction factor, $f\,(=\tau_w/(1/2)\rho V^2)$, equation (3.54) becomes:

$$f = \frac{2}{\rho V^2}\, m' \left(\frac{8V}{D}\right)^{n'} \tag{3.55}$$

Now a Reynolds number may be defined so that in the laminar flow regime, it is related to f in the same way as is for Newtonian fluids, i.e.

$$f = \frac{16}{Re_{MR}} \tag{3.56}$$

from which

$$Re_{MR} = \frac{\rho V^{2-n'} D^{n'}}{8^{n'-1} m'} \tag{3.57}$$

Since Metzner and Reed (1955) seemingly were the first to propose this definition of the generalized Reynolds number, and hence the subscripts 'MR' in Re_{MR}.

It should be realized that by defining the Reynolds number in this way, the same friction factor chart can be used for Newtonian and time-independent non-Newtonian fluids in the laminar region. In effect, we are writing

$$Re_{MR} = \frac{\rho VD}{\mu_{eff}} \tag{3.58}$$

Thus, the flow curve provides the value of the effective viscosity μ_{eff} where $\mu_{eff} = m'(8V/D)^{n'-1}$. It should be noted that the terms apparent and effective viscosity have been used to relate the behaviour of a non-Newtonian fluid to an equivalent property of a hypothetical Newtonian fluid. The apparent viscosity is the point value of the ratio of the shear stress to the shear rate. The effective viscosity is linked to the macroscopic behaviour $(Q - (-\Delta p))$ characteristics, for instance, and is equal to the Newtonian viscosity which would give the same relationship. It will be seen in Chapter 8 that this approach has also been quite successful in providing a reasonable basis for correlating much of the literature data on power consumption for the mixing of time-independent non-Newtonian fluids. The utility of this approach for reconciling the friction factor data for all time-independent fluids including shear-thinning and viscoplastic fluids, has been demonstrated by Metzner and Reed (1955) and subsequently by numerous other workers. Indeed, by writing a force balance on an element of fluid flowing in a circular pipe, it can readily be shown that equation (3.56) is also applicable for visco-elastic fluids. Figure 3.7 confirms this expectation for the flow of highly shear-thinning inelastic and visco-elastic polymer solutions in the range $0.28 \leq n' \leq 0.92$. (Chhabra *et al.*, 1984). Griskey and Green (1971) have shown that the same approach may be adopted for the flow of shear-thickening materials, in the range $1.15 \leq n' \leq 2.50$. Experimental evidence suggests that stable laminar flow prevails for time-independent non-Newtonian fluids for Re_{MR} up to about 2000–2500; the transition from laminar to turbulent flow as well as the friction factor – Reynolds number characteristics beyond the laminar region are discussed in detail in the next section.

Before concluding this section, it is useful to link the apparent power-law index n' and consistency coefficient m' (equation (3.54)) to the true power-law constants n and m, and to the Bingham plastic model constants τ_0^B and μ_B. This is accomplished by

Figure 3.7   *Typical friction factor data for laminar flow of polymer solutions (Chhabra et al., 1984)*

noting that $\tau_w = (D/4)(-\Delta p/L)$ always gives the wall shear stress for steady, incompressible and fully developed flow and the corresponding value of the wall shear rate $\dot{\gamma}_w(= dV_z/dr)_w$ can be evaluated using the expressions for velocity distribution in a pipe presented in Sections 3.2.1 and 3.2.2.

For the laminar flow of a power-law fluid in a pipe, the velocity distribution over the pipe cross-section is given by equation (3.10):

$$\frac{V_z}{V} = \left(\frac{3n+1}{n+1}\right)\left\{1 - \left(\frac{r}{R}\right)^{(n+1)/n}\right\} \tag{3.10}$$

Differentiating it with respect to r gives the expression for the velocity gradient (or shear rate) at a radial location r as:

$$-\frac{dV_z}{dr} = \frac{3n+1}{n}\left(\frac{r}{R}\right)^{1/n}\frac{V}{R} \tag{3.59}$$

Now substituting $r = R$ to get the shear rate at the wall as:

$$\left(-\frac{dV_z}{dr}\right)_{r=R} = \dot{\gamma}_w = \left(\frac{3n+1}{n}\right)\frac{V}{R} = \left(\frac{3n+1}{4n}\right)\left(\frac{8V}{D}\right)$$

The corresponding value of the shear stress at the wall τ_w for a power-law fluid is obtained by substituting this value of $\dot{\gamma}_w$ in equation (3.3):

$$\begin{aligned}
\tau_w = \tau_{rz}|_{r=R} &= m\left[\left(-\frac{dV_z}{dr}\right)_{r=R}\right]^n \\
&= m\left\{\left(\frac{3n+1}{4n}\right)\left(\frac{8V}{D}\right)\right\}^n
\end{aligned} \qquad (3.60)$$

Now comparing this equation with equation (3.54) for a power-law fluid gives:

$$n' = n; \quad m' = m\left(\frac{3n+1}{4n}\right)^n \qquad (3.61)$$

Therefore in this case n' and m' are both constants and independent of shear rate. Similarly, it can be shown that the Bingham model parameters τ_0^B and μ_B are related to m' and n' as (Skelland, 1967):

$$n' = \frac{1 - \frac{4}{3}\phi + \frac{\phi^4}{3}}{1 - \phi^4} \qquad (3.62)$$

and

$$m' = \tau_w\left[\frac{\mu_B}{\tau_w\left(1 - \frac{4}{3}\phi + \frac{\phi^4}{3}\right)}\right]^{n'} \qquad (3.63)$$

where $\phi = (\tau_0^B/\tau_w)$.

It should be noted that in this case, both m' and n' are not constants and depend on the value of the wall shear stress τ_w.

Before leaving this section, it is worthwhile to emphasize here that a $\tau_w - (8V/D)$ plot itself denotes the laminar flow behaviour of time-independent and visco-elastic fluids provided all due corrections arising from end effects and wall-slip effects have been incorporated. In other words, this plot would be valid for the laminar flow of this substance in any size pipe. Conversely, this curve can directly be used to estimate the wall shear stress (hence the pressure drop) for a given flow rate in a pipe of known diameter and length. This is illustrated in Example 3.5 by considering the flow of a polymer solution in pipes of different diameters.

Example 3.5

The following velocity–pressure drop data has been obtained for a polymer solution using pipes of different diameters and lengths in which the pressure drop was recorded across test sections located in the middle of pipe lengths. Show that these data collapse onto one curve thereby confirming that there are no end or wall-slip effects present in this case. Here the velocity is in metre per second and the pressure drop in kilopascal.

$D = 42.1$ mm; $L = 629$ mm		$D = 52.8$ mm; $L = 750$ mm		$D = 63.08$ mm; $L = 0.98$ m		$D = 80.4$ mm; $L = 1.202$ m	
V	$-\Delta p$	V	$-\Delta p$	V	$-\Delta p$	V	$-\Delta p$
0.04	0.360	0.04	0.332	0.39	1.334	0.26	0.820
0.07	0.555	0.09	0.560	0.46	1.450	0.35	1.01
0.14	0.853	0.18	0.866	0.60	1.760	0.47	1.22
0.27	1.322	0.27	1.080	0.77	2.047	0.64	1.508
0.43	1.742	0.37	1.310	0.91	2.334	0.81	1.76
0.60	2.164	0.52	1.607	1.13	2.668	0.895	1.875
0.77	2.53	0.64	1.822	1.37	2.982	1.0	1.99
0.91	2.80	0.84	2.152	1.57	3.193	1.14	2.167
1.07	3.12	0.98	2.424	1.89	3.554	1.36	2.446
1.36	3.62	1.15	2.653	2.11	3.862	1.49	2.540
1.51	3.84	1.39	3.00	2.46	4.249	1.80	2.87
1.64	4.05	1.49	3.16	2.80	4.60	2.01	3.105
1.86	4.41	1.67	3.40	3.15	4.95	2.54	3.512
2.01	4.65	1.82	3.56	3.40	5.31	3.15	4.014
2.32	5.11	2.00	3.78	3.66	5.56	3.97	4.62
2.65	5.58			3.90	5.77	4.55	5.087
2.81	5.81			4.25	6.11	4.93	5.405
3.04	6.13			4.56	6.28	5.34	5.775
3.28	6.28						

Now, use this (pseudo) flow curve to estimate the values of the frictional pressure drop for the flow of this solution under the following conditions:
(i) $D = 150$ mm and $V = 2$ m/s.
(ii) $D = 300$ mm and $V = 1.4$ m/s. (Data courtesy: Dr V. Fester, Cape Peninsula University of Technology, Capetown).

Solution

First, we must convert the given raw data in terms of the wall shear stress and the apparent wall shear rate. For instance, for the first data pair in the smallest diameter pipe, we have:

$$\tau_w = \left(\frac{D}{4}\right)\left(\frac{-\Delta p}{L}\right) = \left(\frac{42.1 \times 10^{-3}}{4}\right)\left(\frac{0.36 \times 10^3}{629 \times 10^{-3}}\right) = 6.02 \, \text{Pa}$$

Table 3.1   *Summary of Calculations*

D = 42.1 mm		D = 52.8 mm		D = 63.08 mm		D = 80.4 mm	
τ_w (Pa)	$\left(8V/D\right)$ (s^{-1})	τ_w (Pa)	$\left(8V/D\right)$ (s^{-1})	τ_w (Pa)	$\left(8V/D\right)$ (s^{-1})	τ_w (Pa)	$\left(8V/D\right)$ (s^{-1})
6.02	7.6	5.84	6.06	21.47	49.46	13.71	25.87
9.29	13.3	9.86	13.64	23.33	58.34	17.02	34.83
14.27	26.6	15.24	27.27	28.32	76.09	20.40	46.77
22.12	51.3	19.01	40.91	32.94	97.65	25.22	63.68
29.15	81.7	23.06	56.06	37.56	115.4	29.43	80.60
36.21	114	28.28	78.79	42.93	143.3	31.35	89.05
42.33	146.3	32.07	96.97	47.98	173.7	33.28	99.50
46.85	172.9	37.88	127.3	51.38	199.1	36.24	113.4
52.2	203.3	42.66	148.5	57.19	239.7	40.90	135.3
60.57	258.4	46.69	174.2	62.14	267.6	42.47	148.3
64.25	286.9	52.8	210.6	68.37	312.0	48.00	179.1
67.77	311.6	55.62	225.8	74.02	355.1	51.92	200.0
73.79	353.4	59.84	253.0	79.65	399.5	58.73	252.7
77.81	381.9	62.66	275.8	85.45	431.2	67.12	313.4
85.51	440.9	66.53	303.1	89.47	464.2	77.26	395.0
93.37	503.6			92.85	494.6	85.06	452.7
97.21	534.0			98.32	539.0	90.38	490.5
102.6	577.7			101.1	578.3	96.57	531.3
105.1	623.3						

and the corresponding nominal shear rate at the wall is

$$\dot{\gamma}_w = \frac{8V}{D} = \frac{8 \times 0.04}{42.1 \times 10^{-3}} = 7.6\,s^{-1}$$

Similarly, other raw data is converted into this form and a summary is provided in Table 3.1.

Figure 3.8 shows a plot of these results in terms of a pseudo-flow curve. Since the results for different pipe diameters collapse onto one curve, it implies the time-independent fluid behaviour for this solution and that these data are free from slip and end effects. Now, for $D = 150$ mm and $V = 2$ m/s, the value of $(8V/D)$ is $106.7\,s^{-1}$ which is well within the range of laminar flow conditions. From Figure 3.8, for $(8V/D) = 106.7\,s^{-1}$, the value of $\tau_w = 34.04$ Pa and hence the value of $(-\Delta p/L) = 4\tau_w/D = (4 \times 34.04)/0.15 = 907.73$ Pa/m.

Similarly, for $D = 300$ mm and $V = 1$ m/s, the value of $(8V/D)$ is $26.67\,s^{-1}$. The corresponding value of τ_w is 14.55 Pa and that of $(-\Delta p/L) = 194$ Pa/m. □

3.3 Criteria for transition from laminar to turbulent flow

For all fluids, the nature of the flow is governed by the relative importance of the viscous and the inertial forces. For Newtonian fluids, the balance between these forces is

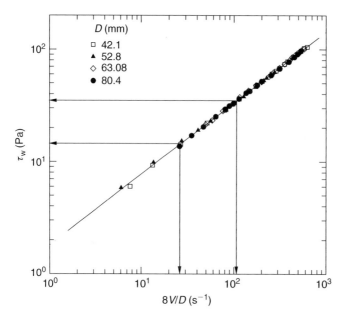

Figure 3.8   *Flow curve in terms of the nominal shear rate at the wall (Example 3.5)*

characterized by the value of the Reynolds number. The generally accepted value of the Reynolds number above which stable laminar flow no longer occurs is 2100 for Newtonian fluids. For time-independent fluids, the critical value of the Reynolds number depends upon the type and the degree of non-Newtonian behaviour. For power-law fluids $(n = n')$, the criterion of Ryan and Johnson (1959) can be used,

$$Re_{MR} = \frac{6464n}{(3n+1)^2}(2+n)^{(2+n)/(1+n)} \tag{3.64}$$

On the other hand, Mishra and Tripathi (1975) argued that the ratio of the average kinetic energy per unit volume of the fluid and the shear stress at the wall remains constant at the point of laminar – turbulent transition for time – independent fluids. Using the value of $Re = 2100$ for Newtonian fluids, they put forward the following expression for the critical Reynolds number for which the flow of power-law fluids is given as:

$$Re_{MR} = \frac{2100(4n+2)(5n+3)}{3(3n+1)^2} \tag{3.65}$$

While for Newtonian fluids, both equations (3.64) and (3.65) predict the critical Reynolds number of 2100, the corresponding limiting values predicted by equation (3.64) increase with the decreasing values of the power-law index, reaching a maximum of about 2400 at $n = 0.4$ and then dropping to 1600 at $n = 0.1$. The latter behaviour is not in line with the experimental results of Dodge and Metzner (1959) who observed laminar flow conditions up to $Re_{MR} \sim 3100$ for a fluid with $n' = 0.38$, a trend also borne out by recent numerical studies (Rudman *et al.*, 2004). On the other hand, equation (3.65) yields values of the critical Reynolds number which increase monotonically with the decreasing value of the power law index, n. In the limiting case of $n = 0$, it predicts a value of 4200. Despite the complex dependence of the limiting Reynolds number on the flow behaviour index embodied in equation (3.64) and the conflicting experimental evidence, it is probably an acceptable approximation to assume that the laminar flow conditions cease to prevail at Reynolds numbers above *ca.* 2000–2500 and, for the purposes of process calculations, the widely accepted figure of 2100 can be used for time-independent fluids characterized in terms of n'. It is appropriate to add here that though the friction factor for visco-elastic fluids in the laminar regime is given by equation (3.56), the limited experimental results available suggest much higher values for the critical Reynolds number. For instance, Metzner and Park (1964) reported that their friction factor data for visco-elastic polymer solutions were consistent with equation (3.56) up to about $Re_{MR} = 10\,000$. However, it is not yet possible to put forward a quantitative criterion for calculating the limiting value of Re_{MR} for visco-elastic fluids.

Several other criteria, depending upon the use of a specific fluid model, are also available in the literature (Hanks, 1963; Govier and Aziz, 1982; Wilson, 1996; Malin, 1997). For instance, Hanks (1963) proposed the following criterion for Bingham plastic fluids:

$$(Re_{B})_{c} = \frac{\rho VD}{\mu_{B}} = \frac{1 - \dfrac{4}{3}\phi_{c} + \dfrac{\phi_{c}^{4}}{3}}{8\phi_{c}} He \qquad (3.66)$$

where the ratio, $\phi_{c} = (\tau_{0}^{B}/\tau_{w_{c}})$ is given by:

$$\frac{\phi_{c}}{(1 - \phi_{c})^{3}} = \frac{He}{16\,800} \qquad (3.67)$$

The Hedström number, He, is defined as:

$$He = \frac{\rho D^{2}\tau_{0}^{B}}{\mu_{B}^{2}} = Re_{B} \times Bi \qquad (3.68)$$

where $Bi = (D\tau_{0}^{B}/\mu_{B}V)$ is the Bingham number. For a given pipe size (D) and Bingham plastic fluid behaviour $(\rho, \mu_{B}, \tau_{0}^{B})$, the Hedström number will be known and the value of ϕ_{c} can be obtained from equation (3.67) which, in turn, facilitates the calculation of $(Re_{B})_{c}$ using equation (3.66), as illustrated in Example 3.6. Subsequent numerical calculations (Malin, 1997) lend further support to the validity of equations (3.66) and (3.67).

Both Wilson (1996) and Slatter (1996) have also re-evaluated the available criteria for the laminar–turbulent transition, with particular reference to the flow of pseudoplastic and yield-pseudoplastic mineral slurries in circular pipes. Wilson (1996) has argued that the larger dissipative micro-eddies present in the wall region result in thicker viscous sub-layers in non-Newtonian fluids which, in turn, produce greater mean velocity, giving a friction factor lower than that for Newtonian fluids, for the same value of the pressure drop across the pipe. For power-law fluids, he was able to link the non-Newtonian apparent viscosity to the viscosity of a hypothetical Newtonian fluid simply through a function of n, the power-law flow behaviour index, such that the same $Q-(-\Delta p)$ relationship applies to both fluids. This, in turn, yields the criterion for laminar–turbulent transition in terms of the critical value of the friction factor as a function of n (power-law index) alone. Note that in this approach, the estimated value of the effective viscosity will naturally depend upon the type of fluid and pipe diameter, D. Similarly, Slatter

(1996) has put forward a criterion in terms of a new Reynolds number for the flow of Herschel–Bulkley model fluids (equation (1.18)) to delineate the laminar–turbulent transition condition. His argument hinges on the fact that the inertial and viscous forces in the fluid are determined solely by that part of the fluid which is undergoing deformation (shearing), and hence he excluded that part of the volumetric flow rate attributable to the unsheared plug of material present in the middle of the pipe. These considerations lead to the following definition of the modified Reynolds number:

$$Re_{mod} = \frac{8\rho V_{ann}^2}{\tau_0^H + m\left(\dfrac{8V_{ann}}{D_{shear}}\right)^n}$$

(3.69)

where

$$V_{ann} = \frac{Q - Q_{plug}}{\pi(R^2 - R_p^2)} \quad \text{and} \quad D_{shear} = 2(R - R_p)$$

Laminar flow conditions cease to exist at $Re_{mod} = 2100$. The calculation of the critical velocity corresponding to $Re_{mod} = 2100$ requires an iterative procedure. For known rheology (ρ, m, n, τ_0^H) and pipe diameter (D), a value of the wall shear stress is assumed which, in turn, allows the calculation of R_p, from equation (3.18), and Q and Q_p from equations (3.41) and (3.40), respectively. Thus, all quantities are then known and the value of Re_{mod} can be calculated. The procedure is terminated when the value of τ_w has been found which makes $Re_{mod} = 2100$, as illustrated in Example 3.6 for the special case of $n = 1$, i.e., for the Bingham plastic model, and in Example 3.7 for a Herschel–Bulkley fluid. Detailed comparisons between the predictions of equation (3.69) and experimental data reveal an improvement in the predictions, though the values of the critical velocity obtained using the criterion $Re_{MR} = 2100$ are only 20–25% lower than those predicted by equation (3.69). Furthermore, the two criteria coincide for power-law model fluids. Subsequently, it has also been shown that while the laminar–turbulent transition in small diameter tubes is virtually unaffected by the value of the yield stress, both the flow behaviour index (n) and the yield stress play increasingly greater roles in determining the transition point with increasing pipe diameter. Thus, it appears that both shear-thinning and yield stress act to stabilize the flow thereby delaying the cessation of the laminar flow conditions. This is also consistent with the available theoretical and experimental studies (Frigaard *et al.*, 1994; Draad *et al.*, 1998; Nouar and Frigaard, 2001; Peixinho *et al.*, 2005; Rudman and Blackburn, 2006). Finally, the scant results obtained with a kaolin slurry and a critical micelle concentration (CMC) solution seem to suggest that the laminar–turbulent transition is not influenced by the pipe roughness (Slatter, 1996, 1997).

Example 3.6

The rheological behaviour of a coal slurry ($1160 \, \text{kg/m}^3$) can be approximated by the Bingham plastic model with $\tau_0^B = 0.5 \, \text{Pa}$ and $\mu_B = 14 \, \text{mPa·s}$. It is to be pumped through a 400 mm diameter pipe at the rate of 188 kg/s. Ascertain the nature of the flow by calculating the maximum permissible velocity for laminar flow conditions. Contrast the predictions of equations (3.66) and (3.69).

Solution

Here, the Hedström number,

$$He = \frac{\rho D^2 \tau_0^B}{\mu_B^2} = \frac{1160 \times 0.4^2 \times 0.5}{(14 \times 10^{-3})^2}$$

i.e., $He = 4.73 \times 10^5$ when substituted in equation (3.67) yields,

$$\frac{\phi_c}{(1 - \phi_c)^3} = \frac{4.73 \times 10^5}{16\,800} = 28.15$$

A trial and error procedure gives $\phi_c = 0.707$. Now substituting for He and ϕ_c in equation (3.66):

$$(Re_B)_c = \frac{1 - \dfrac{4}{3} \times 0.707 + \dfrac{(0.707)^4}{3}}{8 \times 0.707} \times 4.73 \times 10^5$$

or $\quad (Re_B)_c = \dfrac{\rho V_c D}{\mu_B} = 11760$

and the maximum permissible velocity, V_c, therefore is,

$$V_c = \frac{11\,760 \times 14 \times 10^{-3}}{1160 \times 0.4} = 0.354 \,\text{m/s}$$

The actual velocity in the pipe is:

$$V = \left(\frac{188}{1160}\right)\left(\frac{4}{\pi \times 0.4^2}\right) = 1.29 \,\text{m/s}$$

Thus, the flow in the pipe is not streamline.

Alternatively, one can use equation (3.69) to estimate the maximum permissible velocity for streamline flow in the pipe. In this example, $n = 1$, $m = 0.014\,\text{Pa}\,\text{s}$ and $\tau_0^H = 0.5\,\text{Pa}$. As mentioned previously the use of equation (3.69) requires an iterative procedure, and to initiate this method, let us assume a value of $\tau_w = 0.6\,\text{Pa}$.

$$\therefore \phi = \frac{\tau_0^H}{\tau_w} = \frac{0.5}{0.6} = \frac{R_p}{R}, \quad \text{i.e., } R_p = 0.166\,\text{m and } \phi = 0.833$$

Now using equation (3.28), or equation (3.41) for $n = 1$:

$$Q = \pi \times 0.2^3 \times 1\left(\frac{0.6}{0.014}\right)(1 - 0.833)^2$$

$$\times \left[\frac{(1 - 0.833)^2}{4} + \frac{2 \times 0.833(1 - 0.833)}{3} + \frac{0.833^2}{2}\right]$$

$$= 0.0134\,\text{m}^3/\text{s}$$

The plug velocity, V_{zp}, is calculated from equation (3.40) by substituting $R_p/R = \phi = 0.833$, i.e.,

$$V_{zp} = \frac{1 \times 0.2}{(1 + 1)}\left(\frac{0.6}{0.014}\right)(1 - 0.833)^2 = 0.1195\,\text{m/s}$$

$$Q_p = V_p \pi R_p^2 = 0.1195 \times \pi \times 0.166^2 = 0.01035 \, \text{m}^3/\text{s}$$

$$\therefore \quad Q_{ann} = Q - Q_p = 0.0134 - 0.01035 = 0.00305 \, \text{m}^3/\text{s}$$

$$V_{ann} = \frac{Q_{ann}}{\pi(R^2 - R_p^2)} = \frac{0.00305}{\pi(0.2^2 - 0.166^2)} = 0.078 \, \text{m/s}$$

$$D_{shear} = 2(R - R_p) = 2(0.2 - 0.166) = 0.068 \, \text{m}$$

$$\therefore \quad Re_{mod} = \frac{8\rho V_{ann}^2}{\tau_0^H + m\left(\dfrac{8V_{ann}}{D_{shear}}\right)^n} = \frac{8 \times 1160 \times 0.078^2}{0.5 + 0.014\left(\dfrac{8 \times 0.078}{0.068}\right)} = 90$$

which is too small for the flow to be turbulent. Thus, this procedure must be repeated for other values of τ_w to make $Re_{mod} = 2100$. A summary of calculations is presented in the table below.

τ_w (Pa)	Q (m³/s)	Q_p (m³/s)	Re_{mod}
0.6	0.0134	0.01035	90
0.70	0.0436	0.0268	890
0.73	0.0524	0.0305	1263
0.80	0.0781	0.0395	2700
0.77	0.0666	0.0358	1994
0.78	0.0706	0.0370	2233
0.775	0.0688	0.0365	2124

The last entry is sufficiently close to $Re_{mod} = 2100$, and laminar flow will cease to exist at $\tau_w \geq 0.775$ Pa. Also, note that the use of equations (3.40) and (3.41) beyond this value of wall shear stress is incorrect.

$$\therefore \text{Maximum permissible velocity} = \frac{0.0688}{\dfrac{\pi}{4}(0.4)^2} = 0.55 \, \text{m/s}$$

This value is some 40% higher than the previously calculated value of 0.35 m/s. However, even on this count, the flow will be turbulent at the given velocity of 1.29 m/s. □

Example 3.7

Determine the critical velocity for the upper limit of laminar flow for a slurry with the following properties, flowing in a 150 mm diameter pipe.

$$\rho = 1150 \, \text{kg/m}^3, \quad \tau_0^H = 6 \, \text{Pa}, \quad m = 0.3 \, \text{Pa s}^n \quad \text{and} \quad n = 0.4$$

Solution

As in Example 3.6, one needs to assume a value for τ_w and then calculate all other quantities using equations (3.40) and (3.41) which in turn allow the calculation of Re_{mod} using equation (3.69).

A summary of the calculations is presented in the following table:

τ_w (Pa)	Q (m^3/s)	Q_p (m^3/s)	Re_{mod}
6.4	4.72×10^{-5}	4.27×10^{-5}	6.5×10^{-3}
7.4	3.097×10^{-3}	2.216×10^{-3}	26.6
8.4	0.01723	0.01	778
9.3	0.046	0.0224	5257
8.82	0.0287	0.0153	2100

Thus, the laminar–turbulent transition for this slurry in a 150 mm diameter pipe occurs when the wall shear stress is 8.82 Pa and the volumetric flow rate is 0.0287 m^3/s.

$$\therefore \text{Mean velocity at this point} = \frac{Q}{(\pi/4)D^2} = \frac{0.0287}{(\pi/4)(0.15)^2} = 1.62 \text{ m/s}$$

Hence, streamline flow will occur for this slurry in a 150 mm diameter pipe at velocities up to a value of 1.62 m/s. □

3.4 Friction factors for transitional and turbulent conditions

Though turbulent flow conditions are encountered less frequently with polymeric non-Newtonian substances, sewage sludges, coal, china clay and many other mineral suspensions are usually all transported in the turbulent flow regime in large diameter pipes. Therefore, considerable research efforts have been directed at developing a generalized approach for the prediction of the frictional pressure drop for turbulent flow in pipes, especially for purely viscous (power-law, Bingham plastic and Herschel–Bulkley models) fluids. Analogous studies for the flow of visco-elastic and the so-called drag-reducing fluids are somewhat inconclusive. Furthermore, the results obtained with drag-reducing polymer solutions also tend to be strongly dependent on the type and molecular weight of the polymers, the nature of the solvent and on the type of experimental set up used, and it is thus not yet possible to put forward generalized equations for the turbulent flow of such fluids. Therefore, the ensuing discussion is restricted primarily to the turbulent flow of time-independent fluids. However, excellent survey articles on the turbulent flow of visco-elastic and drag-reducing systems are available in the literature (Cho and Hartnett, 1982; Govier and Aziz, 1982; Sellin *et al.*, 1982).

In the same way as there are many equations for predicting friction factor for turbulent Newtonian flow, there are numerous equations for time-independent non-Newtonian fluids; most of these are based on dimensional considerations combined with experimental observations (Govier and Aziz, 1982; Heywood and Cheng, 1984) There is a preponderance of correlations based on the power-law fluid behaviour and additionally some expressions are available for Bingham plastic fluids (Tomita, 1959; Wilson and Thomas, 1985). Here only a selection of widely used and proven methods is presented.

3.4.1 Power-law fluids

In a comprehensive study, Dodge and Metzner (1959) carried out a semi-empirical analysis of the fully developed turbulent flow of power-law fluids in smooth pipes. They used the same dimensional considerations for such fluids, as Millikan (1939) for incompressible Newtonian fluids, and obtained an expression which can be rearranged in terms of the apparent power law index n' (equation 3.54) as follows:

$$\frac{1}{\sqrt{f}} = A(n') \log[Re_{MR} \ f^{(2-n')/2}] + C(n') \tag{3.70}$$

where $A(n')$ and $C(n')$ are two unknown functions of n'. Based on extensive experimental results in the range $2900 \le Re_{MR} \le 36\,000$; $0.36 \le n' \le 1$ for polymer solutions and particulate suspensions, Dodge and Metzner (1959) obtained,

$$A(n') = 4(n')^{-0.75} \tag{3.71}$$

$$C(n') = -0.4(n')^{-1.2} \tag{3.72}$$

Incorporating these values in equation (3.70),

$$\frac{1}{\sqrt{f}} = \frac{4}{(n')^{0.75}} \log[Re_{MR} f^{(2-n')/2}] - \frac{0.4}{(n')^{1.2}} \tag{3.73}$$

and this relation is shown graphically in Figure 3.9.

While a more detailed derivation of equation (3.73) is available in their original paper and elsewhere (Skelland, 1967), it is useful to add here that the limited numerical results available in the literature lend support to the validity of this approach (Rudman et al., 2004). For Newtonian fluids ($n' = 1$), equation (3.73) reduces to the well-known Nikuradse equation as follows:

$$\frac{1}{\sqrt{f}} = 4 \log\left(Re\sqrt{f}\right) - 0.4 \tag{3.74}$$

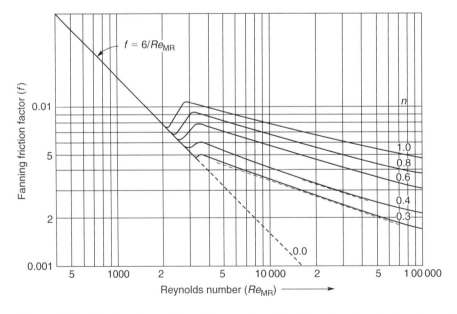

Figure 3.9 *Friction factor–Reynolds number behaviour for time-independent fluids (Dodge and Metzner, 1959)*

Dodge and Metzner (1959) also demonstrated that their data for clay suspensions which did not conform to power-law behaviour, were consistent with equation (3.73) provided that the slope of log τ_w − log(8 V/D) plots was evaluated at the appropriate value of the wall shear stress. It is also important to point out here that equation (3.73) necessitates the values of n' and m' to be evaluated from volumetric flow rate – pressure drop data for laminar flow conditions. Often, this requirement poses significant experimental difficulties. Finally, needless to say, this correlation is implicit in friction factor f (like equation (3.74) for Newtonian fluids) and hence an iterative technique is needed for its solution. The method of Irvine (1988) obviates this difficulty. Based on the Blasius expression for velocity profile for turbulent flow (discussed subsequently) together with modifications based on experimental results, Irvine (1988) proposed the following Blasius-like expression for power-law fluids:

$$f = \{D(n)/Re_{MR}\}^{1/(3n+1)} \tag{3.75}$$

where

$$D(n) = \frac{2^{n+4}}{7^{7n}} \left(\frac{4n}{3n+1} \right)^{3n^2}$$

Note that this cumbersome expression does reduce to the familiar Blasius expression for $n = 1$ and is explicit in friction factor, f. Equation (3.75) was stated to predict the values of friction factor with an average error of ±8% in the range of conditions: $0.35 \leq n \leq 0.89$ and $2000 \leq Re_{MR} \leq 50\,000$. Though this approach has been quite successful in correlating most of the literature data, significant deviations from it have also been observed (Harris, 1968; Quader and Wilkinson, 1980; Heywood and Cheng, 1984; Duffy, 2004); though the exact reasons for such deviations are not immediately obvious but possible visco-elastic effects and erroneous values of the rheological parameters (e.g. n' and m') cannot be ruled out. Example 3.8 illustrates the application of these methods.

Example 3.8

A non-Newtonian polymer solution (density = 1000 kg/m³) is in steady flow through a smooth 300 mm inside diameter 50 m long pipe at the mass flow rate of 300 kg/s. The following data have been obtained for the rheological behaviour of the solution using a tube viscometer. Two tubes, 4 and 6.35 mm in inside diameter and 2 and 3.2 m long, respectively, were used to encompass a wide range of shear stress and shear rate.

D = 4 mm, L = 2 m		D = 6.35 mm, L = 3.2 m	
Mass flow rate (kg/h)	Pressure drop (kPa)	Mass flow rate (kg/h)	Pressure drop (kPa)
33.9	49	18.1	27
56.5	57.6	45.4	36
95	68.4	90.7	44
136	76.8	181	54
153.5	79.5	272	61

Determine the pump power required for this pipeline. How will the power requirement change if the flow rate is increased by 20%?

Solution

First, the tube viscometer data will be converted to give the wall shear stress τ_w and nominal shear rate $(8V/D)$:

$$\tau_w = \frac{D}{4}\left(\frac{-\Delta p}{L}\right) = \frac{4 \times 10^{-3}}{4} \times \frac{49 \times 1000}{2} = 24.5\,\text{Pa}$$

and

$$\frac{8V}{D} = \frac{8}{4 \times 10^{-3}} \times \frac{33.9}{1000 \times 3600} \times \frac{4}{\pi(4 \times 10^{-3})^2} = 1499\,\text{s}^{-1}$$

Similarly the other mass flow rate–pressure drop data can be converted into $\tau_w - (8V/D)$ form as are shown in the following table.

	$D = 4\,$mm		$D = 6.35\,$mm	
τ_w (Pa)	$(8V/D)$ (s^{-1})	τ_w (Pa)	$(8V/D)$ (s^{-1})	
24.5	1499	13.4	200	
28.8	2500	17.86	502	
34.2	4200	21.83	1002	
38.4	6000	26.8	2000	
39.8	6800	30.26	3005	

Note that since L/D for both tubes is 500, entrance effects are expected to be negligible. Figure 3.10 shows the τ_w versus $(8V/D)$ data on log–log coordinates. Obviously, n' is not constant, though there seem to be two distinct power-law regions with parameters:

$$n' = 0.3 \quad m' = 2.74\,\text{Pa s}^{n'} \quad (\tau_w < {\sim}30\,\text{Pa})$$
$$n' = 0.35 \quad m' = 1.82\,\text{Pa s}^{n'} \quad (\tau_w > {\sim}30\,\text{Pa})$$

Also, the overlap in data obtained using tubes of two different diameters confirms the time-independent behaviour of the solution and the absence of wall slip effects.

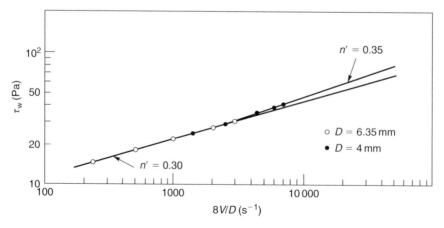

Figure 3.10   *Wall shear stress–apparent wall shear rate plot for data in Example 3.8*

In the large diameter pipe, the mean velocity of flow

$$V = \frac{300}{1000} \times \frac{4}{\pi(0.3)^2} \text{ or } V = 4.1 \text{ m/s}$$

Let us calculate the critical velocity, V_c, for the end of the streamline flow by setting $Re_{MR} = 2100$, i.e.,

$$\frac{\rho V^{2-n'} D^{n'}}{8^{n'-1} m'} = 2100$$

Substituting values,

$$\frac{1000 \times V_c^{2-0.3} \times (0.3)^{0.3}}{8^{0.3-1} \times 2.74} = 2100$$

Solving, $V_c = 1.47$ m/s which is lower than the actual velocity of 4.1 m/s and hence, the flow in the 300 mm pipe is likely to be turbulent. Note that in this case, equations (3.64) and (3.65) predict the critical values of the Reynolds number as 2344 and 2792, respectively. These correspond to the transition velocities of 1.6 and 1.74 m/s, respectively. Thus, the value of 1.47 m/s used in this example is conservative.

Since the value of the wall shear stress in the large pipe is not known, a trial and error procedure is required here. Initially, let us assume that the wall shear stress in the large pipe would be <30 Pa, i.e. $n' = 0.3$ and $m' = 2.74$ Pa s^n can be used for calculating the value of the Re_{MR}, equation (3.57),

$$Re_{MR} = \frac{\rho V^{2-n'} D^{n'}}{8^{n'-1} m'} = \frac{1000 \times 4.1^{2-0.3} \times (0.3)^{0.3}}{8^{0.3-1} \times 2.74}$$

$$= 12\,230$$

Now for $n' = 0.3$ and $Re_{MR} = 12\,230$, from Figure 3.9, $f \approx 0.0033$ (equation (3.75) gives $f = 0.0036$). Next the frictional pressure gradient $(-\Delta p/L)$ is calculated as follows:

$$\left(\frac{-\Delta p}{L}\right) = \frac{2 f \rho V^2}{D} = \frac{2 \times 0.0033 \times 1000 \times 4.1^2}{0.3} = 364 \text{ Pa/m}$$

The corresponding value of $\tau_w = (D/4)(-\Delta p/L) = (0.3 \times 364)/4 = 27.7$ Pa is within the range of the first power-law region and hence no further iteration is needed. The pump power is $Q \cdot \Delta p$, i.e., $(300/1000) \times 364 \times 50 = 5460$ W.

For the case when the flow rate has been increased by 20%, i.e., the new mass flow rate in the large pipe is 360 kg/s.

Mean velocity of flow

$$V = \frac{360}{1000} \times \frac{4}{\pi(0.3)^2} = 4.92 \text{ m/s}$$

Based on the previous calculation, it is reasonable to assume that the new value of the wall shear stress would be greater than 30 Pa and therefore, one should use $n' = 0.35$ and $m' = 1.82$ Pa s$^{n'}$.

The Reynolds number

$$Re_{MR} = \frac{1000 \times 4.92^{2-0.35} \times (0.3)^{0.35}}{8^{0.35-1} \times 1.82}$$

$$= 19\,410$$

For $n' = 0.35$ and $Re_{MR} = 19410$, from Figure 3.9, $f \approx 0.0032$ (while equation (3.75) also yields the same value). The frictional pressure gradient $(-\Delta p/L)$ is:

$$\frac{-\Delta p}{L} = \frac{2f\rho V^2}{D} = \frac{2 \times 0.0032 \times 1000 \times 4.92^2}{0.3} = 511\,\text{Pa/m}$$

Checking:

$$\tau_w = \frac{D}{4}\left(\frac{-\Delta p}{L}\right) = \frac{0.3 \times 511}{4} = 39\,\text{Pa}$$

This value is just within the range of laminar flow data.

$$\text{Pump power} = \frac{360}{1000} \times 511 \times 50 = 9200\,\text{W.}\ \square$$

3.4.2 Viscoplastic fluids

Despite the fact that equation (3.73) is applicable to all kinds of time-independent fluids, numerous workers have presented expressions for turbulent flow friction factors for specific fluid models. For instance, Tomita (1959) applied the concept of the Prandtl mixing length and put forward modified definitions of the friction factor and Reynolds number for the turbulent flow of Bingham Plastic fluids in smooth pipes so that the Nikuradse equation, i.e., equation (3.73) with $n' = 1$, namely, equation (3.74), could be used. Though he tested the applicability of his method using his own data in the range $2000 \leq Re_B(1 - \phi)^2(3 - \phi) \leq 10^5$, the validity of this approach has not been established using independent experimental data.

In contrast, the semi-empirical equations due to Darby *et al.* (1992) obviate these difficulties due to their explicit form which is as follows:

$$f = (f_L^b + f_T^b)^{1/b} \tag{3.76}$$

where f_L is the solution of equation (3.36), or equation (3.38) and f_T is a function of the Reynolds number and Hedström number expressed as:

$$f_T = 10^{a_0}\,Re_B^{-0.193} \tag{3.77}$$

where

$$a_0 = -1.47[1 + 0.146\exp(-2.9 \times 10^{-5}\,He)] \tag{3.78}$$

and

$$b = 1.7 + \frac{40\,000}{Re_B} \tag{3.79}$$

This method has been shown to yield satisfactory values of pressure drop under turbulent conditions for $D < 335\,\text{mm}$, $Re_B \leq 3.4 \times 10^5$ and $1000 \leq He \leq 6.6 \times 10^7$. Example 3.9 illustrates the application of this method.

Example 3.9

A 18% iron oxide slurry (density $= 1170\,\text{kg/m}^3$) behaves as a Bingham plastic fluid with $\tau_0^B = 0.78\,\text{Pa}$ and $\mu_B = 4.5\,\text{mPa\,s}$. Estimate the wall shear stress as a function of the nominal wall shear rate ($8V/D$) in the range $0.4 \leq V \leq 1.75\,\text{m/s}$ for flow in a 79 mm diameter pipeline.

Over the range of the wall shear stress encountered, the slurry can also be modelled as a power-law fluid with $m = 0.16\,\mathrm{Pa\,s}^n$ and $n = 0.48$. Contrast the predictions of the two rheological models.

Solution

(a) Bingham model

Here $\tau_0^B = 0.78\,\mathrm{Pa}$; $\mu_B = 4.5\,\mathrm{mPa\,s}$;
$\rho = 1170\,\mathrm{kg/m^3}$ and $D = 79 \times 10^{-3}\,\mathrm{m}$

The value of Hedström number is calculated first:

$$He = \frac{\rho D^2 \tau_0^B}{\mu_B^2} = \frac{1170 \times 0.079^2 \times 0.78}{(4.5 \times 10^{-3})^2} = 2.81 \times 10^5$$

From equation (3.78):

$$a_0 = -1.47[1 + 0.146\exp(-2.9 \times 10^{-5} \times 2.81 \times 10^5)] = -1.47$$

A sample calculation is shown here for $V = 0.4\,\mathrm{m/s}$.

Apparent shear rate at wall, $\dfrac{8V}{D} = \dfrac{8 \times 0.4}{0.079} = 40.5\,\mathrm{s}^{-1}$

The Reynolds number of flow, Re_B is:

$$Re_B = \frac{\rho VD}{\mu_B} = \frac{1170 \times 0.4 \times 0.079}{4.5 \times 10^{-3}} = 8216$$

$\therefore$ The index b is given by equation (3.79):

$$b = 1.7 + \frac{40\,000}{Re_B} = 1.7 + \frac{40\,000}{8216} = 6.57$$

The value of f_L, i.e., the friction factor in streamline flow, is calculated using equation (3.36) as:

$$f_L = \frac{16}{Re_B}\left[1 + \left(\frac{1}{6}\right)\frac{He}{Re_B} - \left(\frac{1}{3}\right)\frac{He^4}{f_L^3 Re_B^7}\right]$$

For $Re_B = 8216$, and $He = 2.81 \times 10^5$, this equation is solved to get $f_L = 0.0131$. The value of f_T is calculated using equation (3.77):

$$f_T = 10^{-1.47}(8216)^{-0.193} = 0.00595$$

Finally, the actual friction factor is estimated from equation (3.76):

$$f = (0.0131^{6.57} + 0.00595^{6.57})^{1/6.57} = 0.0131$$

and the wall shear stress, $\tau_w = \tfrac{1}{2}f\rho V^2 = \tfrac{1}{2} \times 0.0131 \times 1170 \times 0.4^2$, i.e., $\tau_w = 1.22\,\mathrm{Pa}$.

This procedure is repeated for the other values of the average velocity, and a summary of results is shown in the following table.

V (m/s)	$(8V/D)$ (s^{-1})	Re_{B} $(-)$	b $(-)$	f_{T} $(-)$	f_{L} $(-)$	f $(-)$	τ_{w} (Pa)
0.4	40.5	8216	6.57	0.00595	0.0131	0.0131	1.22
0.6	60.8	1.23×10^4	4.95	0.0055	0.00625	0.0068	1.44
0.8	81	1.64×10^4	4.14	0.00521	0.00377	0.00551	2.62
1.0	101.3	2.05×10^4	3.65	0.00499	0.00257	0.00510	2.99
1.25	126.6	2.57×10^4	3.26	0.00477	0.00176	0.00483	4.42
1.50	152	3.08×10^4	3.00	0.00461	0.00131	0.00465	6.12
1.75	177.2	3.6×10^4	2.81	0.00447	0.00102	0.0045	8.06

(b) Power-law model

Likewise, one sample calculation is shown here for $V = 0.4\,\text{m/s}$.
The Metzner–Reed Reynolds number, Re_{MR} is given by equation (3.57):

$$Re_{\text{MR}} = \frac{\rho V^{2-n} D^n}{8^{n-1} m \left(\dfrac{3n+1}{4n}\right)^n}$$

$$= \frac{1170 \times 0.4^{2-0.48} \times 0.079^{0.48}}{8^{0.48-1} \times 0.16 \times \left(\dfrac{3 \times 0.48 + 1}{4 \times 0.48}\right)^{0.48}} = 1407 < 2100$$

$\therefore$ the flow is streamline and $f = 16/Re_{\text{MR}}$,

$$f = 0.0114 \quad \text{and} \quad \tau_{\text{w}} = \tfrac{1}{2} f \rho V^2 = \tfrac{1}{2} \times 0.0114 \times 1170 \times 0.4^2$$

or $\quad \tau_{\text{w}} = 1.06\,\text{Pa}$

For $V \geq 0.6\,\text{m/s}$, the value of Re_{MR} ranges from 2607 for $V = 0.6\,\text{m/s}$ to 1.33×10^4 for $V = 1.75\,\text{m/s}$. Therefore, one must use either equation (3.73) or (3.75) under these conditions; the latter is used here owing to its explicit form. A summary of calculations is shown in the following table:

V (m/s)	$(8V/D)$ (s^{-1})	Re_{MR} $(-)$	f $(-)$	τ_{w} (Pa)
0.4	40.5	1407	0.0114	1.06
0.6	60.8	2607	0.00911	1.92
0.8	81	4037	0.00761	2.85
1.0	101.3	5667	0.00662	3.87
1.25	126.6	7956	0.00577	5.27
1.50	152	1.05×10^4	0.00515	6.77
1.75	177.2	1.33×10^4	0.00468	8.37

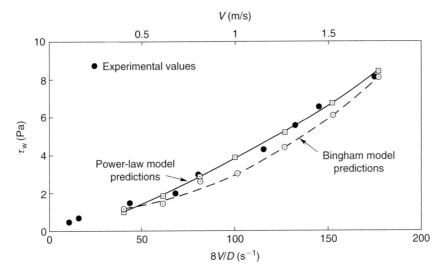

Figure 3.11 $\tau_w - (8V/D)$ *plot for Example 3.9*

The predicted values of τ_w according to the power-law and Bingham plastic models are plotted in Figure 3.11 together with the experimental values. While the maximum discrepancy between the two predictions is of the order of 20%, in this particular example, the actual values are seen to be closer to the predictions of the power-law model than that of the Bingham model. □

3.4.3 Bowen's general scale-up method

Bowen (1961), on the other hand, proposed that for turbulent flow of a *particular* fluid (exhibiting time-independent behaviour), the wall shear stress, τ_w, could be expressed as:

$$\tau_w = A \frac{V^b}{D^c} \tag{3.80}$$

where A, b and c are constants for the fluid and may be determined from experimental measurements in small diameter tubes. While it is difficult to offer a sound justification for the form of equation (3.80), the following simple approach is helpful in this regard. For the turbulent flow of Newtonian fluids in pipes, the well-known Blasius equation is written as:

$$f = 0.079 Re^{-0.25} \tag{3.81}$$

For $n' = 1$, equation (3.75) also reduces to this form. Now using the definition of the friction factor, $f = 2\tau_w/\rho V^2$, and expressing the Reynolds number as $Re = \rho VD/\mu$, it is possible to rewrite equation (3.81) as:

$$\tau_w = 0.0395 \rho^{0.75} \mu^{0.25} \frac{V^{1.75}}{D^{0.25}} \tag{3.82}$$

For a given Newtonian fluid, both ρ and μ are constant and therefore, one can rewrite equation (3.82) as:

$$\tau_w \propto \frac{V^{1.75}}{D^{0.25}} \tag{3.83}$$

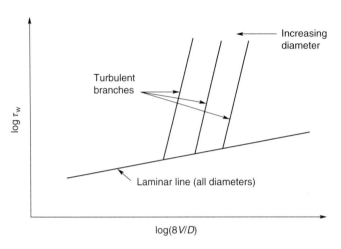

Figure 3.12   *Schematics of $\tau_w - (8V/D)$ behaviour for different diameter tubes in laminar and turbulent regions*

Thus, equation (3.80) due to Bowen (1961) can be seen as a generalization of this equation for the turbulent flow of time-independent non-Newtonian fluids. For laminar flow, wall shear stress (τ_w) and apparent wall shear stress $(8V/D)$ data for different diameter tubes coincide provided these have been corrected for end effects and wall-slip. However, for turbulent conditions, as shown schematically in Figure 3.12, diameter appears to be an additional independent parameter. Because of the inclusion of a diameter effect explicitly, this method is to be preferred for scaling up the results of small-scale pipe experiments to predict frictional pressure drops for the same fluid in large diameter pipes. Note that this method does not offer a generalized scheme of calculating pressure drop in contrast to the method of Metzner and Reed (1955) and Dodge and Metzner (1959).

The three turbulent branches shown in Figure 3.12, which all have the same slope, correspond to three different pipe diameters, whilst a single line suffices for all diameters for laminar flow in the absence of wall-slip effects. The Reynolds number at the transition can be calculated from the value of $(8V/D)$ at which the turbulent branch deviates from the laminar line. In practice, however, the laminar–turbulent transition occurs over a range of conditions rather than abruptly as shown in Figure 3.12. Strictly speaking, only turbulent flow data for small-scale pipe should be used to scale up to turbulent flow in larger pipes. However, this is likely to pose problems because of the extremely high velocities and pressure drops which would frequently be required to achieve turbulence in small diameter pipes. But nevertheless measurements in the streamline and transitional regimes facilitate the delineation of the critical Reynolds number. Due to the fluctuating nature of flow, the operation of pipe lines in the transition region should be avoided as far as possible. Finally, this method can be summarized as follows:

(i) Obtain $\tau_w - (8V/D)$ data using at least two pipe sizes under both laminar and turbulent flow conditions if possible. Plot these data on log–log coordinates as shown schematically in Figure 3.12. The laminar flow data should collapse on to one line and there should be one branching for turbulent flow line for each pipe diameter.

(ii) The index of the velocity term, b, in equation (3.80) is the slope of the turbulent branch. Ideally, all branches would have identical slopes but in practice b should be evaluated for each value of D and the mean value used.

(iii) Next, plot $LV^b/(-\Delta p)$ against D on log–log coordinates for each turbulent flow data point; the slope of this line will be $(1 + c)$ and hence c can be evaluated. This is easily seen by expressing τ_w as $(-\Delta p/L)(D/4)$ in equation (3.80).

(iv) The remaining constant, A, is evaluated by calculating its value for each turbulent flow data point as $\tau_w D^c/V^b$ and again the mean value of A should be used for the purpose of scale-up to large diameter pipes.

Equation (3.80) can now be used directly to give the pressure drop for any pipe diameter if the flow is turbulent. Alternatively, this approach can be used to construct a wall shear stress (τ_w) – apparent wall shear rate $(8V/D)$ turbulent flow line for any pipe diameter. This method is particularly suitable when either the basic rheological measurements for laminar flow are not available or it is not possible to obtain a satisfactory fit of such data. The application of this method is illustrated in Example 3.10.

Example 3.10

The following flow rate–pressure drop data for a 0.2% aqueous carbopol solution (density = $1000\,\text{kg/m}^3$) have been reported for two capillary tubes and three pipes of different diameters. (Data from: D.W. Dodge, PhD Thesis, University of Delaware, Newark, 1958).

Capillary data				Pipeline data					
$D = 0.84\,\text{mm}$ $L = 155.5\,\text{mm}$		$D = 0.614\,\text{mm}$ $L = 203.8\,\text{mm}$		$D = 12.7\,\text{mm}$		$D = 25.4\,\text{mm}$		$D = 50.8\,\text{mm}$	
$\left(\dfrac{8V/D}{(s^{-1})}\right)$	τ_w (Pa)	$\left(\dfrac{8V/D}{(s^{-1})}\right)$	τ_w (Pa)	$\left(\dfrac{8V/D}{(s^{-1})}\right)$	τ_w (Pa)	$\left(\dfrac{8V/D}{(s^{-1})}\right)$	τ_w (Pa)	$\left(\dfrac{8V/D}{(s^{-1})}\right)$	τ_w (Pa)
213.2	4.73	820.5	12.85	472.5	7.76	205	4.53	74	2.19
397.1	7.59	1456	19.50	753.2	10.93	301	6.16	114	2.90
762.8	12.16	2584	29.61	1121	14.94	418	7.88	151	3.87
1472	19.67	4691	46.24	1518	19.68	504	9.42	177	4.86
1837	23.27	8293	69.71	1715	26.45	562	11.83	194*	5.64
2822	31.39	14 420	104.50	1849*	34.45	602	14.24	212	7.61
4000	41.90	25 140	157	1989	40.12	691*	19.93	236.8	9.33
5237	49.08	43 310	231.7	2250	49.41	825	27.03	264	11.40
9520	77.33	76 280	343.4	2642	64.11	1004	37.21	303	14.14
10 300	81.1	111 900	496	3043	81.11	1188	49.08	348	17.61
19 010	125.35	–	–	3485	101	1473	69.76	395	21.62
22 400	142.54	–	–	4047	128.6	1727	90.30	453	27
35 490	191	–	–	4610	160.6	2071	120.40	517	33.52
50 400	246	–	–	5192	196	2482	162.1	582	40.64
59 170	272	–	–	6106	260.5	2990	222.7	671	51.33
94 260	365	–	–	6899	320	3482	286.3	776	64.78
95 600	358	–	–	7853	399	4261	407	907	82.64

Using the methods of Bowen (1961) and of Dodge and Metzner (1959), construct the wall shear stress–apparent shear rate plots for turbulent flow of this material in 101.6 and 203.2 mm diameter pipes. Also, calculate the velocity marking the end of the streamline flow.

Solution

(a) Bowen's method

Figure 3.13 shows the flow behaviour data tabulated above. As expected, all the laminar flow data are independent of tube diameter and collapse on to one line, with $n' = 0.726$ and $m' = 0.0981\,\mathrm{Pa\,s}^n$. Also, the same value of n' is seen to cover the entire range of measurements.

On the other hand, three turbulent branches are obtained with, each corresponding to a particular pipe diameter. The slopes of these lines are remarkably similar at 1.66, 1.66 and 1.65 for the 12.7, 25.4 and 50.8 mm diameter pipes, respectively.

$\therefore b = 1.66$

Note that the first few data points in each case seem to lie in laminar and transitional range as can be seen in Figure 3.13. These are also indicated in the tabulated data by an asterisk.

Now using this value of b, the quantity $(LV^b/ -\Delta p)$ is evaluated for each turbulent data point, and the resulting mean values for $(LV^b)/(-\Delta p)$ are 0.000535, 0.001183 and 0.002713 for the 12.7, 25.4 and 50.8 mm diameter pipes, respectively. These values are plotted against pipe diameter in Figure 3.14. Again, as expected, a straight line results with a slope of 1.18.

$\therefore 1 + c = 1.18$, i.e., $c = 0.18$

Finally, the value of A is calculated by evaluating the quantity $(\tau_w D^c / V^b)$ for each *turbulent* flow data point.

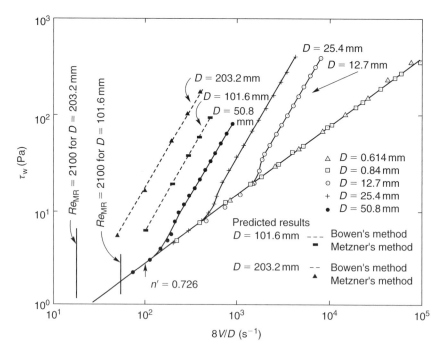

Figure 3.13 *Turbulent flow behaviour of 0.2% carbopol solution (Example 3.10)*

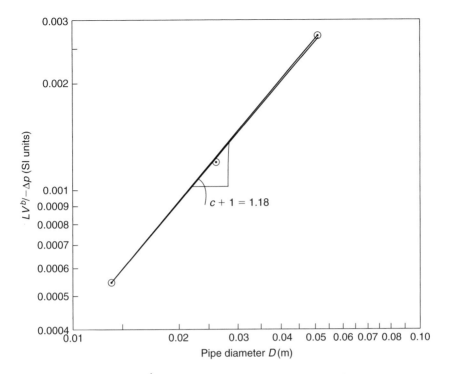

Figure 3.14   *Plot of $(LV^b/-\Delta p)$ versus pipe diameter for evaluating c*

For 12.7 mm diameter pipe, $2.67 \leq (\tau_w D^c/V^b) \leq 2.8$

25.4 mm diameter pipe, $2.75 \leq (\tau_w D^c/V^b) \leq 2.86$

50.8 mm diameter pipe, $2.75 \leq (\tau_w D^c/V^b) \leq 2.86$

The resulting mean value of A is 2.78. Therefore, for this carbopol solution in turbulent flow, the wall shear stress is given as:

$$\tau_w = 2.78 \frac{V^{1.66}}{D^{0.18}}$$

where all quantities are in SI units. Now, the predicted values for $D = 101.6$ and 203.2 mm pipes are shown in the following tables, and are also included in Figure 3.13 as broken lines.

Since the values of n' and m' remain constant over the entire range of wall shear stress, one can readily apply the method of Metzner and Reed (1955) to calculate the wall shear stress in large diameter pipes here. For instance, for $V = 1.27$ m/s, $D = 101.6$ mm

$$\mathrm{Re}_{MR} = \frac{\rho V^{2-n'} D^{n'}}{8^{n'-1} m'} = \frac{1000 \times 1.27^{2-0.726} \times 0.1016^{0.726}}{8^{0.726-1} \times 0.0981}$$

$$= 4645$$

From Figure 3.9, $f \approx 0.0077$.

Predicted values of $\tau_w - (8V/D)$ using the methods of Bowen and of Metzner and co-workers

V (m/s)	$(8V/D)$ s^{-1}	Re_{MR}	f	τ_w (Pa)	
				Bowen method	Metzner and Reed (1955)
$D = 101.6$ mm					
1.27	100	4645	0.0077	6.2	6.21
2.54	200	11 234	6.012×10^{-3}	19.65	19.39
3.81	300	18 831	5.348×10^{-3}	38.51	38.82
5.08	400	27 168	4.695×10^{-3}	62	60.58
6.35	500	36 100	4.5×10^{-3}	90	90.73
$D = 203.2$ mm					
1.27	50	7684	0.00668	5.5	5.39
2.54	100	18 582	0.00513	17.34	16.55
5.08	200	44 937	0.0043	54.8	55.5
7.62	300	75 325	0.0037	107.5	107.4
10.16	400	1 08 670	0.0035	173	180.6

Alternatively equation (3.75) may be used with $n = n' = 0.726$

Then

$$D(n) = \frac{2^{4.726}}{7^{7 \times 0.726}} \left(\frac{4 \times 0.726}{3 \times 0.726 + 1} \right)^{3 \times 0.726^2} = 0.001164$$

$$\therefore f = \left(\frac{0.001164}{4645} \right)^{1/(3 \times 0.276 + 1)} \approx 0.00837$$

The two values differ by about 8%. Using the value of $f = 0.0077$:

$$\tau_w = f \cdot \tfrac{1}{2} \rho V^2 = 0.0077 \times \frac{1000}{2} \times 1.27^2 = 6.21 \text{ Pa}$$

Similarly, one can calculate the values of the wall shear stress for the other values of the average velocity. These results are also summarized in the above table where a satisfactory correspondence is seen to exist between the two predicted values of the wall shear stress.

Finally, the critical values of the velocity marking the limit of streamline flow in 101.6 mm and 203.2 mm diameter pipes are calculated by setting $Re_{MR} = 2100$. These values are found to be $V = 0.68$ m/s in the 101.6 mm diameter pipe and $V = 0.37$ m/s for the other pipe. These values are also shown in Figure 3.13. □

Many other correlations for power-law, Bingham plastic and yield-pseudoplastic fluid models are available in the literature (Govier and Aziz, 1982; Hanks, 1986; Thomas and Wilson, 1987; Darby, 1988) but unfortunately their validity has been established only for a limited range of data and hence they are too tentative to be included here. In an exhaustive comparative study, Heywood and Cheng (1984) evaluated the relative performance of seven such correlations. They noted that the predictions differed widely and that the uncertainty was further compounded by the inherent difficulty in unambiguously evaluating the rheological properties to be used under turbulent flow conditions. Indeed in some cases, the estimated values of friction factor varied by up to ±50%, which is in stark contrast to Newtonian turbulent flow where most predictive expressions yield values of the friction factor within ±5% of each other. Thus, they recommended that as many predictive methods as possible should be used to calculate the value of f, so that an engineering judgement can be made taking into account the upper and lower bounds on the value of f for the particular application. However, in subsequent extensive evaluations of turbulent and transitional flow data (Garcia and Steffe, 1987; Hartnett and Kostic, 1990), it is recommended that the method of Metzner and Reed (1955) can be used to calculate frictional losses in straight sections of smooth pipes, i.e., Figure 3.9, or equation (3.73) or equation (3.75). The method of Bowen (1961) is recommended for scaling up small-scale turbulent flow data to turbulent flow in large diameter pipes. However, it should be reiterated here that the method of Metzner and Reed (1955) necessitates using the values of m' and n' evaluated at the prevailing wall shear stress; this usually means relying on $\tau_w - (8V/D)$ data obtained in laminar region. For a true power law fluid, $n' = n$ and there is no difficulty provided this value of n is applicable all the way up to the wall shear stress levels under turbulent flow conditions. However, since the pressure drop (hence wall shear stress) is usually unknown, a trial and error solution is required, except for the case when the pressure loss drop is known and the flow rate is to be calculated. Thus, first a value of $(-\Delta p/L)$ is assumed and based on this, m' and n' are evaluated from the plot of τ_w versus $(8V/D)$ obtained for laminar conditions. From these parameters, the value of Re_{MR} is calculated and f is found either from Figure 3.9 or equation (3.73) or (3.75). If the value of $(-\Delta p/L)$ calculated using this value of f is appreciably different from that obtained using the assumed value of $(-\Delta p/L)$, other iterations are carried out till the two values are fairly close to each other. The calculation usually converges quickly. However, there does not appear to be a simple way of determining *a priori* the shear stress range which must be covered in the laminar flow rheological tests to ensure that these will include the values of τ_w likely to be encountered under turbulent flow conditions. However, it is usually recommended that the approximate $(8V/D)$ range over which the pipeline is to operate can be used as a guide for performing laminar flow tests. On the other hand, the method of Bowen (1961), while free from all these complications, relies on the use of the same fluid in turbulent flow in both small and large scale pipes.

3.4.4 Effect of pipe roughness

Considerable confusion exists regarding the effect of the pipe wall roughness on the value of friction factor in the turbulent flow region, though the effect is qualitatively similar to that for Newtonian fluids (Slatter, 1997; Slatter and van Sittert, 1997; Slatter *et al.*, 1997). Thus, Torrance (1963) and Szilas *et al.* (1981) have incorporated a small correction to account for pipe roughness in their expressions for the turbulent flow of power-law and Bingham plastic fluids; under fully turbulent conditions, it is assumed that the value of the friction factor is determined only by pipe roughness and rheological properties, and is independent of the Reynolds number. However, in view of the fact that laminar sub-layers tend to be somewhat thicker for non-Newtonian fluids than that for Newtonian liquids, the effect of pipe roughness is likely to be small for time-independent fluids (Bowen, 1961; Hemeida, 1993; Wojs, 1993). Despite this, Govier and Aziz (1982) recommend the use of the same function of relative roughness for time-independent fluids as for Newtonian systems. In any case, this will yield conservative estimates.

3.4.5 Velocity profiles in turbulent flow of power-law fluids

No exact mathematical analysis of turbulent flow has yet been developed for power-law fluids, though a number of semi-theoretical modifications of the expressions for the shear stress in Newtonian fluids at the walls of a pipe have been proposed.

   The shear stresses within the fluid are responsible for the frictional force at the walls and the velocity distribution over the cross-section of the pipe. A given assumption for the shear stress at the walls therefore implies some particular velocity distribution. In line with the traditional concepts that have proved the value for Newtonian fluids, the turbulent flow of power-law fluids in smooth pipes can be considered by dividing the flow into three zones, as shown schematically in Figure 3.15.

(i) *Laminar sub-layer*

This represents a thin layer $(0 \leq y \leq y_L)$ next to the pipe wall in which the effects of turbulence are assumed to be negligible. Assuming the no-slip boundary condition at the wall, $y = 0$, the fluid in contact with the surface is at rest. Furthermore, all the fluid close to the surface is moving at a very low velocity and therefore any changes in its momentum as it flows in the z direction must be extremely small. Consequently, the net shear force acting on any element of fluid in this zone must be negligible, the retarding force at its lower boundary being balanced by the accelerating force at its upper boundary. Thus, the shear stress in the fluid near the surface must approach a constant value which implies that the shear rate in this layer must also be constant or conversely, the velocity variation must be linear in the laminar sub-layer.

(ii) *Transition zone*

This region separates the so-called viscous or laminar sub-layer and the fully turbulent core prevailing in the middle portion of the pipe, and it extends over $y_L \leq y \leq y_T$, and finally,

(iii) *Turbulent core* $(y_T \leq y \leq R)$

A fully turbulent region comprising the bulk of the fluid stream where momentum transfer is attributable virtually entirely to random eddies and the effects of viscosity are negligible.

Figure 3.15   *Schematic representation of three zone model of turbulent flow*

A brief derivation of the turbulent velocity profile for Newtonian fluids in smooth pipes will first be presented and then extended to power-law fluids. The shear stress at any point in the fluid, at a distance y from the wall, is made of 'viscous' and 'turbulent' contributions, the magnitudes of which vary with distance from the wall. Expressing shear stress in terms of a dynamic viscosity and an eddy momentum diffusivity (or eddy kinematic viscosity), E,

$$\tau_{yz} = \left(\frac{\mu}{\rho} + E\right)\frac{d}{dy}(\rho V_z) \tag{3.84}$$

Prandtl postulated that E could be expressed as:

$$E = l^2 \left|\frac{dV_z}{dy}\right| \tag{3.85}$$

where the so-called 'mixing' length l (analogous to the mean free path of the molecules) is assumed to be directly proportional to the distance from the wall (i.e., $l = ky$). Now recalling that the shear stress varies linearly (equation (3.2)) with the radial distance r, and this can be rewritten in terms of the wall shear stress τ_w as:

$$\frac{\tau_{rz}}{\tau_w} = \frac{r}{R} \tag{3.2}$$

Now note that $r = R - y$ and $\tau_{rz} = \tau_{yz}$. Thus, equation (3.84) can be rewritten as:

$$\tau_w\left(\frac{R-y}{R}\right) = \left\{\frac{\mu}{\rho} + k^2 y^2 \frac{dV_z}{dy}\right\}\frac{d}{dy}(\rho V_z) \tag{3.86}$$

In the laminar sub-layer (small values of y), $E \simeq 0$ and:

$$\tau_w \simeq \frac{\mu}{\rho}\frac{d}{dy}(\rho V_z) \tag{3.87}$$

which upon integration with $V_z = 0$ at $y = 0$ to V_z at y yields the expected linear velocity profile (discussed earlier):

$$V_z = \frac{\tau_w}{\mu} y \tag{3.88}$$

It is customary to introduce the friction velocity $V^* = \sqrt{\tau_w/\rho}$ and to express equation (3.88) in dimensionless form:

$$V^+ = y^+ \tag{3.89}$$

where

$$V^+ = V_z/V^*, \quad y^+ = \frac{yV^*\rho}{\mu} \tag{3.90}$$

In the turbulent core, but yet close to the wall $y/R \ll 1$, (μ/ρ) is small compared with E. In addition, (dV_z/dy) will be positive close to the wall and therefore the modulus signs can be omitted in equation (3.85) and equation (3.86) becomes:

$$\tau_w \simeq \rho k^2 y^2 \left(\frac{dV_z}{dy} \right)^2 \tag{3.91}$$

Substituting for $V^* = \sqrt{(\tau_w/\rho)}$ and integration leads to:

$$V_z = \frac{V^*}{k} \ln y + B_0 \tag{3.92}$$

Now introducing the non-dimensional velocity and distance, V^+ and y^+,

$$V^+ = A \ln y^+ + B \tag{3.93}$$

where all constants have been absorbed in A and B. It might be expected that since equation (3.93) has been based on the approximation that $y/R \ll 1$, it should be valid only near the wall, but obviously for $y > y_T$. In fact, it has been found in pipe flow to correlate experimental data well over most of the turbulent core, except close to the centre of the pipe. Experimental values of $A = 2.5$ and $B = 5.5$ have been obtained from a very wide range of experimental data. B is a function of the relative roughness for rough pipes and A which is independent of roughness is readily seen to be equal to $1/k$.

No such simple analysis is possible in the transition zone nor is it possible analytically to delineate the transition boundaries for the three regions of flow. Based on experimental results, it is now generally believed that the laminar sub-layer extends up to $y^+ \approx 5$ and the turbulent core begins at $y^+ \approx 30$. The following empirical correlation provides an adequately approximate velocity distribution in the transition layer $(5 \le y^+ \le 30)$ in smooth pipes:

$$V^+ = 5 \ln y^+ - 3.05 \tag{3.94}$$

Equation (3.94) is a straight line on semi-logarithmic coordinates joining the laminar sub-layer values at $y^+ = 5$ and equation (3.93) for the turbulent core at $y^+ = 30$. Finally, it should be noted that equation (3.93) does not predict the expected zero velocity gradient at the centre of the pipe but this deficiency has little influence on the volumetric flow rate–pressure drop relationship.

Numerous attempts (Dodge and Metzner, 1959; Bogue and Metzner, 1963; Shenoy and Talathi, 1985; Wilson and Thomas, 1985; Shenoy, 1988) have been made at developing analogous expressions for velocity profiles for the steady turbulent flow of power-law fluids in smooth pipes; most workers have modified the definitions of y^+ and V^+, but Brodkey *et al.* (1961) used a polynomial approximation for the velocity profile. Figure 3.16 shows the velocity profiles derived on this basis for power-law fluids; the transition region, shown as dotted lines, is least understood.

The velocity distribution within the laminar sub-layer can be derived by the same reasoning as for a Newtonian fluid that the velocity varies linearly with distance from the wall (equation 3.88)

$$\frac{dV_z}{dy} = \frac{\tau_w}{\mu} = \frac{V_z}{y} \tag{3.88}$$

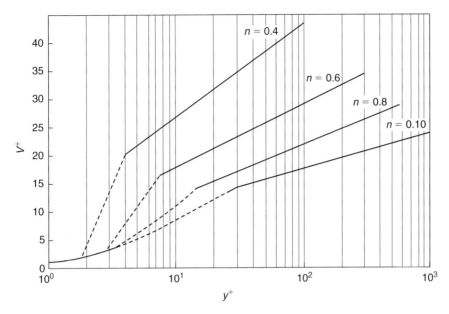

Figure 3.16 *Typical velocity profiles in the three zone representation of turbulent flow*

For a power-law fluid, the shear stress in this layer can be written as,

$$\tau_w \left(1 - \frac{y}{R}\right) = m \left(\frac{dV_z}{dy}\right)^n$$

or

$$\tau_w \simeq m \left(\frac{dV_z}{dy}\right)^n \tag{3.95}$$

which upon integration with the no-slip condition at the wall ($V_z = 0$ when $y = 0$) gives:

$$V_z = \left(\frac{\tau_w}{m}\right)^{1/n} y \tag{3.96}$$

Introducing the friction velocity, V^*, and rearrangement leads to:

$$V^+ = (y^+)^{1/n} \tag{3.97}$$

where

$$y^+ = \frac{\rho (V^*)^{2-n} y^n}{m} \tag{3.98}$$

It should be noted that for $n = 1$, equations (3.97) and (3.98) reduce to equations (3.89) and (3.90), respectively.

For the turbulent core, Dodge and Metzner (1959) used a similar approach to that given above for Newtonian fluids and proposed the following expression for power-law fluids:

$$
V^+ = \frac{5.66}{n^{0.75}} \log y^+ - \frac{0.566}{n^{1.2}} + \frac{3.475}{(n)^{0.75}}
$$
$$
\times \left\{ 1.96 + 0.815n - 1.628n \log\left(\frac{3n+1}{n}\right) \right\} \tag{3.99}
$$

It should be noted that in the limit of $n = 1$, equation (3.99) also reduces to equation (3.93) with $A = 2.47$ and $B = 5.7$; the slight discrepancy in the values of the constants arises from the fact that experimental Q–$(-\Delta p)$ data have been used to obtain the values of the constants rather than velocity measurements. The detailed derivation of equation (3.99) has been given by Skelland (1967). Subsequently, Bogue and Metzner (1963) used point velocity measurements to modify equation (3.99) to give:

$$
V^+ = \frac{5.57}{n} \log y^+ + C(y^*, f) + I(n, Re_{MR}) \tag{3.100}
$$

where

$$
C(y^*, f) = 0.05 \frac{\sqrt{2}}{f} \exp\left\{ \frac{-(y^* - 0.8)^2}{0.15} \right\}
$$

The friction factor, f, is calculated using equation (3.73) and the typical values of the function $I(n, Re_{MR})$ are presented in Table 3.2. Bogue and Metzner (1963) also noted that the velocity distributions for Newtonian and power-law fluids were virtually indistinguishable from each other if plotted in terms of (V_z/V) versus y/R instead of $V^+ - y^+$ coordinates, as shown in Figure 3.17. Finally, attention is drawn to the fact that both Dodge and Metzner (1959) and Bogue and Metzner (1963) have implicitly neglected the transition layer. Clapp (1961), on the other hand, has combined the Prandtl and von Karman approaches to put forward the following expressions for velocity distribution in the transition and turbulent zones:

$$
V^+ = \frac{5}{n} \ln y^+ - 3.05 \quad (5^n \le y^+ \le y_T^+) \tag{3.101}
$$

Table 3.2   *Values of* $I(n, Re_{MR})$ *in equation (3.100)*

n	Re_{MR}			
	5000	10000	50000	10^5
1	5.57	5.57	5.57	5.57
0.8	6.01	5.92	5.69	5.58
0.6	6.78	6.51	5.89	5.60
0.4	8.39	7.70	6.27	5.60

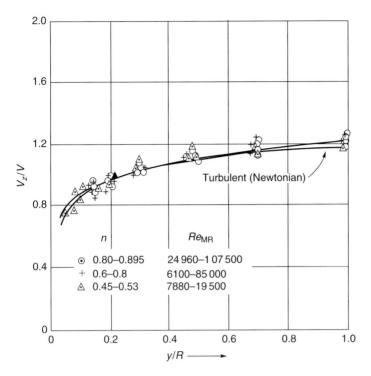

Figure 3.17   *Measured velocity profiles in circular pipes for turbulent flow of power-law fluids (data from Bogue and Metzner, 1963)*

$$V^+ = \frac{2.78}{n} \ln y^+ + \frac{3.8}{n} \quad (y^+ > y_T^+) \tag{3.102}$$

where y_T^+ is evaluated as the intersection point of equations (3.101) and (3.102). The laminar sub-layer region, equation (3.97), extends up to a value of 5^n. The numerical constants in these equations were evaluated using point velocity measurements in the range $0.7 \leq n \leq 0.81$. For $n = 1$, these equations yield $y_T^+ = 22$ which is somewhat lower than the generally accepted value of 30 for Newtonian fluids. This form of velocity profiles is also in line with the limited experimental observations (Park *et al.*, 1989; Peixinho *et al.*, 2005) and numerical predictions (Rudman *et al.*, 2004). Finally, all the above-mentioned velocity distributions fail to predict $(dV_z/dy) = 0$ at $y = R$.

Extensive experimental work indicates that the form of the turbulent velocity profiles established for non-elastic fluids without yield stress is very similar to that found for Newtonian fluids. This supports the contention, implicit in the discussion on friction factors, that it is the properties of the fluid in the wall region which are most important. This is particularly so with shear-thinning materials, for which the apparent viscosity in this region is lower than that in the bulk. While the velocity profiles are expected and observed to be symmetrical about the pipe axis in both laminar and turbulent flow conditions, asymmetric velocity distributions have been observed for shear-thinning and viscoplastic fluids in the transitional zone (Escudier *et al.*, 2005).

A potential and interesting application of this idea is to use pulsating flow in pipelines transporting shear-thinning and viscoplastic materials. The super-imposition of a small oscillating component on the bulk velocity has the effect of raising the average shear rate, thereby lowering the apparent viscosity. It appears advantageous to use some of the pumping energy in this way enabling the use of smaller pumps than would otherwise be required (Edwards *et al.*, 1972; Phan-Thien and Dudek, 1982a,b; Kajiuchi and Saito, 1984; Nakamura and Swada, 1987).

Before concluding this section, it is appropriate to mention briefly that some work is available on velocity distributions in turbulently flowing visco-elastic and drag-reducing polymer solutions in circular pipes. Qualitatively similar velocity distributions have been recorded for such systems as those mentioned in the preceding section for Newtonian and power-law fluids. However, the resulting equations tend to be more complex owing to the additional effects arising from visco-elasticity. Most of these studies have been critically reviewed by Shenoy and Talathi (1985) and Tam *et al.* (1992).

Aside from the foregoing discussion for the fully established flow, the problems involving transient flows of non-Newtonian fluids in circular pipes have also been investigated (Edwards *et al.*, 1972; Brown and Heywood, 1991).

One important area of application is the transportation of waxy crude oils. While the degree of thixotropy varies from one crude oil to another, it is generally believed that the time-dependent effects are significant only in the transient operations, especially during the restart of these pipelines following a period of shut down. Naturally, the values of their plastic viscosity, yield stress and the extent of time dependence are directly influenced by the degree of crystallization at low temperatures in sub-sea and arctic regions. Furthermore, the dissolved gases are also released due to the lowering of their solubility at low temperatures thereby imparting a degree of compressibility to crude oils under these conditions. Therefore, the transient (start up) flow in a pipeline filled with a waxy crude oil is governed by a large number of parameters including Reynolds number, Bingham number, length-to-diameter ratio of the pipe, compressibility number, the power-law index if the crude is also shear thinning, and the rate of wax appearance, etc. From a practical standpoint, one of the key questions is to be able to predict the value of the start-up pressure required to restart a frozen pipeline for given rheological and physical properties of the crude and pipe dimensions (diameter and length). A comprehensive model accounting for all these features is not yet available.

However, some approximate one-dimensional and pseudo-two dimensional flow models incorporating the effects of thixotropy and compressibility, with and without temperature dependent yield stress and plastic viscosity are available in the literature (Cawkwell and Charles, 1987; Sestak *et al.*, 1987; Vinay *et al.*, 2005, 2006, 2007). Broadly, it appears that the start-up of the flow in a pipe is facilitated even by the slight compressibility of the crude. On the other hand, it is, however, not easy to quantify the role of rheological parameters on the restart process. This is so in part due to the fact that the apparent yield stress due to thixotropy (at low temperatures) could be as high as 100 times the steady value of the yield stress of a thoroughly sheared sample. Also, the fact that the temperature will be lowest at the wall of the pipe coupled with a poor heat transfer in the radial direction adds to the difficulties of start-up of such pipelines. Suffice it to say here that it is not uncommon to use start up pressures of 5–10 times the steady state values of the pressure gradient.

Similarly, some guidelines for handling time-dependent thixotropic materials are also available in the literature (Govier and Aziz, 1982; Wardhaugh and Boger, 1987, 1991; Brown and Heywood, 1991).

3.5 Laminar flow between two infinite parallel plates

The steady flow of an incompressible power-law fluid between two parallel plates extending to infinity in x- and z-directions as shown schematically in Figure 3.18 will now be considered. The mid-plane between the plates will be taken as the origin with the flow domain extending from $y = -h$ to $y = +h$. The force balance on the fluid element ABCD situated at distance $\pm y$ from the mid-plane can be set up in a similar manner to that for flow through pipes.

$$p2Wy - (p + \Delta p)2Wy = \tau_{yz}\, 2WL \tag{3.103}$$

i.e.,

$$\tau_{yz} = \left(\frac{-\Delta p}{L}\right) y \tag{3.104}$$

The shear stress is thus seen to vary linearly, from zero at the mid-plane to a maximum value at the plate surface, as in the case of pipe flow. The system is symmetrical about the mid-plane ($y = 0$) and equation (3.104) needs to be solved only for $0 < y < h$. Because dV_z/dy is negative in this region, the shear stress for a power-law fluid is given by:

$$\tau_{yz} = m\left(-\frac{dV_z}{dy}\right)^n \tag{3.105}$$

Now combining equations (3.104) and (3.105) followed by an integration with respect to y yields for the velocity distribution:

$$V_z = -\left(\frac{n}{n+1}\right)\left\{\frac{1}{m}\left(\frac{-\Delta p}{L}\right)\right\}^{1/n} y^{(n+1)/n} + \text{constant} \tag{3.106}$$

At the walls of the channel (i.e., when $y = \pm h$), the velocity V_z must be zero to satisfy the condition of no-slip. Substituting the value $V_z = 0$, when $y = h$:

$$\text{constant} = \left(\frac{n}{n+1}\right)\left\{\frac{1}{m}\left(\frac{-\Delta p}{L}\right)\right\}^{1/n} h^{(n+1)/n} \tag{3.107}$$

and therefore:

$$V_z = \left(\frac{n}{n+1}\right)\left\{\frac{1}{m}\left(\frac{-\Delta p}{L}\right)\right\}^{1/n} h^{(n+1)/n}\left\{1 - \left(\frac{y}{h}\right)^{(n+1)/n}\right\} \tag{3.108}$$

Figure 3.18   *Laminar flow between parallel plates*

The velocity distribution is seen to be parabolic for a Newtonian fluid and becomes progressively blunter as the value of n decreases below unity, and sharper for shear-thickening fluids, similar to that seen in Figure 3.3 for laminar flow in a circular tube. The maximum velocity occurs at the centre-plane and its value is obtained by putting $y = 0$ in equation (3.108):

$$\text{Maximum velocity} = V_{max} = \left(\frac{n}{n+1}\right)h^{(n+1)/n}\left\{\frac{1}{m}\left(\frac{-\Delta p}{L}\right)\right\}^{1/n} \tag{3.109}$$

The total rate of flow of fluid between the plates is obtained by calculating the flow through two laminae of thickness dy and located at a distance y from the centre-plane and then integrating. Flow through laminae:

$$dQ = 2W\,dy\,V_z$$

Substituting for V_z from equation (3.108),

$$dQ = 2W\left(\frac{n}{n+1}\right)h^{(n+1)/n}\left\{\frac{1}{m}\left(\frac{-\Delta p}{L}\right)\right\}^{1/n}\left\{1-\left(\frac{y}{h}\right)^{(n+1)/n}\right\}dy \tag{3.110}$$

Then, on integrating between the limits of y from 0 to h:

$$Q = 2hW\left(\frac{n}{2n+1}\right)\left\{\frac{1}{m}\left(\frac{-\Delta p}{L}\right)\right\}^{1/n}h^{(n+1)/n} \tag{3.111}$$

The average velocity of the fluid:

$$V = \frac{Q}{2hW}$$
$$= \left(\frac{n}{2n+1}\right)\left\{\frac{1}{m}\left(\frac{-\Delta p}{L}\right)\right\}^{1/n}h^{(n+1)/n} \tag{3.112}$$

A similar procedure can, in principle, be used for other rheological models by inserting an appropriate expression for shear stress in equation (3.104).

The analogous result for the laminar flow of Bingham plastic fluids in this geometry is given here. The velocity distribution V_z in the fluid-like zone is given as:

$$V_z = \frac{h^2}{2\mu_B}\left(\frac{-\Delta p}{L}\right)\left(1-\frac{y^2}{h^2}\right)-\frac{\tau_0^B h}{\mu_B}\left(1-\frac{y}{h}\right) \tag{3.113}$$

The uniform plug velocity V_{zp} in the unsheared material is given as:

$$V_{zp} = \frac{h^2}{2\mu_B}\left(\frac{-\Delta p}{L}\right)(1-\phi)^2 \tag{3.114}$$

and finally, the volumetric flow rate, Q:

$$Q = \frac{2Wh^2}{3\mu_B}\left(\frac{-\Delta p}{L}\cdot h\right)\left\{1-\frac{3}{2}\phi+\frac{1}{2}\phi^3\right\} \tag{3.115}$$

where $\phi = \tau_0^B \Big/ \left(\dfrac{-\Delta p}{L} \cdot h \right) = \tau_0^B / \tau_w$

Example 3.11

Calculate the volumetric flow rate per unit width at which a 0.5% polyacrylamide solution will flow down on a wide inclined surface (30° from horizontal) as a 3 mm thick film. The shear stress–shear rate behaviour of this polymer solution may be approximated by the Ellis fluid model, with the following values of the model parameters: $\mu_0 = 9\,\text{Pa s}$; $\tau_{\frac{1}{2}} = 1.32\,\text{Pa}$; $\alpha = 3.22$ and the solution has a density of $1000\,\text{kg/m}^3$. Assume the flow to be laminar.

Solution

A general equation will be derived first for the flow configuration shown in Figure 3.19 by writing a force balance on a differential element of the fluid.

In a liquid flowing down a surface (of width W), a velocity distribution will be established with the velocity increasing from zero at the surface itself ($y = 0$) to a maximum at the free surface ($y = H$). For viscoplastic fluids, it can be expected that plug-like motion may occur near the free surface. The velocity distribution in the film can be obtained in a manner similar to that used previously for pipe flow, bearing in mind that the driving force here is that due to gravity rather than a pressure gradient, which is absent everywhere in the film.

In an element of fluid of length dz, the gravitational force acting on that part of the liquid which is at a distance greater than y from the surface is given by:

$$(H - y)W\,dz\rho g \sin \beta$$

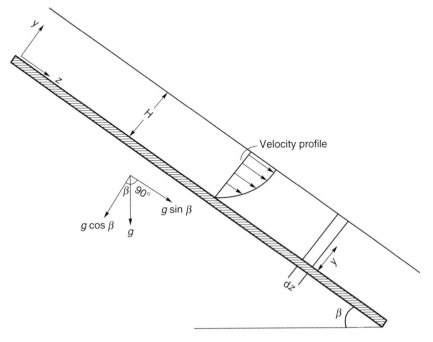

Figure 3.19   *Schematics of flow on an inclined plate*

If the drag force at the free surface is negligible, the retarding force for flow will be attributable to the shear stress prevailing in the liquid at the distance y from the surface and this will be given by:

$$\tau_{yz} W \, dz$$

At equilibrium therefore:

$$\tau_{yz} W \, dz = (H - y)W \, dz \, \rho g \sin \beta$$

or $$\tau_{yz} = \rho g (H - y) \sin \beta \tag{3.116}$$

The shear stress is seen to vary linearly from a maximum value at the solid surface to zero at the free surface.

Since dV_z/dy is positive here, the shear stress for an Ellis model fluid is given by:

$$\tau_{yz} = \frac{\mu_0 \left(\dfrac{dV_z}{dy} \right)}{1 + \left(\dfrac{\tau_{yz}}{\tau_{1/2}} \right)^{\alpha - 1}}$$

or

$$\mu_0 \frac{dV_z}{dy} = \tau_{yz} \left\{ 1 + \left(\frac{\tau_{yz}}{\tau_{1/2}} \right)^{\alpha - 1} \right\} \tag{3.117}$$

Substitution of equation (3.116) into equation (3.117), followed by an integration with respect to y and using the no-slip boundary condition at the solid surface ($V_z = 0$, when $y = 0$) yields:

$$V_z = \frac{\rho g \sin \beta}{\mu_0} \left(Hy - \frac{y^2}{2} \right) + \frac{(\rho g \sin \beta)^{\alpha} H^{\alpha + 1}}{\mu_0 (\alpha + 1) \tau_{1/2}^{\alpha - 1}} \left\{ 1 - \left(1 - \frac{y}{H} \right)^{\alpha + 1} \right\} \tag{3.118}$$

The volumetric flow rate of liquid down the surface can now be calculated as:

$$Q = W \int_0^H V_z \, dy \tag{3.119}$$

Substituting equation (3.118) in equation (3.119) and integrating:

$$Q = \frac{\rho g H^3 W \sin \beta}{3 \mu_0} + \frac{W (\rho g \sin \beta)^{\alpha} H^{\alpha + 2}}{\mu_0 (\alpha + 2)(\tau_{1/2})^{\alpha - 1}} \tag{3.120}$$

For a Newtonian fluid, $\tau_{1/2} \to \infty$, and both equations (3.118) and (3.120) reduce to the corresponding Newtonian expressions.

The maximum velocity occurs at the free surface, and its value is obtained by putting $y = H$ in equation (3.118):

$$\text{Maximum velocity} = V_{\max} = \frac{\rho g \sin \beta H^2}{2 \mu_0} + \frac{(\rho g \sin \beta)^{\alpha} H^{\alpha + 1}}{\mu_0 (\alpha + 1)(\tau_{1/2})^{\alpha - 1}} \tag{3.121}$$

For a vertical surface, $\sin \beta = 1$.

For the numerical example, $\beta = 30°$, i.e., $\sin 30 = 1/2$; $H = 3\,\text{mm}$ and substituting the other values in equation (3.120):

$$\frac{Q}{W} = \frac{1000 \times 9.81 \times (3 \times 10^{-3})^3 \left(\frac{1}{2}\right)}{3 \times 9}$$

$$+ \frac{\left(1000 \times 9.81 \times \frac{1}{2}\right)^{3.22} (3 \times 10^{-3})^{5.22}}{9(3.22 + 2)(1.32)^{3.22-1}}$$

$$= 4.91 \times 10^{-6} + 5.95 \times 10^{-4}$$

$$= 6 \times 10^{-4} \text{ m}^3/\text{s per metre width.} \quad \square$$

3.6 Laminar flow in a concentric annulus

The flow of non-Newtonian fluids through concentric and eccentric annuli represents an idealization of several industrially important processes. One important example is in oil well drilling where a heavy drilling mud is circulated through the annular space around the drill pipe in order to carry the drilling debris to the surface. These drilling muds are typically either Bingham plastic or power-law type fluids. Other examples include the extrusion of plastic tubes and pipes in which the molten polymer is forced through an annular die, and the flow in double-pipe heat exchangers. In all these applications, it is often required to predict the frictional pressure gradient to sustain a fixed flow rate or vice versa. In this section, the isothermal, steady and fully developed flow of power-law and Bingham plastic fluids in concentric annulus is analysed and appropriate expressions and/or charts are presented, which permit the calculation of pressure gradient for a given application.

The calculation of the velocity distribution and the mean velocity of a fluid flowing through an annulus of outer radius R and inner radius σR is more complex than that for flow in a pipe or between two parallel planes (Figure 3.20), though the force balance on an element of fluid can be written in a manner similar to that has been used in previous sections. If the pressure changes by an amount Δp as a consequence of friction over a length L of annulus, the resulting force can be equated to the shearing force acting on the fluid.

Consider the flow of the fluid situated at a distance not greater than r from the centre-line of the pipe. The shear force acting on this fluid comprises two parts: one is the drag on its outer surface ($r = R$) which can be expressed in terms of the shear stress in the fluid at that location; the other contribution is the drag occurring at the inner (solid) boundary of the annulus, i.e., at $r = \sigma R$. This component cannot be estimated at present, however. Alternatively, this difficulty can be obviated by considering the equilibrium of a thin ring of fluid of radius r and thickness dr (Figure 3.20). The pressure force acting on this fluid element is:

$$2\pi r \, dr\{p - (p + \Delta p)\}$$

The only other force acting on the fluid element in the z-direction is that arising from the shearing on both surfaces of the element. Note that, not only will the shear stress change from r to $r + dr$ but the surface area over which shearing occurs will also depend upon the value of r. The net force can be written as:

$$(2\pi r L \cdot \tau_{rz}|_{r+dr} - 2\pi r L \tau_{rz}|_r)$$

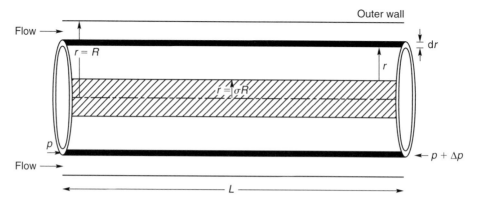

Figure 3.20   *Flow in a concentric annulus*

At equilibrium therefore:

$$2\pi r dr \left(\frac{-\Delta p}{L}\right) L = 2\pi L \{r\tau_{rz}|_{r+dr} - r\tau_{rz}|_r\}$$

or

$$\frac{r\tau_{rz}|_{r+dr} - r\tau_{rz}|_r}{dr} = r\left(\frac{-\Delta p}{L}\right)$$

Now taking limits as $dr \to 0$, it becomes

$$\frac{d}{dr}(r\tau_{rz}) = r\left(\frac{-\Delta p}{L}\right) \tag{3.122}$$

The shear stress distribution across the gap is obtained by integration:

$$\tau_{rz} = \frac{r}{2}\left(\frac{-\Delta p}{L}\right) + \frac{C_1}{r} \tag{3.123}$$

where C_1 is a constant of integration.

Because of the no-slip boundary condition at both solid walls, i.e., at $r = \sigma R$ and $r = R$, the velocity must be maximum at some intermediate point, say at $r = \lambda R$. Then, for a fluid without a yield stress, the shear stress must be zero at this position and for a viscoplastic fluid, there will be a plug moving *en masse*. Thus, the constant of integration, C_1, can be evaluated using the condition of $\tau_{rz} = 0$ at $r = \lambda R$, and hence,

$$C_1 = -\frac{\lambda^2 R^2}{2}\left(-\frac{\Delta p}{L}\right) \tag{3.124}$$

and this allows equation (3.123) to be rewritten as:

$$\tau_{rz} = \left(\frac{-\Delta p}{L}\right)\frac{R}{2}\left(\xi - \frac{\lambda^2}{\xi}\right) \tag{3.125}$$

where $\xi = r/R$, the dimensionless radial coordinate.

3.6.1 Power-law fluids

For this flow, since the velocity gradient changes sign at $r = \lambda R$, the power-law fluid can be written as:

$$\tau_{rz} = -m \left| \frac{dV_z}{dr} \right|^{n-1} \left(\frac{dV_z}{dr} \right) \tag{3.126}$$

It is important to write the equation in this form whenever the sign of the velocity gradient changes within the flow field. In this case, (dV_z/dr) is positive for $\sigma \leq \xi \leq \lambda$ and negative for $\lambda \leq \xi \leq 1$. Now equation (3.126) can be substituted in equation (3.125) and integrated to obtain

$$V_{z_i} = R \left(\frac{-\Delta p}{L} \cdot \frac{R}{2m} \right)^{1/n} \int_{\sigma}^{\xi} \left(\frac{\lambda^2}{x} - x \right)^{1/n} dx \tag{3.127a}$$
$$\sigma \leq \xi \leq \lambda$$

$$V_{z_o} = R \left(\frac{-\Delta p}{L} \cdot \frac{R}{2m} \right)^{1/n} \int_{\xi}^{1} \left(x - \frac{\lambda^2}{x} \right)^{1/n} dx \tag{3.127b}$$
$$\lambda \leq \xi \leq 1$$

where subscripts 'i' and 'o' denote the inner ($\sigma \leq \xi \leq \lambda$) and outer ($\lambda \leq \xi \leq 1$) regions, respectively, and x is a dummy variable of integration. The no-slip boundary conditions at $\xi = \sigma$ and $\xi = 1$ have been incorporated in equation (3.127). Clearly, the value of λ is evaluated by setting $V_{z_i} = V_{z_o}$ at $\xi = \lambda$, i.e.,

$$\int_{\sigma}^{\lambda} \left(\frac{\lambda^2}{x} - x \right)^{1/n} dx = \int_{\lambda}^{1} \left(x - \frac{\lambda^2}{x} \right)^{1/n} dx \tag{3.128}$$

The volumetric flow rate of the fluid, Q, is obtained as:

$$Q = 2\pi \int_{\sigma R}^{R} rV_z \, dr = 2\pi R^2 \int_{\sigma}^{1} \xi V_z \, d\xi$$
$$= 2\pi R^3 \left(\frac{-\Delta p}{L} \cdot \frac{R}{2m} \right)^{1/n} \left[\int_{\sigma}^{\lambda} \xi V_{z_i} \, d\xi + \int_{\lambda}^{1} \xi V_{z_o} \, d\xi \right] \tag{3.129}$$

Clearly, equations (3.128) and (3.129) must be solved and integrated simultaneously to eliminate λ and to evaluate the volumetric rate of flow of liquid, Q. Analytical solutions are possible only for integral values of $(1/n)$, i.e., for $n = 1, 0.5, 0.33, 0.25$, etc. Thus, Fredrickson and Bird (1958) evaluated the integral in equation (3.129) for such values of n and, by interpolating the results for the intermediate values of power-law index, they presented a chart relating non-dimensional flow rate, pressure drop, σ and n. However, the accuracy of their results deteriorates rapidly with the decreasing values of n and/or $(1 - \sigma) \ll 1$, i.e., with narrowing annular region. Subsequently, Hanks and Larsen (1979) were able to evaluate the volumetric flow rate, Q, analytically and their final expression is:

$$Q = \frac{n\pi R^3}{(3n + 1)} \left(\frac{-\Delta p}{L} \frac{R}{2m} \right)^{1/n} \{ (1 - \lambda^2)^{(n+1)/n} - \sigma^{(n-1)/n} (\lambda^2 - \sigma^2)^{(n+1)/n} \} \tag{3.130}$$

Table 3.3   *Values of λ Computed from equation (3.128) (Hanks and Larson, 1979)*

n	Value of σ								
	0.10	0.20	0.30	0.40	0.50	0.60	0.70	0.80	0.90
0.10	0.3442	0.4687	0.5632	0.6431	0.7140	0.7788	0.8389	0.8954	0.9489
0.20	0.3682	0.4856	0.5749	0.6509	0.7191	0.7818	0.8404	0.8960	0.9491
0.30	0.3884	0.4991	0.5840	0.6570	0.7229	0.7840	0.8416	0.8965	0.9492
0.40	0.4052	0.5100	0.5912	0.6617	0.7259	0.7858	0.8426	0.8969	0.9493
0.50	0.4193	0.5189	0.5970	0.6655	0.7283	0.7872	0.8433	0.8972	0.9493
0.60	0.4312	0.5262	0.6018	0.6686	0.7303	0.7884	0.8439	0.8975	0.9494
0.70	0.4412	0.5324	0.6059	0.6713	0.7319	0.7893	0.8444	0.8977	0.9495
0.80	0.4498	0.5377	0.6093	0.6735	0.7333	0.7902	0.8449	0.8979	0.9495
0.90	0.4572	0.5422	0.6122	0.6754	0.7345	0.7909	0.8452	0.8980	0.9495
1.00	0.4637	0.5461	0.6147	0.6770	0.7355	0.7915	0.8455	0.8981	0.9496

The only unknown now remaining is λ, which locates the position where the velocity is maximum. Table 3.3 presents the values of λ for a range of values of σ and n which have been obtained by solving equation (3.128) numerically.

Example 3.12

A polymer solution exhibits power-law behaviour with $n = 0.5$ and $m = 3.2\,\mathrm{Pa\,s}^{0.5}$. Estimate the pressure gradient required to maintain a steady flow of $0.3\,\mathrm{m^3/min}$ of this polymer solution through the annulus between a 10 mm and a 20 mm diameter tube.

Solution

Here,

$$R = \frac{20}{2} \times 10^{-3} = 0.01\,\mathrm{m}$$

$$\sigma R = \frac{10}{2} \times 10^{-3} = 0.005\,\mathrm{m}$$

or

$$\sigma = 0.5$$

From Table 3.3, for $\sigma = 0.5$ and $n = 0.5$, $\lambda = 0.728$.
Substituting these values in equation (3.130)

$$\left(\frac{0.3}{60}\right) = \frac{(0.5)(3.14)(0.01)^3}{(3 \times 0.5 + 1)}\left(\frac{-\Delta p}{L}\right)^2\left(\frac{0.01}{2 \times 3.2}\right)^2$$
$$\times \{(1 - 0.728^2)^{(0.5+1)/0.5} - 0.5^{(0.5-1)/0.5}(0.728^2 - 0.5^2)^{(0.5+1)/0.5}\}$$

and solving:

$$\frac{-\Delta p}{L} = 169\,\mathrm{kPa/m}\;\square$$

3.6.2 Bingham plastic fluids

The laminar axial flow of Bingham plastic fluids through a concentric annulus has generated even more interest than that for the power-law fluids, e.g., see Laird (1957),

Figure 3.21 *Schematics of velocity profiles for Bingham plastic and power-law fluids in a concentric annulus*

Fredrickson and Bird (1958), Bird *et al.* (1983) and Fordham *et al.* (1991). The main feature which distinguishes the flow of a Bingham plastic fluid from that of a power-law fluid is the existence of a plug region in which the shear stress is less than the yield stress. Figure 3.21 shows qualitatively the salient features of the velocity distribution in an annulus; the corresponding profile for a fluid without a yield stress (e.g., power-law fluid) is also shown for the sake of comparison.

In principle, the velocity distribution and the mean velocity of a Bingham plastic fluid flowing through an annulus can be deduced by substituting for the shear stress in equation (3.125) in terms of the Bingham plastic model, equation (3.19). However, the signs of the shear stress (considered positive in the same sense as the flow) and the velocity gradients in the two flow regions need to be treated with special care. With reference to the sketch shown in Figure 3.21, the shearing force on the fluid is positive ($\sigma R \leq r \leq \lambda_- R$) where the velocity gradient is also positive. Thus, in this region:

$$\tau_{rz} = \tau_0^B + \mu_B \left(\frac{dV_z}{dr} \right) \tag{3.131}$$

On the other hand, in the region $\lambda_+ R \leq r \leq R$, the velocity gradient is negative and the shearing force is also in the negative r-direction and hence,

$$-\tau_{rz} = \tau_0^B + \mu_B \left(-\frac{dV_z}{dr} \right) \tag{3.132}$$

Equations (3.131) and (3.132) can now be substituted in equation (3.125) and integrated to deduce the velocity distributions. The constants of integration can be evaluated by using the no-slip boundary condition at both $r = \sigma R$ and $r = R$. However, the boundaries of the plug existing in the middle of the annulus are not yet known; nor is the plug velocity known. The unknown boundaries are evaluated by applying the following three conditions, namely, the continuity of velocity at $r = \lambda_- R$ and $r = \lambda_+ R$, the velocity gradient is also zero at these boundaries and finally, the force balance on the plug of fluid is:

$$2\pi R(\lambda_+ + \lambda_-)\tau_0^B = \left(\frac{-\Delta p}{L} \right) \cdot \pi \left\{ (\lambda_+ R)^2 - (\lambda_- R)^2 \right\} \tag{3.133}$$

Unfortunately, the algebraic steps required to carry out the necessary integrations and the evaluation of the constants are quite involved and tedious. Thus, these are not presented here and readers are referred to the original papers (Laird, 1957) or to the book by Skelland (1967) for detailed derivations. Instead consideration is given here to the practical problem of estimating the necessary pressure gradient to maintain a fixed flow rate of a Bingham plastic fluid or vice versa. Fredrickson and Bird (1958) organized their numerical solutions of the equations presented above in terms of the following dimensionless parameters:

$$\text{Dimensionless velocity: } V_z^* = \frac{2\mu_B V_z}{R^2 \left(\frac{-\Delta p}{L}\right)} \tag{3.134}$$

$$\text{Dimensionless yield stress: } \phi_0 = \frac{2\tau_0^B}{R\left(\frac{-\Delta p}{L}\right)} \tag{3.135}$$

$$\text{Dimensionless flowrate: } \Omega = \frac{Q}{Q_N} \tag{3.136}$$

where Q_N is the flow rate of a Newtonian liquid of viscosity, μ_B. Thus,

$$Q_N = \frac{\pi R^4}{8\mu_B}\left(\frac{-\Delta p}{L}\right) \tag{3.137}$$

Fredrickson and Bird (1958) presented three charts (Figures 3.22–3.24) showing relationships between V_z^*, ϕ_0, Ω, σ and λ_+.

For given values of the rheological constants (μ_B, τ_0^B), pressure gradient ($-\Delta p/L$) and the dimensions of the annulus (σ, R), the values of λ_+ and the plug velocity V_{zp}^* can be read from Figure 3.22 and the value of Ω from Figure 3.23 from which the volumetric rate of flow, Q, can be estimated. For the reverse calculation, the group (Ω/ϕ_0) is independent of the pressure gradient and one must use Figure 3.24 to obtain the value of ϕ_0 and thus evaluate the required pressure gradient ($-\Delta p/L$).

Example 3.13

A molten chocolate (density = 1500 kg/m³) flows through a concentric annulus of inner and outer radii 10 and 20 mm, respectively, at 30°C at the constant flow rate of 0.03 m³/min. The steady-shear behaviour of the chocolate can be approximated by a Bingham plastic model with $\tau_0^B = 35$ Pa and $\mu_B = 1$ Pa s.

(a) Estimate the required pressure gradient to maintain the flow, and determine the velocity and the size of the plug.
(b) Owing to a pump malfunction, the available pressure gradient drops by 25% of the value calculated in (a), what will be the new flow rate?

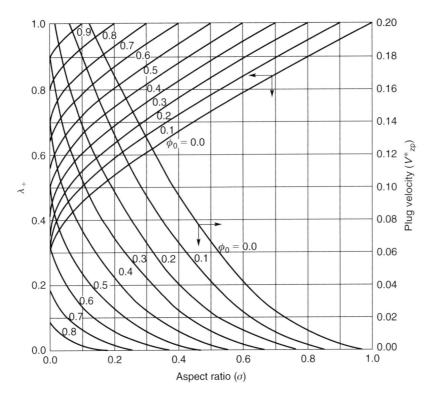

Figure 3.22 *Dimensionless plug velocity and plug size for laminar Bingham plastic flow in an annulus*

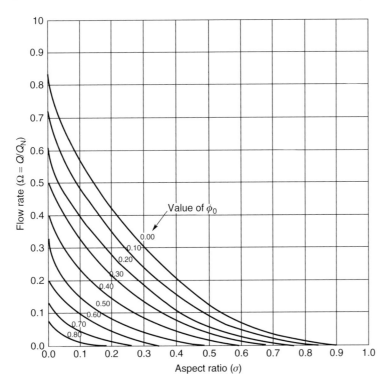

Figure 3.23 *Dimensionless flow rate for Bingham plastic fluids in laminar flow through an annulus*

Figure 3.24   *Chart for the estimation of pressure gradient for laminar flow of Bingham plastic fluids in an annulus*

Solution

Part (a):

In this case, $\tau_0^B = 35$ Pa, $\mu_B = 1$ Pa s

$$Q = 0.03\,\mathrm{m^3/min} = \frac{0.03}{60}\,\mathrm{m^3/s}$$

$$\sigma = \frac{10}{20} = 0.5;\ R = 20 \times 10^{-3}\,\mathrm{m}$$

Since the pressure gradient $(-\Delta p/L)$ is unknown, one must use Figure 3.24.

$$\therefore\ \frac{\Omega}{\phi_0} = \frac{4\mu_B Q}{\pi R^3 \tau_0^B} = \frac{4 \times 1 \times (0.03/60)}{3.14 \times (20 \times 10^{-3})^3 \times 35} = 2.28$$

For $\Omega/\phi_0 = 2.28$ and $\sigma = 0.5$, Figure 3.24 gives:

$$\phi_0 = {\sim}0.048$$

$$\therefore\ \left(\frac{-\Delta p}{L}\right) = \frac{2\tau_0^B}{R\phi_0} = \frac{2 \times 35}{20 \times 10^{-3} \times 0.048}$$

$$= 73\,000\,\mathrm{Pa/m}$$

$$= 73\,\mathrm{kPa/m}$$

Now from Figure 3.22, $\phi_0 = 0.048$ and $\sigma = 0.5$,

$$V_{zp}^* = \sim 0.05$$
$$\lambda_+ = 0.76$$

From the definition of V_z^* (equation 3.134), for plug velocity, we have:

$$V_{zp}^* = \frac{2\mu_B V_{zp}}{R^2\left(\dfrac{-\Delta p}{L}\right)}$$

$$0.05 = \frac{2 \times 1 \times V_{zp}}{(20 \times 10^{-3})^2(73\,000)}$$

$$\therefore V_{zp} = 0.73\,\text{m/s}$$

i.e., the plug in the central region has a velocity of 0.73 m/s (compared with the mean velocity of $Q/\pi R^2 (1 - \sigma^2)$, i.e., 0.53 m/s).

From equation (3.133):

$$(\lambda_+ - \lambda_-)\frac{R}{2}\left(\frac{-\Delta p}{L}\right) = \tau_0^B$$

Substitution of values of λ_+, R, τ_0^B and $(-\Delta p/L)$ gives $\lambda_- = 0.71$. Thus the plug region extends from $\lambda_- R$ to $\lambda_+ R$, i.e., from 14.2 to 15.2 mm. These calculations assume the flow to be laminar. As a first approximation, one can define the corresponding Reynolds number based on the hydraulic diameter, D_h.

$$D_h = \frac{4 \times \text{flow area}}{\text{wetted perimeter}} = \frac{4\pi R^2(1 - \sigma^2)}{2\pi R(1 + \sigma)} = 2R(1 - \sigma)$$

$$= 2 \times 20 \times 10^{-3}(1 - 0.5) = 0.02\,\text{m}$$

$$\text{Reynolds number, } Re = \frac{\rho V D_h}{\mu_B} = \frac{1500 \times 0.53 \times 0.02}{1} = 16$$

The flow is thus likely to be streamline.

Part (b):

In this case, the available pressure gradient is only 75% of the value calculated above,

$$\frac{-\Delta p}{L} = 73 \times 0.75 = 54.75\,\text{kPa/m}$$

We can now evaluate ϕ_0:

$$\phi_0 = \frac{2\tau_0^B}{R\left(\dfrac{-\Delta p}{L}\right)} = \frac{2 \times 35}{20 \times 10^{-3} \times 54.75 \times 1000} = 0.064$$

From Figure 3.23, for $\phi_0 = 0.064$ and $\sigma = 0.5$,

$$\Omega = \frac{Q}{Q_N} = \sim 0.1$$

$$\therefore Q = 0.1 \times Q_N = 0.1 \times \frac{\pi R^4}{8\mu_B}\left(\frac{-\Delta p}{L}\right)$$

$$= \frac{0.1 \times 3.14 \times (20 \times 10^{-3})^4}{8 \times 1} \times 54.75 \times 1000\,\text{m}^3/\text{s}$$

$$= 0.000344\,\text{m}^3/\text{s} \quad \text{or} \quad 0.0206\,\text{m}^3/\text{min}$$

Two observations can be made here. The 25% reduction in the available pressure gradient has lowered the flow rate by 31%. Secondly, in this case the flow rate is only one-tenth of that of a Newtonian fluid of the same viscosity as the plastic viscosity of the molten chocolate! □

3.6.3 Herschel–Bulkley Fluids

Many viscoplastic fluids display shear-thinning characteristics at the shear stress values greater than the yield stress. Typical examples include kaolin slurries, laponite and bentonite suspensions and many mine tailings. In principle, one can follow exactly an identical reasoning as that for the Bingham plastic fluids in the preceding section. However, owing to the non-linear dependence of shear stress on the rate of shear for stress levels above the yield stress, an analytical treatment leading to closed form expressions for the volumetric flow rate Q, plug velocity V_p, etc. are not possible in this case. However, Hanks (1979) has shown that it is possible to develop a series of graphical charts which can be used to carry out process design calculations such as to predict the pressure drop for a given flow system, or, conversely to estimate the volumetric flow rate for a given pipe–pump combination, for the laminar flow of Herschel–Bulkley fluids in a concentric annulus. By dimensional considerations, it can readily be shown that, for a fully developed and steady flow, the dimensionless pressure drop (friction factor) is a function of the Reynolds number, a Bingham number, power-law index and the aspect ratio of the annulus. By a judicious regrouping of these parameters, Hanks (1979) found that it was possible to express these results by using the following dimensionless groups:

- Dimensionless yield stress: $\phi_0 = \dfrac{\tau_0^H}{\tau_w}$ (3.138)

- Dimensionless flow rate: $\Omega = \left(\dfrac{m}{\tau_0^H}\right)^{1/n}\left(\dfrac{Q}{\pi R^3}\right)$ (3.139)

- Dimensionless pressure gradient: $P^+ = \left(\dfrac{Q}{\pi R^3}\right)\left\{\left[\left(\dfrac{-\Delta p}{L}\right)\left(\dfrac{R}{2m}\right)\right]\right\}^{-1/n}$ (3.140)

which can also be expressed in terms of the yield stress as:

$$P^+ = \left(\dfrac{Q}{\pi R^3}\right)\left\{\phi_0\left(\dfrac{m}{\tau_0^H}\right)\right\}^{1/n}$$ (3.141)

Figures 3.25 to 3.30 show the plots of Ω (ϕ_0, σ, n) and $P^+(\phi_0, \sigma, n)$, respectively. Thus, for a given material (known ρ, m, n, τ_0^H) and pipe dimensions (σ, R, L), it is possible to use Figures 3.28–3.30 to calculate the volumetric flow rate for a known pressure gradient. On the other hand, Figures 3.25–3.27 must be used to estimate the unknown pressure gradient for a given flow rate. The utility of these plots is illustrated in Example 3.14.

Example 3.14

A clay slurry ($\rho = 1500\,\text{kg/m}^3$) is to be pumped through the annular space between a 289 mm (inside diameter) of outer pipe and of 219 mm (outer diameter) of inner pipe. The line is 150 m long

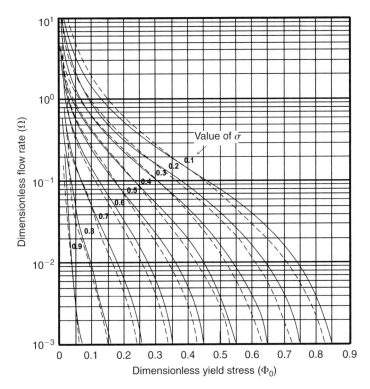

Figure 3.25 *Dimensionless flow rate – yield stress plots, (———) n = 1; (-----) n = 0.8 (replotted from Hanks, 1979)*

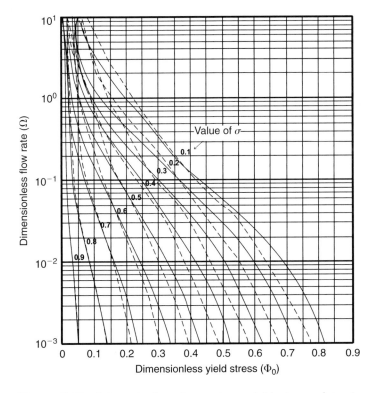

Figure 3.26 *Dimensionless flow rate – yield stress plots, (———) n = 0.7; (-----) n = 0.5 (replotted from Hanks, 1979)*

Figure 3.27 *Dimensionless flow rate – yield stress plots, (———) n = 0.4; (-----) n = 0.2 (replotted from Hanks, 1979)*

Figure 3.28 *Dimensionless pressure drop – yield stress plots, (———) n = 1; (-----) n = 0.7 (replotted from Hanks, 1979)*

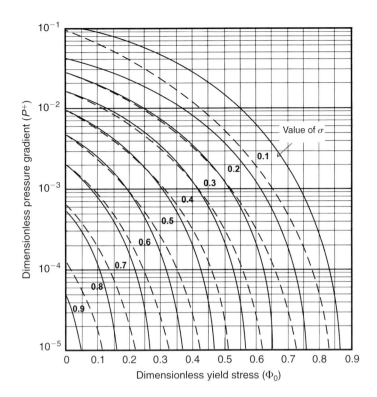

Figure 3.29 *Dimensionless pressure drop – yield stress plots, (———) n = 0.6; (-----) n = 0.4 (replotted from Hanks, 1979)*

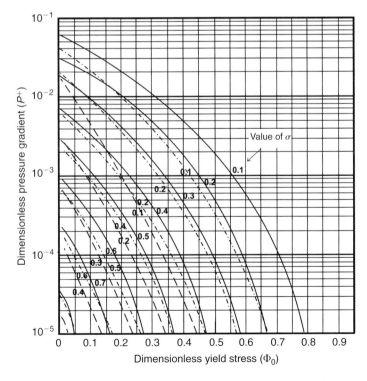

Figure 3.30 *Dimensionless pressure drop – yield stress plots, (———) n = 0.3; (– ·– ·) n = 0.2; (-----) n = 0.1 (replotted from Hanks, 1979)*

and a pump capable of producing a head of 345 kPa is available. The slurry behaves as a Herschel–Bulkley model fluid with the values of constants: $n = 0.5$, $\tau_0^H = 17$ Pa and $m = 0.83$ Pa·s$^{0.5}$. Estimate the volumetric flow rate which can be achieved with this pump.

Due to an increased demand, it is necessary to increase the volumetric flow rate by 20% through this installation, what will be the new value of the pressure drop?

Solution

From the two pipe diameters,

$$\sigma = \frac{219}{289} = 0.758$$

Since the available pressure drop is known in this case (and hence $\tau_w = (D/4)(-\Delta p/L)$), we can calculate the value of ϕ_o using equation (3.138) as:

$$\phi_0 = \frac{2 \times 17}{(289/2) \times 10^{-3} \times (345 \times 10^3/150)} = 0.102$$

Therefore, for $\sigma = 0.758$, $n = 0.5$ and $\phi_0 = 0.102$, Figure 3.29 gives $P^+ = \sim 1.6 \times 10^{-4}$ and from equation (3.141), the volumetric flow rate Q is given as:

$$Q = P^+(\pi R^3)\left(\phi_0 m/\tau_0^H\right)^{-1/n}$$

Substituting values,

$$Q = 1.6 \times 10^{-4} \times 3.14 \times \left(\frac{289}{2} \times 10^{-3}\right)^3 \left(\frac{0.102 \times 0.83}{17}\right)^{-1/0.5} = 0.0611 \, \text{m}^3/\text{s}$$

Therefore, the maximum achievable flow rate with the existing pump is 0.0611 m³/s, or 220 m³/h. Now, with a 20% increase in the flow rate, the new flow rate is:

$$Q_{new} = 1.2Q = 1.2 \times 0.0611 = 0.0733 \, \text{m}^3/\text{s}$$

In this case, the pressure drop is not known and therefore, one can now calculate the value of Ω using equation (3.139):

$$\Omega = \left(\frac{0.83}{17}\right)^{1/0.5}\left(\frac{0.0733}{3.14 \times \left(\frac{289}{2} \times 10^{-3}\right)^3}\right) = 0.0184$$

Using Figure 3.26 for $\Omega = 0.0184$, $\sigma = 0.758$ and $n = 0.5$, one gets

$$\phi_0 = \sim 0.095$$

and therefore, the new value of the pressure drop is given by equation (3.138) as:

$$(-\Delta p) = \frac{2\tau_0^H L}{R\phi_0} = \frac{2 \times 17 \times 150}{\left(\frac{289}{2} \times 10^{-3}\right) \times 0.095} = 392 \times 10^3 \, \text{Pa} = 392 \text{kPa},$$

i.e., only an increase of 8% over the previous value.

For a power-law fluid, $\Delta p \propto Q^{0.5}$ and therefore a 20% increase in the flow rate will cause only about 10% increase in pressure drop, as seen above. □

This section is concluded by noting that analogous treatments for the concentric and eccentric annular flow of fluids described by many other viscosity models are also available in the literature (Uner *et al.*, 1988; Fordham *et al.*, 1991; Walton and Bittleston, 1991; Gücüyener and Mehmeteoglu, 1992; Nouri *et al.*, 1993; Escudier *et al.*, 1995, 2002a,b).

3.7 Laminar flow of inelastic fluids in non-circular ducts

Analytical solutions for the laminar flow of time-independent fluids in non-axisymmetric conduits are not possible. Numerous workers have obtained approximate and/or complete numerical solutions for specific flow geometries including square, rectangular and triangular pipes (Schechter, 1961; Wheeler and Wissler, 1965; Miller, 1972; Mitsuishi and Aoyagi, 1969, 1973). On the other hand, semi-empirical attempts have also been made to develop methods for predicting pressure drop for time-independent fluids in ducts of non-circular cross-section. Perhaps the most systematic and successful friction factor analysis is that provided by Kozicki *et al.* (1966, 1967). It is useful to recall here that the equation (3.52) is a generalized equation for the laminar flow of time-independent fluids in a tube and it can be slightly rearranged as:

$$\dot{\gamma}_{\mathrm{w}} = \left(-\frac{\mathrm{d}V_z}{\mathrm{d}r}\right)_{\mathrm{w}} = \mathrm{f}(\tau_{\mathrm{w}}) = \frac{1}{4}\tau_{\mathrm{w}}\frac{\mathrm{d}(8V/D)}{\mathrm{d}\tau_{\mathrm{w}}} + \frac{3}{4}\left(\frac{8V}{D}\right) \tag{3.142}$$

Similarly, one can parallel this approach for the fully developed laminar flow of time-independent fluids in a thin slit (Figure 3.18) to derive the following relationship:

$$\dot{\gamma}_{\mathrm{w}} = \left(-\frac{\mathrm{d}V_z}{\mathrm{d}r}\right)_{\mathrm{w}} = \mathrm{f}(\tau_{\mathrm{w}}) = \tau_{\mathrm{w}}\frac{\mathrm{d}(V/h)}{\mathrm{d}\tau_{\mathrm{w}}} + 2\left(\frac{V}{h}\right) \tag{3.143}$$

In order to develop a unified treatment for ducts of various cross-sections, it is convenient to introduce the usual hydraulic diameter D_{h} (defined as four times the area for flow/wetted perimeter) into equations (3.142) and 3.143).
For a circular pipe, $D_{\mathrm{h}} = D$ and hence equation (3.142) becomes:

$$\dot{\gamma}_{\mathrm{w}} = \frac{1}{4}\tau_{\mathrm{w}}\frac{\mathrm{d}(8V/D_{\mathrm{h}})}{\mathrm{d}\tau_{\mathrm{w}}} + \frac{3}{4}\left(\frac{8V}{D_{\mathrm{h}}}\right) \tag{3.144}$$

For the slit shown in Figure 3.18, the hydraulic diameter $D_{\mathrm{h}} = 4h$, and thus equation (3.143) is rewritten as:

$$\dot{\gamma}_{\mathrm{w}} = \frac{1}{2}\tau_{\mathrm{w}}\frac{\mathrm{d}(8V/D_{\mathrm{h}})}{\mathrm{d}\tau_{\mathrm{w}}} + \left(\frac{8V}{D_{\mathrm{h}}}\right) \tag{3.145}$$

By noting the similarity between the form of the Rabinowitsch–Mooney equations for the flow of time-independent fluids in circular pipes (equation (3.144)) and that in between two plates (equation (3.145)), they suggested that it could be extended to the ducts having a constant cross-section of arbitrary shape as follows:

$$\left(-\frac{dV_z}{dr}\right)_w = f(\bar{\tau}_w) = a\bar{\tau}_w \frac{d(8V/D_h)}{d\bar{\tau}_w} + b\left(\frac{8V}{D_h}\right) \tag{3.146}$$

where a and b are two geometric parameters characterising the cross-section of the duct ($a = 1/4$ and $b = 3/4$ for a circular tube, and $a = 1/2$ and $b = 1$ for the slit) and $\bar{\tau}_w$ is the mean value of shear stress at the wall, and is related to the pressure gradient as:

$$\bar{\tau}_w = \frac{D_h}{4}\left(\frac{-\Delta p}{L}\right) \tag{3.147}$$

For constant values of a and b, equation (3.146) is an ordinary differential equation of the form $(dy/dx) + p(x)y = q(x)$ which can be integrated to obtain the solution as:

$$y = e^{-\int p(x)dx} \int e^{\int p(x)dx} q(x) dx + C_0 \tag{3.148}$$

Now identifying $y = (8V/D_h)$ and $x = \bar{\tau}_w$, $p(x) = (b/a\ \bar{\tau}_w)$ and $q(x) = (f(\bar{\tau}_w)/a\bar{\tau}_w)$, the solution to equation (3.146) is given as:

$$\left(\frac{8V}{D_h}\right) = \frac{1}{a}(\bar{\tau}_w)^{-b/a} \int_0^{\bar{\tau}_w} \tau^{(b/a)-1} f(\tau) d\tau \tag{3.149}$$

where τ is a dummy variable of integration. The constant C_0 has been evaluated by using the condition that when $V = 0$, $\bar{\tau}_w = 0$ and therefore, $C_0 = 0$.

It should be noted that for a circular pipe of diameter D, $D_h = D$; $a = 1/4$ and $b = 3/4$; equation (3.149) then reduces to equation (3.46). For the flow of a power-law fluid, $f(\tau) = (\tau/m)^{1/n}$ and integration of (3.149) yields:

$$\bar{\tau}_w = m\left\{\frac{8V}{D_h}\left(b + \frac{a}{n}\right)\right\}^n \tag{3.150}$$

which can be rewritten in terms of the friction factor, $f = 2\bar{\tau}_w/\rho V^2$ as:

$$f = \frac{16}{Re_g} \tag{3.151}$$

where the generalized Reynolds number,

$$Re_g = \frac{\rho V^{2-n} D_h^n}{8^{n-1} m\left(b + \frac{a}{n}\right)^n} \tag{3.152}$$

The main virtue of this approach lies in its simplicity and the fact that the geometric parameters a and b can be deduced from the behaviour of Newtonian fluids in the same flow geometry. Table 3.4 lists values of a and b for a range of flow geometries commonly encountered in process applications. A typical comparison between predicted

Table 3.4   *Values of a and b to be used in equations (3.151) and (3.153)*

1. Concentric annuli

$\sigma = R_i/R_o$	a	b
0.00	0.2500	0.7500
0.01	0.3768	0.8751
0.03	0.4056	0.9085
0.05	0.4217	0.9263
0.07	0.4331	0.9383
0.10	0.4455	0.9510
0.20	0.4693	0.9737
0.30	0.4817	0.9847
0.40	0.4890	0.9911
0.50	0.4935	0.9946
0.60	0.4965	0.9972
0.70	0.4983	0.9987
0.80	0.4992	0.9994
0.90	0.4997	1.0000
1.00	0.5000	1.0000

2. Elliptical ducts

$\varepsilon = b'/a'$	a	b
0.00	0.3084	0.9253
0.10	0.3018	0.9053
0.20	0.2907	0.8720
0.30	0.2796	0.8389
0.40	0.2702	0.8107
0.50	0.2629	0.7886
0.60	0.2575	0.7725
0.70	0.2538	0.7614
0.80	0.2515	0.7546
0.90	0.2504	0.7510
1.00	0.2500	0.7500

3. Rectangular ducts

$E = H/W$	a	b
0.00	0.5000	1.0000
0.25	0.3212	0.8182
0.50	0.2440	0.7276
0.75	0.2178	0.6866
1.00	0.2121	0.6766

(*Continued*)

Table 3.4   *(Continued)*

---

4. Isosceles triangular ducts

2α	a	b
10°	0.1547	0.6278
20°	0.1693	0.6332
40°	0.1840	0.6422
60°	0.1875	0.6462
80°	0.1849	0.6438
90°	0.1830	0.6395

5. Regular polygonal ducts

N	A	b
4	0.2121	0.6771
5	0.2245	0.6966
6	0.2316	0.7092
8	0.2391	0.7241

and experimental values of friction factor for rectangular ducts is shown in Figure 3.31. Similar agreement has been reported by, among others, Mitsuishi *et al.* (1972) and, subsequently, by Xie and Hartnett (1992) for visco-elastic fluids in rectangular ducts and more recently by Escudier and Smith (2001) for pipes of square cross-section. Kozicki *et al.* (1966) argued that equation (3.73) can be generalized to include the turbulent flow in non-circular ducts by recasting it in the form:

$$\frac{1}{\sqrt{f}} = \frac{4}{n^{0.75}} \log_{10} (Re_g f^{(2-n)/2}) - \frac{0.4}{n^{1.2}} + 4n^{0.25} \log \left[\frac{4(a+bn)}{(3n+1)} \right] \tag{3.153}$$

Note that since for a circular tube, $a = 1/4$ and $b = 3/4$, equation (3.153) is consistent with that for circular pipes, equation (3.73). The limited data available on turbulent flow in triangular (Irvine Jr, 1988), rectangular (Kostic and Hartnett, 1984) and square ducts (Escudier and Smith, 2001) conforms to equation (3.153). In the absence of any definite information, Kozicki and Tiu (1988) suggested that the Dodge–Metzner criterion, $Re_g \leq 2100$, can be used for predicting the limit of laminar flow in non-circular ducts.

Some further attempts have been made to simplify and/or improve upon the two geometric parameter method of Kozicki *et al.* (1966, 1967). Delplace and Leuliet (1995) revisited the definition of the generalized Reynolds number (equation (3.152)) and argued that while the use of a and b accounts for the non-circular cross-sections of the ducts, but the factor 8^{n-1} appearing in the denominator is strictly applicable for the flow in circular ducts only. Their reasoning hinges on the fact that for the laminar flow of a Newtonian fluid, the product $(f \cdot Re)$ is a function of the conduit shape only. Thus, they wrote:

$$\beta = \frac{48}{(f \cdot Re)} \tag{3.154}$$

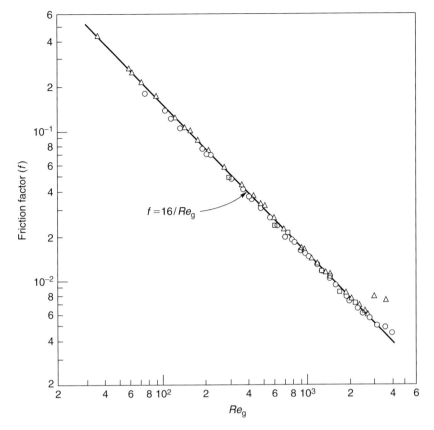

Figure 3.31   *Experimental friction factor values for power-law fluids in laminar regime in rectangular channels. ○, Wheeler and Wissler (1965); △, Hartnett et al. (1986); □, Hartnett and Kostic (1985)*

where both the (Fanning) friction factor and the Reynolds number are based on the use of the hydraulic diameter, D_h and the mean velocity of the flow, V. Furthermore, they were able to link the geometric parameters a and b with the new parameter β as follows:

$$a = \frac{1}{1 + \beta} \tag{3.155}$$

$$b = \frac{\beta}{1 + \beta} \tag{3.156}$$

and finally, the factor of 8^{n-1} in the denominator in equation (3.152) is replaced $(24/\beta)^{n-1}$. With these modifications, one can use the relationship $f = (16/Re_g)$ to estimate the pressure gradient for the laminar flow of a power-law fluid in a non-circular duct for which the value of β is known either from experiments or from numerical results. Therefore, this approach necessitates the knowledge of only one parameter (β) as opposed to the two geometric parameters, namely, a and b in the method of Kozicki et al. (1966) and Kozicki and Tiu (1967), albeit a similar suggestion was also made by

Table 3.5   *Values of $(f \cdot Re)$ for laminar flow in non-circular ducts*

Geometry	Value of $f \cdot Re$	Remarks
Circular	16	–
Elliptical $\left[\alpha = \left(\dfrac{b}{a}\right) < 1\right]$	$2(1 + \alpha^2)\left\{\dfrac{\pi}{E(m)}\right\}^2$ where $m = 1 - \alpha^2$	$E(m)$ is the complete elliptic integral of the second kind
Parallel plates	24	–
Rectangular $\left[\alpha = \left(\dfrac{b}{a}\right) < 1\right]$	$8.968\left(\dfrac{1 + \alpha}{\sqrt{\alpha}}\right)^{0.75}$	$\alpha \geq 0.125$
Semi-circular	15.767	–
Triangular		
Equi-lateral	$\dfrac{40}{3}$	–
Isosceles	$\dfrac{12(B + 2)(1 - \tan^2 \phi)}{(B - 2)(\tan \phi + \sec \phi)^2}$ where $B = \left\{4 + \dfrac{5}{2}(\cot^2 \phi - 1)\right\}^{1/2}$	2ϕ is the apex angle

Miller (1972) and Liu (1983). Finally, for the limiting case of a circular pipe, evidently $\beta = 3$ thereby leading to $a = (1/4)$ and $b = (3/4)$ and the two definitions of the Reynolds number coincide, as expected. The values of β for a few standard duct shapes are summarized in Table 3.5. While in laminar flow, these two methods give almost identical predictions, the applicability of the modified method of Delplace and Leuliet (1995) has not been checked in the transitional and turbulent flow regions.

Scant analytical and experimental results suggest that visco-elasticity in a fluid may induce secondary motion in non-circular conduits, even under laminar conditions. However, measurements reported to date indicate that the friction factor–Reynolds number behaviour is little influenced by such secondary flows (Hartnett and Kostic, 1989).

Example 3.15

A power-law fluid ($m = 0.3\,Pa\,s^n$ and $n = 0.72$) of density $1000\,kg/m^3$ is flowing in a series of ducts of the same flow area but different cross-sections as listed below:

(i)   concentric annulus with $R = 37\,mm$ and $\sigma = 0.40$
(ii)   circular pipe

(iii) rectangular, $(H/W) = 0.5$

(iv) elliptical, $b'/a' = 0.5$

(v) isosceles triangular with half-apex angle, $\alpha = 20°$.

Estimate the pressure gradient required to maintain an average velocity of 1.25 m/s in each of these channels. Use the geometric parameter method. Also, calculate the value of the generalized Reynolds number as a guide to the nature of the flow.

Solution

(i) For a concentric annulus, for $\sigma = 0.4$, and from Table 3.4:
$$a = 0.489;\ b = 0.991$$

The hydraulic diameter, $D_h = 2R(1 - \sigma)$
$$= 2 \times 37 \times 10^{-3} \times (1 - 0.4)$$
$$= 0.044\,\text{m}$$

Reynolds number, $\quad Re_g = \dfrac{\rho V^{2-n} D_h^n}{8^{n-1} m \left(b + \dfrac{a}{n}\right)^n}$ (3.152)

$$= \dfrac{1000 \times 1.25^{2-0.72} \times 0.044^{0.72}}{8^{0.72-1} \times 0.3 \times \left(0.991 + \dfrac{0.489}{0.72}\right)^{0.72}}$$

$$= 579$$

Thus, the flow is laminar and the friction factor is estimated as:

$$\therefore\ f = \dfrac{16}{579} = 0.0276$$

and

$$\dfrac{-\Delta p}{L} = \dfrac{2f\rho V^2}{D_h} = \dfrac{2 \times 0.0276 \times 1000 \times 1.25^2}{0.044} = 1963\,\text{Pa/m}$$

For the sake of comparison, equation (3.130) yields a value of 1928 Pa/m which is remarkably close to the value calculated here.

(ii) For a circular tube, the area of flow
$$= \pi R^2 (1 - \sigma^2) = 3.14 \times 0.037^2\,(1 - 0.4^2) = 0.003613\,\text{m}^2$$

which corresponds to the pipe radius of 0.0339 m or diameter of 0.0678 m.
For a circular pipe, $a = 0.25$, $b = 0.75$, $D_h = D = 0.0678$ m.

$$\therefore\ Re_g = \dfrac{\rho V^{2-n} D_h^n}{8^{n-1} m \left(b + \dfrac{a}{n}\right)^n} = \dfrac{(1000)(1.25)^{2-0.72}(0.0678)^{0.72}}{8^{0.72-1}(0.3)\left(0.75 + \dfrac{0.25}{0.72}\right)^{0.72}}$$

$$= 1070$$

Thus, the flow is laminar and $f = 16/1070 = 0.01495$ and the pressure gradient,

$$\dfrac{-\Delta p}{L} = \dfrac{2f\rho V^2}{D_h} = \dfrac{2 \times 0.01495 \times 1000 \times 1.25^2}{0.0678}$$
$$= 689\,\text{Pa/m}$$

(iii) For a rectangular duct with $H/W = 0.5$, $H = 0.0425$ m and $W = 0.085$ m (for the same area of flow), and from Table 3.4:

$a = 0.244$, $b = 0.728$

The hydraulic diameter, $D_h = \dfrac{4 \times \text{flow area}}{\text{wetted perimeter}}$

$$= \frac{4 \times A}{2(H + W)} = \frac{4 \times 0.003613}{2(0.0425 + 0.085)}$$

$$= 0.0567 \text{ m}$$

$$\therefore Re_g = \frac{\rho V^{2-n} D_h^n}{8^{n-1} m \left(b + \dfrac{a}{n} \right)^n} = \frac{(1000) \times (1.25)^{2-0.72} (0.0567)^{0.72}}{8^{0.72-1} \times 0.3 \times \left(0.728 + \dfrac{0.244}{0.72} \right)^{0.72}}$$

$$= 960$$

Again, the flow is laminar and $f = 16/960 = 0.0167$

The pressure gradient, $\dfrac{-\Delta p}{L} = \dfrac{2f\rho V^2}{D_h}$

$$= \frac{2 \times 0.0167 \times 1000 \times 1.25^2}{0.0567} = 919 \text{ Pa/m}$$

(iv) The cross-sectional area of an elliptic pipe with semi-axes a' and b' is $\pi a' b'$, while $a = 0.2629$ and $b = 0.7886$ from Table 3.4 corresponding to $b'/a' = 0.5$

$$\therefore \pi a' b' = 0.003613$$

Solving it together with $b'/a' = 0.5$, $a' = 0.04795$ m and $b' = 0.024$ m. The hydraulic diameter D_h is calculated next as:

$$D_h = \frac{4 \times \text{flow area}}{\text{wetted perimeter}}$$

No analytical expression is available for the perimeter of an ellipse; however, it can be approximated by $2\pi((a'^2 + b'^2)/2^{1/2})$

$$\therefore \quad D_h = \frac{4 \times 0.003613}{2\pi \left(\dfrac{0.04795^2 + 0.024^2}{2} \right)^{1/2}} = 0.0607 \text{ m}$$

$$Re_g = \frac{1000 \times 1.25^{1.28} \times 0.0607^{0.72}}{8^{-0.28} \times 0.3 \times \left(0.7886 + \dfrac{0.2629}{0.72} \right)^{0.72}} = 953$$

and $f = 16/Re_g = 16/953 = 0.0168$

$$\therefore \quad \frac{-\Delta p}{L} = \frac{2f\rho V^2}{D_h} = \frac{2 \times 0.0168 \times 1000 \times 1.25^2}{0.0607} = 864 \text{ Pa/m}$$

(v) For the pipe of triangular cross-section, with $\alpha = 20°$, $a = 0.184$ and $b = 0.6422$.
Let the base of the triangle be x

$$\therefore \text{ Height of the triangle} = \frac{x}{2\tan\alpha} = \frac{x}{2\tan 20}$$
$$= 1.37x$$

$$\therefore \text{ Area for flow} = \tfrac{1}{2} \times x \times 1.37x = 0.003613$$
$$\text{or } x = 0.0725 \text{ m}$$

The hydraulic diameter is calculated next:

$$D_h = \frac{4 \times 0.003613}{x + 2\left(\dfrac{x}{2\sin\alpha}\right)} = \frac{4 \times 0.003613}{0.0725\left(1 + \dfrac{1}{\sin 20}\right)}$$
$$= 0.0508 \text{ m}$$

The Reynolds number of flow,

$$Re_g = \frac{1000 \times 1.25^{1.28} \times 0.0508^{0.72}}{8^{-0.28} \times 0.3 \times \left(0.6422 + \dfrac{0.184}{0.72}\right)^{0.72}}$$
$$= 1006$$

$$\therefore f = \frac{16}{1006} = 0.016$$

and
$$\frac{-\Delta p}{L} = \frac{2f\rho V^2}{D_h} = \frac{2 \times 0.016 \times 1000 \times 1.25^2}{0.0508}$$
$$= 980 \text{ Pa/m}$$

This example clearly shows that the pressure drop is a minimum in the case of circular pipes, followed by the elliptic, rectangular and triangular cross-sections and the concentric annulus for flow. On the other hand, if one were to maintain the same hydraulic diameter in each case, the corresponding pressure gradients range from 2500 to 4000 Pa/m. □

3.8 Miscellaneous frictional losses

In the analysis of pipe networks, one is usually concerned either with how much power is required to deliver a set flow rate through an existing flow system or with the optimum pipe diameter for a given pump and duty. All such calculations involve determining the frictional pressure losses in the systems, both in the region of fully established flow (as has been assumed so far), and in the associated sudden changes in cross-section (expansions and contractions) and other fittings such as bends, elbows, valves, etc. It is also necessary to establish whether the flow is laminar or turbulent. For Newtonian flow, the magnitudes of these losses are well known, and, although the theory is far from complete, the established design procedures are usually quite satisfactory. The analogous results for non-Newtonian (mainly time-independent type) are presented here, although the experimental results are scant in this field.

3.8.1 Sudden enlargement

When the cross-section of a pipe enlarges gradually, the streamlines follow closely the contours of the duct and virtually no extra frictional losses are incurred. On the other hand, whenever the change is sudden, additional losses arise due to the eddies formed as the fluid enters the enlarged cross-section. The resulting head loss for laminar flow can be evaluated by applying the mechanical energy balance in conjunction with the integral momentum balance. Consider the flow configuration shown in Figure 3.32 in which the section '2' is located immediately after the end of the smaller pipe. By suitable choice of plane '1', the frictional pressure loss may be assumed to be negligible between planes '1' and '2' and hence $p_1 \sim p_2$; the latter acts over the whole cross-section (πR_2^2). Also, immediately following the expansion at section '2', the streamlines will be nearly parallel to the axis and the velocity profile at section '2' will be similar to the fully developed profile at section '1'. If section '3' is sufficiently far downstream, the velocity profile again be fully established. On applying the integral momentum balance over the control volume (broken line) as shown in Figure 3.32:

$$\Sigma F_z = (p_2 - p_3)\pi R_2^2 = -\int_0^{R_1} 2\pi r \rho V_z^2 \, dr + \int_0^{R_2} 2\pi r \rho V_z^2 \, dr \qquad (3.157)$$

The integration in equation (3.157) can be carried out after insertion of the velocity profiles for the appropriate viscosity model to obtain the pressure loss $(p_2 - p_3)$. For power-law fluids, this procedure leads to:

$$\frac{p_2 - p_3}{\rho} = \left(\frac{3n+1}{2n+1}\right)\frac{Q^2}{A_1^2}\left[\left(\frac{A_1}{A_2}\right)^2 - \frac{A_1}{A_2}\right] \qquad (3.158)$$

where $A_1 = \pi R_1^2$ and $A_2 = \pi R_2^2$. Applying the mechanical energy balance equation between points '1' and '3' (neglecting friction over the sections 1–2 and 2–3):

$$\frac{p_1}{\rho} + \frac{V_1^2}{2\alpha} + gz_1 = \frac{p_3}{\rho} + \frac{V_3^2}{2\alpha} + gz_3 + \Sigma F_{\text{exp}}$$

or $\hspace{9cm}$ (3.159)

$$\Sigma F_{\text{exp}} = \frac{p_1 - p_3}{\rho} + (z_1 - z_3)g + \frac{V_1^2 - V_3^2}{2\alpha}$$

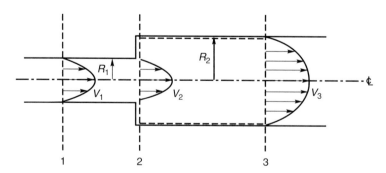

Figure 3.32   *Schematics of laminar flow through a sudden expansion in a tube*

For a horizontal system, $z_1 = z_3$, putting $p_1 \approx p_2$ and substituting for α from equation (3.44), equations (3.158) and (3.159) yield the following expression for the head loss, h_e:

$$h_e = \frac{\Sigma F_{exp}}{g} = \frac{1}{g}\left(\frac{Q}{A_1}\right)^2 \left(\frac{3n+1}{2n+1}\right)$$

$$\times \left[\frac{(n+3)}{2(5n+3)}\left(\frac{A_1}{A_2}\right)^2 - \left(\frac{A_1}{A_2}\right) + \frac{3(3n+1)}{2(5n+3)}\right] \qquad (3.160)$$

If n were equal to zero, the velocity would be uniform across the pipe cross-section ($\alpha = 1$) and equation (3.160) would reduce to

$$h_e = \frac{V_1^2}{2g}\left(1 - \frac{A_1}{A_2}\right)^2 \qquad (3.161)$$

This agrees with the expression for turbulent Newtonian flow when the velocity profile is assumed to be approximately flat.

3.8.2 Entrance effects for flow in tubes

The previous discussion on flow in pipes has been restricted to the fully developed flow where the velocity at any position in the cross-section is independent of distance along the pipe. In the entrance and exit sections of the pipe, this will no longer be true. Since exit effects are much less significant than entrance effects, only the latter are dealt with in detail here.

For all fluids entering a small pipe from either a very much larger one or from a reservoir, the initial velocity profile will be approximately flat and will then undergo a progressive change until fully developed flow is established, as shown schematically in Figure 3.33.

The thickness of the boundary layer is theoretically zero at the entrance and increases progressively along the tube. The retardation of the fluid in the wall region must be accompanied by a concomitant acceleration in the central region in order to satisfy the equation of continuity. When the velocity profile has reached its final shape, the flow is fully developed and the boundary layers may be considered to have converged at the

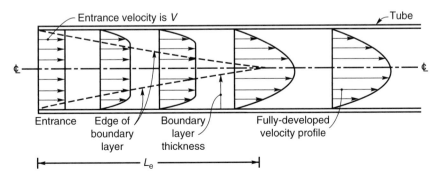

Figure 3.33 *Development of the boundary layer and velocity profile for laminar flow in the entrance region of a pipe*

centre line. It is customary to define an entry length, L_e, as the distance from the inlet at which the centre-line velocity is 99% of that for the fully developed flow. The pressure gradient in this entry region is different from that for the fully developed flow and is a function of the initial velocity profile. There are two factors influencing the pressure gradient in the entry region: first, some pressure energy is converted into kinetic energy as the fluid in the central core accelerates and secondly, the higher velocity gradients in the wall region result in greater frictional losses. It is important to estimate both the pressure drop occurring in the region before the flow has been fully developed and the extent of this entrance length. This situation is amenable to analysis by repeated use of the mechanical energy balance equation. Consider the schematics of the flow shown in Figure 3.34. The stations '1' and '3' are well removed from the tube entrance, '2' is in the plane of entrance while '3' is situated in the fully developed region. The frictional pressure loss between points '1' and '3' can be expressed as:

$$\rho \Sigma F = \Delta p_{fd(1-2)} + \Delta p_{fd(2-3)} + \Delta p_{ex(1-2)} + \Delta p_{ex(2-3)} \tag{3.162}$$

where the subscripts 'fd' and 'ex', respectively, denote the pressure drops over the regions of fully developed flow and the extra or additional pressure drop due to the acceleration of the fluid. Because $V_1 \ll V_3$, the fully developed pressure loss between points '1' and '2' is assumed to be negligible and that between '2' and '3' can be expressed in terms of the wall shear stress in the smaller tube as:

$$\Delta p_{fd(2-3)} = 2\tau_w \left(\frac{L}{R} \right) \tag{3.163}$$

where L is the length of the pipe between '2' and '3'. Thus, the extra frictional loss between '1' and '3' arising from the fact that flow is developing, $\Delta p_{entrance}$, can be written as:

$$\Delta p_{entrance} = (p_1 - p_3) - \Delta p_{fd(2-3)} = p_1 - p_3 - 2\tau_w \left(\frac{L}{R} \right) \tag{3.164}$$

Applying the mechanical energy balance between points '1' and '3', noting $z_1 = z_3$ and $V_1 \ll V_3$, and substituting for $\rho \Sigma F$ from equations (3.162) and (3.163):

$$(p_1 - p_3) = \frac{\rho V_3^2}{2\alpha_3} + \rho \Sigma F = \frac{\rho V_3^2}{2\alpha_3} + 2\tau_w \left(\frac{L}{R} \right) + \Delta p_{ex(1-2)} + \Delta p_{ex(2-3)}$$

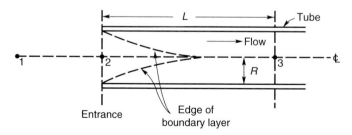

Figure 3.34   *Schematics for calculation of entrance effects*

Table 3.6 *Values of C_1 (n) and C_2 (n) (Boger, 1987)*

N	C_1 (n)	C_2 (n)
1	2.33	0.58
0.9	2.25	0.64
0.8	2.17	0.70
0.7	2.08	0.79
0.6	1.97	0.89
0.5	1.85	0.99
0.4	1.70	1.15
0.3	1.53	1.33

or

$$\Delta p_{\text{entrance}} = p_1 - p_3 - 2\tau_{\text{w}}\left(\frac{L}{R}\right) = \frac{\rho V_3^2}{2\alpha_3} + \Delta p_{\text{ex}(1-2)} + \Delta p_{\text{ex}(2-3)} \tag{3.165}$$

Noting $(1/2)\rho V_3^3 = \tau_{\text{w}}/f$ and that $f = 16/Re_{\text{MR}}$ in the laminar region, it is customary to rearrange equation (3.165) as:

$$\frac{\Delta p_{\text{entrance}}}{2\tau_{\text{w}}} = C_1(n)\frac{Re_{\text{MR}}}{32} + C_2(n) \tag{3.166}$$

where

$$C_1(n) = \left[\frac{1}{\alpha_3} + \frac{\Delta p_{\text{ex}(2-3)}}{(1/2)\rho V_3^2}\right] \tag{3.167}$$

and

$$C_2(n) = \frac{\Delta p_{\text{ex}(1-2)}}{2\tau_{\text{w}}} \tag{3.168}$$

For the laminar flow of power-law fluids, it has been found that both C_1 and C_2 (also known as Couette correction) are functions of the power law index, n alone. Obviously, C_2 representing the loss between points '1' and '2' would be strongly dependent on the geometrical details of the system, more gradual or smooth the entrance, smaller will be the value of C_2. However, to date, its values have been computed only for an abrupt change. Table 3.6 summarizes the predicted values of C_1 and C_2 for a range of values of n. It should be noted that $C_1(n)$ decreases with the increasing degree of shear-thinning behaviour while $C_2(n)$ shows the exacdy opposite type of dependence on n. That is, the contribution of the excess pressure drop between points '1' and '2' increases with the decreasing value of n. Based on extensive comparisons between predictions and experimental data, this approach has been found to be reliable for estimating the value of $\Delta p_{\text{entrance}}$ for linear contraction ratios greater than 2 and downstream Reynolds number ($\rho V^{2-n} D^n/m) > 5$ (Boger, 1987). Figure 3.35 shows the entry length L_e, required to attain fully developed flow in tubes; excellent agreement is seen to exist between predictions and limited experimental data. Based on extensive numerical results, Poole and Ridley (2007) proposed the following equation for L_e for the laminar flow of power law fluids in circular tubes:

$$\frac{L_e}{D} = \left[(0.25n^2 - 0.675n + 1.03)^\alpha + (0.0567\, Re_{\text{MR}})^\alpha\right]^{1/\alpha} \tag{3.169}$$

where $\alpha = 1.6$.

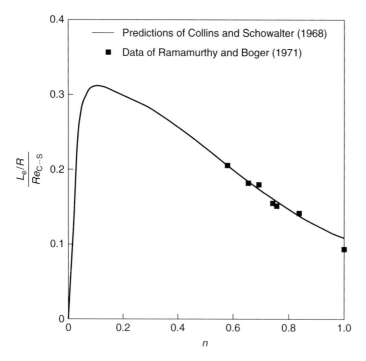

Figure 3.35   *Entrance length for power-law fluids*

This equation was stated to be applicable in the range $0.4 \leq n \leq 1.5$ and $Re_{MR} \leq 1000$.

While the 99% velocity criterion is adequate for Newtonian and shear-thinning fluids, but due to the plug-flow in the centre of the pipe, it is not applicable to the flow of visco-plastic fluids. This is because the plug velocity at the centre of the tube evolves very slowly. Thus, Ookawara *et al.* (2000) defined the entry length as the axial distance (from the inlet of the pipe) at which the velocity at $r = 0.95R_p$ reaches 99% of its ultimately expected value of the plug velocity. Based on this criterion they correlated their numerical results as follows for $Re^* < 10$:

$$\frac{L_e}{D} = 0.379 \exp(-0.148\ Re^*) + 0.0550\ Re^* + 0.260 \tag{3.170}$$

where the modified Reynolds number Re^* is defined as:

$$Re^* = Re_B \left(\frac{\phi^4 - 4\phi + 3}{3} \right) \left[\frac{9}{5} \frac{(5 + 6\phi - 11\phi^2)}{(3 + 2\phi + \phi^2)^2} \right] \tag{3.171}$$

and $\phi = R_p/R$. Evidently, for a Newtonian fluid, $\phi = 0$, and the modified Reynolds number reduces to the usual definition for Newtonian fluids.

In equations (3.169) and (3.171), the constant term (or which is a function of the power-law index only) denotes the increasing role of molecular diffusion in the limit of diminishing Reynolds number. While it is of little importance in traditional engineering applications, but it may become significant in micro-fluidic devices. From a practical standpoint, the currently available body of information suggests that the entrance length,

L_e, is of the order of 40 pipe diameters for inelastic fluids and about $110D$ for visco-elastic fluids (Cho and Hartnett, 1982) in streamline flow. The literature on this subject has been critically reviewed by Boger (1987).

There is little information on either the entrance length or the additional pressure drop for fully developed turbulent flow. Dodge and Metzner (1959) indicated that both the entrance length and the extra pressure loss for inelastic fluids were similar to those for Newtonian fluids under these conditions.

3.8.3 Minor losses in fittings

Little reliable information is available on the pressure drop for the flow of non-Newtonian fluids through pipe fittings. The limited work carried out on the laminar flow of inelastic polymer solutions and of suspensions (Urey, 1966; Steffe *et al.*, 1984; Edwards *et al.*, 1985; Das *et al.*, 1991; Banerjee *et al.*, 1994; Turian *et al.*, 1998; Martinez-Padilla and Linares Garcia, 2001; Polizelli *et al.*, 2003; ESDU, 2004; Bandyopadhyay and Das, 2007; Fester *et al.*, 2007) and on the turbulent flow of magnesia and titania slurries (Weltmann and Keller, 1957; Cheng, 1970) through a range of fittings including elbows, tees, bends, unions, plug, butterfly and globe valves suggests that the shear dependence of viscosity exerts little influence on such minor losses and therefore values for Newtonian fluids can be used. However, the work of Slatter (1997) on the flow of a kaolin slurry and a polymer solution through sudden contractions and expansions and 90° elbows suggests that pressure drops are much larger than that for Newtonian fluids. Therefore, great care must be exercised in estimating such losses, though it is always preferable to carry out some tests with the fluid under consideration. The corresponding losses for visco-elastic fluids are likely to be greater if their extensional viscosities are high. Additional uncertainties arise because the critical Reynolds number for the laminar–turbulent transition is not well defined and is strongly influenced by the design of the particular fitting. Thus, Edwards *et al.* (1985) reported the critical value of Re_{MR} as 900 for an elbow and 12 for globe valves! While overall there appears to be little definite information for the minor losses in various fittings in the laminar flow regime, most of the data is correlated in the form of $KV^2/2g$ and for a fixed fitting design, in turn, the coefficient K is related with the Reynolds number as $K = ARe_{MR}^B$, the most commonly used value of B is typically –1 or close to it, but the values of A differ widely from one fitting to another. For fully open butterfly and gate valves, $A \approx 760$–860 whereas for bends it can range from ~500 to ~1000 as one goes from 45° to 180° bends. Overall there appears to be little conclusive information for the minor losses in various fittings.

Some guidelines for the design and selection of valves for pseudoplastic materials have been developed by De Haven (1959) and Beasley (1992).

3.8.4 Flow measurement

Little information is available on suitable flow measurement devices. It has already been shown that pressure drop across a straight length of pipe is relatively insensitive to flow rate in laminar region of highly shear-thinning materials and therefore it is an unsuitable parameter for flow measurement. In principle, the devices which depend upon the conversion of pressure into kinetic energy (e.g., orifice and venturi metres, rotameters) can be used but they need calibration for each fluid. The problem is compounded in practice because the fluid may not have a constant rheology and/or may display visco-elastic or time-dependent behaviour. Though many investigators have studied the flow of polymer solutions through orifices, most of this work has not been directed towards flow measurement but has related either to the measurement of extensional viscosity or to the extrusion behaviour. Both Harris and Magnall (1972) and Edwards *et al.* (1985) have examined the suitability of orifice and venturi metres for measuring the flow rates of inelastic polymer solutions and kaolin suspensions. While Edwards *et al.* (1985) were unable to bring together data for different aspect ratios, Harris and Magnall (1972) reported satisfactory correlation for the discharge coefficient in the turbulent regime. Despite all these difficulties, the measurement of the flow of shear-thinning oil–water emulsions using orifice and venturi metres has also been the subject of another study (Pal, 1993). The discharge coefficients were successfully correlated with the generalized Reynolds number (Re_{MR}). However, extrapolation to other types of non-Newtonian materials, e.g., polymer solutions, slurries, etc., must be treated with reserve. In view of these difficulties, indirect methods of flow measurement are generally preferred, the electromagnetic flow meter being the most common choice (Dodge and Metzner, 1959; Quader and Wilkinson, 1980; Chhabra *et al.*, 1984) covering a wide variety of materials including polymer solutions, kaolin, anthracite and titania slurries. This method is, however, limited to substances possessing some degree of electrical conductivity. In addition, both Skelland (1967) and Liptik (1967) have suggested rotary type volumetric displacement flow metres for non-Newtonian viscous fluids but have given virtually no practical information on their performance. Ginesi (1991) has found magnetic and coriolis type flow metres to be reliable for slurries and viscous fluids, as also confirmed by a recent study (Fyrippi *et al.*, 2004). In particular, Fyrippi *et al.* (2004) concluded that the ultrasonic flow meters performed poorly due to their extreme sensitivity to the existing velocity distributions for the flow of non-Newtonian fluids in pipes. A recent study also demonstrates the feasibility of using pre-calibrated weirs and notches to meter the flow of inelastic fluids in open channels, as encountered in mining industry (Haldenwang *et al.*, 2007). Brown and Heywood (1992) have discussed instrumentation for the measurement of both flow rate and density for non-Newtonian slurries.

Example 3.16

A kaolin slurry (density $= 1200\,kg/m^3$; $n = 0.2$ and $m = 25\,Pa\,s^n$) flows under gravity from reservoir A to B both of which are of large diameter, as shown in Figure 3.36. Estimate the flow rate of slurry through the 50 mm diameter connecting pipe of total length of 75 m.

Solution

The mechanical energy balance can be applied between points '1' and '2' shown in Figure 3.36.

$$\frac{V_1^2}{2\alpha_1 g} + \frac{p_1}{\rho g} + z_1 = \frac{V_2^2}{2\alpha_2 g} + \frac{p_2}{\rho g} + z_2 + \Sigma h_L$$

Here $p_1 = p_2 = p_{atm}$. Because the two reservoirs are of large diameter, V_1 and V_2 are both approximately zero. With these simplifications, the mechanical energy equation simplifies to:

$$\Sigma h_L = z_1 - z_2 = 40 - 5 = 35\,m$$

The friction losses are associated with the flow in a 75 m length of 50 mm pipe including one gate valve, one globe valve, two short curvature elbows, a contraction (at A) and an expansion at B.

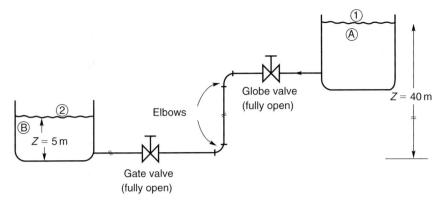

Figure 3.36 *Schematics of flow for Example 3.16*

At this stage, neglecting the contraction and expansion losses, one can express the loss term in terms of the unknown velocity V as:

$$\Sigma h_{\mathrm{L}} = \frac{2fLV^2}{gD} + K\frac{V^2}{2g} + K\frac{V^2}{2g} + 2K\frac{V^2}{2g} = 35\,\mathrm{m}$$

(gate   (globe   (elbows)
valve)   valve)

Edwards *et al.* (1985) have published the following values of K for various fittings:

For a 2 inch (50 mm) fully open gate valve, $K = \dfrac{273}{Re_{\mathrm{MR}}}$ $Re_{\mathrm{MR}} < 120$

For a 2 inch (50 mm) fully open globe valve, $K = \dfrac{384}{Re_{\mathrm{MR}}}$ $Re_{\mathrm{MR}} < 15$

$$K = 25.4 \quad Re_{\mathrm{MR}} > 15$$

For a 2 inch (50 mm) short-curvature elbow, $K = \dfrac{842}{Re_{\mathrm{MR}}}$ $Re_{\mathrm{MR}} < 900$

$$K = 0.9 \quad Re_{\mathrm{MR}} > 900$$

Assuming the flow to be laminar and $Re_{\mathrm{MR}} > 15$. The total head loss is then:

$$\frac{V^2}{2g}\left[\frac{64L}{DRe_{\mathrm{MR}}} + \frac{273}{Re_{\mathrm{MR}}} + 25.4 + 2\times\frac{842}{Re_{\mathrm{MR}}}\right] = 35\,\mathrm{m}$$

(straight pipe   (1 gate   (1 globe   (2 elbows)
length)   valve)   valve)

where

$$Re_{\mathrm{MR}} = \frac{\rho V^{2-n}D^n}{8^{n-1}m\left(\dfrac{3n+1}{4n}\right)^n} \tag{3.17}$$

Here, $\rho = 1200\,\mathrm{kg/m^3}$; $D = 50\times10^{-3}$ m; $m = 25\,\mathrm{Pas}^n$; $L = 75$ m; $n = 0.2$.

$$\therefore Re_{\mathrm{MR}} = \frac{(1200)V^{2-0.2}(50\times10^{-3})^{0.2}}{(8^{0.2-1})(25)\left(\dfrac{3\times0.2+1}{4\times0.2}\right)^{0.2}} = 121.13\,V^{1.8}$$

Substituting for Re_{MR} and other values:

$$V^2 \left\{ \frac{64 \times 75}{50 \times 10^{-3} \times 121.13 V^{1.8}} + \frac{273}{121.13 V^{1.8}} + 25.4 + \frac{2 \times 842}{121.13 V^{1.8}} \right\}$$
$$= 35 \times 9.81 \times 2$$

or $808.69 V^{0.2} + 25.4 V^2 = 686.7$

A trial and error method gives, $V = 0.43 \, \text{m/s}$. Check the value of $Re_{MR} = 121.13 \times 0.43^{1.8} = 26.52$ which is in the laminar range and is also larger than the value of 15 assumed above. □

3.9 Selection of pumps

Non-Newtonian characteristics, notably shear-dependent viscosity and yield stress, strongly influence the choice of a suitable pump and its performance. While no definite quantitative information is available on this subject, general features of a range of pumps commonly used in industry are briefly described here. In particular, consideration is given to positive-displacement, centrifugal, and screw pumps.

3.9.1 Positive-displacement pumps

Reciprocating pumps

Difficulties experienced in initiating the flow of pseudoplastic materials (owing to their high apparent viscosities and/or the yield stress effects) are frequently countered by the use of one of the various types of positive-displacement pumps, see e.g., Coulson and Richardson (1999). Non-Newtonian fluids which are sensitive to breakdown, particularly agglomerates in suspensions, are best handled with pumps which subject the liquid to a minimum of shearing. Diaphragm pumps are then frequently used, but care must be taken that the safe working pressure for the pump and associated pipe network is not exceeded. The use of a hydraulic drive and a pressure-limiting relief valve fitted to the pump ensures that the system is protected from damage. Such an arrangement is shown schematically in Figure 3.37, which depicts general features of all systems using positive-displacement pumps.

Rotary pumps

The selection of a suitable rotary positive-displacement pump for a viscoplastic material has been discussed by Steffe and Morgan (1986). Such pumps operate on the principle of using mechanical means to transfer small elements or 'packages' of fluid from the low pressure (inlet) side to the high pressure (delivery) side. There is a wide range of designs available for achieving this end. The general characteristics of the pumps are similar to those of reciprocating piston pumps but the delivery is more even because of the fluid stream is broken down into much smaller elements. The pumps are capable of delivering to a high pressure, and the pumping rate is approximately proportional to the speed of the pump and is not greatly influenced by the pressure against which it is delivering. Again, it is necessary to provide a pressure relief system to ensure that the safe operating pressure is not exceeded.

Figure 3.37   *A hydraulic drive to protect a positive-displacement pump*

One of the commonest forms of the pump is the gear pump in which one of the gear wheels is driven and the other turns as the teeth engage; two versions are illustrated in Figures 3.38 and 3.39. The liquid is carried around in the spaces between consecutive gear teeth and the outer casing of the pump, and the seal between the high and low pressure sides of the pump is formed as the gears come into mesh and the elements of fluid are squeezed out. Gear pumps are extensively used for both high-viscosity Newtonian liquids and non-Newtonian fluids. The lobe pump (Figure 3.40) is similar, but the gear teeth is replaced by two or three lobes and both axles are driven; it is therefore possible to maintain a small clearance between the lobes, and wear is reduced.

Another form of positive-acting rotary pump is the single-screw extruder pump typified by the Mono pump, illustrated in Figures 3.41 and 3.42. A specially shaped metal helical-rotor revolves eccentrically within a resilient rubber or plastic double-helix, thus creating a continuous forming cavity which progresses towards the discharge of the pump. A continuous seal is created and, the higher the delivery pressure, the greater is the required number of turns, and hence the longer the stator and rotor. This type of pump is suitable for pumping slurries and pastes, whether Newtonian or non-Newtonian in character.

3.9.2 Centrifugal pumps

The most common type of pump used in the chemical industry is the centrifugal pump, though its performance deteriorates rapidly with increasing viscosity of fluids even with Newtonian fluids. The underlying principle is the conversion

Figure 3.38   *Gear pump*

Figure 3.39   *Internal gear pump*

of kinetic energy into a static pressure head. For a pump of this type, the distribution of shear within the pump will vary with throughput. Considering Figure 3.43 where the discharge is completely closed off, the highest degree of shearing is in the gap between the rotor and shell, i.e., at point B. Within the vanes of the rotor (region A) there will be some circulation as sketched in Figure 3.44, but in the discharge line C, the fluid will be essentially static. If the fluid is moving through

Figure 3.40 *Lobe pump*

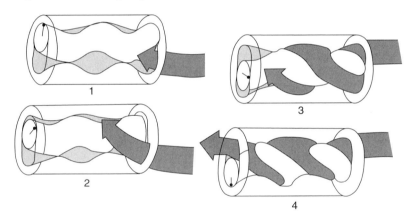

Figure 3.41 *The mode of operation of a Mono pump*

Figure 3.42 *Section of Mono pump*

Figure 3.43   *Zones of differing shear in a centrifugal pump*

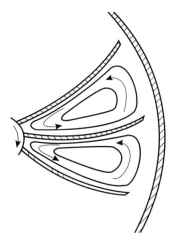

Figure 3.44   *Circulation within a centrifugal pump impeller*

the pump, there will still be differences between these shear rates but they will be less extreme. For a pseudoplastic material, the effective viscosity will vary in these different regions, being less at B than at A and C, while a shear-thickening material will exhibit the opposite behaviour. Under steady conditions, the pressure developed in the rotor produces a uniform flow through the pump. However, there may be problems on starting, when the very high effective viscosities of the fluid as the system starts from rest might result in the overloading of the motor. At this time too, the apparent viscosity of the liquid in the delivery line is at its maximum value, and the pump may take an inordinately long time to establish the required flow. Many pseudoplastic materials (such as food stuffs, pharmaceutical formulations) are damaged and degraded by prolonged shearing, and such a pump would be unsuitable.

As mentioned previously, it is generally accepted that the performance of a centrifugal pump deteriorates increasingly as the extent of non-Newtonian characteristics increases. Both head and particularly efficiency are adversely affected, and the performance of small pumps is impaired to the greatest extent. Severe erosion of the impeller and pump casing is also encountered especially with particulate suspensions. Some guidelines for using charts based on water for non-Newtonian materials by selecting a suitable value of the effective viscosity are, however, available (Duckham, 1971; Walker and Goulas, 1984). Despite these deficiencies, centrifugal pumps are used almost exclusively in the pipelining of scores of mineral slurries. Based on extensive practical experience with wide ranging materials and with impeller diameters up to 380 mm, it is believed that the de-rating factor for the head produced by centrifugal pumps is only of the order of 10% whereas the corresponding factor for pump efficiency could be as high as 25% depending upon the degree of shear-thinning and/or the value of the yield stress of the slurry being pumped (Cooke, 2007). Carter and Lambert (1972), on the other hand, have found helical gear pumps more suitable for viscous Newtonian and non-Newtonian fluids

3.9.3 Screw pumps

Screw extruders, as used in the polymer processing and food industries, form a most important class of pumps for handling highly viscous non-Newtonian materials. Extruders are used for forming simple and complex sections (rods, tubes, etc). The shape of section produced for a given material is dependent only on the profile of the dies through which the fluid is forced just before it cools and solidifies, though additional complications may arise due to die-swell (exhibited by visco-elastic fluids) whereby the diameter of the extrudate may be larger than that of the die.

The basic function of the screw pump or extruder is to shear the fluid in the channel between the screw and the wall of the barrel, as shown schematically in Figure 3.45. The mechanism that generates the pressure can be visualized in terms of a model consisting of an open channel covered by a moving plane surface (Figure 3.46).

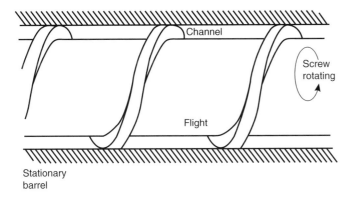

Figure 3.45   *Section of a screw pump*

Figure 3.46   *Planar model of part of a screw pump*

This planar simplification of a stationary screw with rotating barrel is not unreasonable, provided that the depth of the screw channel is small compared with the barrel diameter. The distribution of centrifugal forces will, of course, be different according to whether the rotating member is the wall or the screw; this distinction must be drawn before a detailed force balance can be undertaken, but in any event the centrifugal (inertial) forces are generally far smaller than the viscous forces.

If the upper plate is moved along in the direction of the channel then a velocity profile is established, giving an approximately linear velocity distribution between two walls which are moving parallel to each other. If it moves at right angles to the channel axis, however, a circulation pattern is developed in the gap, as shown in Figure 3.47. In fact, the relative movement is somewhere between these two extremes and is determined by the pitch of the screw.

The fluid path in a screw pump is therefore of a complex helical form within the channel section. The velocity components along the channel depend on the pressure generated and the resistance at the discharge end. If there is no resistance, the velocity profile in the direction of the channel will be of the Couette type, as depicted in Figure 3.48a. With a totally closed discharge end, the net flow would be zero, but the velocity components at the walls would not be affected. As a result, the flow field necessarily would be of the form shown in Figure 3.48b.

Viscous forces within the fluid will always prevent a completely unhindered discharge, but in extrusion-practice the die-head provides additional resistance which generates back-flow and mixing, thus creating a more uniform product. Under these conditions, the flow profile along the channel is of some intermediate form such as that shown in Figure 3.48c.

It must be stressed here that flow in a screw pump is produced as a result of viscous forces and hence the pressures achieved at the outlet with low viscosity materials are small. The screw pump is thus not a modification of the Archimedes screw used in antiquity to raise water–that was essentially a positive-displacement device using a deep cut helix, not running full, mounted at an angle to the horizontal. In any detailed analysis of the flow in a screw pump, it is also necessary to consider the small leakage flow that

Figure 3.47   *Fluid displacement resulting from movement of plane surface*

(a)   (b)   (c)

Figure 3.48   *Velocity profile produced between screw pump surfaces with (a) no resistance on fluid flow, (b) no net flow, (c) partially restricted discharge* will occur between the flight of the screw and the wall. With the large pressures generated in a polymer extruder (~ 100 bar), the flow through this gap (typically 2% of the barrel internal diameter) can be significant because the pressure drop over a single pitch length may be of the order of 10 bar. Once in this region, the viscous fluid is subject to a high rate of shear (the rotational speed of about 2 Hz) and an appreciable part of the fine-scale mixing and viscous heat generation occurs in this part of the extruder. It is thus important to bear in mind that with the very high viscosity materials generally involved, heat generation can be very large and so the temperature of the fluid (and hence its rheological properties) may be a strong function of the power input to the extruder. One must thus solve the coupled momentum and energy equations, including the viscous dissipation term. General descriptions of extrusion technology are available in several books (McKelvey, 1962; Tadmor and Gogos, 1979).

Further reading

Bird, R.B., Armstrong, R.C. and Hassager, O., *Dynamics of Polymeric Liquids. Fluid Dynamics (2nd edn).*, Vol. 1, Wiley, New York (1987).
Carreau, P.J., Dekee, D. and Chhabra, R.P., *Rheology of Polymeric Systems: Principles and Applications*, Hanser, Munich (1997).
Darby, R., *Visco-elastic Fluids*, Marcel–Dekker, New York (1976).
Govier, G.W. and Aziz, K., *The Flow of Complex Mixtures in Pipes*, Krieger, Malabar, FL (1982).

References

Bandyopadhyay, T.K. and Das, S.K., *J. Pet. Sci. Eng.* **55** (2007) 156.
Banerjee, T.K., Das, M. and Das, S.K., *Can. J. Chem. Eng.* **72** (1994) 207.
Beasley, M.E., *Chem. Eng.* **99** (September, 1992) 112.
Bird, R.B., Dai, G.C. and Yarusso, B.J., *Rev. Chem. Eng.* **1** (1983) 1.
Bird, R.B., Armstrong, R.C. and Hassager, O., *Dynamics of Polymeric Liquids. Fluid Dynamics (2nd edn).*, Vol. 1, Wiley, New York (1987).
Boger, D.V., *Ann. Rev. Fluid Mech.* **19** (1987) 157.
Bogue, D.C. and Metzner, A.B., *Ind. Eng. Chem. Fundam.* **2** (1963) 143.
Bowen, R.L., *Chem. Eng.* **68** (July, 1961) 143.
Brodkey, R.S., Lee, J. and Chase, R.C., *AIChE J.* **7** (1961) 392.
Brown, N.P. and Heywood, N.I., Eds, *Slurry Handling: Design of Solid-Liquid Systems*, Elsevier, Amsterdam (1991).
Brown, N.P. and Heywood, N.I., *Chem. Eng.* **99** (Sep, 1992) 106.
Carter, G. and Lambert, D.J., *The Chem. Engr.* (Sep, 1972) 355.
Cawkwell, M.G. and Charles, M.E., *J. Pipelines* **7** (1987) 41.
Cheng, D.C.-H., *Proc. Hydrotransport* 1 (1970) 77.
Chhabra, R.P., Farooqi, S.I. and Richardson, J.F., *Chem. Eng. Res. Des.* **62** (1984) 22.
Cho, Y.I. and Hartnett, J.P., *Adv. Heat Transf.* **15** (1982) 59.
Clapp, R.M., *Int. Dev. Heat Transf,* Part III, 625, D159 & D211–215, ASME, New York (1961).
Collins, M. and Schowalter, W.R., *AIChE J.* **9** (1968) 98.
Cooke, R., *Paterson and Cooke Consulting Engineers*, Private Communication, Capetown (2007).
Coulson, J.M. and Richardson, J.F., *Chemical Engineering (6th edn).*, Vol. 1, Butterworth–Heinemann, Oxford (1999).
Darby, R., *Encyclopedia of Fluid Mechanics* (edited by Cheremisinoff, N.P.), volume **7**, Gulf, Houston, (1988) p. 19.
Darby, R., Mun, R. and Boger, D.V., *Chem. Eng.* **99** (1992) 116.
Das, S.K., Biswas, M.N. and Mitra, A.K., *Chem. Eng. J.* **45** (1991) 165.
De Haven, E.S., *Ind. Eng. Chem.* **51** (July, 1959) 59A.

Delplace, F. and Leuliet, J.C., *Chem. Eng. J.* **56** (1995) 33.

Dodge, D.W. and Metzner, A.B., *AIChE J.* 5 (1959) 189, (corrections ibid **8** (1962) 143).

Draad, A.A., Kuiken, G.D.C. and Nieuwstadt, F.T.M., *J. Fluid Mech.* **377** (1998) 267.

Duckham, C.B., *Chem. Proc. Eng.* **52** (July, 1971) 66.

Duffy, G.G., *Chem. Eng. Commun.* **191** (2004) 182.

Edwards, M.F., Nellist, D.A. and Wilkinson, W.L., *Chem. Eng. Sci.* 27 (1972) 545, *Also see ibid* **27** (1972) 295.

Edwards, M.F., Jadallah, M.S.M. and Smith, R., *Chem. Eng. Res. Des.* **63** (1985) 43.

Escudier, M.P., Gouldson, I.W. and Jones, D.M., *Exp. Fluids* **18** (1995) 225.

Escudier, M.P., Oliveira, P.J. and Pinho, F.T., *Int. J. Heat Fluid Flow* **23** (2002a) 52.

Escudier, M.P., Oliveira, P.J., Pinho, F.T. and Smith, S., *Exp. Fluids* **33** (2002b) 101.

Escudier, M.P., Poole, R.J., Presti, F., Dales, C., Nourar, C., Desaubry, C., Graham, L. and Pullum, L., *J. Non-Newt. Fluid Mech.* **127** (2005) 143.

Escudier, M.P. and Smith, S., *Proc. Royal Soc.* 457A (2001) 911.

ESDU (Engineering Sciences Data Unit), ESDU 69022 (2004).

Fester, V.G., Kazadi, D.M., Mbiya, B.M. and Slatter, P.T., *Chem. Eng. Res. Des.* **85** (2007) 1314.

Fordham, E.J., Bittleston, S.H. and Tehrani, M.A., *Ind. Eng. Chem. Res.* **30** (1991) 517.

Fredrickson, A.G. and Bird, R.B. *Ind. Eng. Chem.* **50** (1958), 347. Also see *ibid* **50** (1958) 1599 & *Ind. Eng. Chem. Fundam.* **3** (1964) 383.

Frigaard, I., Howison, S.D. and Sobey, I., *J. Fluid Mech.* **263** (1994) 133.

Fyrippi, I., Owen, I. and Escudier, M.P., *Flow Meas. Instrum.* **15** (2004) 131.

Garcia, E.J. and Steffe, J.F., *J. Food Process Eng.* **9** (1987) 93.

Ginesi, D., *Chem. Eng.* 98 (May, 1991) 146.

Govier, G.W. and Aziz, K., *The Flow of Complex Mixtures in Pipes*, Krieger, Malabar, FL (1982).

Griskey, R.G. and Green, R.G., *AIChE J.* **17** (1971) 725.

Gücüyener, H.I. and Mehmeteoglu, T., *AIChE J.* **38** (1992) 1139.

Haldenwang, R., Erasmus, R., Slatter, P. and Chhabra, R.P., *Hydrotransport* **17** (2007) 301.

Hanks, R.W., *AIChE J.* **9** (1963) 306.

Hanks, R.W., *Ind. Eng. Chem., Proc. Des. Dev.* **18** (1979) 488.

Hanks, R.W., *Encyclopedia of Fluid Mechanics* (edited by Cheremisinoff, N.P.), volume **5**, Gulf, Houston, p. 213 (1986).

Hanks, R.W. and Larsen, K.M., *Ind. Eng. Chem. Fundam.* **18** (1979) 33.

Harris, J., *Rheol. Acta* **7** (1968) 228.

Harris, J. and Magnall, A.N., *Trans Inst. Chem. Engrs.* **50** (1972) 61.

Hartnett, J.P. and Kostic, M., *Int. J. Heat Mass Transf.* **28** (1985) 1147.

Hartnett, J.P. and Kostic, M., *Adv. Heat Transf.* **19** (1989) 247.

Hartnett, J.P. and Kostic, M., *Int. Commun. Heat Mass Transf.* **17** (1990) 59.

Hartnett, J.P., Kwack, E.Y. and Rao, B.K., *J. Rheol.* **30** (1986) S45.

Hedstrom, B.O.A., *Ind. Eng. Chem.* **44** (1952) 651.

Hemeida, A.M., *J. Pet. Sci. Eng.* **10** (1993) 163.

Heywood, N.I. and Cheng, D.C.-H., *Trans. Inst. Meas. Control* **6** (1984) 33.

Irvine, Jr., T.F., *Chem. Eng. Commun.* **65** (1988) 39.

Kajiuchi, T. and Saito, A., *J. Chem. Eng. Jpn.* **17** (1984) 34.

Kostic, M. and Hartnett, J.P., *Int. Comm. Heat Mass Transf.* **11** (1984) 345.

Kozicki, W., Chou, C.H. and Tiu, C., *Chem. Eng. Sci.* **21** (1966) 665.

Kozicki, W., and Tiu, C. *Can, J., Chem. Eng.* **45** (1967) 127.

Kozicki, W. and Tiu, C., *Encyclopedia of Fluid Mechanics* (edited by Cheremisinoff, N.P.), volume **7**, Gulf, Houston, p. 199 (1988).

Laird, W.M., *Ind. Eng. Chem.* **49** (1957) 138.

Liptik, B.G., *Chem. Eng.* **70** (Jan 30, 1967), 133; also see 70 (Feb 13, 1967) 151.

Liu, T.-J., *Ind. Eng. Chem. Fundam.* **22** (1983) 183.

Malin, M.R., *Int. Comm. Heat Mass Transf.* **24** (1997) 793.

Martinez-Padilla, L.P. and Linares Garcia, J.A., *J. Food Proc. Eng.* **24** (2001) 135.

McKelvey, J.M., *Polymer Processing*, Wiley, New York (1962).

Metzner, A.B., *Adv. Chem. Eng.* **1** (1956) 79.

Metzner, A.B. and Reed, J.C., *AIChE J.* **1** (1955) 434.

Metzner, A.B. and Park, M.G., *J. Fluid Mech.* **20** (1964) 291.

Miller, C., *Ind. Eng. Chem. Fundam.* **11** (1972) 524.

Millikan, C.B., *Proc. 5th Int. Cong. Applied Mech.* (1939) 386.
Mishra, P. and Tripathi, G., *Chem. Eng. Sci.* **26** (1971) 915.
Mitsuishi, N. and Aoyagi, Y., *Chem. Eng. Sci.* **24** (1969) 309.
Mitsuishi, N. and Aoyagi, Y., *J. Chem. Eng. Jpn.* **6** (1973) 402.
Mitsuishi, N., Aoyagi, Y. and Soeda, H., *Kagaku–Kogaku.* **36** (1972) 182.
Nakamura, M. and Swada, T., *J. Non-Newt. Fluid Mech.* **22** (1987) 191.
Nouar, C. and Frigaard, I., *J. Non-Newt. Fluid Mech.* **100** (2001) 127.
Nouri, J.M., Umur, H. and Whitelaw, J.H., *J. Fluid Mech.* **253** (1993) 617.
Ookawara, S., Ogawa, K., Dombrowski, N., Amooie-Foumeny, E. and Riza, A., *J. Chem. Eng. Jpn.* **33** (2000) 675.
Pal, R., *Ind. Eng. Chem. Res.* **32** (1993) 1212.
Park, J.T., Mannheimer, R.J., Grimley, T.A. and Morrow, T.B., *J. Fluids Eng. (ASME)* **111** (1989) 331.
Peixinho, J., Nouar, C., Desaubry, C. and Theron, B., *J. Non-Newt. Fluid Mech.* **128** (2005) 172.
Phan-Thien, N. and Dudek, J., *Nature* **296** (1982a) 843.
Phan-Thien, N. and Dudek, *J., Non-Newt. Fluid Mech.* **11** (1982b) 147.
Polizelli, M.A., Menegalli, F.C., Telis, V.R.N. and Telis-Romero, J., *Braz. J. Chem. Eng.* **20** (2003) 455.
Poole, R.J. and Ridley, B.S., *J. Fluids Eng. (ASME)* **129** (2007) 1281.
Quader, A.K.M.A. and Wilkinson, W.L., *Int. J. Multiphase Flow* **6** (1980) 553.
Ramamurthy, A.V. and Boger, D.V., *Trans. Soc. Rheol.* **15** (1971) 709.
Rudman, M. and Blackburn, H.M., *Appl. Math. Modelling* **30** (2006) 1229.
Rudman, M., Blackburn, H.M., Graham, L.J.W. and Pullum, L., *J. Non-Newt. Fluid Mech.* **118** (2004) 33.
Ryan, N.W. and Johnson, M.M., *AIChE J.* **5** (1959) 433.
Schechter, R.S., *AIChE J.* **7** (1961) 445.
Sestak, J., Charles, M.E., Cawkwell, M.G. and Houska, M., *J. Pipelines* **6** (1987) 15.
Shenoy, A.V., *Encyclopedia of Fluid Mechanics* (edited by Cheremisinoff, N.P.), volume 7, Gulf, Houston, (1988) p. 479 .
Shenoy, A.V. and Talathi, M.M., *AIChE J.* **31** (1985) 520.
Skelland, A.H.P., *Non-Newtonian Flow and Heat Transfer*, Wiley, New York (1967).
Slatter, P.T., *Hydrotransport* **13** (1996) 97.
Slatter, P.T. (1997). *9th Int. Conf. Transport & Sedimentation of Solid Particles*, Cracow, p. 547.
Slatter, P.T. and van Sittert, F.P.(1997). *9th Int. Conf. on Transport & Sedimentation of Solid Particles*, Cracow, p. 621.
Slatter, P.T., Pienaar, V.G. and Petersen, F.W. (1997) *9th Int. Conf. on Transport & Sedimentation of Solid Particles*, Cracow, p. 585.
Steffe, J.F., Mohamed, I.O. and Ford, E.W., *Trans. Amer. Soc. Agri. Engrs.* **27** (1984) 616.
Steffe, J.F. and Morgan, R.G., *Food Technol.* **40** (Dec 1986) 78.
Szilas, A.P., Bobok, E. and Navratil, L., *Rheol. Acta* **20** (1981) 487.
Tadmor, Z. and Gogos, G., *Principles of Polymer Processing*, Wiley, New York (1979).
Tam, K.C., Tiu, C. and Keller, R.J., *J. Hyd. Res.* **30** (1992) 117.
Thomas, A.D. and Wilson, K.C., *Can. J. Chem. Eng.* **65** (1987) 335.
Tomita, Y., *Bull. Jpn. Soc. Mech. Engrs.* **2** (1959) 10.
Torrance, B.McK., *South African Mech. Engr.* **13** (1963) 89.
Turian, R.M., Ma, T.-W., Hsu, F.-L.G., Sung, M.D.-J. and Plackmann, G.W., *Int. J. Multiphase Flow* **24** (1998) 225; also see *ibid* 242.
Uner, D., Ozgen, C. and Tosun, I., *Ind. Eng. Chem. Res.* **27** (1988) 698.
Urey, J.F., *J. Mech. Eng. Sci.* **8** (1966) 226.
Vinay, G., Wachs, A. and Agassant, J.-F., *J. Non-Newt. Fluid Mech.* **128** (2005) 144, Also see *ibid* 136 (2006) 93.
Vinay, G., Wachs, A. and Frigaard, I., *J. Non-Newt. Fluid Mech.* **143** (2007) 141.
Walker, C.I. and Goulas, A., *Proc. Inst. Mech. Engrs, Part A* **198** (1984) 41.
Walton, I.C. and Bittleston, S.H., *J. Fluid Mech.* **222** (1991) 39.
Wardhaugh, L.T. and Boger, D.V., *Chem. Eng. Res. Des.* **65** (1987) 74.
Wardhaugh, L.T. and Boger, D.V., *AIChE J.* **37** (1991) 871.
Weltmann, R.N. and Keller, T.A., *NACA Technical Note 3889* (1957).
Wheeler, J.A. and Wissler, E.H., *AIChE J.* **11** (1965) 207.
Wilson, K.C. *Hydrotransport* **13** (1996) 61–74. *BHR Group* Cranfield, UK.
Wilson, K.C. and Thomas, A.D., *Can. J. Chem. Eng.* 63 (1985) 539.

Wojs, K., *J. Non-Newt. Fluid Mech.* 48 (1993) 337.

Xie, C. and Hartnett, J.P., *Ind. Eng. Chem. Res.* 31 (1992) 727.

Nomenclature

		Dimensions in **M, L, T**
a, b	geometric parameters for non-circular ducts ($-$)	$\mathbf{M^0 L^0 T^0}$
h	half gap between two parallel plates (m)	$\mathbf{L}$
$Bi = D\tau_0^B / V\mu_B$	Bingham number ($-$)	$\mathbf{M^0 L^0 T^0}$
D	pipe diameter (m)	$\mathbf{L}$
D_h	hydraulic diameter (m)	$\mathbf{L}$
E	eddy diffusivity (m²/s)	$\mathbf{L^2 T^{-1}}$
f	friction factor ($-$)	$\mathbf{M^0 L^0 T^0}$
g	acceleration due to gravity (m/s²)	$\mathbf{L T^{-2}}$
H	film thickness in flow over inclined surface (m)	$\mathbf{L}$
$He = \rho D^2 \tau_0^B / \mu_B^2$	Hedström number ($-$)	$\mathbf{M^0 L^0 T^0}$
h_l	frictional head loss (m)	$\mathbf{L}$
L	pipe length (m)	$\mathbf{L}$
l	Prandtl mixing length (m)	$\mathbf{L}$
m	power law consistency coefficient (Pa · s^n)	$\mathbf{M L^{-1} T^{n-2}}$
m'	apparent power law consistency coefficient (Pa · s$^{n'}$)	$\mathbf{M L^{-1} T^{n'-2}}$
n	power-law flow behaviour index ($-$)	$\mathbf{M^0 L^0 T^0}$
n'	apparent power-law flow behaviour index ($-$)	$\mathbf{M^0 L^0 T^0}$
p	pressure (Pa)	$\mathbf{M L^{-1} T^{-2}}$
$-\Delta p$	pressure drop (Pa)	$\mathbf{M L^{-1} T^{-2}}$
Q	volumetric flow rate (m³/s)	$\mathbf{L^3 T^{-1}}$
r	radial coordinate (m)	$\mathbf{L}$
R	pipe radius/annulus outer radius (m)	$\mathbf{L}$
R_p	radius of plug in centre of pipe (m)	$\mathbf{L}$
Re	Reynolds number for Newtonian fluids ($-$)	$\mathbf{M^0 L^0 T^0}$
Re_B	Reynolds number based on Bingham plastic viscosity ($-$)	$\mathbf{M^0 L^0 T^0}$
Re_{C-S}	modified Reynolds number $(= \rho V^{2-n} D^n / m)(-)$	$\mathbf{M^0 L^0 T^0}$
Re_g	generalized Reynolds number for non-circular conduits, equation 3.152 ($-$).	$\mathbf{M^0 L^0 T^0}$
Re_{MR}	Metzner–Reed definition of Reynolds number ($-$)	$\mathbf{M^0 L^0 T^0}$
Re_{mod}	modified Reynolds number, equation (3.69) ($-$)	$\mathbf{M^0 L^0 T^0}$
$Re*$	modified Reynolds number, equation (3.171) ($-$)	$\mathbf{M^0 L^0 T^0}$
V	mean velocity of flow (m/s)	$\mathbf{L T^{-1}}$
V_z	point velocity of flow in z-direction (m/s)	$\mathbf{L T^{-1}}$
V_z^*	non-dimensional point velocity ($-$)	$\mathbf{M^0 L^0 T^0}$

$V^* = \sqrt{\tau_w/\rho}$	friction velocity (m/s)	$\mathbf{LT}^{-1}$
V^+	non-dimensional velocity, equation 3.89 (−)	$\mathbf{M}^0\mathbf{L}^0\mathbf{T}^0$
W	width (m)	$\mathbf{L}$
y	distance from wall (m)	$\mathbf{L}$
y^+	non-dimensional distance from wall, equation 3.89 (−)	$\mathbf{M}^0\mathbf{L}^0\mathbf{T}^0$
z	axial coordinate in the flow direction (m)	$\mathbf{L}$

Greek letters

α	parameter in Ellis fluid model or kinetic energy correction factor (−)	$\mathbf{M}^0\mathbf{L}^0\mathbf{T}^0$
β	factor defined in equation (3.154) (−)	$\mathbf{M}^0\mathbf{L}^0\mathbf{T}^0$
λ	radial coordinate corresponding to zero shear stress in annulus (−)	$\mathbf{M}^0\mathbf{L}^0\mathbf{T}^0$
λ_+, λ_-	boundaries of plug of Bingham plastic fluid in annular flow (−)	$\mathbf{M}^0\mathbf{L}^0\mathbf{T}^0$
μ	apparent viscosity (Pa s)	$\mathbf{ML}^{-1}\mathbf{T}^{-1}$
μ_0	zero shear viscosity (Pa s)	$\mathbf{ML}^{-1}\mathbf{T}^{-1}$
μ_B	Bingham plastic viscosity (Pa s)	$\mathbf{ML}^{-1}\mathbf{T}^{-1}$
$\xi = r/R$	non-dimensional radial coordinate (−)	$\mathbf{M}^0\mathbf{L}^0\mathbf{T}^0$
ρ	density (kg/m³)	$\mathbf{ML}^{-3}$
σ	non-dimensional inner radius of annulus (−)	$\mathbf{M}^0\mathbf{L}^0\mathbf{T}^0$
τ_{rz}	shear stress in fluid (Pa)	$\mathbf{ML}^{-1}\mathbf{T}^{-2}$
τ_w	shear stress at pipe wall (Pa)	$\mathbf{ML}^{-1}\mathbf{T}^{-2}$
$\tau_{1/2}$	parameter in Ellis fluid model (Pa)	$\mathbf{ML}^{-1}\mathbf{T}^{-2}$
τ_0^B	yield stress in Bingham plastic model (Pa)	$\mathbf{ML}^{-1}\mathbf{T}^{-2}$
$\phi = \tau_0^B/\tau_w$	non-dimensional ratio (−)	$\mathbf{M}^0\mathbf{L}^0\mathbf{T}^0$
Ω	non-dimensional flow rate (−)	$\mathbf{M}^0\mathbf{L}^0\mathbf{T}^0$

Chapter 4

Flow of multi-phase mixtures in pipes

4.1 Introduction

The flow problems considered in the previous chapter have concerned either single phases or pseudo-homogeneous fluids such as emulsions and suspensions of fine particles in which little or no separation occurs. Attention will now be focussed on the far more complex problem of the flow of multi-phase systems in which the composition of the mixture may show spatial variation over the cross-section of the pipe or channel. Furthermore, the two components may have different *in situ* velocities as a result of which there is 'slip' between the two phases and *in situ* holdups which are different from those in the feed or exit stream. Consequently, the residence times of the two phases will be different.

Multi-phase flow is encountered in many chemical and process engineering applications, and the behaviour of the material is influenced by the properties of the components, such as their Newtonian or non-Newtonian characteristics or the size, shape and concentration of particulates, the flow rate of the two components and the geometry of the system. In general, the flow is so complex that theoretical treatments, which tend to apply to highly idealized situations, have proved to be of little practical utility. Consequently, design methods rely very much on analyses of the behaviour of such systems in practice. While the term 'multi-phase flows' embraces the complete spectrum of gas–liquid, liquid–liquid, gas–solid, liquid–solid, gas–liquid–solid and gas–liquid–liquid systems, the main concern here is to illustrate the role of non-Newtonian rheology of the liquid phase on the nature of the flow. Attention is concentrated on the simultaneous cocurrent flow of a gas and a non-Newtonian liquid and the transport of coarse solids in non-Newtonian liquids.

Multi-phase mixtures may be transported horizontally, vertically, or at an inclination to the horizontal in pipes and, in the case of liquid–solid mixtures, in open channels. Although there is some degree of similarity between the hydrodynamic behaviour of the various types of multi-phase flows, the range of physical properties is so wide that each system must be considered separately even when the liquids are Newtonian. Liquids may have densities up to three orders of magnitude greater than gases, but they are virtually incompressible.

The liquids themselves may range from simple Newtonian fluids, such as water, to highly viscous non-Newtonian liquids, such as polymer solutions, fine particle slurries. Indeed, because of the large differences in density and viscosity, the flow of gas–liquid mixtures and liquid–solid (coarse) mixtures must, in practice, be considered separately. For all multi-phase flow systems, it is, however, essential to understand the nature of the interactions between the phases and how these affect the flow patterns, including the way in which the two phases are distributed over the cross-section of the pipe or channel. Notwithstanding the importance of the detailed kinematics of flow, the ensuing discussion is mainly concerned with the overall hydrodynamic behaviour, with particular reference to the following features: flow patterns, average holdup of the individual phases and the frictional pressure gradient. Flow patterns are strongly influenced by the difference in the density of the two phases. In gas–liquid systems, it is always the gas which is the lighter phase and in solid–liquid systems, it is more often than not the liquid. The orientation of the pipe may also play a role. In vertical upward flow, for instance, it will be the lighter phase which will tend to rise more quickly. For liquid–solid mixtures, the slip velocity is of the same order as the terminal settling velocity of the particles but, in a gas–liquid system, it depends upon the flow pattern in a complex manner.

Unlike vertical flow, horizontal flow does not exhibit axial symmetry and the flow pattern is more complex because the gravitational force acts normally to the direction of flow, causing asymmetrical distribution over the pipe cross-section.

In practice, many other considerations affect the design of an installation. For example, wherever particulates are involved, pipe blockage may occur with consequences that can be as serious that it is always necessary to operate under conditions which minimize the probability of its occurring. Abrasive solids may give rise to the undue wear, and high velocities may need to be avoided.

Though the main emphasis in this chapter is on the effects of the non-Newtonian rheology, it is useful to draw analogies with the simpler cases of Newtonian liquids, details of which are much more readily available.

4.2 Two-phase gas non-Newtonian liquid flow

4.2.1 Introduction

This section deals with the most important characteristics of the flow of a mixture of a gas or vapour and a Newtonian or non-Newtonian liquid in a round pipe. Despite large differences in rheology, two phase flow of gas–liquid mixtures exhibits many common features whether the liquid is Newtonian or shows inelastic pseudoplastic behaviour. Applications in the chemical, food and processing industries range from the flow of mixtures of crude oil (which may exhibit non-Newtonian characteristics) and gas from oil well heads to that of vapour–liquid mixtures in boilers and evaporators. Further examples are found in the use of vacuum sewers, in gas–condensate pipelines and in the flow of greases (containing air) in pipes, etc.

The nature of the flow of gas–liquid mixtures is complex, and the lack of knowledge concerning local velocities of the individual phases makes it difficult to develop any method of predicting the velocity distribution. In many instances, the gas (or vapour) phase may be flowing considerably faster than the liquid and continually accelerating as a result of its expansion, as the pressure falls. Either phase may be in laminar or in turbulent flow, albeit that laminar flow (especially for a gas), does not have such a clear cut meaning as in the flow of single fluids. In practice, Newtonian liquids will most often be in turbulent flow, whereas the flow of non-Newtonian liquids is more often streamline because of their high apparent viscosities.

Additional complications arise if there is heat transfer from one phase to another such as that encountered in the tubes of a condenser or boiler. Under these conditions, the mass flow rate of each phase is progressively changing as a result of the vapour condensing or the liquid vaporizing. However, this phenomenon is of little relevance to the flow of gas and non-Newtonian liquid mixtures.

Consideration will now be given in turn to three particular aspects of gas–liquid flow which are of practical importance (i) flow patterns or regimes, (ii) holdup and (iii) frictional pressure gradient.

4.2.2 Flow patterns

For two phase cocurrent gas–liquid flow, there is the wide variety of possible flow patterns which are governed principally by the physical properties (density, surface tension, viscosity of gas, rheology of liquid), input fluxes of the two phases and the size and the orientation of the pipe. Since the mechanisms responsible for holdup and momentum transfer (or frictional pressure drop) vary from one flow pattern to another, it is essential to have a method of predicting the conditions under which each flow pattern may occur. Before developing suitable methods for the prediction of flow pattern, it is important to briefly define the flow patterns generally encountered in gas–liquid flows. Horizontal and vertical flows will be discussed separately as there are inherent differences in the two cases.

Horizontal flow

The classification proposed by Alves (1954) encompasses all the major and easily recognizable flow patterns encountered in horizontal pipes. These are sketched in Figure 4.1 and are described below.

(i) *Bubble flow*

This type of flow, sometimes referred to as dispersed bubble flow, is characterized by a train of discrete gas bubbles moving mainly close to the upper wall of the pipe, at almost the same velocity as the liquid. As the liquid flow rate is increased, the bubbles become more evenly distributed over the cross-section of the pipe.

(ii) *Plug flow*

At increased gas throughput, bubbles interact and coalesce to give rise to large bullet shaped plugs occupying most of the pipe cross-section, except for a thin liquid film at the wall of the pipe which is thicker towards the bottom of the pipe.

(iii) *Stratified flow*

In this mode of flow, the gravitational forces dominate the gas phase flows in the upper part of the pipe. At relatively low flow rates, the gas–liquid interface is smooth, but becomes ripply or wavy at higher gas rates thereby giving rise to the so-called 'wavy flow'. As the distinction between the smooth and wavy interface is often ill-defined, it is usual to refer to both flow patterns as stratified–wavy flow.

(iv) *Slug flow*

In this type of flow, frothy slugs of liquid phase carrying entrained gas bubbles alternate with gas slugs surrounded by thin liquid films. Although plug and slug flow are both well defined, as shown in Figure 4.1, in practice it is often not easy to distinguish between them.

(v) *Annular flow*

In this type of flow, most of the liquid is carried along the inner wall of the pipe as a thin film, while the gas forms a central core occupying a substantial portion of the pipe cross-section. Some liquid is usually entrained as fine droplets within the gas core. Sometimes the term 'film flow' is also used to describe this flow pattern.

Figure 4.1 *Flow patterns in horizontal and vertical two phase flow*

(vi) *Mist flow*

Mist flow is said to occur when a significant amount of liquid becomes transferred from the annular film to the gas core; at high gas flow rates nearly all of the liquid is entrained in the gas. Thus, some workers regard mist flow to be an extreme case of annular flow.

The above classification of flow patterns is useful in developing models for the flow, but it is important to note that the distinction between any two flow patterns is far from clear cut, especially in the case of bubble, plug and slug flow patterns. Consequently, these latter three flow patterns are often combined and described as 'intermittent flow'. On this basis, the chief flow patterns adopted in this work are dispersed, intermittent, stratified, wavy and annular–mist flow. Detailed descriptions of flow patterns and of the experimental methods for their determination are available in the literature (e.g. see Hewitt, 1978, 1982; Ferguson and Spedding, 1995). Although most of the information on flow patterns has been gathered from experiments on gas and Newtonian liquids in cocurrent flow, the limited experimental work reported to date with inelastic shear-thinning materials suggests that they give rise to qualitatively similar flow patterns (Chhabra and Richardson, 1984) and therefore the same nomenclature will be adopted here.

4.2.3 Prediction of flow patterns

For the flow of gas and Newtonian liquid mixtures, several, mostly empirical, attempts have been made to formulate flow pattern maps (Govier and Aziz, 1982; Hetsroni, 1982; Chisholm, 1983). The regions over which the different types of flow patterns can occur are conveniently shown as a 'flow pattern map' in which a function of the liquid flow rate is plotted against a function of the gas flow rate and boundary lines are drawn to delineate the various regions. Not only is the distinction between any two flow patterns poorly defined, but the transition from one flow pattern to another may occur over a range of conditions rather than abruptly as suggested in all flow pattern maps. Furthermore,

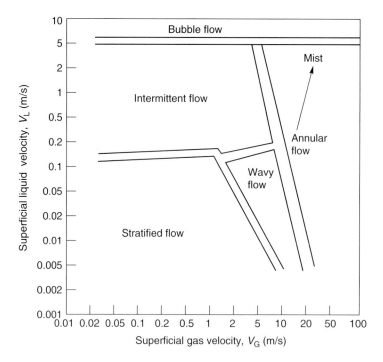

Figure 4.2   *Modified flow pattern map (Chhabra and Richardson, 1984)*

because the flow patterns are usually identified by visual observations of the flow, there is a large element of subjectivity in the assessment of the boundaries.

Most of the data used for constructing such maps have been obtained with the air–water system at or near atmospheric conditions. Although, intuitively, one might expect the physical properties of the two phases to play important roles in determining the transition from one flow pattern to another, it is now generally recognized that they indeed have very little effect (Mandhane *et al.*, 1974; Weisman *et al.*, 1979; Chhabra and Richardson, 1984). Even shear-thinning behaviour seems to play virtually no role in governing the transition from one flow pattern to another (Chhabra and Richardson, 1984; Luo and Ghiaasiaan, 1997; Xu *et al.*, 2007). Based on these considerations and taking into account the extensive experimental study of Weisman *et al.* (1979), Chhabra and Richardson (1984) modified the widely used flow pattern map of Mandhane *et al.* (1974) as shown in Figure 4.2. On the basis of a critical evaluation of the literature on the flow of mixtures of gas and shear-thinning fluids (aqueous polymer solutions, particulate suspensions of china clay, coal and limestone), this scheme was shown to reproduce about 3700 data points on flow patterns with 70% certainty. The range of experimental conditions which have been used in the compilation of this flow pattern map are: $0.021 \leq V_L \leq 6.1$ m/s; $0.01 \leq V_G \leq 55$ m/s; $6.35 \leq D \leq 207$ mm and $0.1 \leq n' \leq 1$. For the systems considered in the preparation of the flow pattern map (Figure 4.2), apparent viscosities of the non-Newtonian liquids at a shear rate of $1 s^{-1}$ varied from 10^{-3} Pas (water) to 50 Pas. Such few experimental results as are available for visco-elastic polymer solutions are also correlated well by this flow pattern map (Chhabra and Richardson, 1984). Though some attempts have been made to improve the prediction of flow patterns, but these have met with only limited success. On the other hand, new data on flow patterns are also consistent with the flow pattern map shown in Figure 4.2 (Luo and Ghiaasiaan, 1997; Xu *et al.*, 2007).

Vertical upward flow

In vertical flow, gravity acts in the axial direction giving symmetry across the pipe cross-section. Flow patterns tend to be somewhat more stable, but with slug flow, oscillations in the flow can occur as a result of sudden changes in pressure as liquid slugs are discharged from the end of the pipe. This effect is also present in horizontal flow.

The flow patterns observed in vertical upward flow of a gas and Newtonian liquid are similar to those shown in Figure 4.1 and are described in detail elsewhere (Barnea and Taitel, 1986). Taitel *et al.* (1980) have carried out a semi-theoretical study of the fundamental mechanisms responsible for each flow pattern, and have derived quantitative expressions for the transition from one regime to another. This analysis shows a strong dependence on the physical properties of the two phases and on the pipe diameter. Figure 4.3 shows their map for the flow of air–water mixtures in a 38 mm diameter pipe along with the more recent data of Dziubinski *et al.* (2004) for $D = 40.6$ mm.

Only a few studies are available on the simultaneous flow of a gas and non-Newtonian fluids in vertical pipes (Khatib and Richardson, 1984; Welsh *et al.*, 1999; Dziubinski *et al.*, 2004). Both Khatib and Richardson (1984) and Dziubinski *et al.* (2004) studied cocurrent upward flow of air and several shear-thinning fluids in pipes ranging from 25 to 50 mm

Figure 4.3 *Experimental and predicted (Taitel et al., 1980) flow patterns for upward flow of air and china clay suspensions (D = 38 mm, Khatib and Richardson, 1984; D = 40.6 mm, Dziubinski et al., 2004)*

in diameter. Khatib and Richardson (1984) reported that their results compared closely with the predictions of Taitel *et al.* (1980) for air–water mixtures and this suggests that the transition boundaries between the various flow pattern are largely unaffected by the rheology of the liquid and that Figure 4.3 can be used when the liquid, is shear-thinning. Interestingly, the results of Dziubinski *et al.* (2004) also seem to be in line with the predictions of Taitel *et al.* (1980), as shown in Figure 4.3. In spite of this, they argued that some improvement is possible by using $V_L \sqrt{\rho_L/\rho_W}$ and $(V_G/V_L)(\rho_G \rho_W/\rho_{air}\rho_L)^{0.5}$ coordinates in lieu of simply the superficial velocities V_L and V_G, as used in Figure 4.3. A detailed examination of their results revealed that over the range of conditions encompassed in their study $\rho_G \approx \rho_{air}$ and $\rho_L \approx \rho_W$ and therefore their method is really equivalent to plotting V_L against (V_G/V_L). Similarly, Welsh *et al.* (1999) concluded that the flow patterns in counter-current flow of a gas and shear-thinning solutions were little influenced by slight inclinations of pipe from vertical. However, no such information is available for viscoelastic liquids.

4.2.4 Holdup

Because of the considerable differences in the physical properties (particularly viscosity and density) of gases and liquids, the gas always tends to flow at a higher average velocity than the liquid. Sometimes, this can also occur if the liquid preferentially wets the surface of the pipe and therefore experiences a greater drag. In both cases, the volume fraction (holdup) of liquid at any point in a pipe will be greater than that in the mixture entering or leaving the pipe. Furthermore, if the pressure falls significantly along the pipeline, the holdup of liquid will progressively decrease as a result of the expansion of the gas.

If α_L and α_G are the holdups for liquid and gas, respectively, it follows that:

$$\alpha_L + \alpha_G = 1 \tag{4.1}$$

Similarly for the input volume fractions:

$$\lambda_L + \lambda_G = 1 \tag{4.2}$$

λ_L and λ_G may be expressed in terms of the flow rates Q_L and Q_G, at a given point in the pipe, as:

$$\lambda_L = \frac{Q_L}{Q_L + Q_G} = \frac{V_L}{V_L + V_G} \tag{4.3}$$

$$\lambda_G = \frac{Q_G}{Q_L + Q_G} = \frac{V_G}{V_L + V_G} \tag{4.4}$$

where V_L and V_G are the superficial velocities of the two phases. Only under the limiting conditions of no-slip between the two phases and of no significant pressure drop along the pipe will α and λ be equal. Liquid (or gas) holdup along the length of the pipe must be known for the calculation of two phase pressure drop.

Experimental determination

The experimental techniques available for measuring holdup fall into two categories, namely, direct and indirect methods. The direct method of measurement involves suddenly isolating a section of the pipe by means of quick-acting valves and then determining the quantity of liquid trapped. Good reproducibility may be obtained, as shown by widespread use of this technique for both Newtonian and non-Newtonian liquids (Hewitt *et al.*, 1963; Oliver and Young-Hoon, 1968; Mahalingam and Valle, 1972; Chen and Spedding, 1983). It yields a volume average value of holdup. Although the method is, in principle, simple, it has two main drawbacks. Firstly, the valves cannot operate either instantaneously or exactly simultaneously. This must lead to inaccuracies and, after each measurement, ample time must be allowed for the flow to reach a steady state. Secondly, it is not practicable to use this method for high temperature and pressure situations and/or when either the gas or the liquid or both are of hazardous nature.

The indirect non-intrusive methods have the advantage of not disturbing the flow. The underlying principle is to measure a physical or electrical property that is strongly dependent upon the composition of the gas–liquid mixture. Typical examples include the measurement of γ-ray or X-ray attenuation (Petrick and Swanson, 1958; Pike *et al.*, 1965; Shook and Liebe, 1976), or of change in impendence (Gregory and Mattar, 1973; Shu *et al.*, 1982) or of change in conductivity (Fossa, 1998). Such methods, however, require calibration and yield values (averaged over the cross-section) at a given position in the pipeline. The γ-ray attenuation method has been used extensively to measure liquid holdup for two phase flow of mixtures of air and non-Newtonian liquids such as polymer solutions and particulate suspensions in horizontal and vertical pipelines (Heywood and Richardson, 1979; Farooqi and Richardson, 1982; Chhabra *et al.*, 1984; Khatib and Richardson, 1984).

Predictive methods for horizontal flow

Methods available for the prediction of the average value of liquid holdup fall into two categories: those methods which are based on models which utilize information implicit in the flow pattern and those which are entirely empirical. Taitel *et al.* (1980) have developed a semi-theoretical expression for the average liquid holdup and the two phase pressure gradient for the stratified flow of mixtures of air and Newtonian liquids. Although such analyses attempt to give some physical insight into the flow mechanism, they inevitably entail gross simplifications and empiricism. For instance, Taitel *et al.* (1980) assumed the interface to be smooth and the interfacial friction factor to be the same as that for the gas, but this model tends to underestimate the two phase pressure drop. This methodology has subsequently been extended to the stratified flow of a gas and power-law liquids (Heywood and Charles, 1979). Similar idealized models are available for the annular flow of gas and power-law liquids in horizontal pipes (Eissenberg and Weinberger, 1979) but most of them assume the liquid to be in streamline flow and the gas turbulent.

The second category of methods includes purely empirical correlations which disregard the flow patterns and are applicable over stated ranges of the variables. Although such an approach contributes little to our understanding, it does provide the designer with the vital information of a known degree of accuracy and reliability. Numerous empirical expressions are available in the literature for the prediction of the liquid holdup when the liquid is Newtonian and these have been critically evaluated (Mandhane *et al.*, 1975; Govier and Aziz, 1982; Spedding and Chen, 1986). The simplest and perhaps most widely used correlation is that of Lockhart and Martinelli (1949) which utilizes

Figure 4.4 *Lockhart–Martinelli correlation for liquid holdup and representative experimental results*

the pressure drop values for single phase flow to define a so-called Lockhart–Martinelli parameter, χ which is:

$$\chi = \left(\frac{-\Delta p_L /L}{-\Delta p_G /L}\right)^{1/2}$$

(4.5)

where $(-\Delta p_L/L)$ and $(-\Delta p_G/L)$ are, respectively, the pressure gradients for the flow of liquid and gas alone at the same volumetric flow rates as in the two phase flow. Although it is based on experimental data for the flow of air–water mixtures in small diameter tubes ($\sim$25 mm) at near atmospheric pressure and temperature, this correlation has proved to be quite successful when applied to other fluids and for tubes of larger diameters. The original correlation, shown in Figure 4.4, consistently overestimates the value of the liquid holdup (α_L) in horizontal flow of two phase gas–Newtonian liquid mixtures. This can be seen in Figure 4.4 which shows the comprehensive data (Chen and Spedding, 1983) for air–water mixtures, of Farooqi (1981) and Farooqi and Richardson (1982) for the flow of air with aqueous glycerol solutions of various compositions. Taken together, the experimental results shown in Figure 4.4 cover a range of four orders of magnitude of the Lockhart–Martinelli parameter, χ, and average liquid holdups from 0.5% to $\sim$100%. These data cover all the major flow patterns and flow regimes, e.g. nominal streamline and turbulent flow of both gas and liquid. The available experimental results are well represented by the following empirical expressions, as shown in Figure 4.4:

$$\alpha_L = 0.24\chi^{0.8} \qquad 0.01 \leq \chi \leq 0.5$$

(4.6a)

$$\alpha_L = 0.175\chi^{0.32} \qquad 0.5 \leq \chi \leq 5$$

(4.6b)

$$\alpha_L = 0.143\chi^{0.42} \qquad 5 \leq \chi \leq 50 \qquad\qquad (4.6c)$$

$$\alpha_L = \frac{1}{0.97 + (19/\chi)} \qquad 50 \leq \chi \leq 500 \qquad\qquad (4.6d)$$

Furthermore, these equations predict values of α_L to within $\pm 1\%$ at the values of χ marking the changeover point between equations. The overall average error is of the order of 7% and the maximum error is about 15%.

Gas–non-Newtonian systems

Because of the widely different types of behaviour exhibited by non-Newtonian fluids, it is convenient to deal with each flow regime separately, depending upon whether the liquid flowing on its own at the same flow rate would be in streamline or turbulent flow. While it is readily conceded that streamline flow does not have as straightforward a meaning in two phase flows as in the flow of single fluids, for the purposes of correlating experimental results, the same criterion is used to delineate the type of flow for non-Newtonian fluids as discussed in Chapter 3 (Section 3.3), and it will be assumed here that the flow will be streamline for $Re_{MR} < 2000$, prior to the introduction of gas.

Streamline flow of liquid

The predictions from equation (4.6) will be compared first with the experimental values of average liquid holdup for cocurrent two phase flow of a gas and shear-thinning liquids. For a liquid of given rheology (m and n), the pressure gradient $(-\Delta p_L/L)$ may be calculated using the methods presented in Chapter 3, but only the power-law model will be used here.

Figure 4.5 shows representative experimental results for average values of liquid holdup α_L, as a function of the parameter χ, together with the predictions of equation (4.6). The curves refer to a series of aqueous china clay suspensions in cocurrent flow with air in a 42 mm diameter horizontal pipe. In addition, Figure 4.6 clearly shows the influence of the liquid superficial velocity (V_L) on the average values of liquid holdup. The results in Figures 4.5 and 4.6 show a similar functional relationship between α_L and χ to that predicted by equation (4.6), but it is seen that this equation overestimates the value of the average liquid holdup, and distinct curves are obtained for each value of the power-law index (n) and the superficial velocity of the liquid. A detailed examination of the voluminous experimental results reported by different investigators (Oliver and Young-Hoon, 1968; Farooqi and Richardson, 1982; Chhabra and Richardson, 1984) reveals the following features:

(i) For a given value of the power-law index (n), the lower the value of the liquid superficial velocity (V_L), the lower is the average liquid holdup (see Figure 4.6).
(ii) The average liquid holdup decreases as the liquid becomes more shear-thinning (i.e. lower value of n), and the deviation from the Newtonian curve becomes progressively greater.

This suggests that any correction factor which will cause the holdup data for shear-thinning fluids to collapse onto the Newtonian curve, must become progressively smaller

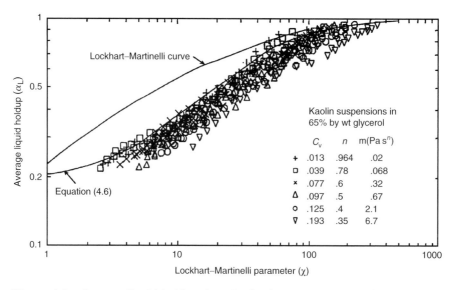

Figure 4.5   *Average liquid holdup data for kaolin suspensions in 65% aqueous glycerol solution in streamline flow (D = 42 mm)*

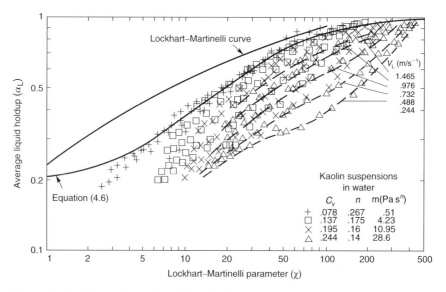

Figure 4.6   *Effect of superficial liquid velocity an average liquid holdup*

as the liquid velocity increases and the flow behaviour index, n, decreases. Based on such intuitive and heuristic considerations, Farooqi and Richardson (1982) proposed a correction factor, J, to be applied to the Lockhart–Martinelli parameter, χ, so that a modified parameter χ_{mod} is defined as:

$$\chi_{\mathrm{mod}} = J\chi \tag{4.7}$$

where $J = \left(\dfrac{V_L}{V_{L_c}}\right)^{1-n}$ (4.8)

and the average liquid holdup is now given simply by replacing χ with χ_{mod} in equation (4.6), viz.:

$$\alpha_L = 0.24(\chi_{mod})^{0.8} \qquad 0.01 \leq \chi_{mod} \leq 0.5 \tag{4.9a}$$

$$\alpha_L = 0.175(\chi_{mod})^{0.32} \qquad 0.5 \leq \chi_{mod} \leq 5 \tag{4.9b}$$

$$\alpha_L = 0.143(\chi_{mod})^{0.42} \qquad 5 \leq \chi_{mod} \leq 50 \tag{4.9c}$$

$$\alpha_L = \dfrac{1}{0.97 + (19/\chi_{mod})} \qquad 50 \leq \chi_{mod} \leq 500 \tag{4.9d}$$

V_{L_c} is the critical velocity for the transition from laminar to turbulent flow. For a given power-law liquid (i.e. known m and n), density and pipe diameter, D, the value of V_{L_c} may be estimated simply by setting the Reynolds number (equation (3.8b)) equal to 2000, i.e.,

$$Re_{MR} = \dfrac{\rho V_{L_c}^{2-n} D^n}{8^{n-1} m \left(\dfrac{3n+1}{4n}\right)^n} = 200 \tag{4.10}$$

For both, $V_L = V_{L_c}$ and/or $n = 1$, the correction factor $J = 1$.

In Figure 4.7, the experimentally determined values of average liquid holdup, α_L, are plotted against the modified parameter χ_{mod} for suspensions of kaolin in aqueous glycerol

Figure 4.7   *Average liquid holdup as a function of modified parameter χ_{mod}*

solutions (same data as shown in Figure 4.5) and it will be seen that they are now well correlated by equation (4.9).

Equally good correlations are obtained for the experimental data for two phase flow of air and nitrogen with aqueous and non-aqueous suspensions of coal (Farooqi *et al.*, 1980) and china clay particles and aqueous solutions of a wide variety of chemically different polymers (Chhabra *et al.*, 1984). A wide range has been covered ($0.14 \leq n \leq 1$; $0.1 \leq \chi_{mod} \leq {\sim}200$) but most data have been obtained in relatively small diameter (3–50 mm) pipes (Chhabra and Richardson, 1986).

Little is known about the influence of visco-elastic properties of the liquid phase on liquid holdup (Chhabra and Richardson, 1986). However, Chhabra *et al.* (1984) used aqueous solutions of polyacrylamide (Separan AP-30) as model visco-elastic liquids and a preliminary analysis of these results indicated that equation (4.9) consistently underestimated the value of liquid holdup. Infact the experimental results for visco-elastic liquid and air lie between those predicted by equations (4.9) and (4.6). It is thus necessary to introduce an additional parameter to account for visco-elastic effects. For this purpose, a Deborah number was defined as:

$$De = \frac{\lambda_f V_M}{D} \tag{4.11}$$

where λ_f, the fluid characteristic time, is deduced from the measurement of the primary normal stress difference N_1. Like viscosity, it is generally possible to approximate the variation of N_1 with the shear rate over a limited range by a power law, e.g. equation (1.39) i.e.,

$$N_1 = m_1(\dot{\gamma})^{p_1} \tag{4.12}$$

which, in turn, allows the fluid characteristic time λ_f to be defined by equation (1.38):

$$\lambda_f = \left(\frac{m_1}{2m}\right)^{1/(p_1 - n)} \tag{4.13}$$

Although the use of the no-slip mixture velocity, V_M, in equation (4.11) is quite arbitrary, it does account for the enhanced shearing of the liquid brought about by the introduction of gas into the pipeline. Over the range of conditions ($0.3 \leq De \leq 200$ and $2 \leq \chi_{mod} \leq 160$), the following simple expression provides a reasonably satisfactory correlation of the available data for α_{LV}, the average value of liquid holdup for visco-elastic liquids:

$$\alpha_{LV} = \alpha_L \left(1 + 0.56 \frac{De^{0.05}}{\chi_{mod}^{0.5}} \right) \tag{4.14}$$

α_L is the value of holdup from equation (4.9) in the absence of visco-elastic effects. Although, equation (4.14) does reduce to the limit of $\alpha_{LV} = \alpha_L$ as $De \to 0$, extrapolation outside the range quoted above must be carried out with caution.

Transitional and turbulent flow of liquids ($Re_{MR} > 2000$)

When the non-Newtonian liquid is no longer in streamline flow (prior to the addition of gas), i.e. $Re_{MR} > 2000$, the experimental results for average liquid holdup agree well with those predicted by equation (4.6) and the original Lockhart–Martinelli parameter χ

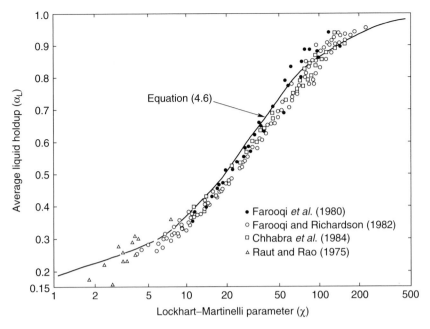

Figure 4.8 *Experimental and predicted (equation (4.6)) values of liquid holdup under turbulent conditions for liquid*

may therefore be used. This is confirmed by the data shown in Figure 4.8 for a variety of shear-thinning liquids including polymer solutions, chalk–water slurries, china clay and coal suspensions (Raut and Rao, 1975; Farooqi *et al.*, 1980; Farooqi and Richardson, 1982; Chhabra *et al.*, 1984). Scant results available in the literature suggest that equation (4.6) underpredicts the value of holdup for visco-elastic liquids in turbulent flow (Rao, 1997).

Predictive methods for upward vertical flow

The previous discussion on holdup related only to horizontal flow of gas – non-Newtonian liquid mixtures. Very few experimental results are available for holdup in vertical upward flow with shear-thinning liquids (Khatib and Richardson, 1984). These authors used a γ-ray attenuation method to measure the average as well as instantaneous values of liquid holdup for shear-thinning suspensions of china clay and air flowing upwards in a 38 mm diameter pipe. The average values of liquid holdup in streamline flow are in line with the predictions from equation (4.9).

Thus, in summary, average liquid holdup can be estimated using equation (4.6) for Newtonian liquids under all flow conditions, and for non-Newtonian liquids in transitional and turbulent regimes ($Re_{MR} > 2000$).

For the streamline flow of shear-thinning fluids ($Re_{MR} < 2000$), it is necessary to use equation (4.9). A further correction must be introduced (equation (4.14)) for visco-elastic liquids. Though most of the correlations are based on horizontal flow, preliminary results indicate that they can also be applied to the vertical upward flow of mixtures of gas and non-Newtonian liquids.

4.2.5 *Frictional pressure drop*

Generally, methods for determining the frictional pressure drop begin by using a physical model of the two phase system, and then applying an approach similar to that for single phase flow. Thus, in the so-called separated flow model, the two phases are first considered to be flowing separately and allowance is then made for the effect of interfacial interactions. Irrespective of the type of flow and the rheology of the liquid phase, the total pressure gradient $(-\Delta p_{TP}/L)$ in horizontal flow consists of two components which represent the frictional and acceleration contribution respectively, i.e.

$$\left(-\frac{\Delta p_{TP}}{L} \right) = \left(-\frac{\Delta p_f}{L} \right) + \left(-\frac{\Delta p_a}{L} \right) \qquad (4.15)$$

Both a momentum balance and an energy balance for two phase flow through a horizontal pipe may be written as expanded forms of those for single phase flow. The difficulty of proceeding in this manner is that local values of important variables such as *in situ* velocities and holdups of the individual phases are not known and cannot readily be predicted. Some simplification is possible if it is assumed that each phase flows separately in the channel and occupies a fixed fraction of the pipe, but there are additional complications stemming from the difficulty of specifying interfacial conditions and the effect of gas expansion along the pipe length. As in the case of single phase flow of a compressible medium, the shear stress is no longer simply linked to the pressure gradient because the expansion of the gas results in the acceleration of the liquid phase. However, as a first approximation, it may be assumed that the total pressure drop can be expressed simply as the sum of a frictional and acceleration components:

$$(-\Delta p_{TP}) = (-\Delta p_f) + (-\Delta p_a) \qquad (4.16)$$

For upward flow of gas–liquid mixtures, an additional term $(-\Delta p_g)$ attributable to the hydrostatic pressure, must be included on the right hand side of equation (4.16), and this depends on the liquid holdup which therefore must be estimated or measured.

Thus, complete analytical solutions for the equations of motion are not possible (even for Newtonian liquids) because of the difficulty of defining the flow pattern and of quantifying the precise nature of the interactions between the two phases. Furthermore, rapid fluctuations in flow frequently occur and these cannot be readily incorporated into analysis. Consequently, most developments in this field are based on dimensional considerations aided by data obtained from experimental measurements. Great care must be exercised, however, when using these methods outside the limits of the experimental work.

Good accounts of idealized theoretical developments in this field are available for mixtures of gas and Newtonian liquids (Govier and Aziz, 1982; Hetsroni, 1982; Chisholm, 1983) and the limited literature on mixtures of gas and non-Newtonian liquids has also been reviewed elsewhere (Mahalingam, 1980; Bishop and Deshpande, 1986; Chhabra and Richardson, 1986).

Practical methods for estimating pressure loss

Over the years, several empirical correlations have been developed for the estimation of the two phase pressure drop for the flow of gas–liquid mixtures, with and without heat transfer. Most of these, however, relate to Newtonian liquids, though some have

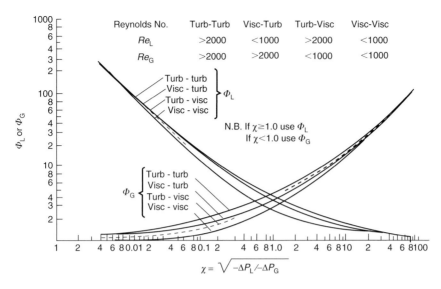

Figure 4.9 *Two phase pressure drop correlation of Lockhart and Martinelli (1949)*

been extended to include shear-thinning liquid behaviour. As the pertinent literature for Newtonian fluids and for non-Newtonian fluids has been reviewed extensively (see references above), attention here will be confined to the methods which have proved to be most reliable and have therefore gained wide acceptance.

Gas–Newtonian liquid systems

The most widely used method for estimating the pressure drop due to friction is that proposed by Lockhart and Martinelli (1949) and subsequently improved by Chisholm (1967). It is based on a physical model of separated flow in which each phase is considered separately and then the interaction effect is introduced. In this method, the two phase pressure drop due to friction $(-\Delta p_{TP})$, is expressed in terms of dimensionless drag ratios, ϕ_L^2 or ϕ_G^2 defined by the following equations:

$$\phi_L^2 = \frac{-\Delta p_{TP}/L}{-\Delta p_L/L} \tag{4.17}$$

$$\phi_G^2 = \frac{-\Delta p_{TP}/L}{-\Delta p_G/L} \tag{4.18}$$

These equations, in turn, are expressed as functions of the Lockhart–Martinelli parameter χ, defined earlier in equation (4.5). Obviously, the drag ratios are inter-related since $\phi_G^2 = \chi^2 \phi_L^2$. Furthermore, Lockhart and Martinelli (1949) used a flow classification scheme depending upon whether the gas–liquid flow is nominally in the laminar–laminar, the laminar–turbulent, the turbulent–laminar or the turbulent–turbulent regime. Notwithstanding the inherently fluctuating nature of two phase flows and the dubious validity of such a flow classification, the regime is ascertained by calculating the value of the Reynolds number (based on superficial velocity) for each phase. The flow is said to be laminar if this Reynolds number is smaller than 1000 and turbulent if it is greater than 2000, with mixed type of flow in the intermediate zone. Figure 4.9 shows the original correlation of Lockhart and Martinelli (1949) who suggested that the $\phi_L - \chi$ curve should be used for $\chi \geq 1$ and the $\phi_G - \chi$ curve for $\chi \leq 1$. Even though this correlation is based on data for the air–water system in relatively small diameter pipes, it has proved to be of value for the flow of other gas–liquid systems in pipes of diameters up to 600 mm. The predictions are well within $\pm 30\%$, but in some cases, errors upto 100% have also been reported. It is paradoxical that this method has been found to perform poorly for the simplest geometric system, namely stratified flow! The main virtue of this method lies, however, in its simplicity and in the fact that no prior knowledge of the flow pattern is needed. This is in contrast to the theoretical models which invariably tend to be flow-pattern dependent.

Chisholm (1967) has developed an algebraic form of relation between ϕ_L^2 and χ:

$$\phi_L^2 = 1 + \frac{C_0}{\chi} + \frac{1}{\chi^2} \tag{4.19}$$

where, for air–water mixtures, the values of C_0 are as follows:

Gas	Liquid	C_0
Laminar	Laminar	5
Laminar	Turbulent	10
Turbulent	Laminar	12
Turbulent	Turbulent	20

Further correction is needed if the densities of the two phases are appreciably different from those of air and water (Chisholm, 1967). Extensive comparisons between the predictions of equation (4.19) and experimental values embracing all the four regimes show satisfactory agreement (Chhabra and Richardson, 1986).

Gas–non-Newtonian liquid systems

As remarked earlier, analytical treatments of two phase flow are of limited value and this applies equally for non-Newtonian liquids. The relatively simple flow patterns, of annular and stratified flow, for power-law liquids have received some attention in the literature. For annular flow, some workers have assumed that the thin liquid film at the wall behaves like a laminar film flowing between two parallel plates (Mahalingam and Valle, 1972) while others (Oliver and Young-Hoon, 1968; Eissenberg and Weinberger, 1979) have approximated the flow area to be an annulus, with no inner wall. Likewise, Heywood and Charles (1979), Bishop and Deshpande (1986) and Xu *et al.* (2007) have modified the Taitel *et al.* (1980) idealized model for stratified flow to include power-law fluids. In most cases, the interface is assumed to be smooth (free from ripples) and the interfacial friction factor has been approximated by that for a gas flowing over a solid surface. These and other simplifications account for the fact that values of pressure drop may deviate from the experimental data by a factor of up to four. On the other hand, experimental work in this field has yielded results which can be used to predict pressure drops over a wide range of conditions. The results obtained with the liquid in laminar flow ($Re_{MR} < 2000$) or in turbulent flow ($Re_{MR} > 2000$) prior to the introduction of gas

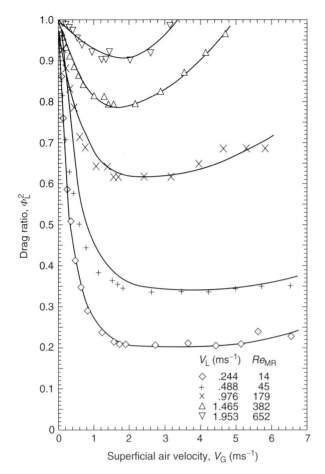

Figure 4.10   *Drag ratio data for a 24.4% kaolin-in-water suspension in a 42 mm diameter pipe*

will now be treated separately. Shear-thinning fluids are found to exhibit completely different behaviour from Newtonian liquids in streamline conditions whereas in turbulent flow, the non-Newtonian properties appear to be of little consequence.

Laminar conditions

When a gas is introduced into a shear-thinning fluid in laminar flow, the frictional pressure drop may, in some circumstances, actually be reduced below the value for the liquid flowing alone at the same volumetric rate. As the gas flow rate is increased, the two phase pressure drop decreases, then passes through a minimum (maximum drag reduction, MDR) and finally increases again and eventually exceeds for the flow of liquid alone. This effect which has been observed with flocculated suspensions of fine kaolin and anthracite coal and with shear-thinning polymer solutions occurs only where the flow of liquid on its own would be laminar. A typical plot of drag ratio (ϕ_L^2) as a function of superficial air velocity is shown in Figure 4.10, for a range of values of the liquid superficial velocity, and the corresponding values of its Reynolds number Re_{MR}

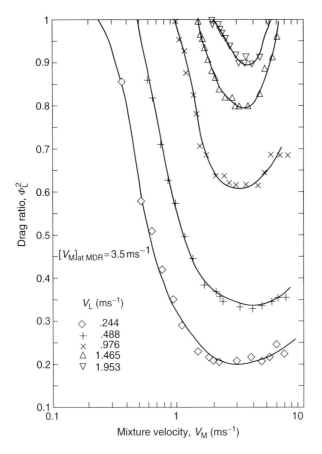

Figure 4.11 *Drag ratio data for a 24.4% (by volume) kaolin-in-water suspension as a function of mixture velocity*

are given. The liquid is a 24.4% (by volume) kaolin suspension in water which exhibits power-law rheology. An analysis of a large number of experimental results identifies the following salient features:

(i)   For a liquid with known values of m, n, ρ, the value of the minimum drag ratio $(\phi_L^2)_{min}$ decreases as the liquid velocity is lowered.

(ii) The superficial gas (air) velocity needed to achieve MDR increases as the liquid velocity decreases.

(iii) The higher the degree of shear-thinning behaviour (i.e. the smaller the value of n), the greater is the extent of the drag reduction obtainable.

If, for a given liquid (i.e. fixed values of m and n for a power-law fluid), the drag ratio is plotted against the no-slip mixture velocity, $V_M (= V_L + V_G)$, as opposed to the superficial gas velocity, it is found that the minima in ϕ_L^2 all occur at approximately the same mixture velocity, irrespective of the liquid flow rate (see Figure 4.11). Furthermore, this minimum always occurs when Re_{MR} (based on V_M rather than V_L) is approximately 2000, corresponding to the upper limit of streamline flow. This suggests that the value of

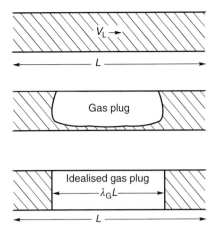

Figure 4.12   *Idealized plug flow model*

ϕ_L^2 continues to fall progressively until the liquid is no longer in streamline flow. Thus, at low flow rates of liquid, more air can be injected before this point is reached.

At first sight, it seems rather anomalous that on increasing the total volumetric throughput by injection of air, the frictional pressure drop can actually be lower than that for the flow of liquid alone. Also, the magnitude of the effect can be very large, with values of ϕ_L^2 as low as 0.2 (obtained with highly shear-thinning china clay suspensions), i.e. the two phase pressure drop can be reduced by a factor of 5 by air injection. The mechanism by which this can occur may be illustrated using a highly idealized model. Suppose that the gas and liquid form a series of separate plugs, as depicted schematically in Figure 4.12.

For the two phase flow, the total pressure drop will be approximately equal to the sum of the pressure drops across the individual liquid slugs, the pressure drop across the gas slugs being negligible in comparison with that for the liquid slugs. For a power-law fluid in laminar flow at a velocity of V_L in a pipe of length L, the pressure drop $(-\Delta p_L)$ is given by (equation (3.9), Chapter 3):

$$-\Delta p_L = A_1 L V_L^n \tag{4.20}$$

where A_1 is a constant for a given pipe (D) and fluid (m and n).

The addition of gas has two effects: the length of pipe in contact with liquid is reduced, and the velocity of the liquid plug is increased. If λ_L (defined by equation (4.3)) is the input volume fraction of liquid, then in the absence of slip, the wetted length of pipe is reduced to $L\lambda_L$ and the velocity of the liquid plug is increased to V_L/λ_L. The two phase pressure drop is:

$$-\Delta p_{TP} = A_1 (L\lambda_L)(V_L/\lambda_L)^n \tag{4.21}$$

The drag ratio, ϕ_L^2, is obtained as:

$$\phi_L^2 = \frac{-\Delta p_{TP}}{-\Delta p_L} = (\lambda_L)^{1-n} \tag{4.22}$$

For a shear-thinning fluid $n < 1$ and $\lambda_L < 1$, the drag ratio ϕ_L^2 must be less than unity, and hence a reduction in pressure drop occurs as a result of the presence of the air. The lower the value of n and the larger the value of λ_L, the greater will the effects be, and this situation is qualitatively consistent with experimental observations. It should be noted that any effects due to the expansion of the gas in the pipeline have not been considered here.

This simple model is likely to under-estimate the magnitude of $(-\Delta p_{TP})$ and ϕ_L^2 because the liquid and gas will not form idealized plugs and there will be some slip between the two phases. It has been found experimentally that equation (4.22) does apply at low air velocities ($<\sim 1\,m/s$), but at higher gas flow rates the model holds progressively less well. In the limiting case of a Newtonian liquid ($n = 1$), equation (4.22) yields $\phi_L^2 = 1$, for all values of λ_L and the two phase pressure drop would be unaffected by air injection. In practice, because gas will always disturb the flow, there will be additional pressure losses, and the two phase pressure drop will always increase, with the introduction of a gas.

Drag reduction can also occur with a fluid exhibiting an apparent yield stress (Farooqi *et al.*, 1980). Similar drag reduction has also been reported for the cocurrent horizontal flow of air and extremely shear-thinning lubricating greases ($n \sim 0.1-0.2$) and for farm waste slurries without, however, reaching the conditions corresponding to the MDR (Bjerkholt *et al.*, 2005; Delgado *et al.*, 2005; Ruiz–Viera *et al.*, 2006).

Maximum drag reduction

As noted earlier, for a given liquid and pipe, the minimum value of ϕ_L^2 occurs at a constant value of the no-slip mixture velocity which corresponds approximately to $Re_{MR} \sim 1000-2000$. This implies that ϕ_L^2 attains a minimum value $(\phi_L^2)_{min}$ when the flow in the liquid plug no longer remains streamline. Values of $(\phi_L^2)_{min}$ have been correlated against the correction factor $J = (V_L/V_{L_c})^{1-n}$ introduced earlier in connection with the prediction of liquid holdup. Thus:

$$(\phi_L^2)_{min} = J^{0.205} \qquad\qquad 0.6 \leq J \leq 1 \qquad\qquad (4.23a)$$

$$(\phi_L^2)_{min} = 1 - 0.0315J^{-2.25} \quad 0.35 \leq J \leq 0.6 \qquad\qquad (4.23b)$$

$$(\phi_L^2)_{min} = 1.9J \qquad\qquad 0.05 \leq J \leq 0.35 \qquad\qquad (4.23c)$$

Figure 4.13 compares the predictions from equation (4.23) with representative experimental results for both aqueous polymer solutions and particulate suspensions in pipes of diameters up to 200 mm. It will be noted that equation (4.23) is particularly useful in estimating, *a priori*, the minimum achievable drag ratio as it requires a knowledge only of the properties of liquid (ρ, m, n) and the operating conditions (D, V_L); the corresponding gas velocity is calculated as $(V_{L_c} - V_L)$.

General method for estimation of two phase pressure loss

The discussion so far has related to the drag reduction occurring when a gas is introduced into a shear-thinning fluid initially in streamline flow. A more general method is required for the estimation of the two phase pressure drop for mixtures of gas and non-Newtonian liquids. The well-known Lockhart–Martinelli (1949) method will now be extended to encompass shear-thinning liquids, first by using the modified Lockhart–Martinelli parameter, χ_{mod} (equation (4.8)). Figure 4.14 shows a comparison between the Lockhart–Martinelli correlation and typical experimental measurements of two phase pressure drop

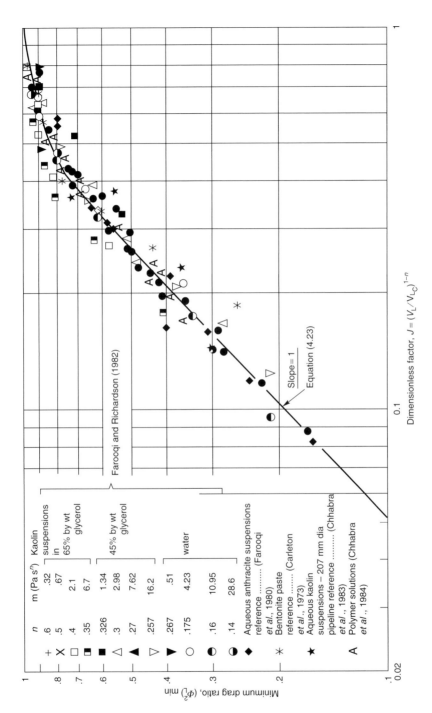

Figure 4.13 *Correlation for minimum drag ratio*

Figure 4.14   *Drag ratio versus modified Lockhart–Martinelli parameter for cocurrent flow of air and a china clay suspension in a 42 mm diameter pipe*

for air and a china clay suspension ($m = 0.67 \, \text{Pa} \, \text{s}^n$, $n = 0.50$) flowing cocurrently in a 42 mm diameter horizontal pipe. The conditions are such that the gas would be in turbulent flow and the kaolin suspension in streamline flow if each phase were flowing on its own. The value of J ranges from 0.16 to 0.65 for these conditions. Evidently, as the liquid velocity increases, the experimental values of the drag ratio move towards the correlation of Lockhart and Martinelli (1949), approximated here by equation (4.19) with χ replaced by χ_{mod}. Indeed, when a large amount of data culled from various sources in the literature is analysed in this fashion, the deviations from the predictions of equation (4.19) range from $+60\%$ to -800%, the experimental values being generally overestimated. Based on these observations, Dziubinski and Chhabra (1989) empirically modified the drag ratio to give

$$(\phi_L^2)_{\text{mod}} = \phi_L^2 / J \tag{4.24}$$

The results shown in Figure 4.14 are re-plotted in Figure 4.15 using the modified drag ratio, $(\phi_L^2)_{\text{mod}}$ and the modified Lockhart–Martinelli parameter, χ_{mod}; data points are now seen to straddle the original correlation of Lockhart and Martinelli (1949), i.e. equation (4.19). Indeed, this approach reconciles nearly 1500 data points relating to the streamline flow of liquid with an error of $\pm 40\%$ which is comparable with the uncertainty associated with the original correlation for Newtonian liquids. The validity of this approach has been tested over the following ranges of conditions as: $0.10 \leq n \leq 0.96$; $2.9 \leq D \leq 207 \, \text{mm}$; $0.17 \leq V_L \leq 2 \, \text{m/s}$; $0.11 \leq V_G \leq 23 \, \text{m/s}$.

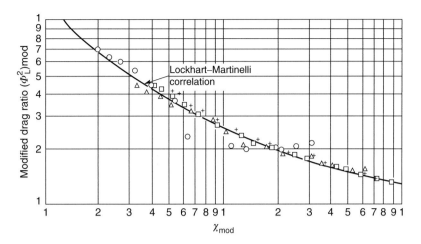

Figure 4.15 *Modified drag ratio versus modified Lockhart–Martinelli parameter (same data as shown in Figure 4.14)*

It is emphasized, however, that because the adaptation of the correlation for non-Newtonian fluids is entirely empirical and that the same factor J appears in both abscissa and ordinate, great caution must be exercised in using this method outside the limits of the variables employed in its formulation.

Turbulent flow

For both Newtonian and non-Newtonian liquids in turbulent flow, the addition of gas always results in an increase in the pressure drop and gives values of drag ratio, ϕ_L^2, in excess of unity. Using χ, both the graphical correlation of Lockhart and Martinelli in Figure 4.9 and equation (4.19) satisfactorily represent the data, as illustrated in Figure 4.16 for turbulent flow of both gas and liquid, as also argued by Rao (1997). More recent data also lend further support to this inference (Bjerkholt *et al.*, 2005; Jinming and Jingxuan, 2006).

On the other hand, Dziubinski (1995) has put forward an alternative formulation for the prediction of the two phase pressure drop for a gas and shear-thinning liquid mixture in the intermittent flow regime. By analogy with the flow of single phase fluids, he introduced a loss coefficient Λ defined as:

$$\Lambda = \frac{\tau_w \rho D^2}{\mu^2} \tag{4.25}$$

For power-law fluids in streamline flow, Dziubinski's expression for the drag ratio is:

$$\phi_L^2 = \frac{1 + 1.036 \times 10^{-4} (Re_{TP})^{1.235}}{1 + 1.036 \times 10^{-4} (Re_L)^{1.235}} \lambda_L^{1-n} \tag{4.26}$$

where λ_L is the input liquid fraction, equation (4.3), and the two Reynolds numbers in equation (4.26) are defined as:

$$Re_{TP} = \frac{\rho V_M^{2-n} D^n}{8^{n-1} m \left(\dfrac{3n+1}{4n}\right)^n} \tag{4.27}$$

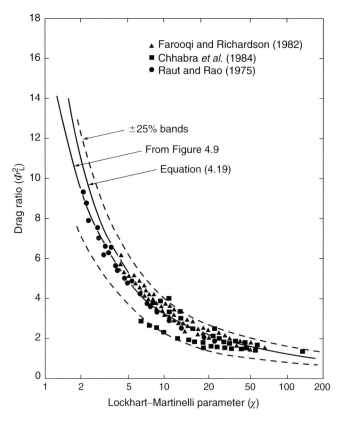

Figure 4.16   *Experimental and predicted (equation (4.19)) values of drag ratio for turbulent flow of polymer solutions and suspensions*

and Re_L is based on the superficial velocity of the liquid. Similarly, for the turbulent flow of gas/pseudoplastic liquid mixtures ($Re_{TP} > 2000$) his expression in terms of the loss coefficient Λ_{TP} is:

$$\Lambda_{TP} = 0.0131\lambda_L \Delta^{-5} \exp(1.745\Delta - 0.643\lambda_L)(Re_{TP}^*)^{1.75} \tag{4.28}$$

where $\Delta = \dfrac{3n+1}{4n}$,   and $Re_{TP}^* = \Delta^2 Re_{TP}$

Attention is drawn to the fact that the values of m and n for use in turbulent region are deduced from the data in the laminar range at the values of $(8V_L/D)$ which is only the *nominal* shear rate at the tube wall for streamline flow, and thus this aspect of the procedure is completely empirical. Dziubinski (1995) stated that equation (4.26) reproduced the same experimental data as those referred to earlier with an average error of $\pm 15\%$, while equation (4.28) correlated the turbulent flow data with an error of $\pm 25\%$. Notwithstanding the marginal improvement over the method of Dziubinski and Chhabra (1989), it is reiterated here that both methods are of an entirely empirical nature and therefore the extrapolation beyond the range of experimental conditions must be treated with reserve.

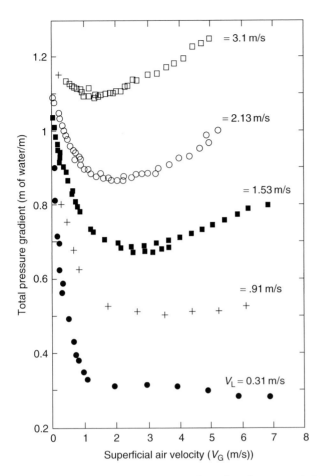

Figure 4.17   *Total pressure gradient for the upward flow of air–water mixtures in a 38 mm diameter pipe*

Subsequently, based on the notion of the fractional pipe surface in contact with the liquid, Kaminsky (1998) has developed a new method for the prediction of the two phase pressure drop for the flow of a mixture of gas and a power-law fluid. This method is implicit in pressure gradient (and therefore requires an iterative solution) and also necessitates additional information about the fraction of the pipe surface in contact with liquid which is not always available. It is appropriate to add here that new experimental data for highly viscous and shear-thinning greases suggest that neither equation (4.22) nor equation (4.26) is entirely satisfactory (Ruiz–Viera *et al.*, 2006) and hence, the latter authors have put forward a new empirical correlation for the prediction of two-phase pressure drop in these systems.

Vertical (upward) flow

The interpretation of results for vertical flow is more complicated since they are strongly dependent on the *in situ* liquid holdup which, in turn, determines the hydrostatic component of the pressure gradient. Khatib and Richardson (1984) reported measured values of the two phase pressure drop and liquid holdup for the vertical upward cocurrent flow of air and aqueous china clay suspensions in a 38 mm diameter pipe. Representative results are shown in Figures 4.17 and 4.18 for air–water and air–china clay suspensions, respectively.

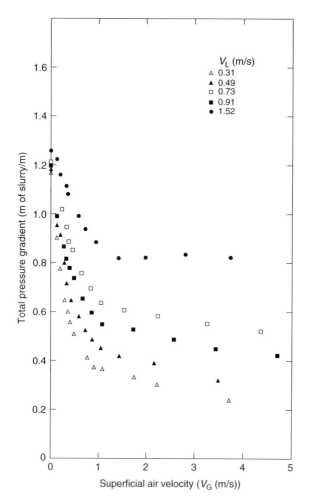

Figure 4.18   *Total pressure gradient for the upward flow of air and 18.9% (by volume) aqueous kaolin suspension*

In all cases, as the air flow rate is increased, the total pressure gradient decreases, passes through a minimum and then rises again. Although the minimum pressure gradient occurs at about the same value of the no-slip mixture velocity as in horizontal flow, it has no connection with the laminar–turbulent transition. There is a minimum in the curve because, as the gas flow rate is increased, the holdup of liquid decreases, but the frictional pressure gradient increases due to the higher liquid velocities.

The frictional pressure drop may be estimated by subtracting the hydrostatic component (calculated from the holdup) from the total pressure gradient, as shown in Figure 4.19. It will be seen that under certain conditions, particularly at low liquid flow rates, the frictional component appears to approach zero. For the flow of air–water mixtures, 'negative friction losses' are well documented in the literature. This anomaly arises because not all of the liquid present in the pipe contributes to the hydrostatic pressure, because some liquid may form a film at the pipe wall. This liquid is sometimes flowing downwards and most of its weight is supported by an upward shear force at the wall. The drag exerted by the gas on the liquid complements the frictional force at the pipe wall.

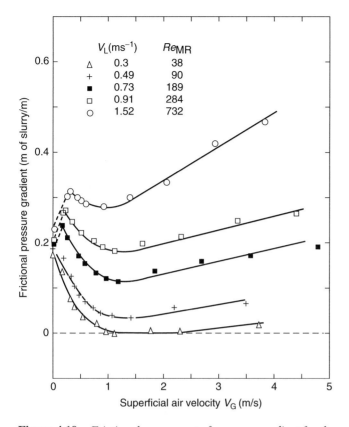

Figure 4.19   *Frictional component of pressure gradient for the results shown in Figure 4.18*

It is, in principle, possible to split the liquid holdup, measured at a given position, into two components; one associated predominantly with liquid slugs and the other with liquid films at the walls which do not contribute to the hydrostatic pressure. Then, using the effective hydrostatic pressure, a more realistic value for the frictional pressure drop can be obtained. Because it is not possible to make sufficiently accurate measurements of the separate components of the liquid holdup, there will be large errors in the amended friction terms calculated in this fashion.

4.2.6 Practical applications and optimum gas flow rate for maximum power saving

Drag reduction offers the possibility of lowering both the pressure drop and the power requirements in slurry pipelines, as is being used in the pumping of waste molasses from a sugar making plant (Dziubinski and Fidos, 1992). Air injection can be used, in practice, in two ways:

(i) To reduce the pressure drop, and hence the upstream pressure in a pipeline, for a given flow rate of shear-thinning liquid.
(ii) To increase the throughput of liquid for a given value of pressure drop.

Air injection can also be beneficial because it may be easier to re-start pumping after a shutdown as the pipe will not be completely full of slurry. On the other hand, if the pipeline follows an undulating topography, difficulties can arise from air collecting at the high points.

Air injection may sometimes be an alternative to deflocculation before pumping. In general, less power is required to pump deflocculated slurries but highly consolidated sediments may form on shutdown and these may be difficult to resuspend when pumping is resumed. Furthermore, deflocculating agents are expensive and may be undesirable contaminants of the product.

However, additional energy will be required to compress the air to a pressure in excess of the upstream pressure. Thus, the circumstances under which there will be a net saving of power for pumping will be strongly dependent on the relative efficiencies of the slurry pump and the gas compressor, and on the specific plant layout. Dziubinski and Richardson (1985) have addressed this problem and the salient features of their study are summarized here. They introduced a power saving coefficient, ψ, defined as:

$$\psi = \frac{N_L - N_{TP} - N_G}{N_L} \qquad (4.29)$$

where N_L is the power needed for pumping the slurry on its own; N_{TP} is the power for pumping the two-phase mixture and N_G is the power for compressing the gas prior to injection into the liquid.

It can be readily seen that the power saving coefficient can be expressed as a function of the efficiencies of the pump and the compressor, the superficial velocities of the gas and the slurry and the two phase pressure drop. Furthermore, Dziubinski and Richardson (1985) noted that the conditions for the MDR coincide approximately with the range of conditions over which the simple plug model applies. Thus, they expressed the two phase pressure drop and the the $(-\Delta p_L)$ term appearing in the expression for ψ in terms of V_L and V_G. Finally, for a given slurry (ρ, m, n), pump and compressor efficiencies, pipe dimensions (D, L) and the slurry flow rate (V_L), they obtained the optimum value of V_G by setting $d\psi/dV_G = 0$. Based on extensive computations, they concluded that, unless the efficiency of the compressor exceeds that of the pump, there will be no net power saving. Also, the liquid flow rate must be well within the laminar regime and it must be moderately shear-thinning ($n < {\sim}0.5$). Finally, the maximum power saving occurs at a much lower gas velocity than that needed for the MDR. Reference must be made to their original paper for a detailed treatment of this aspect of two phase flow.

Example 4.1

Air is injected into a 50 m long horizontal pipeline (of 42 mm diameter) carrying a china clay slurry of density 1452 kg/m³. The rheological behaviour of the slurry follows the power-law model, with $m = 5.55$ Pa s^n and $n = 0.35$. The volumetric flow rates of air and liquid are 7.48 and 1.75 m³/h, respectively. The air is introduced into the pipeline at 20°C and at a pressure of 1.2 bar. Ascertain the flow pattern occurring in the pipeline. Estimate (a) the average liquid holdup at the midpoint, (b) the pressure gradient for the two phase flow, (c) the maximum achievable drag reduction and the air velocity to accomplish it.

Solution

Cross-sectional area of pipe $= \dfrac{\pi}{4} D^2$

$$= \frac{\pi}{4}(42 \times 10^{-3})^2 = 1.38 \times 10^{-3}\,\mathrm{m}^2$$

Superficial liquid velocity, $V_L = \dfrac{1.75}{3600} \times \dfrac{1}{1.38 \times 10^{-3}}$

$$= 0.35\,\mathrm{m/s}$$

Superficial gas velocity, $V_G = \dfrac{7.48}{3600} \times \dfrac{1}{1.38 \times 10^{-3}}$

$$= 1.5\,\mathrm{m/s}$$

From Figure 4.2, it can be seen that the flow pattern is likely to be of the intermittent type under these conditions.

For liquid flowing on its own, the power-law Reynolds number,

$$Re_{MR} = \frac{\rho V_L^{2-n} D^n}{8^{n-1} m \left(\dfrac{3n + 1}{4n}\right)^n} \qquad \text{(eq. (3.17))}$$

$$= \frac{1452 \times (0.35)^{2-0.35}(42 \times 10^{-3})^{0.35}}{8^{0.35-1}(5.55)\left(\dfrac{3 \times 0.35 + 1}{4 \times 0.35}\right)^{0.35}} = 51.6$$

therefore the flow is streamline and the Fanning friction factor is given by equation (3.16),

$$f = \frac{16}{Re_{MR}} = \frac{16}{51.6} = 0.31$$

$$-\frac{\Delta p_L}{L} = \frac{2f \rho V_L^2}{D} = \frac{2 \times 0.31 \times 1452 \times (0.35)^2}{0.042} = 2630\,\mathrm{Pa/m}$$

For air alone, one can follow the same method to estimate $(-\Delta p_G/L)$. The viscosity of air at $20°C$ is $1.8 \times 10^{-5}\,\mathrm{Pa\,s}$ and the density is estimated by assuming it to be an ideal gas, at mean pressure of 1.1 bar:

$$\rho_G = \frac{p M}{RT} = \frac{1.1 \times 1.013 \times 10^5 \times 29}{8314 \times 293} = 1.33\,\mathrm{kg/m}^3$$

$$\therefore \text{Reynolds number, } Re_G = \frac{\rho_G V_G D}{\mu_G} = \frac{1.33 \times 1.5 \times 42 \times 10^{-3}}{1.8 \times 10^{-5}}$$

$$= 4460$$

and the friction factor is calculated using the Blasius formula

$$f = 0.079 Re_G^{-0.25} = 0.079(4460)^{-0.25} = 0.0096$$

$$\therefore -\frac{\Delta p_G}{L} = \frac{2f \rho_G V_G^2}{D} = \frac{2 \times 0.0096 \times 1.33 \times 1.5^2}{0.042} = 1.35\,\mathrm{Pa/m}$$

The Lockhart–Martinelli parameter, χ, is evaluated as:

$$\chi = \sqrt{\frac{-\Delta p_L/L}{-\Delta p_G/L}} = \sqrt{\frac{2630}{1.35}} = 44.2$$

Since the liquid is in laminar flow, the correction factor J must be calculated. The critical liquid velocity corresponding to $Re_{MR} = 2000$ is estimated from the relation:

$$Re_{MR} = \frac{\rho\left(V_{L_c}\right)^{2-n} D^n}{8^{n-1} m\left(\dfrac{3n+1}{4n}\right)^n} = 2000$$

Substituting for ρ, D, m and n, solving for V_{L_c}

$$V_{L_c} = \left[\frac{(2000)(8)^{0.35-1}(5.55)\left(\dfrac{3\times 0.35+1}{4\times 0.35}\right)^{0.35}}{1452\times(0.042)^{0.35}}\right]^{1/(2-0.35)}$$

$$= 3.21\,\text{m/s}$$

$$\therefore J = \left(V_L/V_{L_c}\right)^{1-n} = \left(\frac{0.35}{3.21}\right)^{1-0.35} = 0.237$$

$$\therefore \chi_{mod} = \chi J = 44.2 \times 0.237 = 10.46$$

The average liquid holdup is given by equation (4.9c):

$$\alpha_L = 0.143\chi_{mod}^{0.42} = 0.143 \times 10.46^{0.42} = 0.38$$

that is, on average 38% of the pipe cross-section is filled with liquid. This fraction will, however, continually change along the pipe length as the pressure falls.

The two phase pressure gradient is estimated using equation (4.19) in terms of χ_{mod} and $\phi_{L_{mod}}^2$. Since the liquid is in streamline flow and the gas is in turbulent regime, $C = 12$ and

$$\phi_{L_{mod}}^2 = 1 + \frac{12}{\chi_{mod}} + \frac{1}{(\chi_{mod})^2} = 1 + \frac{12}{10.46} + \frac{1}{(10.46)^2} = 2.16$$

or $\phi_L^2 = J\phi_{L_{mod}}^2 = 0.237 \times 2.16 = 0.51$

Hence, $-\Delta p_{TP}/L = 0.51 \times (-\Delta p_L/L) = 0.51 \times 2630 = 1340\,\text{Pa/m}$.

$\therefore$ Total pressure drop over 50 m pipe length $= 1340 \times 50 = 67\,\text{kPa}$.

It is also of interest to contrast this value with the prediction of equation (4.26).

Here, $V_M = V_L + V_G = 0.35 + 1.50 = 1.85 \, \text{m/s}$.

$$\therefore Re_{TP} = \frac{1452 \times 1.85^{2-0.35} \times 0.042^{0.35}}{8^{0.35-1} \times 5.55 \times \left(\frac{3 \times 0.35 + 1}{4 \times 0.35} \right)^{0.35}} = 805$$

and $\quad \lambda_L = \frac{V_L}{V_G + V_L} = \frac{0.35}{1.5 + 0.35} = 0.189$

From equation (4.26),

$$\phi_L^2 = \left\{ \frac{1 + 1.036 \times 10^{-4} \times 805^{1.235}}{1 + 1.036 \times 10^{-4} \times (51.6)^{1.235}} \right\} (0.189)^{1-0.35} = 0.47$$

therefore $(-\Delta p_{TP})/L = 0.47 \times 2630 = 1230 \, \text{Pa/m}$ which is about 10% lower than the value calculated above. However, both these values of 1340 and 1230 Pa/m compare well with the corresponding experimental value of 1470 Pa/m.

The maximum achievable drag reduction is calculated by using equation (4.23c):

$$\phi_{L_{min}}^2 = 1.9J = 1.9 \times 0.237 = 0.45$$

The corresponding air velocity is obtained simply by subtracting the value of V_L from V_{L_c}, i.e. $3.21 - 0.35 = 2.86 \, \text{m/s}$.

However, for the relative efficiencies of the slurry pump and the compressor in the ratio of 1 to 2, the approach of Dziubinski and Richardson (1985) yields the optimum gas velocity to be 0.2 m/s, which is much smaller than the value of 2.86 m/s for MDR. □

4.2.7 Two phase flow of drag reducing polymers

In many multi-phase oil–gas production systems, the distance from the well to the gas–liquid separation facility is typically of the order of 10–30 km, but it can be as much as 100 km. A high pressure loss due to the flow of multi-phase mixtures in these pipes can sometimes create a significant back pressure on the oil wells thereby reducing their production capacity. So it is desirable to reduce the frictional pressure loss in such pipelines as far as possible (Manfield *et al.*, 1999). Beside this, there is an intrinsic interest in improving the pumping efficiency of long distance pipelines by reducing the frictional pressure gradients. Following the significant drag reductions achieved under turbulent conditions by using polymeric additives in parts per million concentrations in single phase flows (Sellin *et al.*, 1982), there has been a growing interest to extend this idea to two phase and even three phase flows. At the outset, it needs to be recognized that this type of drag reduction occurs under turbulent conditions and therefore the underlying mechanism is completely different from that of the drag reduction in the laminar flow of shear-thinning fluids, discussed in Section 4.2.5.

Numerous experimental studies reporting varying extents of drag reduction using a range of polymeric additives (including polyethylene oxide, high molecular weight poly-alpha-olefin polymers, polyacrylamide, commercial proprietary polymers) are available in the literature. Indeed, the reduction in the two phase pressure drop for the flow of air–water mixtures with the addition of polymers up to 65–70% have been obtained with reference to the pressure drop without the addition of polymers (Rosehart *et al.*, 1972; Otten and Fayed, 1976; Sylvester and Brill, 1976; Kang *et al.*, 1997; Al-Sarkhi and Hanratty, 2001; Soleimani *et al.*, 2002; Baik and Hanratty, 2003; Fernandes *et al.*, 2004; Al-Sarkhi *et al.*, 2006). While the exact mechanism for drag reduction in these systems is not completely understood, it is believed that the injection of polymeric additives suppresses the level of wall turbulence, or reduces the amplitude of interfacial waves in stratified flow thereby reducing the entrainment of droplets from the liquid film into the gas phase. There is also some evidence which suggests that the transition to slug flow is delayed, especially at high gas flow rates. For each class of polymers, there appears to be an optimum dose for the MDR which is also dependent on the pipe diameter. For instance, Al-Sarkhi and Hanratty (2001) achieved ~48% drag reduction in a 95 mm diameter pipe with 10–15 ppm of a commercial polymer (Percol 727) while they reported a drag reduction of 63% in a 25 mm diameter pipe. They attributed a part of this drag reduction to the transition from an annular to a stratified flow pattern. Similarly, drag reduction of about 50% has been observed in oil–water pipelines (Al-Wahaibi *et al.*, 2007). Additional complications arise from the fact that dilute polymer solutions are prone to mechanical degradation. Owing to the lack of sufficient information, it is not yet possible to put forward definitive methods to estimate the two phase flow pressure drop in these systems.

4.3 Two phase liquid–solid flow (hydraulic transport)

Hydraulic transport is the conveyance of particulate matter in liquids. Although most of the earlier applications of the technique used water as the carrier medium (and hence the term hydraulic), there are now many industrial plants, particularly in the minerals, mining and power generation industries, where particles are transported in a variety of liquids which may exhibit either Newtonian or non-Newtonian flow behaviour. Additional applications are found in horizontal oil well processes wherein sand is transported in a viscous oil (Gillies *et al.*, 1999), in drilling operations where debris and rock cuttings are conveyed in viscous drilling muds in laminar flow conditions, and in thermal processing of foods (see Section 4.3.2). Transport may be in vertical or horizontal or inclined pipes, but in the case of long pipelines, it may follow the undulations of the land over which the pipeline is installed. The diameter and length of the pipeline and its inclination, the properties of the solids (size, shape and density) and of the liquid (density, viscosity, Newtonian or non-Newtonian), and the flow rates all influence the nature of the flow and the pressure gradient. Design methods are, in general, not very reliable, especially for the transportation of coarse particles; therefore, the estimated values of the pressure drop and power should be treated with caution. In practice, it is more desirable and important to ensure that the system operates reliably, and without the risk of blockage and without excessive erosion, than to achieve optimal operating conditions in relation to power requirements.

Another important application of solid–liquid flows in pipes is in the sterilizing of foods by heat treatment to enable them to be transported and stored safely. In canning, the container is filled and sealed before it is heated. Owing to the restricted circulation space and rather low thermal conductivity of foodstuffs, this necessitates not only a significant processing time, but it also leads to non-uniform product quality. For instance, in order to ensure the necessary degree of sterility at the centre of the can (mainly

Table 4.1 *Important variables in slurry pipelines*

Component	Parameters
Solids	Shape, size and size distribution, density, strength, abrasiveness, texture, shear sensitivity
Liquid	Type of liquid (Newtonian, pseudoplastic, viscoplastic), corrosive nature, density, rheological properties and their temperature dependence, stability
Pipeline	Diameter and length, its orientation, fittings, valves, material of construction
Operating conditions	Flow rates of liquid and solids, concentration, type of pump, etc.

by conduction), the material near the wall may become over-cooked. Thus, the quality of the product, particularly flavour and taste, may suffer. Therefore, recently there has been a growing interest in developing continuous processes for the thermal treatment of foodstuffs. While, in principle, there is a large degree of similarity with the liquid–solid flows in minerals transport and in food processing applications, the two also differ in significant ways. Firstly, in slurry pipelines, the key design variable is the frictional pressure gradient and how it is influenced by the other system variables (listed in Table 4.1). On the other hand, the key concerns in food processing applications are the residence time distributions (RTD) of the large food particles and of the suspending medium which may be either highly viscous Newtonian or shear-thinning or viscoplastic in behaviour, as the requirement here is to reach a predetermined level of sterility yet minimizing over-cooking in order to meet quality standards. Thus, the prediction of the frictional pressure gradient is of secondary importance. Secondly, there is an upper limit to the mixture velocity which may or may not be adequate for fully suspended flow of large and/or heavy particles. Thus, the sliding bed is an accepted mode of transportation of minerals in slurry pipelines. In contrast, since most food particles have density close to that of water (typically 1010–1100 kg/m^3), there is very little gravity-induced separation. Furthermore, it would not be desirable to have a bed of particles sliding against the pipe wall as this will invariably lead to a non-uniform degree of processing and/or variable product quality. Finally, it is not uncommon to process food particles as large as 25 mm in a 50 mm diameter pipe at solid fractions as high as 60%. In contrast, slurry pipelines are seldom designed for such large particle-to-pipe diameter ratios. In food processing, the liquid flow is usually laminar and therefore, the radial velocity gradients are steeper than for turbulent flow, except near the walls. Notwithstanding the fact that both the frictional pressure drop and RTDs are influenced by the same variables (Table 4.1), but owing to the inherently different objectives, these two situations are dealt with separately in the following sections.

4.3.1 Pressure drop in slurry pipe lines

It is customary to divide suspensions into two broad categories – fine particle suspensions in which the particles are reasonably uniformly distributed in the liquid with little separation;

and coarse suspensions in which particles, if denser than the liquid, tend to separate out and to travel predominantly in the lower part of a horizontal pipe (at a lower velocity than the liquid); in a vertical pipe the solids may have an appreciably lower velocity than the liquid. Although, this is obviously not a very clear cut classification and is influenced by the flow rate and concentration of solids, it does nonetheless provide a convenient initial basis for classifying the flow behaviour of liquid–solid mixtures.

Fine particles usually form fairly homogeneous suspensions which do not separate to any significant extent during flow. In high concentration suspensions, settling velocities of the particles are small in comparison with the liquid velocities (under normal operating conditions) and in turbulent flow, the eddies in the liquid phase keep particles suspended. In practice, turbulent conditions will prevail, except when the liquid has a very high viscosity (such as in coal–oil slurries) or exhibits non-Newtonian behaviour. In addition, concentrated flocculated suspensions are frequently conveyed in streamline flow when they behave essentially as single phase shear-thinning liquids (e.g. flocculated kaolin and coal suspensions).

Depending upon its state, the suspension may exhibit Newtonian or non-Newtonian behaviour. It is often a good approximation to treat it as a pseudo-homogeneous single phase systems by ascribing to it an effective density and viscosity. Thus, one can use the methods outlined in Chapter 3 to estimate the pressure gradient in terms of the flow rate. Attention will now be focused on the transportation of coarse particles in non-Newtonian carrier media which offer two advantages when the flow is streamline: firstly, the effective or apparent viscosity of a shear-thinning fluid is a maximum at the centre of the pipe and this facilitates the suspension of the particles (though some of this effect may be offset by the propensity of particles for migration across streamlines and the enhanced settling velocities in sheared fluids); secondly, the apparent viscosity will be minimum at the pipe wall, as a result of which the frictional pressure gradient will be low and will increase only relatively slowly as the liquid velocity is raised. Furthermore, if the fluid exhibits a yield stress, suspension of coarse particles in the central part of the pipeline will be further assisted. In practice with the transport of particulate matter of wide size distribution (e.g. coal dust to large lumps), the fine colloidal particles tend to form a pseudo-homogeneous shear-thinning medium of enhanced apparent viscosity and density in which the coarse particles are conveyed. On the other hand, the heavy medium may consist of fine particles of a different solid, particularly one of higher density such as in the transport of cuttings in drilling muds in drilling applications. In such a case, it is necessary to separate and re-cycle the heavy medium. The use of heavy carrier media can be advantageous when the coarse particles are transported in suspension rather than as a sliding bed and may enable operation to be carried out under streamline flow conditions.

In spite of these potential benefits, only a few studies dealing with the transport of coarse particles in heavy media have been reported. Charles and Charles (1971) investigated the feasibility of transporting 216 μm sand particles in highly shear-thinning clay suspensions ($0.24 \leq n \leq 0.35$) and they concluded that energy requirements could be reduced by a factor of six when using heavy media as opposed to water. Similarly, Ghosh and Shook (1990) reported slight reduction in head loss for the transport of 600 μm sand particles in a 52 mm diameter pipe in a moderately shear-thinning carboxymethyl cellulose solution; however, no reduction in head loss was observed for 2.7 mm pea gravel particles, presumably because these large particles were conveyed in the form of a sliding bed. Indeed Duckworth *et al.* (1983, 1986) successfully conveyed coal particles (up to 19 mm) in a pipe of 250 mm diameter in a slurry of fine coal which behaved as an ideal

Bingham plastic fluid. However, in none of these studies has an attempt been made to develop a general method for the prediction of pressure gradient in such applications.

In an extensive experimental study, Chhabra and Richardson (1985) transported coarse gravel particles (3.5, 5.7, 8.1 mm) in a 42 mm diameter horizontal pipe in a variety of carrier fluids, including Newtonian liquids of high viscosity, pseudoplastic china clay suspensions and polymer solutions. However, a majority of the coarse particles were seen to be transported in the form of a sliding bed along the bottom of the pipe while the liquid flow could be either streamline or turbulent. In this mode of hydraulic transport, the resistance to motion of bed of particles due to friction between the solids and the pipe wall is balanced by the force due to the hydraulic pressure gradient and, following the procedure of Newitt *et al.* (1955), a force balance gives:

$$k_1 \gamma C Q(\rho_S - \rho_L)g = i_S Q \rho_L g \qquad (4.30)$$

where γ is the friction coefficient between solids and pipe wall; Q is the total volume of suspension of concentration C in the control volume; k_1 is a system constant and i_S is the hydraulic gradient attributable to the presence of solids. Upon re-arrangement,

$$i_S = k_1 \gamma C \left(\frac{\rho_S - \rho_L}{\rho_L} \right) = k_1 \gamma C(s - 1) \qquad (4.31)$$

$$\text{where} \quad s = (\rho_S/\rho_L) \text{ and } i_S = i - i_L \qquad (4.32)$$

where i and i_L are the hydraulic pressure gradients, respectively, for the flow of mixture (total) and of liquid (heavy medium) alone at the same volumetric rate. Substituting $i_L = 2f_L V^2/gD$ and eliminating i_S, yields:

$$\frac{i - i_L}{i_L} = \frac{k_1 \gamma C(s - 1)}{2f_L V^2/gD}$$

$$\text{or} \quad f_L \frac{(i - i_L)}{i_L} = k_2 \frac{gDC(s - 1)}{V^2} \qquad (4.33)$$

where D is the pipe diameter and C is the volume fraction of coarse solids, and k_2 is a constant to be determined from experimental results. Although, in equation (4.30), C should be the *in situ* concentration but in view of the fact that such measurements are generally not available, it is customary to replace it with the concentration of solids in the discharged mixture.

From equation (4.33), it would be expected that in the moving bed regime, $f_L(i - i_L)/i_L$ would vary linearly with the concentration C in the discharged mixture. Chhabra and Richardson (1985) found that the following modified correlation represented their own data and those of Kenchington (1978) (who transported 750 μm sand particles in 13 and 25 mm pipes in a kaolin suspension) somewhat better than equation (4.33) as seen in Figure 4.20:

$$\left(\frac{i - i_L}{Ci_L} \right) f_L = 0.55 \left[\frac{gD(s - 1)}{V^2} \right]^{1.25} \qquad (4.34)$$

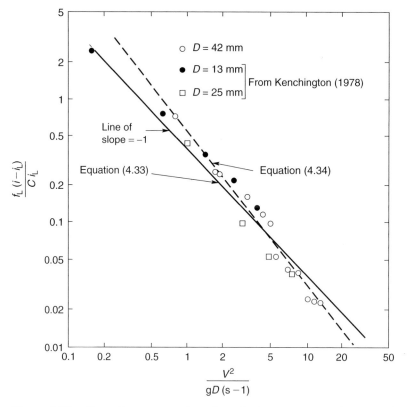

Figure 4.20   *Overall representation of results for transport in heavy media*

There are insufficient reliable results in the literature for expressions to be given for the pressure gradients in other flow regimes.

Example 4.2

A china clay slurry ($m = 9.06\,\text{Pa}\,\text{s}^n$; $n = 0.19$, $\rho_L = 1210\,\text{kg/m}^3$) is used to transport 5 mm gravel particles (nearly spherical) of density $2700\,\text{kg/m}^3$ at a mean mixture velocity of 1.25 m/s in a horizontal pipe of 50 mm diameter. The transport is in the sliding-bed regime and the discharged mixture contains 22% (by volume) gravel particles. Estimate the pressure gradient for the mixture flow.

Solution

First, the value of the friction factor f_L for the flow of liquid alone at the same average velocity is estimated. The Reynolds number is given by equation (3.17):

$$Re_{MR} = \frac{\rho_L V^{2-n} D^n}{8^{n-1} m \left(\dfrac{3n+1}{4n}\right)^n} = \frac{1210 \times (1.25)^{2-0.19}(50 \times 10^{-3})^{0.19}}{(8^{0.19-1})(9.06)\left(\dfrac{3 \times 0.19 + 1}{4 \times 0.19}\right)^{0.19}} = 532$$

As the flow of the china clay slurry is laminar,

$$f = \frac{16}{Re_{MR}} = \frac{16}{532} = 0.03$$

$$i_L = \frac{2f_L V^2}{gD} = \frac{2 \times 0.03 \times 1.25^2}{9.81 \times 50 \times 10^{-3}}$$

$$= 0.192 \text{ m of china clay slurry per m of pipe}$$

Rearranging equation (4.34) and substituting values

$$i = i_L + 0.55 \frac{Ci_L}{f_L} \left[\frac{gD(s-1)}{V^2} \right]^{1.25}$$

$$= 0.192 + \frac{0.55 \times 0.22 \times 0.192}{0.03}$$

$$\times \left[\frac{9.81 \times 50 \times 10^{-3} \left(\frac{2700}{1210} - 1 \right)}{1.25^2} \right]^{1.25}$$

$$= 0.192 + 0.236 = 0.428 \text{ m of china clay slurry/m}$$

or $\left(-\dfrac{\Delta p}{L} \right) = i \times \rho_L \times g = 0.428 \times 1210 \times 9.81 = 5080 \text{ Pa/m}$

Note that about (0.236/0.428), i.e. 55% of the total pressure drop is attributable to the presence of coarse gravel particles which in this flow regime is independent of the gravel size. □

4.3.2 RTD and slip velocity

The heat treatment to which a food particle is subjected is directly influenced by its residence (or contact) time in the plant, especially in the holding section. Therefore, reliable knowledge of RTD is very important, although, the RTD of single phase fluids in continuous sterilization or pasteurization in tube flows is reasonably well understood. For instance, for the laminar flow of Newtonian fluids in a tube, the centerline velocity is twice of the mean velocity. Therefore, the fluid elements in the center of the tube would have the shortest residence time. For highly shear-thinning power-law fluids, the velocity profile is flattened and the ratio of the mean velocity to the centerline velocity is given by equation (3.11), namely, $\{(n + 1)/(3n + 1)\}$. The greater is the degree of shear-thinning (small value of power-law index n), flatter is the velocity profile across the tube cross-section and clearly, this is advantageous in reducing the width of the RTD. Based on experimental results, the estimation of the required length of holding tube is typically based on a maximum (centerline) velocity of twice the mean velocity (as in laminar flow of a Newtonian fluid) to ensure a safe process for all conditions (Lareo *et al.*, 1997b).

On the other hand, however, the corresponding situation for the flow of food suspensions is less clear. Most of the advances made in this field are based on experimental observations aided by empirical and/or phenomenological considerations. Model studies with single food (real and prototype) particles suspended in Newtonian and in shear-thinning vehicles in tube flows have provided useful insights. For instance, Liu *et al.* (1993) investigated the behaviour of 6 mm carrot cubes in water (up to ~35% concentration) in a 44 mm diameter horizontal pipe. Depending upon the mean mixture velocity, they identified six possible flow regimes (with increasing flow rate): stationary particles at the wall, stick-slide type of flow with a variable velocity, sliding along the wall, turning and tumbling but still a great majority of particles in contact with the wall, saltation and suspended. Using dimensional analysis, Liu *et al.* (1993) correlated the ratio of the particle to fluid velocity for single particles with the reciprocal of the particle Froude number, defined as $[V_m / \sqrt{gd(s-1)}]$. For concentrated systems, the particle velocity was lower than that of the single particle, and the data on relative velocity still correlated well with the inverse of the Froude number, but with a significant scatter. One can then infer the mean residence time for particles from a knowledge of the slip velocity.

Several other studies on the RTD of food-like particles in shear-thinning and visco-plastic carriers in horizontal tubes have been reviewed by Lareo *et al.* (1997b), but none of these is conclusive. For instance, the mean and minimum residence times and the standard deviation of their RTDs of particles all show a complex dependence on the particle Reynolds number, power-law behaviour index, particle-to-tube diameter ratio and particle concentration (Sandeep and Zuritz, 1994, 1995). Broadly speaking, the velocity profile flattens with the increasing concentration of solids and, as the particles are nearly neutrally buoyant, most of them travel with an approximate uniform velocity. However, the particles never seem to achieve the maximum expected velocity (based on the assumption of power-law fluid behaviour) even at the axis of the tube (Tucker and Withers, 1994). Under certain conditions, especially at low solid loadings, it is observed that the particle RTDs show two distinct peaks thereby suggesting the formation of two groups of particles moving at different velocities, i.e., segregation of slow moving particles near the wall and of fast moving particles in the middle of the tube. Naturally, in this flow regime, particles near the tube walls will have much longer residence time than their counterparts near the axis of the tube thereby resulting in uneven quality of product.

Similarly, several commercial food processes (i.e. the APV baker ohmic heater) use vertical or inclined pipe work through which solid–liquid food mixtures are passed. Some attempts have been made to predict the RTD of solids and/or their slip velocities in vertical flow. For instance, Lareo *et al.* (1997a), Lareo and Fryer (1998) and Fairhurst *et al.* (1999) have investigated the RTD for real and model food particles in shear-thinning polymer solutions in vertical (upward and downward) flows. Lareo *et al.* (1997a) reported significant migration of particles across streamlines at volume concentrations up to about 10% using 6–10 mm carrot particles in a 44 mm diameter pipe. Over a range of velocities and particle loadings, they observed a plug-like flow wherein the particles segregated in the region $\sim 0.2 \leq (r/R) \leq \sim 0.5$, and were seen to be moving with the approximately same axial velocity. Also, a great majority of the particles were seen to be travelling faster than the mean mixture velocity (presumably due to the segregation in the middle of the tube). Therefore, the mean residence time of the particles was less than that of the mixture, *though not significantly so*. The RTD of particles also appeared to be strongly influenced by whether the liquid was simply shear thinning or also exhibited a yield stress. On the other hand, the minimum and mean residence (or passage) times showed a positive correlation with the concentration in the discharged mixture.

Similarly, preliminary studies have also been reported on RTDs in scraped surface heat exchanger (SSHE) and in helical coils with a view to achieve long contact time, albeit results are strongly dependent on the detailed geometrical design of the equipment (Tucker and Heydon, 1998; Sandeep *et al.*, 2000; Cheng *et al.*, 2005). Although both these geometries offer longer contact times and space economy than a straight tube, but only at the expense of added complexity of the flow.

Thus, in summary, although the advantages of continuous processing of foodstuffs are well recognized, due to the inadequate understanding of solid–liquid flows even in straight tubes, very little definitive information is yet available on sound process design and performance of the equipment used in the thermal treatment of foodstuffs.

Further Reading

Brown, N.P. and Heywood, N.I., (Eds.) *Slurry Handling: Design of Solid–Liquid Systems*, Elsevier, London (1991).

Chhabra, R.P., *Civil Engineering Practice* (edited by Cheremisinoff, P.N., Cheremisinoff, N.P. and Cheng, S.L.), volume 2, Technomic, Lancaster, PA, p. 251 (1988).

Chhabra, R.P. and Richardson, J.F., *Encyclopedia of Fluid Mechanics* (edited by Cheremisinoff, N.P.), volume 3, Gulf, Houston, p. 563 (1986).

Govier, G.W. and Aziz, K., *The Flow of Complex Mixtures in Pipes*, Krieger, Malabar, FL (1980).

Mahalingam, R., *Adv. Transport Process.* **1** (1980) 58.

McKetta, J.J., (Ed.) *Piping Design Handbook*, Marcel–Dekker, New York (1992).

Shook, C.A. and Roco, M.C., *Slurry Flow: Principles and Practice*, Butterworth-Heinemann, Stoneham, MA (1991).

References

Al-Sarkhi, A. and Hanratty, T.J., *Int. J. Multiphase Flow* **27** (2001) 1151, Also see *Chem. Eng. Res. Des.* **79** (2001) 402.

Al-Sarkhi, A., Abu-Nada, E. and Batayneh, M., *Int. J. Multiphase Flow* **32** (2006) 926.

Alves, G.E., *Chem. Eng. Prog.* **50** (1954) 449.

Al-Wahaibi, T., Smith, M. and Angeli, P., *J. Pet. Sci. Eng.* **57** (2007) 334.

Baik, S. and Hanratty, T.J., Int. J. *Multiphase Flow* **29** (2003) 1749.

Barnea, D. and Taitel, Y., *Encyclopedia of Fluid Mechanics* (edited by Cheremisinoff, N.P.), volume 3, Gulf, Houston (1986), Chapter 17.

Bishop, A.A. and Deshpande, S.D., *Int. J. Multiphase Flow* **12** (1986) 977.

Bjerkholt, J.T., Cumby, T.R. and Scotford, I.M., *Biosystems Eng.* **91** (2005) 361.

Carleton, A.J., Cheng, D.C.-H. and French, R.J., *Pneumotransport 2*, BHRA Fluid Eng., Bedford, UK (1973), paper F-2.

Charles, M.E. and Charles, R.A., *Advances in Solid–Liquid Flow and its Applications* (edited by Zandi, I.), Pergamon, Oxford, p. 187 (1971).

Chen, J.J. and Spedding, P.L., *Int. J. Multiphase Flow* **9** (1983) 147.

Cheng, L., Kuznetsov, A.V. and Sandeep, K.P., *Int. J. Numer. Meth. Fluid* **48** (2005) 649.

Chhabra, R.P., Farooqi, S.I. and Richardson, J.F., *Chem. Eng. Res. Des.* **62** (1984) 22.

Chhabra, R.P., Farooqi, S.I., Richardson, J.F. and Wardle, A.P., *Chem. Eng. Res. Des.* **61** (1983) 56.

Chhabra, R.P., and Richardson, J.F., *Can. J. Chem. Eng.* **62** (1984) 449.

Chhabra, R.P. and Richardson, J.F., *Chem. Eng. Res. Des.* **63** (1985) 390.

Chhabra, R.P. and Richardson, J.F., *Encyclopedia of Fluid Mechanics* (edited by Cheremisinoff, N.P.), volume 3, Gulf, Houston, p. 563 (1986).

Chisholm, D., *Int. J. Heat Mass Transf.* **10** (1967) 1767.

Chisholm, D., *Two Phase Flow in Pipelines and Heat Exchangers*, George Goodwin, London (1983).

Delgado, M.A., Franco, J.M., Partal, P. and Gallegos, C., *Chem. Eng. Proc.* **44** (2005) 805.

Duckworth, R.A., Pullum, L. and Lockyear, C.F., *J. Pipelines* **3** (1983) 251.

Duckworth, R.A., Pullum, L., Addie, G.R. and Lockyear, C.F., *Hydrotransport 10*, BHR Group, Bedford (1986), Paper # C2.

Dziubinski, M., *Chem. Eng. Res. Des.* **73** (1995) 528.

Dziubinski, M. and Chhabra, R.P., *Int. J. Eng. Fluid Mech.* **2** (1989) 63.

Dziubinski, M. and Fidos, H., *Zucker Industrie* **117** (1992) 631.

Dziubinski, M. and Richardson, J.F., *J. Pipelines* **5** (1985) 107.

Dziubinski, M., Fidos, H. and Sosno, M., *Int. J. Multiphase Flow* **30** (2004) 551.

Eissenberg, F.G. and Weinberger, C.B., *AICEJ* **25** (1979) 240.

Fairhurst, P.G., Barigou, M., Fryer, P.J. and Pain, J.-P., *Trans. Inst. Chem. Engrs.* 77C (1999) 293.

Farooqi, S.I., *PhD Thesis*, University of Wales, Swansea, U.K. (1981).

Farooqi, S.I. and Richardson, J.F., *Trans. Inst. Chem. Engrs.* **60** (1982) 292 & 323.

Farooqi, S.I., Heywood, N.I. and Richardson, J.F., *Trans. Inst. Chem. Engrs.* **58** (1980) 16.

Ferguson, M.E.G. and Spedding, P.L., *J. Chem. Tech. Biotechnol.* **62** (1995) 262.

Fernandes, R.L.J., Jutte, B.M. and Rodriguez, M.G., *Int. J. Multiphase Flow* **30** (2004) 1051.

Fossa, M., *Flow Meas. Instrum.* 9 (1998) 103.

Ghosh, T. and Shook, C.A., *Freight Pipelines* (edited by Liu, H. and Round, G.F.), Hemisphere, New York, p. 281 (1990).

Gillies, R.G., Hill, K.B., McKibben, M.J. and Shook, C.A., *Powder Technol.* **104** (1999) 269.

Govier, G.W. and Aziz, K., *The Flow of Complex Mixtures in Pipes*, Krieger, Malabar, FL (1982).

Gregory, G.A. and Mattar, L., *J. Can. Pet. Tech.* **12** (1973) 48.

Hetsroni, G., (Ed.) *Handbook of Multiphase Systems*, McGraw Hill, New York (1982).

Hewitt, G.F., *Measurement of Two-Phase Flow Parameters*, Academic, New York (1978).

Hewitt, G.F., *Handbook of Multiphase Systems* (edited by Hetsroni, G.), McGraw Hill, New York, pp. 2–25 (1982).

Hewitt, G.F., King, I. and Lovegrove, P.C., *Brit. Chem. Eng.* **8** (1963) 311.

Heywood, N.I. and Charles, M.E., *Int. J. Multiphase Flow* **5** (1979) 341.

Heywood, N.I. and Richardson, J.F., *Chem. Eng. Sci.* **34** (1979) 17.

Jinming, D. and Jingxuan, Z., *Civil Eng. Environ. Syst.* **23** (2006) 1.

Kaminsky, R.D., *J. Energ. Res. Technol. (ASME)* **120** (1998) 2.

Kang, C., Vancko, R.M., Green, A., Kerr, H. and Jepson, W., *J. Energ. Res. Technol. (ASME)* **120** (1997) 15.

Kenchington, J.M., *Hydrotransport 5*, BHR Group, Bedford, U.K. (1978), Paper # D7.

Khatib, Z. and Richardson, J.F., *Chem. Eng. Res. Des.* **62** (1984) 139.

Lareo, C. and Fryer, P.J., *J. Food Eng.* **36** (1998) 417.

Lareo, C., Branch, C.A. and Fryer, P.J., *Powder Technol.* **93** (1997a) 23, Also see ibid. 35.

Lareo, C., Fryer, P.J. and Barigou, M., *Trans. Inst. Chem. Eng.* 75C (1997b) 73.

Liu, S., Pain, J.-P., Proctor, J., DeAlwis, A.A.P. and Fryer, P.J., *Chem. Eng. Commun.* **124** (1993) 97.

Lockhart, R.W. and Martinelli, R.C., *Chem. Eng. Prog.* **45** (1949) 39.

Luo, D. and Ghiaasiaan, S.M., *Int. Comm. Heat Mass Transf.* **24** (1997) 1.

Mahalingam, R., *Adv. Transport Proc.* **1** (1980) 58.

Mahalingam, R. and Valle, M.A., *Ind. Eng. Chem. Fundam.* **11** (1972) 470.

Mandhane, J.M., Gregory, G.A. and Aziz, K., *Int. J. Multiphase Flow* **1** (1974) 537.

Mandhane, J.M., Gregory, G.A. and Aziz, K., *J. Pet. Technol.* **27** (1975) 1017.

Manfield, P.D., Lawrence, C.J. and Hewitt, G.F., *Multiphase Sci. Technol.* **11** (1999) 197.

Newitt, D.M., Richardson, J.F., Abbott, M. and Turtle, R.B., *Trans. Inst. Chem. Engrs.* **33** (1955) 93.

Oliver, D.R. and Young–Hoon, A., *Trans. Inst. Chem. Engrs.* **46** (1968) 106.

Otten, L. and Fayed, A.S., *Can. J. Chem. Eng.* **54** (1976) 111.

Petrick, P. and Swanson, B.S., *Rev. Sci. Instrum.* **29** (1958) 1079.

Pike, R.W., Wilkins, B. and Ward, H.C., *AIChE J* **11** (1965) 794.

Rao, B.K., *Int. J. Heat Fluid Flow* **18** (1997) 559.

Raut, D.V. and Rao, M.N., *Indian J. Technol.* **13** (1975) 254.

Rosehart, R.G., Scott, D.S. and Rhodes, E., *AIChE J.* **18** (1972) 744.

Ruiz-Viera, M.J., Delgado, M.A., Franco, J.M., Sanchez, M.C. and Gallegos, C., *Int. J. Multiphase Flow* **32** (2006) 232.

Sandeep, K.P. and Zuritz, C.A., *J. Food Sci.* **59** (1994) 1314.

Sandeep, K.P. and Zuritz, C.A., *J. Food Eng.* **25** (1995) 31.

Sandeep, K.P., Zuritz, C.A. and Puri, V.M., *Int. J. Food Sci. Technol.* **35** (2000) 511.

Sellin, R.H.J., Hoyt, J.W. and Scrivener, O., *J. Hyd. Res.* **20** (1982) 29. Also *see ibid* 235.

Soleimani, A., Al-Sarkhi, A. and Hanratty, T.J., *Int. J. Multiphase Flow* **28** (2002) 1911.

Sylvester, N.D. and Brill, J.P., *AIChE J.* **22** (1976) 615.

Tucker, G. and Heydon, C., *Trans. Inst. Chem. Engrs.* 76C (1998) 208.

Tucker, G. and Withers, P.M., *J. Food Proc. Eng.* **17** (1994) 401.

Shook, C.A. and Liebe, J.O., *Can. J. Chem. Eng.* **54** (1976) 118.

Shu, M.T., Weinberger, C.B. and Lee, Y.H., *Ind. Eng. Chem. Fundam.* **21** (1982) 175.

Spedding, P.L. and Chen, J.J., *Encyclopedia of Fluid Mechanics* (edited by Cheremisinoff, N.P.), volume 3, Gulf, Houston (1986), Chapter 18.

Taitel, Y., Barnea, D. and Dukler, A.E., *AIChE J* **26** (1980) 345.

Weisman, J., Duncan, D., Gibson, J. and Crawford, T., *Int. J. Multiphase Flow* **5** (1979) 437.

Welsh, S.A., Ghiaasiaan, S.M. and Abdel-Khalik, S.I., *Ind. Eng. Chem. Res.* **38** (1999) 1083.

Xu, J.-Y., Wu, Y.-X., Shi, Z.-H., Lao, L.-Y. and Li, D.-H., *Int. J. Multiphase Flow* **33** (2007) 948.

Nomenclature

Dimensions in **M, L, T**

C	volumetric concentration of solids in discharged mixture $(-)$	$\mathbf{M^0L^0T^0}$
C_0	constant, equation (4.19) $(-)$	$\mathbf{M^0L^0T^0}$
C_v	volume fraction of solid in suspension $(-)$	$\mathbf{M^0L^0T^0}$
D	pipe diameter (m)	$\mathbf{L}$
De	Deborah number $(-)$	$\mathbf{M^0L^0T^0}$
Fr	particle Froude number $(V_m/\sqrt{gd(s-1)})$ $(-)$	$\mathbf{M^0L^0T^0}$
f	fanning friction factor $(-)$	$\mathbf{M^0L^0T^0}$
g	acceleration due to gravity (m/s²)	$\mathbf{LT^{-2}}$
i	pressure gradient (m of liquid/m of pipe length)	$\mathbf{M^0L^0T^0}$
J	correction factor $(-)$	$\mathbf{M^0L^0T^0}$
L	length of pipe (m)	$\mathbf{L}$
m	power-law consistency coefficient (Pa s^n)	$\mathbf{ML^{-1}T^{n-2}}$
m_1	power-law coefficient for first normal stress difference (Pa s^{p_1})	$\mathbf{ML^{-1}T^{p'-2}}$
m'	apparent power-law consistency coefficient (Pa s$^{n'}$)	$\mathbf{ML^{-1}T^{p'-2}}$
n	power-law flow behaviour index $(-)$	$\mathbf{M^0L^0T^0}$
n'	Apparent power-law index $(-)$	$\mathbf{M^0L^0T^0}$
N	Power (W)	$\mathbf{ML^2T^{-3}}$
N_1	first normal stress difference (Pa)	$\mathbf{ML^{-1}T^{-2}}$
p_1	index in first normal stress difference, equation (4.12) $(-)$	$\mathbf{M^0L^0T^0}$
$-\Delta p/L$	pressure gradient (Pa/m)	$\mathbf{ML^{-2}T^{-2}}$
Q	volumetric flow rate (m³/s)	$\mathbf{L^3T^{-1}}$
Re_{MR}	Metzner–Reed Reynolds number $(-)$	$\mathbf{M^0L^0T^0}$
s	specific gravity of solids $(-)$	$\mathbf{M^0L^0T^0}$
V	superficial velocity (m/s)	$\mathbf{LT^{-1}}$

Greek letters

α	average holdup $(-)$	$\mathbf{M^0L^0T^0}$
γ	coefficient of friction $(-)$	$\mathbf{M^0L^0T^0}$
$\dot{\gamma}$	shear rate$(\mathrm{s^{-1}})$	$\mathbf{T^{-1}}$
λ	input fraction $(-)$	$\mathbf{M^0L^0T^0}$
λ_f	fluid characteristic time, equation (4.13) (s)	$\mathbf{T}$
Λ	loss coefficient $(-)$	$\mathbf{M^0L^0T^0}$
ρ	density (kg/m^3)	$\mathbf{ML^{-3}}$

Dimensions in **M, L, T**

μ	viscosity (Pa s)	$\mathbf{ML^{-1}T^{-1}}$
ϕ^2	drag ratio $(-)$	$\mathbf{M^0L^0T^0}$
χ	Lockhart–Martinelli parameter $(-)$	$\mathbf{M^0L^0T^0}$
ψ	coefficient, equation (4.29) $(-)$	$\mathbf{M^0L^0T^0}$
τ_w	wall shear stress (Pa)	$\mathbf{ML^{-1}T^{-2}}$

Subscripts

a	accelerational contribution
f	frictional contribution
G	gas
g	gravitational
L	liquid
L$_\mathrm{c}$	corresponding to liquid Reynolds number of 2000
L$_\mathrm{v}$	visco-elastic liquid
M	mixture
MDR	maximum drag reduction
min	minimum
mod	modified
s	solid
TP	two phase

Chapter 5

Particulate systems

5.1 Introduction

In many practical applications, we need to know the force required to move a solid object through a surrounding fluid, or conversely, the force that a moving fluid exerts on a solid as the fluid flows past it. Many processes for the separation of particles of various sizes, shapes and materials depend on their behaviour when subjected to the action of a moving fluid, as encountered in settling tanks, hydrocyclones and centrifuges, for instance. Frequently, the liquid phase may exhibit complex non-Newtonian behaviour whose characteristics may be measured using falling-ball viscometry. Furthermore, it is often necessary to calculate the fluid dynamic drag on solid particles in process equipment, e.g. for the design of slurry pipelines, fixed and fluidized beds. Similarly, in the degassing of polymer melts prior to processing, bubbles rise through a still mass of molten polymer. Likewise, the movement of oil droplets and polymer solutions in narrow pores (albeit strongly influenced by capillary forces) occurs in enhanced oil recovery operations. The settling behaviour of a particle is also strongly influenced by the presence of other neighbouring particles as in concentrated suspensions. Furthermore, it is often desirable to keep the active component uniformly suspended, as in many pharmaceutical products, paints, detergents, agro-chemicals, emulsions and foams. Thus, the need to estimate the terminal falling (or rising) velocity of a solid (or fluid) particle arises frequently in process engineering and design calculations.

Frequently, the particles are in the form of clusters (such as in fixed and some fluidized beds) and ensembles as in foams, dispersions and emulsions. However, experience with Newtonian fluids has shown that the hydrodynamics of systems consisting of single particles, drops and bubbles serves as a useful starting point for understanding the mechanics of the more complex multi-particle systems which are not amenable to rigorous analysis. This chapter aims to provide an overview of the developments in the field of non-Newtonian fluid–particle systems. In particular, consideration is given to the drag force, wall effects and settling velocity of single spherical and non-spherical particles, bubbles and drops in various types of non-Newtonian fluids (particularly based on power-law and Bingham plastic models). Flow in packed and fluidized beds and hindered settling are then considered. Detailed accounts and extensive bibliographies on this subject are available elsewhere (Chhabra, 1986, 1993, 2006; Ghosh *et al.*, 1994).

Unlike the flows considered in Chapter 3 which were essentially unidirectional, the fluid flows in particulate systems are either two- or three-dimensional and hence are inherently more difficult to analyse theoretically, even in the creeping (small Reynolds number) flow regime. Secondly, the results are often dependent on the rheological model appropriate to the fluid and a more generalized treatment is not possible. For instance, there is no 'standard non-Newtonian drag curve' even for spheres, and the relevant dimensionless groups depend on the fluid model which is used. Most of the information in this chapter relates to time-independent fluids, with occasional reference to visco-elastic fluids.

5.2 Drag force on a sphere

All bodies immersed in a fluid are subject to a buoyancy force. In a flowing fluid (or in the situation of relative velocity between the fluid and the object), there is an additional force which is made up of two components: the skin friction (or viscous drag) and the form drag (due to the pressure distribution) as shown schematically in Figure 5.1. At low velocities, no separation of the boundary layer takes place, but as the velocity is increased, separation occurs and the skin friction forms a gradually decreasing proportion of the total fluid dynamic drag exerted on the immersed object. The conditions of flow over a spherical particle are characterized by the Reynolds number; the exact form of the latter depends upon the rheological model of the fluid.

The total drag force is obtained by integrating the components of the forces attributable to skin friction and form drag in the direction of fluid motion (z-coordinate) that act on an elemental area of the surface of the sphere, i.e.

$$F_D \quad = \quad F_t \quad + \quad F_n \qquad\qquad (5.1)$$

$$\text{(total drag} \qquad \text{(skin friction} \qquad \text{(form drag}$$
$$\text{force)} \qquad\qquad \text{force)} \qquad\qquad \text{force)}$$

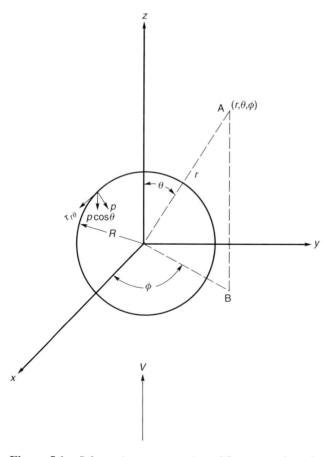

Figure 5.1 *Schematic representation of flow around a sphere*

The total drag force, $F_D = F_t + F_n$, is often expressed using a dimensionless drag coefficient C_D as:

$$C_D = \frac{F_D}{(\frac{1}{2}\rho V^2)(\pi R^2)} \tag{5.2}$$

The shear stress and pressure distributions necessary for the evaluation of the integrals implicit in equation (5.1) can, in principle, be obtained by solving the continuity and momentum equations. In practice, however, numerical solutions are often necessary even at low Reynolds numbers. Since detailed discussions of this subject are available elsewhere (Chhabra, 2006), only the significant results are presented here for power-law and viscoplastic fluid models.

5.2.1 Drag on a sphere in a power-law fluid

A simple dimensional analysis (see Example 5.1) of this flow situation shows that the drag coefficient can be expressed in terms of the Reynolds number and the power-law index, i.e.

$$C_D = f(Re, n) \tag{5.3}$$

Often for the creeping flow region ($Re \ll 1$), the numerical results may be expressed as a deviation factor, $X(n)$, in the relation between drag coefficient and Reynolds number obtained from the Stokes law

$$C_D = \frac{24}{Re} X(n) \tag{5.4}$$

where $Re = \rho V^{2-n} d^n / m$, d being the sphere diameter. The numerical values of $X(n)$ for both shear-thinning and shear-thickening fluid behaviour are listed in Table 5.1 (Gu and Tanner, 1985; Tripathi *et al.*, 1994; Tripathi and Chhabra, 1995). Evidently, shear-thinning behaviour causes drag increase ($X > 1$), and drag reduction ($X < 1$) occurs in shear-thickening ($n > 1$) fluids.

The numerical results presented in Table 5.1 are well correlated by the following analytical expression (Renaud *et al.*, 2004):

$$X(n) = 6^{(n-1)/2} \left\{ \frac{3}{n^2 + n + 1} \right\}^{n+1} \tag{5.5}$$

As expected, equation (5.5) predicts $X = 1$ for Newtonian fluids.

Based on a detailed analysis of experimental results and numerical simulations, the creeping flow occurs in shear-thinning fluids up to about $Re \sim 1 - 2$, even though a visible wake appears only when Re for the sphere reaches the value of 20 (Tripathi *et al.*, 1994; Dhole *et al.*, 2006).

In the case of creeping sphere motion in Newtonian fluids, the skin and form friction contributions are in the proportion of 2 to 1. This ratio continually decreases with the increasing degree of pseudoplasticity, the two components becoming nearly equal at $n \sim 0.4$; these

Table 5.1   *Values of* X(n) *for a sphere*

n	X(n)
1.8	0.261
1.6	0.390
1.4	0.569
1.2	0.827
1.0	1.002
0.9	1.14
0.8	1.24
0.7	1.32
0.6	1.382
0.5	1.42
0.4	1.442
0.3	1.458
0.2	1.413
0.1	1.354

relative proportions, on the other hand, increase with the increasing degree of shear-thickening behaviour.

Numerical predictions of drag on a sphere moving in a power-law fluid are available for the sphere Reynolds number up to 500 (Tripathi *et al.*, 1994; Graham and Jones, 1995; Dhole *et al.*, 2006) and the values of drag coefficient are best represented by the following expression (Dhole *et al.*, 2006):

$$C_\mathrm{D} = \frac{24}{Re}\left[1 + 0.148\,Re^{2.35n/(2.42n+0.918)}\right] \tag{5.6}$$

The empirical constants appearing in equation (5.6) are based on the numerical results of Dhole *et al.* (2006) extending over the range of conditions as: $0.5 \leq n \leq 2$ and $5 \leq Re \leq 500$. The resulting average and maximum deviations are 6.2% and 21%, respectively, which drop to 4.4% and 13%, respectively, if the results for $Re > 200$ are excluded from the database.

Experimental results for the drag on spheres are now available for Reynolds numbers up to about 1000 and for the power-law index in the range $0.38 \leq n \leq 1$ (Chhabra, 1990, 2006). Extensive comparisons between numerical simulations and experimental data show rather poor agreement in the creeping flow regime ($Re < 1$), but this improves somewhat as the Reynolds number increases. While the exact reasons for these discrepancies are not known, they have often been attributed to possible visco-elastic behaviour and the choice of inappropriate values of the power-law constants. On the other hand, the standard drag curve for Newtonian fluids correlates the drag results in power-law liquids in the region $1 \leq Re \leq 1000$ within ±30%, as seen in Figure 5.2; this is comparable with the degree of agreement between the numerical predictions and data in

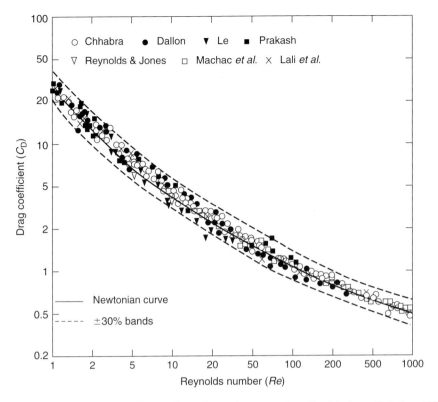

Figure 5.2 *Drag coefficient for spheres in power-law liquids (see Chhabra (1990) for original sources of data)*

the range $1 \leq Re \leq 500$, and confirms that the effect of non-Newtonian viscosity generally diminishes with the increasing Reynolds number.

On the other hand, Renaud *et al.* (2004) and Chhabra (2006) have collated most of the literature data and extended their previous work to estimate the drag on a sphere in power law fluids as:

$$C_D = C_{D0} + 0.44 \chi C_{D0}^{2A_0} B_0 \left\{ \frac{6X(n)b_0}{6X(n)b_0 + C_{D0}} \right\}^{A_0} + 0.44 \left\{ \frac{6X(n)b_0}{6X(n)b_0 + 128C_{D0}} \right\}^{11/12} \tag{5.7}$$

In equation (5.7), C_{D0}, the drag coefficient in the creeping flow regime is given by equation (5.4). The fitted parameters include A_0, B_0 and b_0, which are only functions of the flow behaviour index, are given as follows:

$$b_0 = \exp\{3(C_1 - \ln 6)\} \tag{5.8}$$

$$C_1 = \left\{ 6^{(1-n)/2} X(n) \right\}^{1/n+1} \tag{5.9}$$

$$B_0 = \left\{\frac{3 - C_1}{6C_1}\right\} \exp\left[\frac{1}{2}\left(\frac{3 - C_1}{C_1}\right)\ln 3\right] \tag{5.10}$$

$$A_0 = \frac{11}{48}\sqrt{6}\left[1 - \exp\left\{\left(\frac{3 - C_1}{2C_1}\right)^2 \ln\frac{\sqrt{6} - 1}{\sqrt{6}}\right\}\right] \tag{5.11}$$

Thus, for a known value of the power-law index, the three constants A_0, B_0 and b_0 are known. The remaining parameter χ is the ratio of the surface area to the projected (normal to the direction of settling) area of the particle. It is retained here, as it is shown in Section 5.3 that without introducing any further constants, this equation is also successful in correlating most of the currently available drag results for non-spherical particles. Evidently, for a sphere of diameter, d, the value of the composite shape factor χ is equal to $\pi d^2/(\pi d^2/4)$, i.e. $\chi = 4$. Finally, it needs to be emphasized here that the fitted constants appearing in equations (5.7)–(5.11) are evaluated using *only* the numerical results of Tripathi *et al.* (1994) and of Tripathi and Chhabra (1995) which are limited to $Re \leq 100$. Renaud *et al.* (2004) demonstrated that equation (5.7) reproduced most of the literature data (about 550 data points) with an average error of 15–16%, without any discernable trends. Thus, equation (5.7) does offer significant improvement over the standard Newtonian drag curve, but only at the expense of added complexity.

In addition to the drag behaviour, some information is also available on the detailed flow field and wake formation. At low Reynolds numbers, the flow over a sphere shows fore and after symmetry and the flow is attached to the surface of the sphere, as shown in Figure 5.3 for a Newtonian ($n = 1$), a shear-thinning ($n = 0.5$) and a shear-thickening ($n = 2$) fluid at the Reynolds number of 5. As the Reynolds number is gradually incremented, the flow gets separated from the surface of the sphere and a wake is formed in the rear of the sphere. While the precise values of the Reynolds number denoting the onset of wake formation are not yet known, suffice it to say here that these are likely to be dependent on the value of the flow behaviour index. Thus, for instance, the critical value of the Reynolds number for Newtonian fluids is *ca.* 20 or so. The available scant numerical studies suggest that the wake formation is delayed in shear-thinning fluids and it is advanced to lower values of the Reynolds number in shear-thickening liquids as compared to the value of 20 for Newtonian fluids. Furthermore, the wake appears to be shorter in shear-thinning fluids than that in Newtonian fluids, and shear-thickening fluid behaviour, however, tends to result in longer wakes, as can be seen in Figure 5.4 at $Re = 200$. Unfortunately while no experimental results are available on the size of the wake in power-law liquids, Figure 5.5 shows a comparison between the predicted (Dhole *et al.*, 2006) and observed (Taneda, 1956) sizes of the wake for a sphere in Newtonian fluids at $Re = 73.6$. Not only the comparison is seen to be good, but the wake also seems to be well formed and is about $0.7d$ long in the direction of flow.

Example 5.1

A sphere of diameter d moving at a constant velocity V through a power-law fluid (density, ρ; flow behaviour index, n; and consistency coefficient, m) experiences a drag force, F_D. Obtain the pertinent dimensionless groups of variables for this flow.

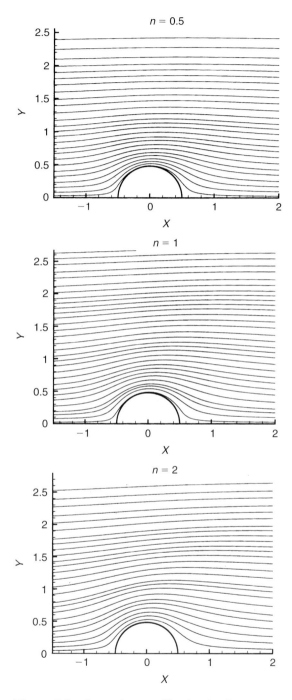

Figure 5.3   *Streamline profiles for the flow (from left to right) of power-law fluids past a sphere at Re = 5 (after Dhole et al., 2006)*

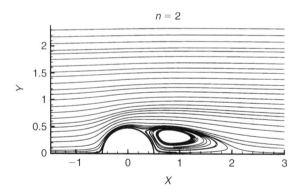

Figure 5.4   *Streamline profiles showing separation (wake) for the flow (from left to right) of power-law fluids past a sphere at Re = 200 (after Dhole et al., 2006)*

Solution

There are six variables here and three fundamental dimensions (**M**, **L**, **T**), and therefore there will be three dimensionless groups. Thus, one can write:

$$F_D = f(\rho, d, m, n, V)$$

Figure 5.5   *Comparison between the experimental (lower half) and predicted (upper half) streamline patterns for the flow of Newtonian fluids past a sphere at Re = 73.6 (based on experiments of Taneda (1956) and computations of Dhole et al. (2006))*

Writing dimensions of each of these variables:

$$F_D \equiv \mathbf{MLT}^{-2} \qquad m \equiv \mathbf{ML}^{-1}\mathbf{T}^{n-2}$$
$$\rho \equiv \mathbf{ML}^{-3} \qquad n \equiv \mathbf{M}^0\mathbf{L}^0\mathbf{T}^0$$
$$d \equiv \mathbf{L} \qquad\qquad V \equiv \mathbf{LT}^{-1}$$

Choosing ρ, d, V as the recurring set, the fundamental dimensions $\mathbf{M}$, $\mathbf{L}$ and $\mathbf{T}$ can be expressed as:

$$\mathbf{M} \equiv \rho d^3; \qquad \mathbf{L} \equiv d; \qquad \mathbf{T} \equiv d/V$$

and the three π-groups can be formed as:

$$F_D/(\rho d^3)(d)(V/d)^2, \qquad \text{i.e.} \qquad F_D/\rho V^2 d^2$$

$$m/(\rho d^3)(d^{-1})(d/V)^{n-2}, \qquad \text{i.e.} \qquad \frac{\rho V^{2-n} d^n}{m}$$

and n. Therefore,

$$\frac{F_D}{\rho V^2 d^2} = \mathrm{f}\left(\frac{\rho V^{2-n} d^n}{m}, n \right)$$

By inspection, $(F_D/\rho V^2 d^2) \propto C_D$.
$\therefore C_D = \mathrm{f}(Re, n)$ which is the same relationship as given by equation (5.3). $\square$

5.2.2 *Drag on a sphere in viscoplastic fluids*

By virtue of its yield stress, a viscoplastic material in an unsheared state will support an immersed particle for an indefinite period of time. In recent years, this property has been successfully exploited in the design of slurry pipelines, as briefly discussed in Section 4.3. Before undertaking an examination of the drag force on a spherical particle in a visco-plastic medium, the question of static equilibrium will be discussed and a criterion will be developed to delineate the conditions under which a sphere will either settle or be held stationary in a liquid exhibiting a yield stress.

(i) *Static equilibrium*
The question of whether or not a sphere will settle in an unsheared viscoplastic mater-ial has received considerable attention in the literature (Chhabra and Uhlherr, 1988; Chhabra, 2006). For the usual case where the sphere is acted upon by gravity, it is con-venient to introduce a dimensionless group, Y, which denotes the ratio of the forces due to the yield stress and due to gravity. Neglecting numerical constants, the simplest defi-nition of Y is:

$$Y = \frac{\tau_0}{gd(\rho_s - \rho)} \tag{5.12}$$

Thus, small values of Y will favour motion of a sphere. The critical values of Y reported by various investigators (Chhabra and Uhlherr, 1988) fall in two categories. One group, with the value of Y in the range 0.06 ± 0.02, includes the numerical predictions (Beris *et al.*, 1985; Beauline and Mitsoulis, 1997; Blackery and Mitsoulis, 1997), observations on the motion/no motion of spheres under free fall conditions (Ansley and Smith, 1967) and the residual force upon the cessation of flow (Brookes and Whitmore, 1968, 1969). More recent experimental results based on the extrapolation of force – velocity data to zero velocity also yield the values in the range $0.048 \leq Y \leq 0.088$ (Jossic and Magnin, 2001; Chafe and de Bruyn, 2005; Merkak *et al.*, 2006). The second group, with $Y \sim 0.2$, relies on the intuitive consideration that the buoyant weight of the sphere is supported by the vertical component of the force due to the yield stress, and on measurements on a fixed sphere held in an unsheared viscoplastic material (Uhlherr, 1986). The large discrepancy between the two sets of values suggests that there is a fundamental differ-ence in the underlying mechanisms inherent in these two approaches. Additional com-plications arise from the fact that the values of yield stress (τ_0) obtained using different methods differ widely (Nguyen and Boger, 1992). Thus, it is perhaps best to establish the upper and lower bounds on the size and/or density of a sphere that will settle in par-ticular circumstances.

(ii) *Flow field*
As in the case of the solid plug-like motion of viscoplastic materials in pipes and slits (discussed in Chapter 3), there again exists a bounded zone of flow associ-ated with a sphere moving in a viscoplastic medium, and beyond this zone, the fluid

experiences elastic deformation, similar to that in elastic solids (Volarovich and Gutkin, 1953; Tyabin, 1953). Indeed the difficulty in delineating the interface between the flow and no flow zones has been the main impediment to obtaining numerical solutions to this problem. Furthermore, even within the cavity of shear deformation, there is unsheared material adhering to parts of the sphere surface and this suggests that the yield stress may not act over the entire surface. The existence of such unsheared material attached to a moving sphere has been observed experimentally, but it is rather difficult to estimate its exact shape and size (Valentik and Whitmore, 1965; Atapattu *et al.*, 1995).

Using the laser speckle photographic method, Atapattu *et al.* (1995) measured point velocities in the fluid near a sphere moving at a constant speed on the axis of a cylindrical tube containing viscoplastic carbopol solutions. Notwithstanding the additional effects arising from the walls of the tube, Figure 5.6 shows the typical size and shape of deformation cavity for a range of values of the sphere to tube diameter ratio and the Bingham number, $Bi(=\tau_0^B d/V\mu_B)$. The slight difference between the size of cavity in the radial and axial directions should be noted, especially for large values of sphere to tube diameter ratio (d/D), but the deformation envelope rarely extends beyond 4–5 sphere radii. Nor has it been possible to identify small caps of solid regions near the front and rear stagnation points. It is appropriate to add here that the experimental results shown in Figure 5.6 are in line with the available numerical predictions (Beauline and Mitsoulis, 1997).

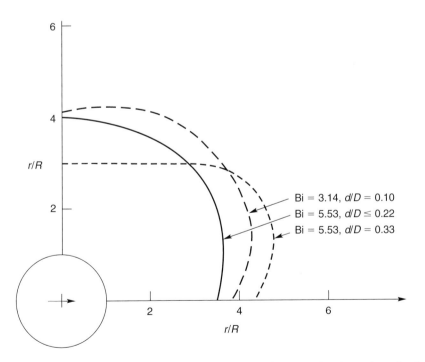

Figure 5.6 *Size of sheared cavity around a sphere moving in a viscoplastic (aqueous carbopol) solution*

Example 5.2

A china clay suspension has a density of $1050\,\text{kg/m}^3$ and a yield stress of $13\,\text{Pa}$. Determine the diameter of the smallest steel ball (density $7750\,\text{kg/m}^3$) which will settle under its own weight in this suspension.

Solution

Here $\rho = 1050\,\text{kg/m}^3;$ $\rho_s = 7750\,\text{kg/m}^3$
       $\tau_0 = 13\,\text{Pa}$

From equation (5.12), the sphere will settle only if $Y < \sim 0.04 - 0.05$
   Substituting values,

$$\frac{\tau_0}{gd(\rho_s - \rho)} = \frac{13}{9.81 \times d \times (7750 - 1050)} \leq 0.04$$

or

$$d = 4.9\,\text{mm}.$$

For a less conservative estimation, $Y = 0.212$ may be used. The use of this criterion gives $d = 0.93\,\text{mm}$. Thus, a 5-mm sphere will definitely settle in this suspension, but there is an element of uncertainty about the 1-mm steel ball. □

(iii) Drag force

The main difficulty in making theoretical estimates of the drag force on a sphere moving in a viscoplastic medium has been the lack of quantitative information about the shape of the sheared cavity. Both Beris *et al.* (1985) and Blackery and Mitsoulis (1997) have used the finite element method to evaluate the total drag on a sphere moving slowly (creeping regime) in a Bingham plastic medium and have reported their predictions in terms of the correction factor, $X\ (=C_D Re_B/24)$ which now becomes a function of the Bingham number, $Bi(=\tau_0^B d/V\mu_B)$ as:

$$X = 1 + a(Bi)^b \tag{5.13}$$

While Beris *et al.* (1985) evaluated the drag in the absence of walls (i.e. $d/D = 0$), Blackery and Mitsoulis (1997) have numerically computed the value of X for a range of diameter ratios $0 \leq d/D \leq 0.5$ and up to $Bi = 1000$. For the case of $(d/D) = 0$ (i.e. no wall effects), $a = 2.93$ and $b = 0.83$. In the range $0 \leq (d/D) \leq 0.5$, the values of a and b vary monotonically in the ranges $1.63 \leq a \leq 2.93$ and $0.83 \leq b \leq 0.95$, respectively. As the Bingham number progressively becomes smaller, X would be expected to approach unity. The higher drag ($X > 1$) in a viscoplastic medium is attributable to the additive effects of viscosity and yield stress.

In addition, many workers have reported experimental correlations of their drag data for spheres falling freely or being towed in viscoplastic media (Chhabra and Uhlherr, 1988; Chhabra, 1993a,b, 2006; Atapattu *et al.*, 1995); most correlations are based on the use of the Bingham model, though some have found the three parameter Herschel–Bulkley fluid model (equation 1.18) to correlate their data somewhat better (Sen, 1984; Atapattu *et al.*, 1995; Beauline and Mitsoulis, 1997). At the outset, it is important to establish the criterion for creeping flow in viscoplastic fluids. For a sphere falling in a Newtonian fluid

$(\tau_0 = 0)$, the creeping flow is assumed to occur up to about $Re \sim 1$. One of the characteristics of creeping flow in a Newtonian fluid is the reciprocal relationship between the Reynolds number and drag coefficient, i.e. $C_D Re = 24$. For Bingham plastic fluids, intuitively this product must be a function of the Bingham number, as can be seen in equation (5.13). Applying this criterion to the available data, the maximum value of the Reynolds number, $Re (= \rho V d / \mu_B)$, for creeping flow is given as (Chhabra and Uhlherr, 1988):

$$Re_{max} \simeq 100 Bi^{0.4} \tag{5.14}$$

Thus, the greater the Bingham number, the higher is the Reynolds number up to which the creeping flow conditions apply for spheres moving in Bingham plastic fluids.

As mentioned previously, the three parameter Herschel–Bulkley fluid model gives a somewhat better fit of the fluid rheology than the Bingham model. Atapattu *et al.* (1995) put forward the following semi-empirical correlation for drag on spheres in Herschel-Bulkley model liquids:

$$C_D = \frac{24}{Re}(1 + Bi^*) \tag{5.15}$$

where the Reynolds number, $Re = \rho V^{2-n} d^n / m$ and the modified Bingham number, $Bi^* = \tau_0^H / m(V/d)^n$. Equation (5.15) covers the ranges: $10^{-5} \leq Re \leq 0.36$; $0.25 \leq Bi^* \leq 280$ and $0.43 \leq n \leq 0.84$; and it also correlates the scant literature data available in the creeping flow region (Sen, 1984; Hariharaputhiran *et al.*, 1998; Chafe and de Bruyn, 2005). These results are also in line with the numerical predictions for Herschel–Bulkley fluids (Beauline and Mitsoulis, 1997).

In the intermediate Reynolds number region, though some predictive expressions have been developed, e.g. see Chhabra (2006), most of these data are equally well in line with the standard drag curve for Newtonian liquids (Machac *et al.*, 1995).

Thus, in summary, the non-Newtonian characteristics seem to be much more important at low Reynolds numbers and their role progressively diminishes as the inertial effects become significant with the increasing Reynolds number. Therefore, in creeping flow region, equations (5.4), (5.13) and (5.15), respectively, should be used to estimate drag forces on spheres moving in power-law, Bingham model or Herschel–Bulkley fluids. On the other hand, at high Reynolds numbers, the application of the standard drag curve for Newtonian fluids yields values of drag on spheres which are about as accurate as the empirical correlations available in the literature. The Reynolds number defined as $\rho V^{2-n} d^n / m$ for power-law fluids, as $\rho V d / \mu_B$ for Bingham plastics and as $(\rho V^{2-n} d^n / m)/(1 + Bi^*)$ for Herschel–Bulkley model fluids must be used in the standard Newtonian drag curve.

5.2.3 Drag in visco-elastic fluids

From a theoretical standpoint, the creeping, steady translation motion of a sphere in a visco-elastic medium has been selected as one of the benchmark problems for the validation of procedures for numerical solutions (Walters and Tanner, 1992; McKinley, 2002; Chhabra, 2006). Unfortunately, the picture which emerges is not only incoherent but also inconclusive. Most simulation studies are based on the creeping flow assumption (zero Reynolds number) and take into account the influence of fluid visco-elasticity on the drag of a sphere in the absence of shear-thinning behaviour. Early studies suggested a slight reduction ($\sim$5–10%) in drag below the Stokes value, with the amount of drag reduction showing a weak dependence on Deborah or Weissenberg number (defined as $\lambda_f V/d$). However, subsequent simulations (Degand and Walters, 1995) and reviews (McKinley, 2002; Chhabra, 2006) suggest that after an initial period of reduction, the drag on a sphere in a visco-elastic medium can exceed that in a Newtonian medium at high values of Deborah number; the latter enhancement is attributed to extensional effects of the fluid. Both drag reduction (up to 25%) and enhancements (up to 200%) compared with the Newtonian value have been observed experimentally (Chhabra, 2006). However there is very little quantitative agreement among various workers between the results of numerical simulations and experimental studies. The former seem to be strongly dependent on the choice of constitutive equation, details of the numerical procedure, mesh size, etc., while the experimental results appear to be very sensitive to the chemical nature, water purity, polymer–solvent interactions (Solomon and Muller, 1996), etc. of the polymer solutions used. It is not yet possible to interpret and/or correlate experimental results of drag in visco-elastic fluids in terms of measurable rheological properties. Aside from these uncertainties, other time-dependent effects have also been observed. For instance, unlike the monotonic approach to the terminal velocity in Newtonian and power-law type fluids (Bagchi and Chhabra, 1991; Chhabra *et al.*, 1998), a sphere released in a visco-elastic liquid could attain a transitory velocity almost twice that of its ultimate falling velocity (Walters and Tanner, 1992; McKinley, 2002). Yet a sphere released in worm-like micellar solutions may exhibit oscillations in its settling velocity and/or never reach a constant falling velocity (Jayaraman and Belmonte, 2003).

On the other hand, the effects of shear-thinning viscosity completely overshadow those of visco-elasticity, at least in the creeping flow region. Indeed, a correlation based on a viscosity model, with zero shear viscosity and/or a characteristic time constant, provides satisfactory representation of drag data when the liquid exhibits both shear-thinning properties and visco-elasticity (Chhabra, 2006).

5.2.4 Terminal falling velocities

In many process design calculations, it is necessary to know the terminal velocity of a sphere settling in a fluid under the influence of the gravitational field. When a spherical particle at rest is introduced into a liquid, it accelerates until the buoyant weight is exactly balanced by the fluid dynamic drag. Although the so-called terminal velocity is approached asymptotically, the effective transition period is generally of short duration for Newtonian and power-law fluids (Chhabra *et al.*, 1998). For instance, in the creeping flow regime, the terminal velocity is attained after the particle has traversed a path of length equal to only a few diameters.

For gravity settling of a sphere at its terminal velocity the drag force on it, F_D, is equal to the buoyant weight, i.e.

$$F_D = \frac{\pi d^3}{6}(\rho_s - \rho)g \tag{5.16}$$

Combining equations (5.4) and (5.16), the terminal velocity of a sphere in a power-law fluid is given as ($Re < 1$):

$$V = \left[\frac{gd^{n+1}(\rho_s - \rho)}{18mX(n)} \right]^{1/n} \tag{5.17}$$

In shear-thinning power-law fluids, therefore, the terminal falling velocity shows a stronger dependence on sphere diameter and density difference than in a Newtonian fluid.

This method of calculation is satisfactory provided it is known *a priori* that the Reynolds number is small (<1). As the unknown velocity appears in both the Reynolds number and the drag coefficient, any approach based on the use of these dimensionless groups requires an iterative procedure to estimate the unknown terminal falling velocity of a sphere of known diameter and density in a quiescent power-law medium of given rheological parameters and its density. It is thus more satisfactory to work in terms of new dimensionless groups which are formed simply by combining the Reynolds number (*Re*), drag coefficient (*C*$_D$) and power law index (*n*) (Kelessidis, 2004; Shah *et al.*, 2007). For instance, following the works of Turton and Clark (1987) and of Haider and Levenspiel (1989) for drag on spheres in Newtonian liquids, Kelessidis (2004) defined dimensionless sphere diameter *d*$^+$ and terminal falling velocity *V*$^+$ in power law fluids as:

$$d^+ = \left[\frac{3}{4} C_D \, Re^{2/(2-n)} \right]^{(2-n)/(2+n)} \tag{5.18}$$

$$V^+ = \left[\frac{Re}{\{(3/4)C_D\}^n} \right]^{1/(2+n)} \tag{5.19}$$

For a sphere falling under the influence of gravity, the drag coefficient *C*$_D$ is given as:

$$C_D = \frac{4}{3} \frac{gd}{V^2} \left(\frac{\rho_s - \rho}{\rho} \right) \tag{5.20}$$

Substituting for *C*$_D$ from equation (5.20) and for $Re = (\rho V^{2-n} d^n)/m$ into equations (5.18) and (5.19):

$$d^+ = d \left\{ \frac{g(\rho_s - \rho)}{\rho} \right\}^{(2-n)/(2+n)} \left(\frac{\rho}{m} \right)^{2/(2+n)} \tag{5.21}$$

and

$$V^+ = V \left\{ \frac{\rho^{n+1}}{g^n m (\rho_s - \rho)^n} \right\}^{1/(2+n)} \tag{5.22}$$

Evidently, the dimensionless diameter *d*$^+$ is independent of the falling velocity, and similarly *V*$^+$ is independent of sphere diameter. It can easily be seen that *d*$^+$ is directly related to the well-known Archimedes or Galileo number used in the literature (Chhabra, 2006; Shah *et al.*, 2007). Finally, the relationship between *d*$^+$ and *V*$^+$ developed by Kelessidis (2004) is written here in a slightly modified form as:

$$\frac{1}{V^+} = \left[\left\{ \frac{18X(n)}{(d^+)^{n+1}} \right\}^{K_2/n} + \left(\frac{0.321}{d^+} \right)^{K_2/2} \right]^{1/K_2} \tag{5.23}$$

Equation (5.23) differs from that of Kelessidis (2004) on two counts: At low Reynolds numbers, the drag coefficient is given by equation (5.4) whereas Kelessidis assumed the value of $X(n) = 1$ for all values of n. Secondly, Kelessidis (2004) suggested the use of $K_2 = 0.824$, the value obtained for Newtonian fluids. Kelessidis (2004) tested the validity of his equation with only 45 data points for power-law fluids ($0.56 \leq n \leq 0.92$), he reported the agreement between predictions and experiments to be satisfactory. Using a much larger database (Renaud *et al.*, 2004) and introducing the correction factor $X(n)$ into his equation, the new value of K_2 is found to be 0.793 which is only slightly smaller than his value of 0.824. The new value of $K_2 = 0.793$ is recommended as it reproduces the literature data on drag in power-law fluids with a slightly lower uncertainty than the original equation of Kelessidis (2004). The utility of this approach is illustrated in Example 5.5.

Example 5.3

For spheres of equal terminal falling velocities, obtain the relationship between their diameters and density difference between particle and fluid for creeping flow in power-law fluids.

Solution

From equation (5.17), the terminal settling velocity of a sphere increases with both its density and size. For two spheres of different diameters d_A, d_B and densities, ρ_{sA} and ρ_{sB}, settling in the same fluid, the factor X is a function of n only (see Table 5.1) and

$$\frac{V_A}{V_B} = \left[\frac{d_A^{n+1}(\rho_{sA} - \rho)}{d_B^{n+1}(\rho_{sB} - \rho)} \right]^{1/n}$$

Thus, for pseudoplastic fluids ($n < 1$), the terminal velocity is more sensitive to both sphere diameter and density difference than in a Newtonian fluid and it should, in principle, be easier to separate closely sized particles. For equal settling velocities,

$$\frac{d_B}{d_A} = \left(\frac{\rho_{sA} - \rho}{\rho_{sB} - \rho} \right)^{1/(n+1)}$$

For $n = 1$, this expression reduces to its Newtonian counterpart. □

Example 5.4

Estimate the terminal settling velocity of a 3.18 mm steel sphere (density = 7780 kg/m^3) in a viscoplastic polymer solution of density 1000 kg/m^3. The flow curve for the polymer solution is approximated by the three parameter Herschel–Bulkley model as:

$$\tau = 3.3 + 3.69(\dot{\gamma})^{0.53}$$

where all values are in SI units.

The settling may be assumed to occur in creeping flow region.

Solution

In the creeping flow region, the drag coefficient is given by equation (5.15), i.e.

$$C_D = \frac{24}{Re}(1 + Bi^*) \tag{5.15}$$

The other dimensionless groups are:

$$C_D = \frac{4}{3}\frac{gd}{V^2}\left(\frac{\rho_s - \rho}{\rho}\right)$$

$$Re = \frac{\rho V^{2-n} d^n}{m}$$

$$Bi^* = \frac{\tau_0^H}{m(V/d)^n}$$

Trial and error solution is needed as the unknown velocity appears in all of these groups. The other values (in S.I. units) are:

$$\tau_0^H = 3.3\,\text{Pa}; \quad m = 3.69\,\text{Pa}\cdot\text{s}^{0.53}; \quad n = 0.53; \quad d = 3.18 \times 10^{-3}\,\text{m}$$
$$\rho_s = 7780\,\text{kg/m}^3; \quad \rho = 1000\,\text{kg/m}^3; \quad g = 9.81\,\text{m/s}^2$$

Substituting these values in equation (5.15):

$$C_D = \frac{0.2813}{V^2} \quad \text{or} \quad V = \sqrt{\frac{0.2813}{C_D}}$$

$$Re = \frac{(1000)(3.18 \times 10^{-3})^{0.53}\,V^{2-0.53}}{3.69} = 12.86V^{1.47}$$

$$Bi^* = \frac{\tau_0^H}{m(V/d)^n} = \frac{3.3 \times (3.18 \times 10^{-3})^{0.53}}{3.69 \times V^{0.53}} = 0.0424\,V^{-0.53}$$

Assume a value of $V = 15\,\text{mm/s} = 15 \times 10^{-3}\,\text{m/s}$.

$$\therefore Bi^* = 0.393; \quad Re = 12.86 \times (15 \times 10^{-3})^{1.47} = 0.0268$$

Now from equation (5.15), the value of C_D:

$$C_D = \frac{24}{Re}(1 + Bi^*) = \frac{24}{0.0268}(1 + 0.393)$$
$$= 1248$$

$$\therefore \text{ the velocity, } V = \sqrt{\frac{0.2813}{C_D}} = \sqrt{\frac{0.2813}{1248}} = 0.015\,\text{m/s} = 15\,\text{mm/s}$$

which matches with the assumed value. Also, in view of the small value of the Reynolds number ($Re_{max} \sim 70$, from equation 5.14), the assumption of the creeping flow is justified. □

Example 5.5

Calculate the terminal falling velocity of the following spherical particles in a power-law fluid ($m = 0.5\,\text{Pa}\cdot\text{s}^n$; $n = 0.65$; $\rho = 1050\,\text{kg/m}^3$):

(i)   plastic spheres ($\rho_s = 1500\,\text{kg/m}^3$) of diameters 1, 5 and 10 mm.
(ii) glass spheres ($\rho_s = 2500\,\text{kg/m}^3$) of diameters 1, 2 and 5 mm.
(iii) steel spheres ($\rho_s = 7780\,\text{kg/m}^3$) of diameters 0.5, 1 and 5 mm.

Compare the predictions of equations (5.7) and (5.23). How much error will be incurred if the original equation of Kelessidis were used?

Solution

For $n = 0.65$, the values of the constants $X(n)$, C_1, A_0, B_0 and b_0 appearing in equation (5.7) are evaluated first. Using equation (5.5):

$$X(n) = 6^{(n-1)/2}\left\{\frac{3}{n^2+n+1}\right\}^{n+1} = 6^{(0.65-1)/2}\left\{\frac{3}{0.65^2+0.65+1}\right\}^{1.65}$$

or

$$X(n) = 1.345$$

Now substituting this value in equation (5.9):

$$C_1 = \left\{6^{(1-0.65)/2}\times(1.345)\right\}^{1/1.65} = 1.447$$
$$b_0 = \exp\{3(C_1 - \ln 6)\} = \exp\{3(1.447 - \ln 6)\}$$

or

$$b_0 = 0.355$$

Substituting value of C_1 in equation (5.10):

$$B_0 = \left\{\frac{3-1.447}{6\times1.447}\right\}\exp\left[\left(\frac{3-1.447}{2\times1.447}\right)\ln 3\right] = 5.109$$

and finally, using equation (5.11):

$$A_0 = \frac{11}{48}\sqrt{6}\left[1-\exp\left\{\left(\frac{3-1.447}{2\times1.447}\right)^2\ln\frac{\sqrt6-1}{\sqrt6}\right\}\right]$$

or

$$A_0 = 0.0787$$

The composite shape factor χ for a sphere:

$$\chi = \frac{\text{surface area}}{\text{projected area}} = \frac{\pi d^2}{\pi d^2/4} = 4$$

Now, re-writing $C_{D0} = (24X(n))/Re$ and substituting these values in equation (5.7):

$$C_D = \frac{24}{Re} \times 1.345 + 0.44 \times 4 \times \left(\frac{24 \times 1.345}{Re}\right)^{2 \times 0.0787} \times 5.109$$

$$\times \left\{\frac{6 \times 1.345 \times 0.355}{6 \times 1.345 \times 0.355 + (24 \times 1.345/Re)}\right\}^{0.0787}$$

$$+ 0.44 \left\{\frac{6 \times 1.345 \times 0.355}{6 \times 1.345 \times 0.355 + (128 \times 24 \times 1.345/Re)}\right\}^{11/12}$$

Simplification yields:

$$C_D = \frac{32.28}{Re} + 15.54 Re^{-0.1574} \left(\frac{2.865}{2.865 + (32.28/Re)}\right)^{0.0787}$$

$$+ 0.44 \left(\frac{2.865}{2.865 + (4131.8/Re)}\right)^{11/12} \tag{5.A}$$

Since the unknown velocity appears in both Reynolds number Re and drag coefficient C_D, an iterative approach must be employed here. Let us assume an initial value of the free falling velocity $V = 50\,\mu m/s$ and the corresponding value of the Reynolds number for 1 mm plastic sphere is calculated as:

$$Re = \frac{\rho V^{2-n} d^n}{m} = \frac{1050 \times (50 \times 10^{-6})^{2-0.65} \times (1 \times 10^{-3})^{0.65}}{0.5}$$

or

$$Re = 3.68 \times 10^{-5}$$

and now substituting this value of $Re = 3.68 \times 10^{-5}$ in equation (5.A):

$$C_D = \frac{32.28}{3.68 \times 10^{-5}} + 15.54 \times (3.68 \times 10^{-5})^{-0.1574}$$

$$\times \left\{\frac{2.865}{2.865 + (32.28/3.68 \times 10^{-5})}\right\}^{0.0787}$$

$$+ 0.44 \left(\frac{2.865}{2.865 + (4131.8/3.68 \times 10^{-5})}\right)^{11/12}$$

$$= 8.772 \times 10^5 + 77.5 \times 0.37 + 0.44 \times 0.996$$

Neglecting the last two terms, $C_D = 8.772 \times 10^5$.

Now using the definition of the drag coefficient as:

$$C_D = \frac{4gd}{3V^2}\left(\frac{\rho_s - \rho}{\rho}\right)$$

Rearranging it in terms of the unknown velocity V as:

$$V = \sqrt{\frac{4gd(\rho_s - \rho)}{3C_D\rho}}$$

Now substituting values for 1 mm plastic sphere:

$$V = \sqrt{\frac{4 \times 9.81 \times 1 \times 10^{-3} \times (1150 - 1050)}{3 \times 8.772 \times 10^5 \times 1050}}$$

Solving for V:

$$V = 3.768 \times 10^{-5} \text{ m/s or } 37.68\,\mu\text{m/s}$$

This value is quite different from our assumed value of $50\,\mu$m/s and therefore further iterations are required. However in this case since the value of the Reynolds number is so small and hence one can now use equation (5.17) to estimate the terminal velocity as:

$$V = \left\{\frac{gd^{n+1}(\rho_s - \rho)}{18mX(n)}\right\}^{1/n}$$

Substituting values:

$$V = \left\{\frac{9.81 \times (1 \times 10^{-3})^{1.65} \times (1150 - 1050)}{18 \times 0.5 \times 1.345}\right\}^{1/0.65}$$

$$V = 2.095 \times 10^{-5} \text{ m/s, i.e., } 20.95\,\mu\text{m/s}$$

Therefore, the terminal falling velocity of this sphere is $20.95\,\mu$m/s.
Now turning our attention to the application of (5.23), the value of d^+ is calculated using equation (5.21):

$$d^+ = 1 \times 10^{-3} \left\{\frac{9.81 \times (1150 - 1050)}{1050}\right\}^{(2-0.65)/(2+0.65)} \left(\frac{1050}{0.5}\right)^{2/(2+0.65)}$$

i.e.

$$d^+ = 0.311$$

Now substituting this value in equation (5.23) and using $X(n) = 1.345$ and $K_2 = 0.793$:

$$\frac{1}{V^+} = \left[\left(\frac{18 \times 1.345}{0.311^{1.65}}\right)^{0.793/0.65} + \left(\frac{0.321}{0.311}\right)^{0.793/2}\right]^{1/0.793}$$

or

$$V^+ = \{(166.5)^{1.22} + 1.013\}^{-1/0.793}$$

i.e.

$$V^+ = 3.814 \times 10^{-4}$$

Now using this value and substituting for the other quantities in equation (5.22):

$$3.814 \times 10^{-4} = V \left[\frac{1050^{1.65}}{9.81^{0.65} \times 0.5 \times (1150 - 1050)^{0.65}} \right]^{1/(2+0.65)}$$

Solving for V:

$$V = 2.09 \times 10^{-5} \, \text{m/s} \quad \text{or} \quad 20.9 \, \mu\text{m/s}$$

Finally, using the standard drag curve for Newtonian fluids which is approximated here by the original equation of Kelessidis (2004), i.e. using $X(n) = 1$ and $K_2 = 0.824$ in equation (5.23), i.e.

$$\frac{1}{V^+} = \left[\left\{ \frac{18}{(d^+)^{1.65}} \right\}^{0.824/0.65} + \left(\frac{0.321}{d^+} \right)^{0.824/2} \right]^{1/0.824}$$

Substituting the value of d^+ and solving for V^+:

$$V^+ = 5.49 \times 10^{-4}$$

This, in turn, yields the value of V (using equation (5.22)):

$$V = 30 \, \mu\text{m/s}$$

This value is about 50% larger than the previously calculated values. However, the differences in various predictions are likely to be significant only at low Reynolds numbers. Similar calculations can now be repeated for the other spheres of different sizes and densities. A summary of results is presented here in Table 5.2.

Some observations can be made at this stage: firstly, at low Reynolds numbers, equations (5.7) and (5.23) yield comparable values, and obviously equation (5.23) is convenient as it obviates the necessity of an iterative solution. Secondly, at high Reynolds numbers, all three predictions begin

Table 5.2 *Summary of calculations (Example 5.5)*

d (mm)	ρ_s (kg/m^3)	Terminal falling velocity (mm/s)			Re^a
		Equation (5.7)	Equation (5.23)	Kelessidis	
1	1150	0.0209	0.0209	0.030	1.13×10^{-5}
5	1150	1.244	1.39	1.89	8.03×10^{-3}
10	1150	7.19	7.12	10.07	0.135
1	2500	1.281	1.48	1.97	2.93×10^{-3}
2	2500	7.43	7.72	10.80	0.0494
5	2500	71.35	60.00	85.00	1.9
0.5	7780	2.34	2.66	3.58	4.2×10^{-3}
1	7780	13.54	13.84	19.5	0.071
5	7780	465	371	481	23.8

[a] Based on the value of V obtained from equation (5.7)

to come close to each other whereas the original equation of Kelessidis over-predicts (by up to 50%) the falling velocity at low Reynolds numbers. □

5.2.5 Effect of container boundaries

The problem discussed so far relates to the motion of a single spherical particle in an unbounded, or effectively infinite, expanse of fluid. If other particles are present in the neighbourhood of the sphere, its settling velocity will be influenced and the effect will become progressively more marked as the concentration of particles increases. There are three contributory factors. First, as the particles settle, the displaced liquid flows upwards. Secondly, the particle experiences increased buoyancy force owing to the higher density of the suspension. Finally, the flow pattern of the liquid relative to the particles will be altered thereby affecting the velocity gradients. The sedimentation of concentrated suspensions in non-Newtonian fluids is discussed in Section 5.2.6 while the effect of the vessel walls is discussed here.

The walls of the vessel containing the liquid exert an extra retarding effect on the terminal falling velocity of the particle. The upward flow of the displaced liquid through the annular region, not only influences the relative velocity, but also sets up a velocity profile in the confined geometry of the tube. This effect may be quantified by introducing a wall factor, f, which is defined as the ratio of the terminal falling velocity of a sphere in a tube, V_m, to that in an unconfined liquid, V, viz.,

$$f = \frac{V_m}{V} \tag{5.24}$$

The experimental determination of the settling velocity in an infinite medium requires the terminal falling velocity of a sphere to be measured in tubes of different diameters and then extrapolating these results to $(d/D) = 0$, as shown in Figure 5.7 for a series of plastic spheres falling in a 0.5% Methocel solution. When the settling occurs in the creeping flow region ($Re < 1$), the measured falling velocity shows a linear dependence on the diameter ratio and can readily be extrapolated to $(d/D) = 0$. The available experimental results in Newtonian and power-law liquids indicate that the wall factor, f, is independent of the sphere Reynolds number (based on the measured velocity, V_m) both at small ($<\sim 1$) and large ($>\sim 1000$) values of the Reynolds number (Chhabra and Uhlherr, 1980; Uhlherr and Chhabra, 1995; Chhabra *et al.*, 1996). Based on an extensive experimental study in the range of conditions $0.5 \leq n \leq 1$; $0.01 \leq Re_m \leq 1000$ and $(d/D) \leq 0.5$, the wall factor can be empirically correlated with the diameter ratio and the sphere Reynolds number as (Chhabra and Uhlherr, 1980):

$$\frac{(1/f) - (1/f_\infty)}{(1/f_0) - (1/f_\infty)} = (1 + 1.3Re_m^2)^{-1/3} \tag{5.25}$$

where f_0 and f_∞, the values of the wall factor in the low and high Reynolds number regions, respectively, are given by:

$$f_0 = 1 - 1.6\frac{d}{D} \tag{5.26}$$

Figure 5.7 *Dependence of terminal falling velocity of spheres in a 0.5% aqueous hydroxyethyl cellulose solution (Re*$_m$ *> 1): • PVC spheres, △ Perspex spheres*

and

$$f_\infty = 1 - 3 \left(\frac{d}{D} \right)^{3.5} \tag{5.27}$$

While it is readily recognized that the creeping flow occurs up to about $Re_m \sim 1$, the critical value of the Reynolds number corresponding to the upper asymptotic value, f_∞, is strongly dependent upon the value of (d/D), e.g. ranging from $Re_m \sim 30$–40 for $(d/D) = 0.1$ to $Re_m \sim 1000$ for $(d/D) = 0.5$. Qualitatively, the additional retardation caused by the walls of the vessel is less severe is power-law fluids than that in Newtonian fluids under otherwise identical conditions; the effect becoming progressively less important with the increasing Reynolds number and/or decreasing diameter ratio and/or the decreasing value of the power-law index. Also in the creeping flow region, the predictions of equation (5.26) are consistent with the available numerical results (Missirlis *et al.*, 2001). However, the numerical results seem to suggest the value of the constant $-$ 1.6 in equation (5.26) – to be weakly dependent on the value of the power-law index, but the currently available data correlate satisfactorily by the single value of 1.6 (Chhabra, 2006). The wall effect is even smaller in visco-elastic liquids (Chhabra, 2006).

Sedimenting particles are also subject to an additional retardation as they approach the bottom of the containing vessel because of the influence of the lower boundary on the flow pattern. No results are available on this effect for non-Newtonian fluids and therefore the corresponding expressions for Newtonian fluids offer the best guide

(Clift *et al.*, 1978), at least for inelastic shear-thinning fluids. For instance, in the creeping flow regime and $d/D < 0.1$, the effect is usually expressed as

$$\frac{V_m}{V} = \frac{1}{1 + 1.65(d/L)} \tag{5.28}$$

where L is the distance from the bottom of the vessel.

5.2.6 Hindered settling

As mentioned earlier, the terminal falling velocity of a sphere is also influenced by the presence of neighbouring particles. In concentrated suspensions, the settling velocity of a sphere is less than the terminal falling velocity of a single particle. For coarse (non-colloidal) particles in mildly shear-thinning liquids ($1 > n \geq 0.8$) (Chhabra *et al.*, 1992), the expression proposed by Richardson and Zaki (1954) for Newtonian fluids applies at values of $Re(= \rho V^{2-n} d^n/m)$ up to about 2:

$$\frac{V_0}{V} = (1 - C)^z \tag{5.29}$$

where V_0 is the hindered settling velocity of a suspension of uniform size spheres at a volume fraction C, V is the terminal falling velocity of a single sphere in the same liquid, Z is a constant which is a function of the Archimedes number and (d/D) and is given as (Coulson and Richardson, 1991):

$$\frac{4.8 - Z}{Z - 2.4} = 0.0365 Ar^{0.57}[1 - 2.4(d/D)^{0.27}] \tag{5.30}$$

where for power-law liquids, the Archimedes number, Ar, is defined as $Ar = C_D \, Re^{2/(2-n)}$.

In visco-elastic fluids, some internal clusters of particles form during hindered settling and the interface tends to be diffuse (Allen and Uhlherr, 1989; Bobroff and Phillips, 1998). For 100–200 μm glass spheres in visco-elastic polyacrylamide solutions, significant deviations from equation (5.30) have been observed. Furthermore, due to severe segregation of particles, it is difficult to locate the clear liquid interface in these systems.

Example 5.6

Estimate the hindered settling velocity of a 25% (by volume) suspension of 200 μm glass beads in an inelastic carboxymethyl cellulose solution ($n = 0.8$ and $m = 2.5 \, \text{Pa} \cdot \text{s}^n$) in a 25 mm diameter tube. The density of glass beads and of the polymer solution are 2500 and 1020 kg/m³, respectively.

Solution

The velocity of a single glass bead is calculated first. In view of the small size and rather high consistency coefficient of the solution, the particle Reynolds number will be low, equation (5.17) can be used. From Table 5.1, $X(n) = 1.24$ corresponding to $n = 0.8$.

Substituting values in equation (5.17):

$$V = \left[\frac{g d^{n+1}(\rho_s - \rho)}{18mX} \right]^{1/n}$$

$$= \left[\frac{9.81 \times (200 \times 10^{-6})^{0.8+1}(2500 - 1020)}{18 \times 2.5 \times 1.24} \right]^{1/0.8}$$

$$= 4.97 \times 10^{-6} \text{ m/s or } 4.97 \,\mu\text{m/s}$$

Check the value of Reynolds number, *Re*:

$$Re = \frac{\rho V^{2-n} d^n}{m} = \frac{1020 \times (4.97 \times 10^{-6})^{2-0.8}(200 \times 10^{-6})^{0.8}}{2.5}$$

$$= 1.93 \times 10^{-7} \ll 1$$

Therefore, the settling occurs in the creeping flow region and the equation (5.17) is valid.
The Archimedes number is given as:

$$Ar = C_{\mathrm{D}} \, Re^{2/(2-n)} = \frac{4}{3} g d^{(2+n)/(2-n)}(\rho_s - \rho)\rho^{n/(2-n)} m^{2/(n-2)}$$

$$= \left(\frac{4}{3} \right) \times 9.81 \times (200 \times 10^{-6})^{(2+0.8)/(2-0.8)}$$

$$\times (2500 - 1020)(1020)^{(0.8)/(2-0.8)}(2.5)^{2/(0.8-2)}$$

$$= 0.0009964$$

The value of *Z* is evaluated from equation (5.30):

$$\frac{4.8 - Z}{Z - 2.4} = 0.0365 \times (0.000\,996\,4)^{0.57} \left[1 - 2.4 \left(\frac{200 \times 10^{-6}}{25 \times 10^{-3}} \right)^{0.27} \right]$$

$$= 0.000\,247$$

and

$$Z \approx 4.8$$

$$\therefore \quad \frac{V_0}{V} = (1 - C)^Z$$

or

$$V_0 = 4.97 \times 10^{-6} \times (1 - 0.25)^{4.8} = 1.25 \times 10^{-6} \text{ m/s}$$

This velocity is about a quarter of the value for a single particle. □

5.3 Flow over a cylinder

The flow of fluids over an infinitely long cylinder oriented normal to the direction of flow has also been studied extensively. It is well known that no solution exists for the creeping (zero Reynolds number) flow of Newtonian fluids over an unconfined cylinder, and it is thus not possible to obtain a formula similar to the Stokes equation for a sphere. This difficulty stems from the fact that the inertial effects do not vanish completely even far away from the cylinder. This is known as the Stokes paradox. However, in spite of this conceptual difficulty, some approximate closed form expressions for drag (Tanner, 1993; Panton,

2005) and extensive numerical results on drag and flow field are now available in the literature, e.g., see Bharti *et al.* (2006). On the other hand, a shear-thinning fluid ($n < 1$) will have a very high apparent viscosity far away from the cylinder where the rate of shear is small. Thus, unlike in the case of a Newtonian fluid, viscous forces would be much greater than the inertial forces far away from the cylinder thereby making the Stokes paradox irrelevant (Tanner, 1993). Thus, a few numerical studies providing the values of drag on a cylinder in power-law fluids are now available in the literature (e.g., see Whitney and Rodin, 2001; Chhabra *et al.*, 2004; Soares *et al.*, 2005; Bharti *et al.*, 2006). Broadly speaking, the drag coefficient for a cylinder shows a similar dependence on the power-law index as that seen for a sphere. Thus, in the creeping flow region shear-thinning fluid behaviour increases the drag on a cylinder as compared to that in Newtonian fluids at equal Reynolds numbers. However, this effect gradually becomes smaller and smaller as the value of the Reynolds number is progressively increased (Chhabra, 2006). Similar to the case of a sphere, at certain Reynolds number, the flow separates from the surface of the cylinder and loses the fore and aft symmetry. For Newtonian fluids, this transition occurs at $Re \sim 4$–5 whereas this transition occurs at slightly higher values of the Reynolds number in shear-thinning fluids ($Re \approx 11$ for $n = 0.3$) and at lower values in shear-thickening fluids ($Re \approx 0.5$ for $n = 1.8$) (Sivakumar *et al.*, 2006). Qualitatively similar numerical results for power-law fluids are also available for cylinders of square and elliptical cross-sections (Dhiman *et al.*, 2006, 2008; Sivakumar *et al.*, 2007).

5.4 Effect of particle shape on terminal falling velocity and drag force

A spherical particle is unique in that it presents the same projected area to the oncoming fluid irrespective of its orientation. For non-spherical particles, on the other hand, the orientation must be specified before the drag force can be calculated. The drag force on spheroidal (oblates and prolates) particles moving in shear-thinning and shear-thickening power-law fluids ($0.4 \leq n \leq 1.8$) have been evaluated for Reynolds numbers up to 100. The values of drag coefficient are given in the original papers (Tripathi *et al.*, 1994; Tripathi and Chhabra, 1995) and the main trends are summarized here. For pseudo-plastic fluids ($n < 1$), creeping flow occurs for Re up to about 1 (based on equal volume sphere diameter) and for dilatant fluids ($n > 1$) up to about 0.2–0.5. For a given Reynolds number and aspect ratio (minor/major axis), the drag on oblates is less than that on a sphere of equal volume whereas for prolate particles, it is higher. The drag force in the creeping flow region is higher for shear-thinning fluids than for Newtonian fluids; this is consistent with the behaviour observed for a sphere. The influence of power-law index, however, diminishes with the increasing particle Reynolds number. The opposite effect is observed with shear-thickening fluids, that is, the drag is lower than that in a Newtonian fluid. Likewise, limited numerical predictions of freely falling disks (flat surface oriented normal to the direction of gravity) in power-law liquids suggest an increase in drag of up to 30–40% due to shear-thinning viscosity (Nitin and Chhabra, 2006). All in all, currently available numerical results are restricted to a very few axisymmetric shapes of particles and of fixed orientation in moving liquids. On the other hand, one often encounters particles of various shapes in industrial practice. Therefore, many workers have measured drag coefficients for particles, including cylinders, rectangular prisms, discs, cones settling at their terminal velocities in power-law fluids. Work in this area has recently been reviewed (Chhabra, 1996, 2006), but it is not yet possible to develop a universally applicable scheme to predict the drag or the terminal falling velocity of a particle of an arbitrary shape and orientation. Unlike for the case of a sphere, a minimum of three attributes are required to describe the settling behaviour of a non-spherical particle; size, shape and orientation. For instance, one needs to specify the length and diameter of a cylinder, or the length of the three sides of a prism, height and radius of a cone, etc. The most commonly used measure of the size of a non-spherical particle is the diameter of an equal volume sphere, d_s. Similarly, while several parameters have been proposed in the literature to characterize the shape of a non-spherical particle, perhaps the most common is the so-called sphericity, ψ. The sphericity of a particle is defined as the surface area of the equal volume sphere to the surface area of the actual particle. Since for a fixed volume, a sphere has the minimum surface area, the value of sphericity, ψ, is always less than unity. Indeed, smaller is the value of ψ, larger is the deviation from the spherical shape. It is useful to recognize here that both the equal volume sphere diameter and the sphericity are fixed for a given particle, and as such do not provide any information about the orientation of the particle during its free fall in a quiescent liquid medium. While there does not appear to be an easy way of quantifying the orientation, one convenient parameter seems to be the ratio of the surface area to the projected (normal) area of the particle, χ. Some authors prefer to express this ratio in terms of (d_s/d_n) where d_n is the diameter of the circle of area equal to the projected area of the particle. Thus, for instance, a cylinder of known size will have constant values of d_s and ψ, but the value of χ will change depending upon its orientation, that is, whether it falls with its axis aligned or normal to the direction of its fall. It is worthwhile to show here that the shape factor χ indeed combines all these attributes into a single parameter. By definition, the composite shape factor, χ:

$$\chi = \frac{\text{surface area of particle}}{\text{projected area of particle}} \tag{5.31}$$

which can be re-written in terms of sphericity ψ as:

$$\chi = \frac{\text{surface area of an equal volume surface}/\psi}{\text{projected area of particle}} \tag{5.32}$$

Now writing the two area terms in terms of d_s and d_n to get:

$$\chi = \frac{\pi d_s^2/\psi}{\dfrac{\pi d_n^2}{4}} = \frac{4}{\psi}\left(\frac{d_s}{d_n}\right)^2 \tag{5.33}$$

The virtue of this parameter lies in its simplicity and in the fact that it combines the three attributes (size, d_s, shape, ψ; orientation, d_n) into one composite shape factor χ. Table 5.3 lists values of χ for typical (regular) shapes of particles.

An inspection of Table 5.3 shows some counter-intuitive results. Since the shape factor χ incorporates the particle shape, size and orientation, it is possible to obtain the same value of χ for two particles of different shapes, for example, see the values for a cube and for a cylinder of $L/D = 1$ falling with its axis aligned with gravity. Similarly, depending upon the particle shape and its orientation, it is possible to obtain values of χ lower than that for a sphere, i.e., 4. Finally, intuitively one would expect the drag of a cone falling with its apex pointing downward to be less than that for a cone settling

Table 5.3 *Typical values of* χ

Shape	Orientation	Sphericity (ψ)	Shape factor (χ)
Sphere	–	1	4
Cube	Axis aligned with vertical	0.806	6
Cylinder $L/D = 1$	$\parallel$	0.873	6
	$\perp$	0.873	4.7
$L/D = 10$	$\parallel$	0.579	42
	$\perp$	0.579	3.3
$L/D = 20$	$\parallel$	0.471	82
	$\perp$	0.471	3.2
Cone	Apex up or down		
$h/d = 1$		0.778	3.24
$h/d = 1.5$		0.793	4.16
$h/d = 2$		0.781	5.12

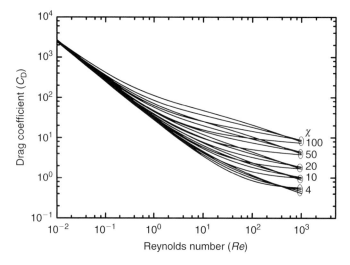

Figure 5.8 *Effect of particle shape and of power-law index on drag coefficient as predicted by equation (5.7)*

with its apex oriented upward, but the approach outlined above does not distinguish between these two cases. This illustrates the difficulty in developing a general approach to account for the orientation of a particle during its free fall under its own weight.

Most of the literature data for drag on non-spherical particles in power-law fluids has been collated recently (Rajitha *et al.*, 2006; Chhabra, 2006; Agarwal and Chhabra, 2007) and it was found that equation (5.7) with sphere diameter replaced by d_s reproduced nearly 1000 individual data points (for cones, cylinders, square plates and disks, prisms, parallelepipeds, etc.) encompassing wide ranges of the flow behaviour index ($0.31 \le n < 1$), of the Reynolds number ($\sim 10^{-5}$–300) and of sphericity ($0.62 \le \psi \le 1$) with a mean error of $\pm 30\%$, albeit the maximum error was of the order of 80%. Figure 5.8

shows the predictions of equation (5.7) for non-spherical particles for a range of values of χ and n. One can readily see that neither shape nor orientation seem to be important at low Reynolds numbers. This is consistent with the fact that particle size plays an important role at low Reynolds numbers (predominantly viscous drag) and shape and orientation matter at high Reynolds number due to a major contribution from the form drag. Example 5.7 illustrates the utility of this approach.

Example 5.7

The viscosity behaviour of a shear-thinning polymer solution (density 1020 kg/m³) is approximated by the simple power-law model and the values of the constants are: $n = 0.5$ and $m = 0.01 \, \text{Pa} \cdot \text{s}^n$. Estimate the terminal falling velocity of the following particles (all made of plastic, $\rho_s = 1200 \, \text{kg/m}^3$ and of the same volume as a 3 mm diameter sphere) in this polymer solution:

(i) Circular cylinders of $L/d = 1$, 2 and 10 falling in both orientations, that is, with their axis aligned with and normal to the direction of gravity.
(ii) Cones of $h/d = 0.5$, 1, 2
(iii) Square cross-section bars of $L/d = 1$, 2 and 10 in both orientations as in (i) above.

Solution

As noted above, equation (5.7) also applies to the free settling of non-spherical particles in stationary power-law media, it will therefore be used here to calculate the terminal falling velocity of the non-spherical particles. We begin with by evaluating the various constants appearing in equation (5.7) for $n = 0.5$.

$$X(n) = 6^{(n-1)/2} \left[\frac{3}{n^2 + n + 1} \right]^{n+1} = 6^{(0.5-1)/2} \left[\frac{3}{0.5^2 + 0.5 + 1} \right]^{0.5+1}$$

or

$$X(n) = 1.434$$

Now substituting this value in equation (5.9):

$$C_1 = \left\{ 6^{(1-0.5)/2} \times 1.434 \right\}^{1/(1+0.5)} = 1.714$$

Similarly, the remaining three constants, namely, b_0, B_0 and A_0 are evaluated next.

$$b_0 = \exp\{3(C_1 - \ln 6)\} = \exp\{3 \times (1.714 - \ln 6)\}$$

or

$$b_0 = 0.792$$

Now using equation (5.10):

$$B_0 = \left\{ \frac{3 - 1.714}{6 \times 1.714} \right\} \exp\left[\left(\frac{3 - 1.714}{2 \times 1.714} \right) \ln 3 \right] = 0.189$$

and finally using equation (5.11):

$$A_0 = \frac{11}{48}\sqrt{6}\left[1 - \exp\left\{\left(\frac{3 - 1.714}{2 \times 1.714}\right)^2 \ln\frac{\sqrt{6} - 1}{\sqrt{6}}\right\}\right]$$

or

$$A_0 = 0.04$$

One sample calculation is demonstrated here for a cylinder with $L/d = 2$ falling with its long axis normal to the direction of gravity.

$$\text{Volume of each particle, } \vartheta_p = \frac{\pi d^3}{6} = \frac{3.14 \times (3 \times 10^{-3})^3}{6}$$

i.e.,

$$\vartheta_p = 1.414 \times 10^{-8}\,\text{m}^3$$

The equal volume sphere diameter of each particle is thus 3 mm, i.e.

$$d_s = 3 \times 10^{-3}\,\text{m}$$

We need to estimate the length (L) and diameter (d) of this cylinder by equating the two volumes, i.e.

$$\frac{\pi}{4}d^2 L = 1.414 \times 10^{-8}$$

For a cylinder with $L/d = 2$, i.e. $L = 2d$ therefore:

$$\frac{\pi}{4}d^2(2d) = 1.414 \times 10^{-8}$$

Solving for d:

Diameter of cylinder, $d = 2.08 \times 10^{-3}\,\text{m}$, and
Length of cylinder, $L = 2 \times d = 4.16 \times 10^{-3}\,\text{m}$

With the long axis oriented normal to the direction of settling, the projected area would be equal to $L \times d = 4.16 \times 10^{-3} \times 2.08 \times 10^{-3} = 8.653 \times 10^{-6}\,\text{m}^2$. The equal projected area circle diameter d_n is thus obtained as:

$$\frac{\pi d_n^2}{4} = 8.653 \times 10^{-6}\,\text{m}^2$$

or, $d_n = 3.32 \times 10^{-3}\,\text{m}$

The sphericity of this cylinder is calculated as:

$$\psi = \frac{\text{Surface area of equal volume sphere}}{\text{Surface area of cylinder}}$$

$$= \frac{\pi d_s^2}{\pi dL + 2 \times (\pi d^2/4)}$$

Substituting values:

$$\psi = \frac{3.14 \times (3 \times 10^{-3})^2}{3.14 \times 2.08 \times 10^{-3} \times 4.16 \times 10^{-3} + (2 \times 3.14 \times (2.08 \times 10^{-3})^2/4)}$$

i.e.

$$\psi = 0.832$$

Now evaluating the composite shape factor, χ:

$$\chi = \frac{4}{\psi}\left(\frac{d_s}{d_n}\right)^2$$

Substituting values:

$$\chi = \frac{4}{0.832}\left(\frac{3 \times 10^{-3}}{3.32 \times 10^{-3}}\right)^2 = 3.926$$

Now re-writing $C_{D0} = (24X(n))/Re$ and substituting values in equation (5.7):

$$C_D = \frac{24 \times 1.434}{Re} + 0.44 \times 3.926 \times \left\{\frac{24 \times 1.434}{Re}\right\}^{2 \times 0.04}$$

$$\times 0.189 \times \left\{\frac{6 \times 1.434 \times 0.792}{6 \times 1.434 \times 0.792 + (24 \times 1.434/Re)}\right\}^{0.04}$$

$$+ 0.44 \times \left\{\frac{6 \times 1.434 \times 0.792}{6 \times 1.434 \times 0.792 + (128 \times 24 \times 1.434/Re)}\right\}^{11/12}$$

Simplification yields:

$$C_D = \frac{34.416}{Re} + 0.433Re^{-0.08} \times \left\{\frac{6.814}{6.814 + (34.416/Re)}\right\}^{0.04}$$

$$+ 0.44 \times \left\{\frac{6.814}{6.814 + (4405.2/Re)}\right\}^{11/12} \tag{5.A}$$

Once again since the unknown V occurs in both Re and C_D, an iterative approach must be used. Let us begin with an initial estimate of $V = 0.15\,\text{m/s}$, and therefore, the corresponding value of the Reynolds number is calculated as:

$$Re = \frac{\rho V^{2-n} d_s^n}{m} = \frac{1020 \times 0.15^{2-0.5} \times (3 \times 10^{-3})^{0.5}}{0.01}$$

i.e. $Re = 324.6$

Substituting this value in equation (5.A):

$$C_D = \frac{34.416}{324.6} + 0.433 \times (324.6)^{-0.08} \times \left\{ \frac{6.814}{6.814 + (34.416/324.6)} \right\}^{0.04}$$

$$+ 0.44 \times \left\{ \frac{6.814}{6.814 + (4405.2/324.6)} \right\}^{11/12}$$

$$= 0.106 + 0.272 + 0.161 = 0.539$$

i.e.

$$C_D = 0.539$$

Now using the definition of the drag coefficient for a free falling particle as:

$$C_D = \frac{4gd_s}{3V^2} \left(\frac{\rho_s - \rho}{\rho} \right)$$

Rearranging it in terms of the unknown velocity V and substituting values as:

$$V = \sqrt{\frac{4gd_s(\rho_s - \rho)}{3C_D\rho}} = \sqrt{\frac{4 \times 9.81 \times 3 \times 10^{-3} \times (1200 - 1020)}{3 \times 0.539 \times 1020}}$$

$$= 0.113 \, \text{m/s}$$

i.e.

$$V = 0.113 \, \text{m/s}$$

This value is quite different from our assumed value of 0.15 m/s and therefore further iterations are required. Using $V = 0.11$ m/s, the new value of the Reynolds number is calculated as:

$$Re = \frac{1020 \times 0.11^{2-0.5} \times (3 \times 10^{-3})^{0.5}}{0.01} = 203.8$$

The corresponding new value of the drag coefficient is obtained from equation (5.A) as:

$$C_D = \frac{34.416}{203.8} + 0.433 \times (203.8)^{-0.08} \times \left\{ \frac{6.814}{6.814 + (34.416/203.8)} \right\}^{0.04}$$

$$+ 0.44 \times \left\{ \frac{6.814}{6.814 + (4405.2/203.8)} \right\}^{11/12}$$

$$= 0.169 + 0.283 + 0.119 = 0.571$$

Now again using this value of C_D, one can calculate the terminal falling velocity V as:

$$V = \sqrt{\frac{4gd_s(\rho_s - \rho)}{3C_D\rho}} = \sqrt{\frac{4 \times 9.81 \times 3 \times 10^{-3} \times (1200 - 1020)}{3 \times 0.571 \times 1020}}$$

Table 5.4   *Summary of calculations (Example 5.7)*

Particle shape	Size	Orientation	d_n (mm)	ψ	χ	Re	V (m/s)
Cylinder	L/d = 1	⊥	2.96	0.874	4.71	188.1	0.104
		‖	2.62	0.874	6	166.7	0.096
	L/d = 2	⊥	3.32	0.832	3.93	203.8	0.110
		‖	2.08	0.832	10	124.2	0.079
	L/d = 10	⊥	4.34	0.579	3.30	219.6	0.116
		‖	1.216	0.579	42	45.86	0.0407
Cone	h/d = 0.5	Apex up or	2.634	0.826	2.414	245.6	0.125
	h/d = 1	down	2.957	0.874	3.236	221.3	0.116
	h/d = 2		3.319	0.832	5.123	180.7	0.102
Square	L/d = 1	⊥	2.728	0.806	6	166.7	0.096
cross-	(cube)	‖	2.728	0.806	6	166.7	0.096
section	L/d = 2	⊥	3.063	0.768	5	183	0.102
bar		‖	2.166	0.768	10	124.2	0.0791
	L/d = 10	⊥	4.005	0.534	4.2	198.4	0.108
		‖	1.266	0.534	42	45.86	0.0407

i.e.

$V = 0.11\,\text{m/s}$

which coincides with our assumed value in the second iteration. Thus we finally have:

$$V = 0.11\,\text{m/s}; \quad Re = 203.8; \quad C_D = 0.571$$

This procedure can be repeated for the other particles specified in this example. A summary of results is presented in Table 5.4.

In order to elucidate the influence of particle shape on the settling velocity, equation (5.23) is used here to calculate the terminal falling velocity of an equivalent volume sphere, i.e. $d = d_s = 3\,\text{mm}$.

The value of d^+ is calculated using equation (5.21):

$$d^+ = 3 \times 10^{-3} \times \left\{ \frac{9.81 \times (1200 - 1020)}{1020} \right\}^{(2-0.5)/(2+0.5)} \left(\frac{1020}{0.01} \right)^{2/(2+0.5)}$$

or

$$d^+ = 42.36$$

Now using this value in equation (5.23) to obtain the value of V^+:

$$\frac{1}{V^+} = \left[\left(\frac{18 \times 1.434}{42.36^{1+0.5}} \right)^{0.793/0.5} + \left(\frac{0.321}{42.36} \right)^{0.793/2} \right]^{1/0.793}$$

$$= (0.0233 + 0.1443)^{1/0.793}$$

i.e.

$$V^+ = 9.506$$

Now using equation (5.22) to calculate V:

$$V = V^+ \left\{ \frac{g^n m (\rho_s - \rho)^n}{\rho^{n+1}} \right\}^{1/(2+n)}$$

$$= 9.506 \times \left\{ \frac{9.81^{0.5} \times 0.01 \times (1200 - 1020)^{0.5}}{1020^{0.5+1}} \right\}^{1/(2+0.5)}$$

$$V = 0.105 \, \text{m/s}.$$

Thus, this approach will always yield the same value of the terminal falling velocity for particle of any shape as long as it has the same volume as a 3 mm diameter sphere.

   An inspection of Table 5.4 shows that depending upon the particle shape and orientation, the results based on the assumption of an equal volume sphere may entail errors up to 60%. On the other hand, equal volume sphere approximation seems to work satisfactorily when the value of χ does not deviate significantly from the value of 4 corresponding to a sphere, albeit this trend is likely to be somewhat dependent on the value of the power-law index, n. □

   The scant experimental and theoretical results available for viscoplastic and visco-elastic fluids have been reviewed elsewhere (Chhabra, 1996, 2006).

5.5 Motion of bubbles and drops

The drag force acting on a gas bubble or liquid droplet will not, in general, be the same as that acting on a rigid particle of the same shape and size because circulating patterns are set up inside bubbles and drops. While the radial velocity at the interface is zero, the angular velocity, shear, and normal stresses are continuous across the interface for fluid particles and the velocity gradient in the continuous phase (hence shear stress and drag force) is therefore less than that for a rigid particle. In the absence of surface tension and inertial effects ($Re \ll 1$), the terminal velocity of a fluid sphere falling in an incompressible Newtonian fluid, as calculated from the Stokes' law (equation (5.17) with $n = 1$, $X = 1$) is increased by a factor Q_1 which accounts for the internal circulation:

$$Q_1 = \frac{3}{2} \left(\frac{1 + \delta}{1 + 1.5\delta} \right) \tag{5.34}$$

where δ is the ratio of the viscosity of the fluid in the sphere to that of the ambient liquid. Clearly, in the limit of $\delta \to \infty$, i.e. for a solid sphere, $Q_1 = 1$ and as $\delta \to 0$, i.e. for a bubble, $Q_1 \to 3/2$. In the intermediate range of viscosity ratios, the internal circulation effects generally decrease with the increasing value of δ. With very small droplets, the surface tension forces nullify the tendency for circulation and the droplet falls at a velocity close to that of a solid sphere of the same size. Likewise, even small amounts of surface active agents tend to immobilize the free surface of small fluid particles, thereby inhibiting internal circulation, and these particles again fall at velocities similar to those of solid spheres.

Drops and bubbles, in addition, undergo deformation because of the differences in the pressures acting on various parts of the surface. Thus, when a drop is falling in a quiescent medium, both the hydrostatic and impact pressures will be greater on the forward than on the rear face; this will tend to flatten the drop. Conversely, the viscous drag will tend to elongate it.

The deformation of the drop is resisted by surface tension forces and very small drops (large surface area) therefore remain spherical, whereas large drops (small surface tension forces) are appreciably deformed and the drag is increased. For droplets above a certain size, the deformation is so great that the drag force increases in almost direct proportion to volume and the terminal velocity is almost independent of size. The literature on the motion of fluid particles in Newtonian fluids has been thoroughly reviewed by Clift *et al.* (1978), the corresponding developments in non-Newtonian fluids are briefly summarized here.

The drag coefficient for freely falling spherical droplets (or rising gas bubbles) of Newtonian fluids in power-law liquids at low Reynolds numbers has been approximately evaluated and, in the absence of surface tension effects, it is given by equation (5.4), i.e.

$$C_D = \frac{24}{Re} X(n, \delta) \tag{5.4}$$

where the Reynolds number is defined as $\rho_c V^{2-n} d^n / m_c$, and the correction factor $X(n, \delta)$ now depends on both the power-law index and the ratio of the viscosities of the dispersed and continuous phases. However, in the case of gas bubbles ($\delta = 0$), the correction factor X is a function of the power-law index alone (Hirose and Moo-Young, 1969):

$$X = 2^n 3^{(n-3)/2} \left[\frac{13 + 4n - 8n^2}{(2n+1)(n+2)} \right] \tag{5.35}$$

This equation is valid for only mildly shear-thinning behaviour ($n \geq 0.6$ or so). For a Newtonian fluid ($n = 1$), equation (5.35) gives $X = 1/Q_1 = 2/3$. This simple expression does not, however, account for either surface tension effects or for the shape and changing size (due to expansion) of a freely rising bubble in a stationary power-law medium. The agreement between the predictions of equation (5.35) and the scant data for approximately spherical gas bubbles ($1 \geq n \geq 0.5$) is reasonable, as seen in Figure 5.9.

No such simple relation exists between the value of the correction factor X and the values of n and δ for droplets, and reference must be made to original or review papers (Nakano and Tien, 1970; DeKee and Chhabra, 1992; DeKee *et al.*, 1996; Kishore *et al.*, 2007; Chhabra, 2006). Droplets are subject to greater drag in power-law shear-thinning fluids than in a Newtonian fluid in the creeping flow region and the effect of shear-thinning becomes progressively less marked with the increasing Reynolds number (Nakano and Tien, 1970).

The usefulness of these studies is severely limited by the various simplifying assumptions which have been made concerning shape and surface tension effects. Indeed, a variety of shapes of drops and bubbles has been observed under free fall conditions in non-Newtonian fluids and these differ significantly from those in Newtonian fluids (Clift *et al.*, 1978; Chhabra, 2006).

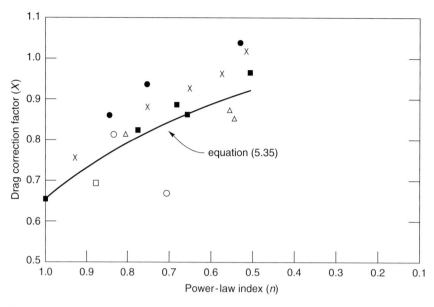

Figure 5.9   *Drag on spherical gas bubbles in power-law fluids in creeping flow region*

In addition, many workers have reported experimental data on various aspects (such as shapes, coalescence, terminal falling velocity) of bubble and drop motion in a non-Newtonian continuous phase (Chhabra, 2006).

The bulk of the data on drag on freely falling drops relate to conditions where the viscosity ratio parameter, δ, rarely exceeds 0.001, thereby suggesting that behaviour is really tantamount to that of gaseous spheres. In spite of these difficulties, many correlating expressions for drag on droplets and bubbles in non-Newtonian liquids are available in the literature, but these are too tentative and restrictive to be included here. By way of illustration, Rodrigue (2002, 2004) asserted that the following correlation predicted the free rising velocity of gas bubbles in quiescent power-law fluids with an average error of $\pm 25\%$:

$$V^* = \frac{F}{12(1 + 0.018F)^{0.75}} \tag{5.36}$$

where the re-scaled free rise velocity V^* and the parameter F are defined as:

$$V^* = V\left(\frac{\rho_c^2 d^2}{\sigma \mu_c}\right)^{1/3} \tag{5.37}$$

$$F = g\left(\frac{d^8 \rho_c^5}{\sigma \mu_c^4}\right)^{1/3} \tag{5.38}$$

The characteristic viscosity of the continuous phase μ_c appearing in equations (5.37) and (5.38) is evaluated at the shear rate value of (V/d). Thus, for instance, for a power-law fluid, $\mu_c = m(V/d)^{n-1}$ is substituted in equations (5.37) and (5.38) which, in turn, leads to the form of equation (5.36) which is implicit in velocity.

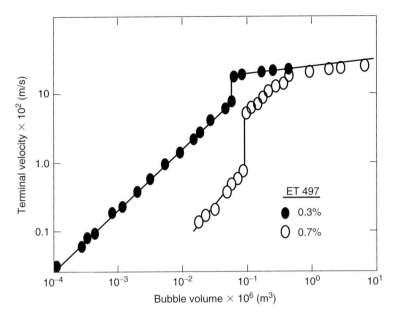

Figure 5.10   *Free rise velocity–bubble volume data showing an abrupt increase in velocity*

A striking feature is the way the free rise velocity of a bubble varies with its size and displays the so-called discontinuity. Indeed, Astarita and Appuzzo (1965) noted a 6- to 10-fold increase in the rise velocity at a critical bubble size, as seen in Figure 5.10 for air bubbles in two polymer solutions. Subsequently, similar, though less dramatic results, summarized in recent reviews (DeKee *et al.*, 1996; Chhabra, 2006), have been reported. The critical bubble radius appears to hover around 2.6 mm, irrespective of the type or the degree of non-Newtonian properties of the continuous phase, and this value is well predicted by the following equation due to Bond and Newton (1928):

$$r = \sqrt{\frac{\sigma}{g(\rho_l - \rho_g)}} \tag{5.39}$$

where σ is the surface tension of the liquid. Indeed, the critical bubble sizes observed experimentally in a variety of inelastic and visco-elastic liquids, and those calculated using equation (5.39) seldom differ by more than 5%. As mentioned earlier, small gas bubbles behave like rigid spheres (no-slip boundary condition at the interface) whereas large ones display a shear-free surface, and this change over in the type of particle behaviour gives rise to a jump of 50% in terminal velocity for Newtonian fluids. For creeping bubble motion in power-law fluids, the ratio of the terminal falling velocity of a bubble to that of a rigid sphere of the same size and density may be expressed as:

$$\frac{V_b}{V_s} = \left(\frac{X_s}{X_b}\right)^{1/n} \tag{5.40}$$

where the subscripts 'b' and 's' refer to the bubble and sphere, respectively. The values of X_s are listed in Table 5.1 and for gas bubbles, they can be estimated from equation (5.35). For $n = 1$, equation (5.40) gives the expected result, namely, $V_b = 1.5V_s$. For a power-law liquid with $n = 0.6$, $X_s = 1.42$ (Table 5.1) and $X_b = 0.93$, and $V_b \simeq 2V_s$. Thus, this mechanism does explain the effect qualitatively, but it does not predict as large an increase in rise velocity as that has been observed experimentally. Furthermore, visco-elasticity possibly contributes to the abrupt jump in rise velocity. In view of this, equation (5.35) may be applied only when the bubbles display a shear-free interface, tentatively when they are larger than 2.5 mm. Recent developments on this issue have been critically re-examined by Rodrigue and Dekee (2002), Chhabra (2006) and others (Herrera-Velarde *et al.*, 2003; Soto *et al.*, 2006; Pilz and Brenn, 2007).

Little is known about the effect of visco-elasticity on the motion of bubbles and drops in non-Newtonian fluids, though a preliminary study suggests that spherical bubbles are subject to a larger drag in a visco-elastic than in an inelastic liquid. Recent surveys clearly reveal the paucity of reliable experimental data on the behaviour of fluid particles in non-Newtonian liquids (Chhabra, 2006; DeKee *et al.*, 1996).

5.6 Flow of a liquid through beds of particles

The problems discussed so far relate to the motion of single particles and assemblies of particles in stationary non-Newtonian media. Consideration will now be given to the flow of non-Newtonian liquids through a bed of particles, as encountered in a variety of processing applications. For instance, the filtration of polymer melts, slurries and sewage sludges using packed beds, or the leaching of uranium from a dilute slurry of ore in a fluidized bed. Further examples are found in enhanced oil recovery by polymer flooding in which non-Newtonian polymer solutions are forced to flow through a porous rock. It is important therefore to be able to predict pressure drop across such beds.

With downward flow of a liquid, no relative movement occurs between the particles except for that arising from their unstable initial packing. In the streamline region, the pressure drop across the bed is directly proportional to the rate of flow for Newtonian liquids and to the rate of flow raised to a lower (less than unity) exponent for a shear-thinning and to a higher (>1) exponent for a shear-thickening fluid. At rates of flow high enough for turbulence to develop in some of the flow channels, the pressure drop will rise more steeply ($\propto Q^{1.8-2}$).

For upward flow through a bed which is not constrained at the top, the frictional pressure drop will be the same as for the downward flow as long as the structure of the bed is not disturbed by the flow. When the upward flow rate has been increased to the point where the frictional drag on the particles becomes equal to their buoyant weight, rearrangement will occur within the bed which then expands in order to offer less resistance to flow. Once the packing of the bed has reached its loosest stable form, any further increase in flow rate causes the individual particles to separate from one another and become freely supported in the liquid stream and the bed is then said to be fluidized. With further increase in flow rate, the particles move further apart and the bed voidage increases while the pressure difference remains approximately equal to the buoyant weight per unit area of the bed, as shown in Figure 5.11. Up to this stage, the system behaves in a similar manner whether the fluid is a gas or a liquid. Liquid fluidized beds continue to expand in a uniform manner, with the degree of agitation of particles

Figure 5.11   *Qualitative pressure drop–flow rate behaviour in fixed and fluidized beds*

increasing progressively. Fluidization is then said to be particulate. A qualitatively similar fluidization behaviour is obtained with non-Newtonian liquids. Since the main thrust here is on the role of liquid rheology, the behaviour of gas fluidized beds (aggregative type) is not covered in this chapter, but attention is drawn to several excellent books on this subject (Davidson *et al.*, 1985; Coulson and Richardson, 1991).

The estimation of the pressure gradient for the flow of a non-Newtonian liquid through a fixed (or packed) bed is addressed in section 5.7, fluidized beds are considered in Section 5.8.

5.7 Flow through packed beds of particles (porous media)

The flow of non-Newtonian liquids through beds of particles is treated in an analogous way to that adopted in Chapter 3 for the flow through ducts of regular cross-section. No complete analytical solution is, however, possible and a degree of empiricism complemented by the use of experimental results is often necessary. Firstly, however, the basic nature and structure of porous media (or beds of particles) will be briefly discussed.

5.7.1 Porous media

The simplest way of regarding a porous medium is as a solid structure with passages through which fluids can flow. Most naturally occurring minerals (sand, limestones) are consolidated having been subjected to compressive forces for long times. Packed beds of glass beads, catalyst particles, Raschig rings, berl saddles, etc. as used in process equipment are unconsolidated. Unconsolidated media generally have a higher permeability and offer less resistance to flow. Packing may be ordered or random according to whether or not there is a discernable degree of order of the particles, though completely random packing hardly ever occurs as 'order' tends to become apparent as the domain of examination is progressively reduced in size. Cakes and breads are good examples of random media!

Porous media may be characterized at two distinct levels: microscopic and macroscopic. At the microscopic level, the structure is expressed in terms of a statistical description of the pore size distribution, degree of inter-connection and orientation of the pores, fraction of dead pores, etc. In the macroscopic approach, bulk parameters are employed which have been averaged over scales much larger than the size of pores. These two approaches are complementary and are used extensively depending upon the objective. Clearly, the microscopic description is necessary for understanding surface phenomena such as adsorption of macromolecules from polymer solutions and the blockage of pores, etc., whereas the macroscopic approach is often quite adequate for process design where fluid flow, heat and mass transfer are of greatest interest, and the molecular dimensions are much smaller than the pore size. Detailed accounts of micro- and macro-level characterization methods frequently used for porous media are available in the literature (Greenkorn, 1983; Dullien, 1992). Of the numerous macroscopic parameters used to quantify porous media, those gaining widest acceptance in the literature for describing the flow of single phase fluids are voidage, specific surface, permeability and tortuosity. Their values can often be inferred from experiments on the streamline flow of single phase Newtonian fluids.

(i) *Voidage*

Voidage, ε, is defined as the fraction of the total volume which is free space available for the flow of fluids, and thus the fractional volume of the bed occupied by solid material is $(1 - \varepsilon)$. Depending upon the nature of the porous medium, the voidage may range from near zero to almost unity. For instance, certain rocks, sandstones, etc. have values of the order of 0.15–0.20 whereas fibrous beds and ring packings may have high values of voidage up to 0.95. Obviously, the higher the value of voidage, the lower is the resistance to flow of a fluid.

(ii) *Specific surface*

In addition, the specific surface, S_B, of the bed affects both its general structure and the resistance it offers to flow. It is defined as the surface area per unit volume of the bed, i.e. m^2/m^3. Hence, S_B can be expressed in terms of the voidage ε and the specific area S of the particles

$$S_B = S(1 - \varepsilon) \tag{5.41}$$

where S is the specific area per unit volume of a particle. Thus for a sphere of diameter d,

$$S = \frac{\pi d^2}{\pi d^3/6} = \frac{6}{d} \tag{5.42}$$

For a given shape, S is inversely proportional to the particle size. Highly porous fibre glasses have specific surface areas in the range $5–7 \times 10^4 \, m^2/m^3$ while compact limestones ($\varepsilon \sim 0.04–0.10$) have specific surface areas in the range $\sim 0.2–2 \times 10^6 \, m^2/m^3$.

(iii) *Permeability*

The permeability of a porous medium may be defined by means of the well-known Darcy's law for the streamline flow of an incompressible Newtonian fluid:

$$\frac{Q}{A} = V_c = \left(\frac{k}{\mu}\right)\left(-\frac{\Delta p}{L}\right) \tag{5.43}$$

where Q is the volume rate of flow of a fluid of viscosity, μ, through a porous medium of area A (normal to flow) under the influence of the pressure gradient $(-\Delta p/L)$, and k is called the permeability of the porous medium.

A porous material is said to have a permeability of 1 darcy if a pressure gradient of 1 atm/cm results in a flow of 1 cm³/s of a fluid having viscosity of 1 cP through an area of 1 cm². In SI units, it is expressed as m² and 1 darcy $\simeq 10^{-12}$ m². Evidently, the lower the permeability, the greater is the resistance to flow. Typical values of permeability range from 10^{-11} m² for fibre glass to 10^{-14} m² for silica powder and limestone.

(iv) Tortuosity

Tortuosity is a measure of the extent to which the path traversed by fluid elements deviates from a straight-line in the direction of overall flow and may be defined as the ratio of the average length of the flow paths to the distance travelled in the direction of flow. Though the tortuosity depends on voidage and approaches unity as the voidage approaches unity, it is also affected by particle size, shape and orientation in relation to the direction of flow. For instance, for plate like particles, the tortuosity is greater when they are oriented normal to the flow than when they are packed parallel to flow. However, the tortuosity factor is not an intrinsic characteristic of a porous medium and must be related to whatever one-dimensional flow model is used to characterize the flow.

5.7.2 Prediction of pressure gradient for flow through packed beds

Many attempts have been made to obtain general relations between pressure drop and mean velocity of flow through porous media or packings, in terms of the bed voidage which is either known or can easily be measured. The following discussion is limited primarily to the so-called capillary tube bundle approach while the other approaches of treating the flow of both Newtonian and non-Newtonian fluids are described in the literature (Happel and Brenner, 1965; Greenkorn, 1983; Dullien, 1992; Chhabra, 2006).

(i) Streamline flow

The interstitial void space in a porous matrix or bed of particles may be envisaged as consisting of tortuous conduits of complex cross-section but having a constant average area for flow. Thus, flow in a porous medium is equivalent to that in a non-circular conduit offering the same resistance to flow. However, the flow passages in a bed of particles will be oriented and inter-connected in an irregular fashion and the elementary capillary models do not account for these complexities. Despite these difficulties, the analogy between flow through a circular tube and through the channels in a bed of particles, shown schematically in Figure 5.12, provides a useful basis for deriving a general flow rate–pressure drop expression. As seen in Chapter 3, such expressions vary according to the flow model chosen for the fluid. Following Kemblowski *et al.* (1987), we begin by re-writing the well-known Hagen–Poiseuille equation, for the streamline flow of an incompressible Newtonian fluid through circular tubes, as follows:

$$V = \frac{D^2}{32\mu}\left(\frac{-\Delta p}{L}\right) \tag{5.44}$$

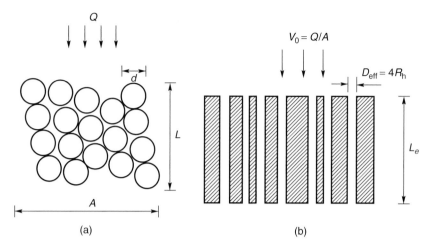

Figure 5.12   *Schematic representation of flow through a bed of uniform spheres (a) and the capillary model idealization (b)*

Then, for flow through a non-circular duct, this may be re-arranged as:

$$V = \frac{D_h^2}{16K_0\mu}\left(\frac{-\Delta p}{L}\right) \tag{5.45}$$

where D_h is the hydraulic mean diameter (4 × pore volume/surface area of particles) and K_0 is a constant which depends only on the shape of the cross-section. For a circular tube of diameter D, for instance $D_h = D$ and $K_0 = 2$. Likewise, for flow in between two plates separated by a distance $2h$, $D_h = 4h$ and $K_0 = 3$. Kemblowski *et al.* (1987) have further rearranged equation (5.45) as:

$$\frac{D_h}{4}\left(\frac{-\Delta p}{L}\right) = \mu\left(\frac{4K_0 V}{D_h}\right) \tag{5.46}$$

For an incompressible Newtonian fluid, $(D_h/4)(-\Delta p/L)$ is the average shear stress at the wall of the flow passage and $(4K_0 V/D_h)$, the shear rate at the wall, which may be regarded as the nominal shear rate for time-independent non-Newtonian fluids. Thus,

$$\langle \tau_w \rangle = \frac{D_h}{4}\left(\frac{-\Delta p}{L}\right) \tag{5.47}$$

and

$$\langle \dot{\gamma}_w \rangle_n = \frac{4K_0 V}{D_h} \tag{5.48}$$

where $\langle \rangle$ denotes the values averaged over the perimeter of the conduit.

Equations (5.47) and (5.48) can be used to obtain expressions for the streamline flow of time-independent fluids through beds of particles, in which case V must be replaced by the mean velocity in the pores or interstices, i.e. V_i, and the length L is replaced by the average length of the tortuous path, L_e, traversed by the fluid elements.

For a bed whose structure is independent of its depth, then L_e and L will be linearly related, i.e.

$$L_e = TL \qquad (5.49)$$

where T is the tortuosity factor.

The interstitial velocity, V_i, is related to the superficial velocity V_0 by the Dupuit relation which is based on the following considerations.

In a cube of side, l, the volume of the voids is εl^3 and the mean cross-sectional area is the free volume divided by the height, i.e. εl^2. The volumetric flow rate through the cube is given by $V_0 l^2$, so that the average interstitial velocity V_i, is given by:

$$V_i = \frac{V_0 l^2}{\varepsilon l^2} = \frac{V_0}{\varepsilon} \qquad (5.50)$$

Although equation (5.50) is a good approximation for random packings, it does not apply to all regular packings. For instance, for a bed of uniform spheres arranged in cubic packing, $\varepsilon = 0.476$, but the fractional area varies continuously from 0.215 in a plane across the diameters to unity between successive layers. Furthermore, equation (5.50) implicitly assumes that an element of fluid moving at a velocity V_i covers a distance L in the same time as a fluid element of superficial velocity V_0 in an empty tube. This implies that the actual interstitial velocity is likely to be somewhat greater than the value given by equation (5.50). Because an element of fluid in a bed actually travels a distance greater than L, in the Kozeny–Carman capillary model, a correction is made for this effect as:

$$V_i = \frac{V_0}{\varepsilon} T \qquad (5.51)$$

Some authors (Comiti and Renaud, 1989) call the velocity given by equation (5.51) as the pore velocity.

Finally, the hydraulic mean diameter D_h must be expressed in terms of the packing characteristics. Thus, for a bed of uniform spheres of diameter d, the hydraulic mean diameter D_h can be estimated as follows:

$$D_h = \frac{4 \times \text{flow area}}{\text{wetted perimeter}} = \frac{4 \times \text{volume of flow channels}}{\text{surface area of packing}}$$

$$= \frac{4 \times \dfrac{\text{volume of flow channels}}{\text{volume of bed}}}{\dfrac{\text{surface area of packing}}{\text{volume of bed}}} = \frac{4\varepsilon}{S_B} \qquad (5.52)$$

Substitution from equations (5.41) and (5.42) gives:

$$D_h = \left(\frac{2}{3}\right) \frac{d\varepsilon}{(1 - \varepsilon)} \qquad (5.53)$$

Note that the wetted surface of the column walls has been neglected, which is justified under most conditions of interest.

Although the early versions of capillary models, namely, the Blake and the Blake–Kozeny models, are based on the use of $V_i = V_0/\varepsilon$ and $L_e = LT$, it is now generally accepted (Dullien, 1992; Chhabra *et al.*, 2001) that the Kozeny–Carman model, using $V_i = V_0 T/\varepsilon$ provides a more satisfactory representation of the flow in beds of particles. Using equations (5.49), (5.51) and (5.53), the average shear stress and the nominal shear rate at the wall of the flow passage (equations 5.47 and 5.48) may now be expressed as:

$$\langle \tau_w \rangle = \frac{d\varepsilon}{6(1 - \varepsilon)T} \left(\frac{-\Delta p}{L} \right) \tag{5.54}$$

and

$$\langle \dot{\gamma}_w \rangle_n = 6K_0 T \left(\frac{1 - \varepsilon}{\varepsilon^2} \right) \frac{V_0}{d} \tag{5.55}$$

For generalized non-Newtonian fluids, Kemblowski *et al.* (1987) postulated that the shear stress at the wall of a pore or 'capillary' is related to the corresponding nominal shear rate at the wall by a power-law type relation:

$$\langle \tau_w \rangle = m'(\langle \dot{\gamma}_w \rangle_n)^{n'} \tag{5.56}$$

where m' and n' are the apparent consistency coefficient and flow behaviour index, respectively, inferred from pressure drop/flow rate data obtained in a packed bed. By analogy with the generalized procedure for streamline flow in circular tubes outlined in equation (3.61), m' and n' can be linked to the actual rheological parameters. For a truly power-law fluid, for instance:

$$n' = n \tag{5.57a}$$

$$m' = m \left(\frac{3n + 1}{4n} \right)^n \tag{5.57b}$$

It should be, however, noted that the Rabinowitsch–Mooney factor of $((3n + 1)/4n)^n$ is strictly applicable only to cylindrical tubes, but the limited results available for non-circular ducts suggest that it is nearly independent of the shape of the conduit cross-section (Miller, 1972; Tiu, 1985). For instance, the values of this factor are within 2–3% of each other for circular tubes and parallel plates over the range $0.1 \leq n \leq 1$.

The cross-section of the channels formed in a bed of spheres would be expected to lie between that of a circular tube and of a plane slit and Kemblowski *et al.* (1987) therefore suggested the use of a mean value of 2.5 for K_0. Considerable confusion also exists in the literature about the value of the tortuosity factor, T. Thus, Carman (1956) proposed a value of $\sqrt{2}$ based on the assumption that the capillaries deviate on average by 45° from the mean direction of flow ($\cos 45 = 1/\sqrt{2}$). On the other hand, if a fluid element follows the surface round the diameter of a spherical particle, the tortuosity factor should equal $\pi/2$. Indeed, the values ranging from ~1 to 1.65 have been used in the literature for Newtonian fluids (Agarwal and O'Neill, 1988; Chhabra *et al.*, 2001). Because T is a function of the geometry of the bed, it has the same value whatever the liquid rheology, provided that it is time-independent, although there is some evidence that T is weakly dependent on flow rate (Dharamadhikari and Kale, 1985). This is not surprising because macromolecules have a tendency to adsorb on the walls of the pores and, if the flow rate is high enough, the shearing forces may overcome the surface forces; thus polymer molecules become detached thereby making more space available for flow. Thus, flow passages blocked at low flow rates may open up to flow again at high flow rates. Such unusual effects observed with non-Newtonian fluids in porous media are briefly discussed in a later section in this chapter.

A dimensionless friction factor may be defined as:

$$f = \left(\frac{-\Delta p}{L}\right)\frac{d}{\rho V_0^2}\left(\frac{\varepsilon^3}{1-\varepsilon}\right) \tag{5.58}$$

Noting $K_0 = 2.5$ and $T = \sqrt{2}$, combining equations (5.54) to (5.58):

$$f = \frac{180}{Re^*} \tag{5.59}$$

where

$$Re^* = \frac{\rho V_0^{2-n} d^n}{m(1-\varepsilon)^n}\left(\frac{4n}{3n+1}\right)^n\left(\frac{15\sqrt{2}}{\varepsilon^2}\right)^{1-n} \tag{5.60}$$

For a Newtonian fluid, $n = 1$, both equations (5.59) and (5.60) reduce to the well-known Kozeny–Carman equation. Equation (5.60) correlates most of the literature data on the flow of power-law fluids through beds of spherical particles up to about $Re^* \sim 1$, though most work to date has been carried out in beds having voidages in the range $0.35 \leq \varepsilon \leq 0.41$ (Kemblowski *et al.*, 1987; Chhabra *et al.*, 2001; Chhabra, 2006).

Bingham plastic fluids

The flow of viscoplastic fluids through beds of particles has not been studied as extensively as that of power-law fluids. However, since the expressions for the average shear stress and the nominal shear rate at the wall, equations (5.54) and (5.55), are independent of fluid model, they may be used in conjunction with any time-independent behaviour fluid model, as illustrated here for the streamline flow of Bingham plastic fluids. The mean velocity for a Bingham plastic fluid in a circular tube is given by equation (3.28):

$$V = \frac{D^2}{32\mu_B}\left(\frac{-\Delta p}{L}\right)\left(1 - \frac{4}{3}\phi + \frac{1}{3}\phi^4\right) \tag{3.28}$$

where $\phi = \tau_0^B/\tau_w$. This equation can be re-arranged in terms of the nominal shear rate and shear stress at the wall of the pore as:

$$(\dot{\gamma}_w)_n = \frac{8V}{D} = \frac{\tau_w}{\mu_B}\left(1 - \frac{4}{3}\phi + \frac{1}{3}\phi^4\right) \tag{5.61}$$

As seen in Chapter 3, the quantity $(8V/D)$ is the nominal shear rate at the wall (also see equation (5.48), for a circular tube, $D_h = D$, $K_0 = 2$). Substituting for the nominal shear rate and wall shear stress from equations (5.53) and (5.54) in equation (5.61), slight re-arrangement gives:

$$f = \frac{180}{Re_B F(\phi)} \tag{5.62}$$

where

$$Re_B = \frac{\rho V_0 d}{\mu_B} \tag{5.63}$$

$$F(\phi) = 1 - \frac{4}{3}\phi + \frac{\phi^4}{3} \tag{5.64}$$

and

$$\phi = \frac{\tau_0^B}{\langle \tau_w \rangle} \tag{5.65}$$

It should be noted that $T = \sqrt{2}$ and $K_0 = 2.5$ have been used in deriving equation (5.62). Again for the special case of Newtonian fluids, $\tau_0^B = 0$ or $\phi = 0$, $F(\phi) = 1$ and equation (5.62) reduces to the Kozeny–Carman equation. The scant experimental data available on pressure drop for the streamline flow for Bingham plastic fluids ($Re_B \sim 1$) is consistent with equation (5.62).

This section is concluded by noting that similar expressions for the friction factor have been derived for a range of purely inelastic fluid models and these have been critically reviewed elsewhere (Chhabra *et al.*, 2001; Chhabra, 1993, 2006).

(ii) *Transitional and turbulent flow*
Because the apparent viscosity of non-Newtonian systems is usually high, flow conditions rarely extend beyond the streamline flow regime. There is no clear cut value of the Reynolds number marking the end of streamline flow. An examination of the available data indicates that the lower the value of the power-law index, the higher the Reynolds number up to which streamline flow occurs. An important factor is that with a range of pore sizes, some can be in laminar flow and others turbulent. Despite these uncertainties, the value of $Re^* \sim 5$–10 is a good approximation for engineering design calculations.

The capillary bundle approach has also been extended for correlating data on pressure drop in packed beds of spherical particles in the transitional and turbulent regions. Both Mishra *et al.* (1975) and Brea *et al.* (1976) proposed the following empirical method for estimating the 'effective viscosity', μ_{eff}:

$$\mu_{eff} = m' \left\{ \frac{12 V_0 (1 - \varepsilon)}{d \varepsilon^2} \right\}^{n-1} \tag{5.66}$$

This is then incorporated in the modified Reynolds number, Re', to give:

$$Re' = \frac{\rho V_0 d}{\mu_{eff}(1 - \varepsilon)} \tag{5.67}$$

They assumed the 'viscous' and 'inertial' components of the pressure drop to be additive and proposed the following relationship between the friction factor and the modified Reynolds number:

$$f = \frac{\alpha}{Re'} + \beta \qquad (5.68)$$

Based on their experimental data for the flow of power-law fluids in packed and fluidized beds of spheres ($0.7 \leq n \leq 1$; $0.01 \leq Re' \leq 1000$; $0.37 \leq \varepsilon \leq 0.95$), Mishra *et al.* (1975) obtained $\alpha = 150$ and $\beta = 1.75$. With these values, equation (5.68) coincides with the well-known Ergun equation for Newtonian fluids (Ergun, 1952). On the other hand, the data of Brea *et al.* (1976) encompass somewhat wider ranges of the power-law index ($0.4 \leq n' \leq 1$) and Reynolds number ($0.01 \leq Re' \leq 1700$) but a limited range of voidage ($0.36 \leq \varepsilon \leq 0.40$) and they proposed $\alpha = 160$ and $\beta = 1.75$. A close scrutiny of equation (5.66) shows that it is tantamount to using $K_0 = 2$ (corresponding to a circular cross-section) and $T = (25/12)$, and thus the lower value of K_0 is compensated for by the higher value of the tortuosity factor, T. Although the original papers give mean deviations of 15–16% between the predictions of equation (5.68) and experimental data, careful inspection of the pertinent graphs reveal maximum deviations of up to 100%.

Based on the re-appraisal of the literature data and new data, the following simplified expression provides a somewhat better representation of the data in packed beds, at least for $\varepsilon \leq 0.41$ and $Re^* < 100$ (Chhabra, 2006):

$$f = \frac{150}{Re^*} + 1.75 \qquad (5.69)$$

Thus, it is suggested that for the flow of shear-thinning fluids in packed beds, equation (5.69) should be used for $Re^* < 100$, and for $\varepsilon > 0.41$ and $Re^* > 100$ equation (5.68) is preferable, with $\alpha = 150$ and $\beta = 1.75$.

Before leaving this section, it is appropriate to add here that some other variants of the capillary model are also available in the literature. Perhaps the most successful of these is due to Comiti and Renaud (1989). Based on the notion that the pressure drop varies proportional to V_0^n in the low Reynolds number region and as $\sim V_0^2$ in the fully turbulent conditions, and by further assuming these two contributions to be additive, Sabiri and Comiti (1995, 1997) presented the following expression to estimate the frictional pressure drop for the flow of power-law fluids in homogeneous beds:

$$f_{\text{pore}} = \frac{16}{Re_{\text{pore}}} + 0.194 \qquad (5.70)$$

where both the friction factor and the Reynolds number are based on the pore velocity given by equation (5.50) and are given as:

$$Re_{\text{pore}} = \frac{\rho (TV_0/\varepsilon)^{2-n}}{2^{n-3} m \{(3n+1)/4n\}^n \{S(1-\varepsilon)/\varepsilon\}^n} \qquad (5.71)$$

and

$$f_{pore} = \frac{2(-\Delta p/L)\varepsilon^3}{\rho V_0^2 T^3 S(1 - \varepsilon)} \tag{5.72}$$

By a suitable choice of the value of the tortuosity factor T and noting that for a bed made up of uniform size spheres of diameter d, the specific surface area $S = 6/d$, the form of equation (5.70) is in line with the other developments in this field, e.g., equations (5.68) and (5.69). On the other hand, for a given bed, it is possible to evaluate the values of T and S by using experimental values of pressure drop and volumetric flow rate for a Newtonian fluid. Thus, while this approach does offer the possibility of accounting for particle shape directly, or for beds of mixed size particles, this advantage is somewhat offset by the necessity of requiring experimental data for Newtonian fluids thereby limiting its utility. The validity of this approach is demonstrated in Example 5.8.

Example 5.8

Estimate the frictional pressure gradient for the flow of a polymer solution $(m = 3.7\,\text{Pa}\cdot\text{s}^n$, $n = 0.5$, density $= 1008\,\text{kg/m}^3)$ at the rate of $0.001\,\text{m}^3/\text{s}$ through a 50 mm diameter column packed with 1.5 mm lead shots. The average voidage of the packing is 0.39.

Solution

Superficial velocity of flow, $V_0 = (0.001/((\pi/4)(50 \times 10^{-3})^2)) = 0.51\,\text{m/s}$
   The Reynolds number of flow:

$$Re^* = \frac{\rho V_0^{2-n} d^n}{m(1 - \varepsilon)^n} \left(\frac{4n}{3n + 1}\right)^n \left(\frac{15\sqrt{2}}{\varepsilon^2}\right)^{1-n}$$

$$= \frac{(1008)(0.51)^{2-0.5}(1.5 \times 10^{-3})^{0.5}}{3.7(1 - 0.39)^{0.5}} \left(\frac{4 \times 0.5}{3 \times 0.5 + 1}\right)^{0.5} \left(\frac{15\sqrt{2}}{0.39^2}\right)^{1-0.5}$$

$$= 52$$

Therefore the flow is in the transitional regime. Equation (5.68) or (5.69) or (5.70) may be used. For equation (5.69),

$$f = \frac{150}{Re^*} + 1.75 = \frac{150}{52} + 1.75 = 4.63$$

The pressure gradient $(-\Delta p/L)$ across the bed is calculated using this value of f in equation (5.58),

$$\frac{-\Delta p}{L} = \frac{f\rho V_0^2}{d}\left(\frac{1 - \varepsilon}{\varepsilon^3}\right) = \frac{(4.63)(1008)(0.51)^2(1 - 0.39)}{(1.5 \times 10^{-3})(0.39)^3}$$

$$= 8\,300\,000\,\text{Pa/m}\quad\text{or}\quad 8.3\,\text{MPa/m}.$$

For the sake of comparison, the value of $(-\Delta p/L)$ using equation (5.68) will also be calculated here. For a power-law fluid,

$$m' = m\left(\frac{3n+1}{4n}\right)^n \quad \text{and} \quad n' = n = 0.5$$

$$\therefore \quad m' = 3.7\left(\frac{3 \times 0.5 + 1}{4 \times 0.5}\right)^{0.5} = 4.14\,\text{Pa}\cdot\text{s}^n$$

$$\therefore \quad \mu_{\text{eff}} = m'\left\{\frac{12V_0(1-\varepsilon)}{d\varepsilon^2}\right\}^{n'-1}$$

$$= 4.14\left\{\frac{12 \times 0.51 \times (1-0.39)}{(1.5 \times 10^{-3})(0.39)^2}\right\}^{0.5-1} = 0.0324\,\text{Pa}\cdot\text{s}$$

The modified Reynolds number, Re', is evaluated using equation (5.67):

$$Re' = \frac{\rho V_0 d}{\mu_{\text{eff}}(1-\varepsilon)} = \frac{(1008)(0.51)(1.5 \times 10^{-3})}{0.0324(1-0.39)} = 39.1$$

This value also suggests that flow is in the transition regime. The corresponding friction factor is estimated as:

$$f = \frac{150}{Re'} + 1.75 = \frac{150}{39.1} + 1.75 = 5.59$$

Again using equation (5.58), $(-\Delta p/L) = 9.6\,\text{MPa/m}$.

This value is only about 15% higher than that calculated previously.

Finally, the value of $(-\Delta p/L)$ will also be calculated here using equation (5.70). Based on extensive data, Comiti and Renaud (1989) proposed the following relationship between T and ε for a bed of uniform spheres:

$$T = 1 - 0.41 \ln \varepsilon$$

For $\varepsilon = 0.39$, this relationship yields $T = 1.386$ which is only slightly smaller than the commonly used value of $\sqrt{2}$. The specific surface area S for spheres is given by $6/d$, i.e.

$$S = \frac{6}{d} = \frac{6}{1.5 \times 10^{-3}} = 4000\,\text{m}^{-1}$$

Now calculating the pore Reynolds number Re_{pore} by substituting values in equation (5.71):

$$Re_{\text{pore}} = \frac{1008 \times (1.386 \times 0.51/0.39)^{2-0.5}}{2^{0.5-3} \times 4.14 \times \{4000(1-0.39)/0.39\}^{0.5}}$$

$$= 42.5$$

This value is quite comparable to the two values obtained above. Using this value in equation (5.70):

$$f_{\text{pore}} = \frac{16}{42.5} + 0.194 = 0.571$$

Now re-arranging equation (5.72) to solve for the unknown pressure gradient $(-\Delta p/L)$:

$$\frac{-\Delta p}{L} = \frac{\rho V_o^2 T^3 S (1 - \varepsilon) f_{pore}}{2\varepsilon^3}$$

Substituting values:

$$\frac{-\Delta p}{L} = \frac{1008 \times 0.51^2 \times 1.386^3 \times 4000(1 - 0.39) \times 0.571}{2 \times 0.39^3}$$

$$= 8.19 \times 10^6 \text{ Pa/m, i.e., } 8.19 \text{ MPa/m}$$

While this value is remarkably close to the one obtained using equation (5.69), it is only about 17% lower than the prediction of equation (5.68). □

The above discussion is limited to the flow of inelastic fluids in unconsolidated beds of particles where the pore size is substantially larger than the characteristic dimensions of the polymer molecules. Interaction effects between the walls of the pore and the polymer molecules are then small. Thus, measuring the relationship between pressure drop and flow rate in a packed bed and in a tube would therefore lead to the prediction of the same rheological properties of the fluid. Visco-elastic effects and other phenomena including blockage of pores, polymer adsorption/retention, etc., observed in beds of low permeability or in consolidated systems will be briefly discussed in Section 5.7.7.

5.7.3 Wall effects

In practice, the influence of the confining walls on the fluid flow will be significant only when the ratio of the particle to column diameters is less than about 30 (Cohen and Metzner, 1981). Particles will pack less closely near the wall, so that the resistance to flow in a bed of smaller diameter may be less than that in a large column at the same superficial velocity. One would expect the effect to be a characteristic of the packing and the column size and information on wall effects for Newtonian fluids can also be used for non-Newtonian fluids, at least for inelastic fluids. Thus, in streamline flow, Coulson (1949) found that, for the given pressure gradient, the superficial velocity of a Newtonian liquid increases by a factor, f_w, defined as:

$$f_w = \left(1 + \frac{1}{2}\frac{S_c}{S}\right)^2 \tag{5.73}$$

where S_c is the surface of the column per unit volume of the bed. Thus, for a cylindrical tube of diameter D packed with spheres of diameter d, equation (5.73) becomes:

$$f_w = \left(1 + \frac{d}{3D}\right)^2 \tag{5.74}$$

clearly as (D/d) increases, the correction factor approaches unity.

In contrast to this simple approach, Cohen and Metzner (1981) have made use of the detailed voidage profiles in the radial direction and have treated wall effects in a rigorous manner for both Newtonian and power-law fluids in streamline flow. From a practical standpoint, this analysis suggests the effect to be negligible in beds with $(D/d) > 30$

for inelastic fluids. Other approaches to account for wall effects have been discussed by Chhabra *et al.* (2001) and Chhabra (2006).

5.7.4 *Effect of particle shape*

The voidage of a bed of particles is strongly influenced by the particle shape, orientation and size distribution. However, voidage correlates quite well with sphericity, as shown in Figure 5.13; values of sphericity for different shape particles have been compiled by Brown *et al.* (1950) and German (1989) amongst others.

For beds of spheres of mixed sizes, the voidage can be significantly less if the smaller size spheres can fill the voids between the larger ones. In general, the voidage of a bed of mixed particles is a function of the volume fractions x of each particle size and the ratio of the sizes. Hence, for a binary-size particle system, the mean voidage of the bed will be a function of d_1/d_2 and x_1, x_2, and the curves of ε versus (d_1/d_2) and (x_1/x_2) will pass through minima. Some predictive expressions for voidage of beds of binary and ternary spherical and non-spherical particle systems are available in the literature (Yu and Standish, 1993; Yu *et al.*, 1993; Chhabra, 2006).

For the flow of Newtonian fluids in packed beds of mixed size spheres, Leva (1957) suggested the use of the volume-mean diameter, whereas the subsequent limited work with Newtonian and power-law fluids (Jacks and Merrill, 1971; Rao and Chhabra, 1993) shows that it is more satisfactory to use volume/surface mean size:

$$d = \frac{\sum n_i d_i^3}{\sum n_i d_i^2} \tag{5.75}$$

where n_i, is the number of spheres of size d_i.

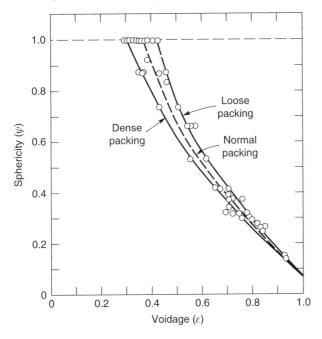

Figure 5.13 *Variation of mean bed voidage with particle sphericity*

For the flow of power-law fluids through packed beds of cubes, cylinders and gravel chips (Machac and Dolejs, 1981; Chhabra and Srinivas, 1991; Sharma and Chhabra, 1992; Sabiri and Comiti, 1995a, b, c; Tiu *et al.*, 1997), the few available data for stream-line flow correlate well with equation (5.56) if the equal volume sphere diameter, d_s, multiplied by sphericity, ψ, is employed as the effective diameter, $d_{eff} = d_s \psi$ in the definitions of the Reynolds number and friction factor. In this regard, the approach of Comiti and Renaud (1989) and modifications thereof, equations (5.70)–(5.72), offers a distinct advantage provided appropriate experimental data on pressure drop–flow rate for a Newtonian fluid are available which can be used to evaluate T and S for a given bed.

5.7.5 Dispersion in packed beds

Dispersion is the general term which denotes the various types of self-induced mixing processes which can arise during the flow of a fluid. The effects of dispersion are impor-tant in packed beds, though they are also present in the simple flow conditions existing in a straight tube. There are two important mechanisms of dispersion – molecular diffusion and mixing arising from the flow pattern within the fluid. An important consequence of dispersion in a packed bed or porous medium is that true plug-flow never occurs (except possibly for viscoplastic materials). Consequently, the performance of oil displace-ment processes using non-Newtonian polymer solutions, and of packed bed reactors for effecting polymerisation reactions, are adversely affected. For laminar flow in a tube, dispersion arises from random molecular motion and is governed by Fick's law accord-ing to which the flux of a component is proportional to the product of its concentration gradient and the molecular diffusivity. Furthermore, the velocity profile for a Newtonian fluid is parabolic with the ratio of the centre-line velocity to the mean velocity equal to 2. For a power-law fluid, this ratio is $(3n + 1)/(n + 1)$ (equation (3.11)). Dispersion occurs since elements of fluid take different times to traverse the length of the pipe, depending upon their radial positions and at the exit, elements with a range of residence times mix together. Thus, if a plug of tracer is injected into the entering liquid, it will first appear in the exit stream after the interval of time taken by the fastest moving fluid element at the axis to travel the length of the pipe. Then tracer will appear later in the liquid issuing at progressively greater distances from the axis of the pipe. Because the fluid velocity approaches zero at the wall, some tracer will still appear after a long period.

In turbulent flow, molecular diffusion is augmented by the presence of turbulent eddies and mixing is more intensive though, due to the flatter velocity profiles in tubes, the role of velocity gradient in dispersion diminishes.

In a bed of particles, the effects of dispersion will generally be greater than in a straight tube, partly because of the successive contractions and expansions in the flow passages. Radial mixing readily occurs in the flow passages or cells because a liquid element enters them with high kinetic energy, much of which is converted into rotational motion within the cells. Also, the continually changing velocity promotes dispersion in a bed of particles. Wall effects can be significant because of channelling through the region of high voidage near the wall.

At low flow rates, molecular diffusion dominates and cell mixing (attributable to the development of rotational motion) contributes relatively little to the overall dispersion. At high flow rates, however, mixing in a packed bed may be modelled by considering it to consist of a series of mixing cells, each being of the same size as the packing itself.

$$\left[2\pi r\,dr\,V_0\left(C+\frac{\partial C}{\partial \ell}d\ell\right)\right.$$
$$+(-)\,2\pi r\,dr\,\varepsilon\,D_L$$
$$\left.\left(\frac{\partial C}{\partial \ell}+\frac{\partial^2 C}{\partial \ell^2}d\ell\right)\right]$$

Flow

$d\ell$

dr

r

$$-2\pi r\,dr\,\varepsilon\,D_R\frac{\partial C}{\partial r}$$

$r+dr$

$$-2\pi\,(r+dr)$$
$$dr\,\varepsilon\,D_R\left(\frac{\partial C}{\partial r}+\frac{\partial^2 C}{\partial r}dr\right)$$

$$2\pi r\,dr\,V_0 C+(-)\,2\pi r\,dr\,\varepsilon\,D_L\frac{\partial C}{\partial \ell}$$

Figure 5.14 *Control volume for the derivation of the differential equation for dispersion*

Irrespective of the actual mechanism, dispersion processes may be characterized by a dispersion coefficient. In packed beds, dispersion is generally anisotropic, except at very low velocities; that is, the dispersion coefficients D_L in the axial (longitudinal) and D_R in the radial directions will generally be different. The process may normally be considered to be linear in so far as the rate of dispersion is proportional to the concentration gradient multiplied by the corresponding dispersion coefficient. However, unlike the molecular diffusion coefficient, D_L and D_R are strongly dependent on the flow regime and the bed geometry. In fact, the dispersion coefficient is analogous to the eddy kinematic viscosity which has been discussed in connection with momentum transfer in turbulent flow (Section 3.6.4).

The differential equation for dispersion in a cylindrical bed of voidage ε can be readily derived by considering the material balance over a differential annular control volume of length dl and of thickness dr, as shown in Figure 5.14. On the basis of a dispersion model, it is seen that the concentration C of a tracer will depend upon the axial position l, radial position r and time t. A material balance over this element gives (Coulson and Richardson, 1991):

$$\frac{\partial C}{\partial t}+\frac{V_0}{\varepsilon}\frac{\partial C}{\partial l}=D_L\frac{\partial^2 C}{\partial l^2}+\frac{1}{r}D_R\frac{\partial}{\partial r}\left(r\frac{\partial C}{\partial r}\right) \tag{5.76}$$

Longitudinal dispersion coefficients can be evaluated by injecting a flat pulse of tracer into the bed so that $\partial C/\partial r = 0$. The values of D_L can be estimated by matching the predicted (equation (5.76)) and the actual (measured) changes in the shape of a pulse of tracer as it passes between two locations (axial direction) in the bed. The results are often expressed in terms of dimensionless groups, namely, Peclet number, Pe, Reynolds number, Re and Schmidt number, Sc.

Wen and Yim (1971) reported a few results on axial dispersion coefficients (D_L) for the flow of two weakly shear-thinning polymer (PEO) solutions ($n = 0.81$ and 0.9) through a bed packed with glass spheres ($d = 4.76$ and $14.3\,\text{mm}$) of voidages 0.4 and 0.5. Over the range ($7 \leq Re_1 \leq 800$), their results did not deviate substantially from the correlation developed by these authors previously for Newtonian fluids:

$$Pe = 0.2 + 0.011(Re_1)^{0.48} \tag{5.77}$$

where

$$Pe = V_0 d/D_L \quad \text{and} \quad Re_1 = \frac{\rho V_0^{2-n} d^n}{m' 8^{n-1}} \tag{5.78}$$

Subsequent experimental work (Payne and Parker, 1973) was carried out with aqueous polyethylene oxide solutions having values of n in the range $0.53 \leq n < 1$, flowing in a packed bed of $350\,\mu\text{m}$ glass beads ($\varepsilon = 0.365$). The Reynolds number was of the order of 10^{-3} thereby rendering the second term on the right hand side of equation (5.77) negligible. However, under these conditions, the Peclet number showed weak dependence on the power-law index, decreasing from $Pe = 0.2$ for $n = 1$ to $Pe \sim 0.1$ for $n = 0.53$, thereby suggesting that dispersion was greater in shear-thinning fluids.

In the only reported study of radial mixing of non-Newtonian rubber solutions in packed beds (Hassell and Bondi, 1965), the quality of mixing was found to deteriorate rapidly with the increasing viscosity.

5.7.6 Mass transfer in packed beds

Little work has been reported on mass transfer between non-Newtonian fluids and particles in a fixed bed. Kumar and Upadhyay (1981) measured the rate of dissolution of benzoic acid spheres and cylindrical pellets in an aqueous carboxymethyl cellulose solution ($n = 0.85$) and, using the plug flow model, proposed the following correlation for mass transfer in terms of the j_m factor:

$$\varepsilon j_m = \frac{0.765}{(Re')^{0.82}} + \frac{0.365}{(Re')^{0.39}} \tag{5.79}$$

where ε is the voidage of the bed; Re' is the modified Reynolds number defined by equations (5.66) and (5.67) and the j_m factor is:

$$j_m = \frac{k_c}{V_0} Sc^{2/3} \tag{5.80}$$

where k_c is the mass transfer coefficient and the Schmidt number, $Sc = \mu_{\text{eff}}/\rho D_{AB}$ with the effective viscosity, μ_{eff}, as defined in equation (5.66). This correlation was found

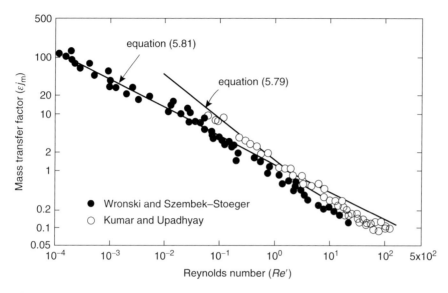

Figure 5.15 *Correlation of liquid–solid mass transfer for power-law fluids in packed beds*

to represent the experimental data with a mean error of $\pm 10\%$ for the range of conditions: $0.1 \le Re' \le 40$ and $800 \le Sc \le 72\,000$, but for a single value of $n = 0.85$. For cylindrical pellets, the characteristic linear dimension was taken as the diameter of the sphere of equal volume multiplied by the sphericity. The validity of equation (5.79) has subsequently been confirmed independently (Wronski and Szembek-Stoeger, 1988) but it does overpredict the j_m factor at low Reynolds numbers ($Re' \le 0.1$), as can be seen in Figure 5.15. The following correlation due to Wronski and Szembek-Stoeger (1988) takes account of most of the data in the literature and offers some improvement over equation (5.79), especially in the low Reynolds number region:

$$\varepsilon j_m = [0.097(Re')^{0.30} + 0.75(Re')^{0.61}]^{-1} \qquad (5.81)$$

Finally, suffice it to add here that the other limited literature data on mass transfer is also consistent with these predictions (Chhabra, 2006).

No analogous heat transfer studies have been reported with non-Newtonian fluids.

5.7.7 *Visco-elastic and surface effects in packed beds*

So far the flow of inelastic non-Newtonian fluids through unconsolidated beds of particles has been considered. The pore size has been substantially larger than the characteristic dimensions of the polymer molecules or of the flocs. In such cases, there is negligible or no interaction between the walls of the pore and the polymer molecules or flocs, and the rheological properties (values of m and n, for instance) may be inferred from pressure drop–flow rate data obtained in packed beds and tubes. Visco-elastic effects and other phenomena such as blockage of pores, polymer adsorption and apparent slip effects, observed in low permeability and consolidated systems will now be briefly discussed.

(i) *Visco-elastic effects*
Visco-elastic effects may be quantified in terms of a Deborah or Weissenberg number (De). For flow in packed beds, De is simply defined as:

$$De = \frac{\lambda_f V_0}{d} = \frac{\lambda_f}{(d/V_0)} \tag{5.82}$$

where λ_f is a characteristic time of the fluid and (d/V_0) is a measure of the time-scale of the flow. The characteristic time, λ_f, is determined from rheological measurements but the time-scale of the process clearly depends upon the kinematic conditions, principally, flow rate and packing size. At low velocities, the time-scale of flow (d/V_0) is large as compared with λ_f and De is therefore small. Thus, a fluid element is able continually to adjust to its changing flow geometry and area, and no effects arise from visco-elasticity. In other-words, the pressure loss through a bed of particles is determined essentially by the effective viscosity of the fluid at low flow rates and Deborah numbers. With gradual increase in flow rate, the value of (d/V_0) decreases progressively (hence the value of De increases) and the fluid elements are no longer able to re-adjust themselves to the rapidly changing flow conditions. The energy expended in squeezing (extensional flow) through 'throats' or 'narrow passages' in the bed is dissipated, as the fluid element has insufficient time ($\lambda_f > (d/V_0)$) to relax or recover its original state; this, in turn, may result in a substantially increased pressure drop. If the loss coefficient $\Lambda(=fRe^*)$ is plotted against Deborah number, or simply (V_0/d), if λ_f is constant for a fluid. Λ remains constant (at 150 or 180) at low values of (V_0/d) and, beyond a critical value of (V_0/d), it begins to rise gradually, as seen in Figure 5.16 which refers to the flow of aqueous solutions of hydroxypropyl guar (a polymer used as a cracking fluid in drilling and enhanced oil recovery operations) through a packed bed of 200–900 μm glass beads (Vorwerk and Brunn, 1994). As polymer concentration is increased, visco-elastic behaviour becomes more marked, and values of $(\Lambda/180)$ begin to deviate from unity at progressively lower values of (V_0/d), as shown in Figure 5.16.

The occurrence of very high pressure drops is well documented in the literature for a range of chemically different polymer melts and solutions (Chhabra, 2006), but there is little evidence concerning the critical value of Deborah number at which visco-elastic effects are significant or the extent to which Λ is increased as a result of visco-elasticity. The nature of the polymer solution and the geometry of the bed interact in a complex manner. For instance, Marshall and Metzner (1967) observed visco-elastic effects at $De \simeq 0.05$–0.06 whereas Michele (1977) reported the critical value of De to be as large as 3! The main difficulty stems from the fact that there is no simple method of calculating the value of λ_f in an unambiguous manner; some workers have inferred its value from steady shear data (first normal stress difference and/or shear viscosity) while others have used transient rheological tests to deduce the value of λ_f. However, polymer solutions certainly undergo substantial extensional deformation, in addition to shearing, in the narrow capillaries in beds of particles and in other porous matrices such as screens, mats and dense rocks. No satisfactory correlations for the frictional pressure gradient for visco-elastic fluids in packed beds are therefore available.

(ii) *Anomalous surface effects*
As mentioned previously, additional complications arise when the size of the pores or flow capillaries in the porous matrix in only slightly larger than the polymer molecules

Figure 5.16 *Visco-elastic effects in packed beds*

themselves. Polymer molecules may then be retained by both adsorption on to the walls of the pores and by mechanical entrapment, thereby leading to partial or even complete blockage of pores and increased pressure drops. The exact cause of the sharp rise in pressure gradient as shown in Figure 5.16 is therefore difficult to pinpoint. The existence of a slip velocity at the wall of the pore may explain many of the observations. However, at present there is no way of assessing *a priori* whether slip effects would be important or not in any new situation. Extensive reviews of the developments in this field are available in the literature (Cohen, 1988; Agarwal *et al.*, 1994).

5.8 Liquid–solid fluidization

5.8.1 Effect of liquid velocity on pressure gradient

As shown schematically in Figure 5.11 for the upward flow of a liquid through a bed of particles, a linear relation is obtained between the pressure gradient and the superficial velocity (on logarithmic coordinates) up to the point where the bed is fluidized and where expansion of the bed starts to occur (A), but the slope of the curve then gradually diminishes as the bed expands. As the liquid velocity is gradually increased, the pressure drop passes through a maximum value (B) and then falls slightly and eventually attains a nearly constant value, independent of liquid velocity (CD). If the velocity of flow is reduced again, the bed contracts until it reaches the condition where the particles are just resting in contact with one another (E); it then has the maximum stable voidage for a fixed bed of the particles in question. No further change in the voidage of the bed occurs as the velocity is reduced, provided it is not shaken or vibrated. The pressure drop (EF) in this re-formed packed bed may then be less than that in the original bed at the same velocity. If the liquid velocity were now to be increased again, the new curve (EF) would normally be re-traced and the slope will suddenly drop to zero at the fluidizing point E; this condition is difficult to produce, however, because the bed tends to become consolidated again as a result of stray vibrations and disturbances.

In an ideal fluidized bed, the pressure drop corresponding to ECD in Figure 5.11 is equal to the buoyant weight of the bed per unit area. In practice, however, deviations from this value may be observed due to channelling and interlocking of particles. Point B is situated above CD because the frictional forces between the particles must be overcome before the rearrangement of particles can occur.

Figure 5.17 shows representative experimental results of $\log(-\Delta p)$ versus $\log(V_0)$ for the flow of an aqueous carboxymethyl cellulose solution $(n = 0.9)$ through a bed of 3.57 mm glass spheres in a 100 mm diameter column (Srinivas and Chhabra, 1991). In spite of the scatter of the data at the onset of fluidization, the regions up to A (fixed bed) and CD (fluidized) can be clearly seen in this case.

The velocity corresponding to the incipient fluidizing point can also be calculated from the relation presented in Section 5.7, by equating the pressure drop across the bed to its buoyant weight per unit area. The voidage at the onset of fluidization should correspond to the maximum value attainable in the fixed bed. Hence, in a fluidized bed, the total frictional force on the particles must be equal to the buoyant weight of the bed. Thus, in a bed of height L and voidage ε:

$$-\Delta p = (\rho_s - \rho)(1 - \varepsilon)Lg \qquad (5.83)$$

where g is the acceleration due to gravity.

There may be some discrepancy between the observed and calculated minimum fluidization velocities. This difference is attributable to channelling or wall effects, or to the agglomeration of small particles. Equation (5.83) applies from the initial expansion of the bed up to high values (*ca.* 0.95) of voidage.

For streamline flow of a power-law fluid through a fixed bed of spherical particles, the relationship between liquid velocity V_0, bed voidage ε and pressure drop $(-\Delta p)$ is given by equation (5.59). Substituting for $(-\Delta p)$ from equation (5.83), equation (5.59) becomes:

$$V_0 = \left[\frac{(\rho_s - \rho)g\varepsilon^3 d^{n+1}}{180m(1 - \varepsilon)^n} \left(\frac{4n}{3n + 1} \right)^n \left(\frac{15\sqrt{2}}{\varepsilon^2} \right)^{1-n} \right]^{(1/n)} \qquad (5.84)$$

Figure 5.17   *Experimental pressure drop–superficial velocity curve and determination of minimum fluidisation velocity*

For $n = 1$, equation (5.84) reduces to the well-known Kozeny–Carman equation:

$$V_0 = 0.005\,55 \frac{(\rho_s - \rho)g\varepsilon^3 d^2}{\mu(1 - \varepsilon)} \qquad (5.85)$$

where μ is the Newtonian fluid viscosity.

5.8.2 Minimum fluidizing velocity

At the point of incipient fluidisation, the bed voidage, ε_{mf}, depends on the shape and size range of the particles, but is approximately equal to 0.4 for isometric particles. The minimum fluidizing velocity V_{mf} for a power-law fluid in streamline flow is then obtained by substituting $\varepsilon = \varepsilon_{mf}$ in equation (5.84). Although this equation applies only at low values of the bed Reynolds numbers (<1), this is not usually a limitation at the high apparent viscosities of most non-Newtonian materials.

At high velocities, the flow may no longer be streamline, and a more general equation must be used for the pressure gradient in the bed, such as equation (5.68) or (5.69) or (5.70). Substituting $\varepsilon = \varepsilon_{mf}$ and $V_0 = V_{mf}$, this equation becomes:

$$f_{mf} = \frac{150}{Re^*_{mf}} + 1.75 \qquad (5.86)$$

Replacement of f_{mf} by the Galileo number eliminates the unknown velocity V_{mf} which appears in Re^*_{mf}. Multiplying both sides of equation (5.86) by $(Re^*_{mf})^{2/(2-n)}$:

$$Ga_{mf} = 150(Re^*_{mf})^{n/(2-n)} + 1.75(Re^*_{mf})^{2/(2-n)} \qquad (5.87)$$

where

where $\quad Ga_{mf} = f_{mf}(Re^*_{mf})^{n/(2-n)}$

$$= \frac{(\rho_s - \rho)gd\varepsilon^3_{mf}}{\rho V^2_{mf}}(Re^*_{mf})^{2/(2-n)} \qquad (5.88)$$

For a given liquid (known m, n, ρ) and particle (ρ_s, d, ε_{mf}) combination, equation (5.87) can be solved for Re^*_{mf} which in turn enables the value of the minimum fluidizing velocity to be calculated, as illustrated in Example 5.9.

Currently available other methods for the estimation of the minimum fluidizing velocity also lead to comparable predictions (Chhabra, 2006).

Example 5.9

A bed consists of uniform glass spheres of size 3.57 mm diameter (density = 2500 kg/m³). What will be the minimum fluidizing velocity in a polymer solution of density, 1000 kg/m³, with power-law constants: $n = 0.6$ and $m = 0.25\,Pa \cdot s^n$? Assume the bed voidage to be 0.4 at the point of incipient fluidization.

Solution

First calculate the value of the Galileo number using equation (5.88) which after substitution for Re^*_{mf} from equation (5.60) becomes:

$$Ga_{mf} = \frac{(\rho_s - \rho)gd\varepsilon^3_{mf}}{\rho}\left[\frac{\rho d^n}{m(1 - \varepsilon_{mf})^n}\left(\frac{4n}{3n + 1}\right)^n\left(\frac{15\sqrt{2}}{\varepsilon^2_{mf}}\right)^{1-n}\right]^{2/(2-n)}$$

Substituting the numerical values:

$$Ga_{mf} = \frac{(2500 - 1000)(9.81)(3.57 \times 10^{-3})(0.4)^3}{1000}$$

$$\times \left[\frac{1000(3.57 \times 10^{-3})^{0.6}}{0.25(1 - 0.4)^{0.6}}\left(\frac{4 \times 0.6}{3 \times 0.6 + 1}\right)^{0.6}\times\left(\frac{15\sqrt{2}}{(0.4)^2}\right)^{1-0.6}\right]^{2/(2-0.6)}$$

$$= 83.2$$

Substituting this value in equation (5.87):

$$150(Re^*_{mf})^{(0.6)/(2-0.6)} + 1.75(Re^*_{mf})^{2/(2-0.6)} = 83.2$$

or

$$150(Re^*_{mf})^{0.43} + 1.75(Re^*_{mf})^{1.43} = 83.2$$

By trial and error procedure, $Re^*_{mf} = 0.25$. Thus,

$$\frac{\rho V^{2-n}_{mf}d^n}{m(1 - \varepsilon_{mf})^n}\left(\frac{4n}{3n + 1}\right)^n\left(\frac{15\sqrt{2}}{\varepsilon^2_{mf}}\right)^{1-n} = 0.25$$

On substituting numerical values and solving for the unknown velocity, $V_{mf} = 0.00236$ m/s = 2.36 mm/s.

   This value compares well with the experimental value of 2.16 mm/s for this system (Srinivas and Chhabra, 1991). □

5.8.3 Bed expansion characteristics

Liquid–solid fluidized systems are generally characterized by the regular expansion of the bed which takes place as the liquid velocity increases from the minimum fluidization velocity to a value approaching the terminal falling velocity of the particles. The general form of relation between velocity and bed voidage is found to be similar for both Newtonian and inelastic power-law liquids. For fluidization of uniform spheres by Newtonian liquids, equation (5.29), introduced earlier to represent hindered settling data, is equally applicable:

$$\frac{V_0}{V} = (1 - C)^Z = \varepsilon^Z \tag{5.29}$$

where V_0 is now the superficial velocity of the liquid in the fluidized bed and V is the terminal settling velocity of a single sphere in the same liquid; Z is a constant related to the particle Archimedes number and to the particle-to-column diameter ratio by equation (5.30).

Figure 5.18 shows a typical bed expansion behaviour for 3.57 mm glass spheres fluidized by aqueous carboxymethyl cellulose solutions *(n = 0.84 and n = 0.9)*, and the behaviour is seen to conform to the form of equation (5.29). Qualitatively similar results have been reported by many workers, and in a recent review (Chhabra, 1993, 2006) it has been shown that equation (5.30) correlates most of the data available for inelastic power-law fluids ($0.6 \leq n \leq 1$; $(d/D) \leq 0.16$). Values calculated from equation (5.30) and experimental values of Z differ by less than 10%. This suggests that the modified Archimedes number (see Example 5.6) takes account of power-law shear-thinning behaviour. On the other hand, much larger values of Z have been reported for fluidization with visco-elastic polymer solutions (Briend *et al.*, 1984; Srinivas and Chhabra, 1991; Chhabra *et al.*, 2001; Machac *et al.*, 2003; Chhabra, 2006), but no systematic study has been made to predict the value of Z for visco-elastic liquids.

5.8.4 Effect of particle shape

Little is known about the influence of particle shape on the minimum fluidizing velocity and bed expansion of liquid fluidized beds even for Newtonian liquids (Couderc, 1985; Flemmer *et al.*, 1993). The available scant data suggests that, if the diameter of a sphere of equal volume is used together with its sphericity factor, satisfactory predictions of the minimum fluidizing velocity are obtainable from the expressions for spherical particles.

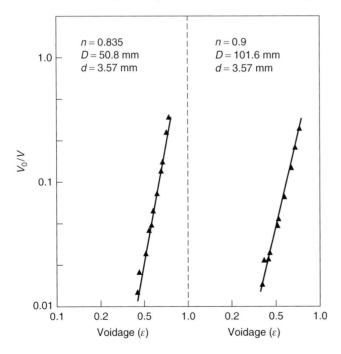

Figure 5.18 *Typical bed expansion data for 3.57 mm glass spheres fluidized by shear-thinning polymer solutions (Srinivas and Chhabra, 1991)*

Only one experimental study has been identified (Sharma and Chhabra, 1992) which is of relevance, and in that gravel chips (of sphericity of 0.6) were fluidized by aqueous carboxymethyl cellulose solutions ($0.78 \leq n \leq 0.91$). If an effective particle size (diameter of a sphere of equal volume multiplied by its sphericity) is used in lieu of sphere diameter in equation (5.87), the values of V_{mf} are predicted with a mean error of 18%. Similarly, the values of Z were within $\pm 10\%$ of the predictions of equation (5.30), although the latter correlation consistently underpredicted the value of Z for gravel chips.

5.8.5 Dispersion in fluidized beds

The material presented earlier on dispersion in packed beds (Section 5.7.5) is also relevant to fluidized beds. However, in the only experimental study reported (Wen and Fan, 1973), longitudinal dispersion coefficient D_{L} was measured by introducing a fluorescein dye into beds of glass and of aluminium spheres fluidized by mildly shear-thinning carboxymethyl cellulose solutions ($n = 0.86$ and $n = 0.89$). The limited results are well correlated by the following modified form of equation (5.77):

$$\chi_0 Pe = 0.2 + 0.011 Re_1^{0.48} \tag{5.89}$$

where $\chi_0 = 1$ for packed beds and $\chi_0 = (V_0/V_{mf})^{2-n}$ for fluidized systems; the Reynolds number is based on the superficial velocity.

5.8.6 Liquid–solid mass transfer in fluidized beds

Kumar and Upadhyay (1980, 1981) have measured the rates of dissolution of spheres and cylindrical pellets of benzoic acid in a bed fluidized by an aqueous carboxymethyl cellulose solution $(n = 0.85)$. Mass transfer coefficients were calculated on the basis of a plug-flow model from the measured rate of weight loss of the particles, and using the log mean value of concentration driving force at the inlet and outlet of the bed. Their data were satisfactorily correlated using equation (5.79) (as can be seen in Figure 5.19), though the results are over-predicted at low values of the Reynolds number. Subsequently, Hwang et al. (1993) have measured concentrations within a fluidized bed of cylindrical pellets of benzoic acid for a number of aqueous solutions of carboxymethyl cellulose ($0.63 \leq < n \leq 0.92$). The longitudinal concentration profile was adequately represented by a model combining plug-flow with axial dispersion. Using equation (5.89) to calculate D_{L}, together with the experimentally determined concentration profiles, mass transfer coefficients were evaluated in terms of the particle Reynolds number ($0.01 \leq Re' \leq 600$) and the Schmidt number. They proposed the following correlation for the j_m factor defined by equation (5.80):

$$\log(\varepsilon j_m) = 0.169 - 0.455 \log Re' - 0.0661 (\log Re')^2 \tag{5.90}$$

The effective viscosity used in the definitions of the Reynolds and Schmidt numbers is estimated from equation (5.66). Figure 5.19 also includes the predictions of equation (5.90) along with the data of Hwang et al. (1993). The scatter is such that, although both equations (5.79) and (5.90) give similar predictions at Reynolds numbers greater than about 1, equation (5.90) seems better in the low Reynolds number region (Chhabra et al., 2001; Chhabra, 2006).

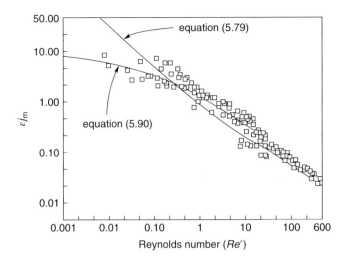

Figure 5.19   *Correlations for liquid–solid mass transfer in beds fluidized by power-law liquids*

Further reading

Chhabra, R.P., *Bubbles, Drops and Particles in Non-Newtonian Fluids*, 2nd edition, CRC Press, Boca Raton, FL (2006).

Clift, R., Grace, J. and Weber, M.E., *Bubbles, Drops and Particles*, Academic, New York (1978).

Coulson, J.M. and Richardson, J.F., *Chemical Engineering. Vol. II*, 4th edition, Butterworth-Heinemann, Oxford (1991).

Davidson, J.F., Clift, R. and Harrison, D., Eds, *Fluidisation*, 2nd edition, Academic, New York (1985).

Dullien, F.A.L., *Porous Media: Fluid Transport and Pore Structure*, 2nd edition, Academic, New York (1992).

References

Agarwal, N. and Chhabra, R.P., *Powder Technol.* **178** (2007) 17.

Agarwal, P.K. and O'Neill, B.K., *Chem. Eng. Sci.* **43** (1988) 2487.

Agarwal, U.S., Dutta, A. and Mashelkar, R.A., *Chem. Eng. Sci.* **49** (1994) 1693.

Allen, E. and Uhlherr, P.H.T., *J. Rheol.* **33** (1989) 627.

Ansley, R.W. and Smith, T.N., *AIChE J.* **13** (1967) 1193.

Astarita, G. and Appuzzo, G., *AIChE J* **11** (1965) 815.

Atapattu, D.D., Chhabra, R.P. and Uhlherr, P.H.T., *J. Non-Newt. Fluid Mech.* **59** (1995) 245.

Bagchi, A. and Chhabra, R.P., *Powder Technol.* **68** (1991) 85.

Beauline, M. and Mitsoulis, E., *J. Non-Newt. Fluid Mech.* **72** (1997) 55.

Beris, A.N., Tsamapolous, J., Armstrong, R.C. and Brown, R.A., *J. Fluid Mech.* **158** (1985) 219.

Bharti, R.P., Chhabra, R.P. and Eswaran, V., *Can. J. Chem. Eng.* **84** (2006) 406.

Blackery, J. and Mitsoulis, E., *J. Non-Newt. Fluid Mech.* **70** (1997) 59.

Bobroff, S. and Phillips, R.J., *J. Rheol.* **42** (1998) 1419.

Bond, W.N. and Newton, D.A., *Phil. Mag.* **5** (1928) 794.

Brea, F.M., Edwards, M.F. and Wilkinson, W.L., *Chem. Eng. Sci.* **31** (1976) 329.

Briend, P., Chavarie, C., Tassart, M. and Hlavacek, B., *Can. J. Chem. Eng.* **62** (1984) 26.

Brookes, G.F. and Whitmore, R.L., *Rheol. Acta.* **7** (1968) 188.

Brookes, G.F. and Whitmore, R.L., *Rheol. Acta.* **8** (1969) 472.

Brown, G.G. and associates, *Unit Operations*, Wiley, New York (1950).

Carman, P.C., *Flow of Gases through Porous Media*, Butterworths, London (1956).

Chafe, N.P. and de Bruyn, J.R., *J. Non-Newt. Fluid Mech.* 131 (2005) 44.

Chhabra, R.P., *Encyclopedia of Fluid Mechanics* (edited by Cheremisinoff, N.P.), Vol. 1, Gulf, Houston, p. 983 (1986).

Chhabra, R.P., *Chem. Eng. & Process* **28** (1990) 89.

Chhabra, R.P., *Adv. Heat Trans.* **23** (1993a) 178.

Chhabra, R.P., *Advances in Transport Processes* **9** (1993b) 501.

Chhabra, R.P., *Handbook of Polymer Processing Technology* (edited by Cheremisinoff, N.P.), Marcel-Dekker, New York (1996), Chapter 1.

Chhabra, R.P., *Bubbles, Drops and Particles in Non-Newtonian Fluids*, 2nd edition CRC Press, Boca Raton, FL (2006).

Chhabra, R.P. and Srinivas, B.K., *Powder Technol.* **67** (1991) 15.

Chhabra, R.P. and Uhlherr, P.H.T., Chem. *Eng. Commun,* 5 (1980) 115.

Chhabra, R.P. and Uhlherr, P.H.T., *Encyclopedia of Fluid Mechanics* (edited by Cheremisinoff, N.P.), Vol. 7, Gulf, Houston, p. 253 (1988).

Chhabra, R.P., Unnikrishnan, A. and Nair, V.R.U., *Can. J. Chem. Eng.* **70** (1992) 716.

Chhabra, R.P., Uhlherr, P.H.T. and Richardson, J.F., *Chem. Eng. Sci.* **51** (1996) 4532.

Chhabra, R.P., Soares, A.A. and Ferreira, J.M., *Can. J. Chem. Eng.* **76** (1998) 1051.

Chhabra, R.P., Comiti, J. and Machac, I., *Chem. Eng. Sci.* **56** (2001) 1.

Chhabra, R.P., Soares, A.A. and Ferreira, J.M., *Acta Mech.* **172** (2004) 1.

Clift, R., Grace, J. and Weber, M.E., *Bubbles, Drops and Particles*, Academic, New York (1978).

Cohen, Y., *Encyclopedia of Fluid Mechanics* (edited by Cheremisinoff, N.P.), volume **7**, Gulf, Houston (1988), Chapter 14.

Cohen, Y. and Metzner, A.B., *AIChE J.* **27** (1981) 705.

Comiti, J. and Renaud, M., *Chem. Eng. Sci.* **44** (1989) 1539.

Couderc, J.-P., *Fluidisation* (edited by Davidson, J.F., Clift, R. and Harrison, D.), 2nd edition, Academic, New York (1985), Chapter 1.

Coulson, J.M., *Trans. Inst. Chem. Engrs.* **27** (1949) 237.

Coulson, J.M. and Richardson, J.F., *Chemical Engineering, Vol. 2*, 4th edition, Butterworth-Heinemann, Oxford (1991).

Davidson, J.F., Clift, R. and Harrison, D., *Fluidisation*, 2nd edition, Academic, New York (1985).

Degand, E. and Walters, K., *J. Non-Newt. Fluid Mech.* **57** (1995) 103.

DeKee, D. and Chhabra, R.P., *Transport Processes in Bubbles, Drops and Particles* (edited by Chhabra, R.P. and DeKee, D.), Hemisphere, New York (1992), Chapter 2.

DeKee, D., Chhabra, R.P. and Rodrigue, D., *Handbook of Polymer Processing Technology* (edited by Cheremisinoff, N.P.), Marcel-Dekker, New York (1996), Chapter 3.

Dharamadhikari, R.V. and Kale, D.D., *Chem. Eng. Sci.* **40** (1985) 427.

Dhiman, A.K., Chhabra, R.P. and Eswaran, V., *Chem. Eng. Res. Des.* **84** (2006) 300.

Dhiman, A.K., Chhabra, R.P. and Eswaran, V., *J. Non-Newt. Fluid Mech.* **148** (2008) 141.

Dhole, S.D., Chhabra, R.P. and Eswaran, V., *Ind. Eng. Chem. Res.* **45** (2006) 4773.

Dhole, S.D., Chhabra, R.P. and Eswaran, V., *Ind. Eng. Chem. Res.* **46** (2007) 939.

Dullien, F.A.L., *Porous Media: Fluid Transport and Pore Structure*, 2nd edition, Academic, New York (1992).

Ergun, S., *Chem. Eng. Prog.* **48** (Feb, 1952) 89.

Flemmer, R.C., Pickett, J. and Clark, N.N., *Powder Technol.* **77** (1993) 123.

German, R.M., *Particle Packing Characteristics*, Metal Powder Industries Federation, Princeton, NJ (1989).

Ghosh, U.K., Upadhyay, S.N. and Chhabra, R.P., *Adv. Heat Trans.* **25** (1994) 251.

Graham, D.I. and Jones, T.E.R., *J. Non-Newt. Fluid Mech.* **54** (1995) 465.

Greenkorn, R.A., *Flow Phenomena in Porous Media*, Marcel-Dekker, New York (1983).

Gu, D. and Tanner, R.I., *J. Non-Newt. Fluid Mech.* **17** (1985) 1.

Haider, A. and Levenspiel, O., *Powder Technol.* **58** (1989) 63.

Happel, J. and Brenner, H., *Low Reynolds Number Hydrodynamics*, Prentice Hall, Englewood Cliffs, NJ (1965).

Hariharaputhiran, M., Shankar Subramanian, R., Campbell, G.A. and Chhabra, R.P., *J. Non-Newt. Fluid Mech.* **79** (1998) 87.

Hassell, H.L. and Bondi, A., *AIChE J.* **11** (1965) 217.

Herrera-Velarde, J.R., Zenit, R., Chehata, D. and Mena, B., *J. Non-Newt. Fluid Mech.* **111** (2003) 199.

Hirose, T. and Moo-Young, M., *Can. J. Chem. Eng.* **47** (1969) 265.

Hwang, S.-J., Lu, C.-B. and Lu, W.-J., *Chem. Eng. J.* **52** (1993) 131.

Jacks, J.-P. and Merrill, R.P., *Can. J. Chem. Eng.* **49** (1971) 699.

Jayaraman, A. and Belmonte, A., *Phys. Rev.* **E67** (2003) 653011.

Jossic, L. and Magnin, A., *AIChE J.* **47** (2001) 2666.

Kelessidis, V.C., *Chem. Eng. Sci.* **59** (2004) 4437.

Kemblowski, Z., Dziubinski, M. and Sek, J., *Advances in Transport Processes* **5** (1987) 117.

Kishore, N., Chhabra, R.P. and Eswaran, V., *Chem. Eng. Sci.* **62** (2007) 2422.

Kumar, S. and Upadhyay, S.N., *Lett. Heat Mass Trans.* **7** (1980) 199.

Kumar, S. and Upadhyay, S.N., *Ind. Eng. Chem. Fundam.* **20** (1981) 186.

Leva, M., *Chem. Eng.* **64** (1957) 245.

Machac, I. and Dolejs, V., *Chem. Eng. Sci.* **36** (1981) 1679.

Machac, I., Ulbrichova, I., Elson, T.P. and Cheesman, D.J., *Chem. Eng. Sci.* **50** (1995) 3323.

Marshall, R.J. and Metzner, A.B., *Ind. Eng. Chem. Fundam.* **6** (1967) 393.

Machac, I., Comiti, J., Brokl, P. and Siska, B., *Chem. Eng. Res. Des.* **81A** (2003) 1217.

McKinley, G.H., *Transport Processes in Bubbles, Drops and Particles* (edited by De Kee, D. and Chhabra, R.P.), IInd ed., Taylor and Francis, New York, p. 338 (2002).

Merkak, O., Jossic, L. and Magnin, A., *J. Non-Newt. Fluid Mech.* **133** (2006) 99.

Michele, H., *Rheol. Acta.* **16** (1977) 413.

Miller, C., *Ind. Eng. Chem. Fundam.* **11** (1972) 524.

Mishra, P., Singh, D. and Mishra, I.M., *Chem. Eng. Sci.* **30** (1975) 397.

Missirlis, K.A., Assimacopoulos, D., Mitsoulis, E. and Chhabra, R.P., *J. Non-Newt. Fluid Mech.* **96** (2001) 459.

Nakano, Y. and Tien, C., *AIChE J.* **16** (1970) 569.

Nguyen, Q.D. and Boger, D.V., *Annu. Rev. Fluid Mech.* **24** (1992) 47.

Nitin, S. and Chhabra, R.P., *J. Colloid Interface Sci.* **295** (2006) 520.

Panton, R.L., *Incompressible Flow*, 3rd edition, Wiley, New York (2005).

Payne, L.W. and Parker, H.W., *AIChE J.* **19** (1973) 202.

Pilz, G. and Brenn, G., *J. Non-Newt. Fluid Mech.* **145** (2007) 124.

Rajitha, P., Chhabra, R.P., Sabiri, N.E. and Comiti, J., *Int. J. Mineral Proc.* **78** (2006) 110.

Rao, P.T. and Chhabra, R.P., *Powder Technol.* **77** (1993) 171.

Renaud, M., Mauret, E. and Chhabra, R.P., *Can. J. Chem. Eng.* **82** (2004) 1066.

Richardson, J.F. and Zaki, W.N., *Trans. Inst. Chem. Engrs.* **32** (1954) 35.

Rodrigue, D., *AIChE J* **47** (2001) 39.

Rodrigue, D., *Can. J. Chem. Eng.* **80** (2002) 289.

Rodrigue, D., *Can. J. Chem. Eng.* **82** (2004) 382.

Rodrigue, D. and De Kee, D., *Transport Processes in Bubbles, Drops and Particles, Chapter 4* (edited by De Kee, D. and Chhabra, R.P.), 2nd edition, Taylor and Francis, New York (2002).

Sabiri, N.E. and Comiti, J., *Chem. Eng. Sci.* **50** (1995a) 1193.

Sabiri, N.E. and Comiti, J., *Chem. Eng. Sci.* **52** (1995b) 3589.

Sivakumar, P., Bharti, R.P. and Chhabra, R.P., *Chem. Eng. Sci.* **61** (2006) 6035.

Sivakumar, P., Bharti, R.P. and Chhabra, R.P., *Chem. Eng. Sci.* **62** (2007) 1682.

Soares, A.A., Ferreira, J.M. and Chhabra, R.P., *Ind. Eng. Chem. Res.* **44** (2005) 5815.

Solomon, M.J. and Muller, S.J., *J. Non-Newt. Fluid Mech.* **62** (1996) 81.

Soto, E., Goujon, C., Zenit, R. and Manero, O., *Phys. Fluids* **18** (2006) 121570.

Sen, S., *MS Dissertation*, Brigham Young University, Provo, Utah (1984).

Shah, S.N., El Fadili, Y. and Chhabra, R.P., *Int. J. Multiphase Flow* **33** (2007) 51.

Sharma, M.K. and Chhabra, R.P., *Can. J. Chem. Eng.* **70** (1992) 586.

Srinivas, B.K. and Chhabra, R.P., *Chem. Eng. and Process* **29** (1991) 121.

Taneda, S., *J. Phys. Soc. Jpn.* **11** (1956) 1104.

Tanner, R.I., *J. Non-Newt. Fluid Mech.* 50 (1993) 217.

Tiu, C., *Developments in Plastics Technology-2* (edited by Whelan, A. and Craft, J.L.), Elsevier, Amsterdam (1985), Chapter 7.

Tiu, C., Zhou, J.Z.Q., Nicolae, G., Fang, T.N. and Chhabra, R.P., *Can. J. Chem. Eng.* **75** (1997) 843.

Tripathi, A. and Chhabra, R.P., *AIChE J.* **42** (1995) 728.

Tripathi, A., Chhabra, R.P. and Sundararajan, T., *Ind. Eng. Chem. Res.* **33** (1994) 403.

Turton, R. and Clark, N.N., *Powder Technol.* **53** (1987) 127.

Tyabin, N.V., *Colloid, J. (USSR)* **19** (1953) 325.

Uhlherr, P.H.T., *Proc. 9th National Conf. on Rheology*, Adelaide, Australia. p. 231 (1986).

Uhlherr, P.H.T. and Chhabra, R.P., *Can. J. Chem. Eng.* **73** (1995) 918.

Valentik, L. and Whitmore, R.L., *Brit. J. Appl. Phys.* **16** (1965) 1197.

Volarovich, M.P. and Gutkin, A.M., *Colloid J. (USSR)* **15** (1953) 153.

Vorwerk, J. and Brunn, P.O., *J. Non-Newt. Fluid Mech.* **51** (1994) 79.

Walters, K. and Tanner, R.I., *Transport Processes in Bubbles, Drops and Particles* (edited by Chhabra, R.P. and DeKee, D.), Hemisphere, New York (1992), Chapter 3.

Wen, C.Y. and Fan, L.-S., *Chem. Eng. Sci.* **28** (1973) 1768.

Wen, C.Y. and Yim, J., *AIChE J.* **17** (1971) 1503.

Whitney, M.J. and Rodin, G.J., *Int. J. Non-linear Mech.* **36** (2001) 947.

Wronski, S. and Szembek-Stoeger, M., *Inzynieria Chem. Proc.* **4** (1988) 627.

Yu, A.-B. and Standish, N., *Powder Technol.* **74** (1993) 205.

Yu, A.-B., Standish, N. and McLean, A., *J. Am. Ceram. Soc.* **76** (1993) 2813.

Nomenclature

		Dimensions in **M, L, T**
A	area of flow (m²)	$\mathbf{L}^2$
A_0	constant, equation (5.11) (−)	$\mathbf{M}^0\mathbf{L}^0\mathbf{T}^0$
Ar	Archimedes number (−)	$\mathbf{M}^0\mathbf{L}^0\mathbf{T}^0$
Bi	Bingham number (−)	$\mathbf{M}^0\mathbf{L}^0\mathbf{T}^0$
Bi*	modified Bingham number (−)	$\mathbf{M}^0\mathbf{L}^0\mathbf{T}^0$
B_0	constant, equation (5.10) (−)	$\mathbf{M}^0\mathbf{L}^0\mathbf{T}^0$
b_0	constant, equation (5.8) (−)	$\mathbf{M}^0\mathbf{L}^0\mathbf{T}^0$
C	volume fraction of solids in a suspension or in a fluidized bed (−)	$\mathbf{M}^0\mathbf{L}^0\mathbf{T}^0$
C_1	constant, equation (5.9) (−)	$\mathbf{M}^0\mathbf{L}^0\mathbf{T}^0$
C_D	drag coefficient (−)	$\mathbf{M}^0\mathbf{L}^0\mathbf{T}^0$
C_DN	drag coefficient in a Newtonian fluid (−)	$\mathbf{M}^0\mathbf{L}^0\mathbf{T}^0$
D	tube or column diameter (m)	$\mathbf{L}$
D_h	hydraulic diameter for a packed bed (m)	$\mathbf{L}$
De	Deborah number (−)	$\mathbf{M}^0\mathbf{L}^0\mathbf{T}^0$
D_L	longitudinal dispersion coefficient (m²/s)	$\mathbf{L}^2\mathbf{T}^{-1}$
D_R	radial dispersion coefficient (m²/s)	$\mathbf{L}^2\mathbf{T}^{-1}$
d	sphere diameter (m)	$\mathbf{L}$
d_n	diameter of an equal projected are a circle (m)	$\mathbf{L}$
d_s	equal volume sphere diameter (m)	$\mathbf{L}$
d^+	dimensionless diameter, equation (5.18) or (5.21) (−)	$\mathbf{M}^0\mathbf{L}^0\mathbf{T}^0$
F	dimensionless parameter, equation (5.38) (−)	$\mathbf{M}^0\mathbf{L}^0\mathbf{T}^0$
F_D	total drag force on a particle (N)	$\mathbf{MLT}^{-2}$
F_n	normal component of drag force due to pressure (N)	$\mathbf{MLT}^{-2}$
F_t	tangential component of drag force due to shear stress (N)	$\mathbf{MLT}^{-2}$
f	friction factor for flow in packed beds (−) or wall correction factor for particle settling in a tube (−)	$\mathbf{M}^0\mathbf{L}^0\mathbf{T}^0$
f_0	value of wall correction factor in streamline region (−)	$\mathbf{M}^0\mathbf{L}^0\mathbf{T}^0$
f_∞	value of wall correction factor in the high Reynolds number region (−)	$\mathbf{M}^0\mathbf{L}^0\mathbf{T}^0$
f_w	wall correction factor for packed beds (−)	$\mathbf{M}^0\mathbf{L}^0\mathbf{T}^0$
g	acceleration due to gravity (m/s²)	$\mathbf{LT}^{-2}$
Ga	Galileo number (−)	$\mathbf{M}^0\mathbf{L}^0\mathbf{T}^0$
He	Hedström number (−)	$\mathbf{M}^0\mathbf{L}^0\mathbf{T}^0$
j_m	*j* factor for mass transfer (−)	$\mathbf{M}^0\mathbf{L}^0\mathbf{T}^0$
K_0	constant dependent on shape of cross-section (−)	$\mathbf{M}^0\mathbf{L}^0\mathbf{T}^0$
K_2	constant, equation (5.23) (−)	$\mathbf{M}^0\mathbf{L}^0\mathbf{T}^0$
k	permeability of porous medium (m²)	$\mathbf{L}^2$
k_c	mass transfer coefficient (m/s)	$\mathbf{LT}^{-1}$
L	height of bed (m)	$\mathbf{L}$
L_e	effective path length traversed by a fluid element (m)	$\mathbf{L}$
m	power law consistency coefficient (Pa sⁿ)	$\mathbf{ML}^{-1}\mathbf{T}^{n-2}$
m'	apparent power law consistency coefficient (Pa $s^{n'}$), equation (5.61)	$\mathbf{ML}^{-1}\mathbf{T}^{n'-2}$

		Dimensions in $\mathbf{M, L, T}$
n	flow behaviour index $(-)$	$\mathbf{M^0L^0T^0}$
n'	apparent flow behaviour index $(-)$	$\mathbf{M^0L^0T^0}$
p	pressure (Pa)	$\mathbf{ML^{-1}T^{-2}}$
$(-\Delta p/L)$	pressure gradient (Pa/m)	$\mathbf{ML^{-2}T^{-2}}$
Pe	Peclet number $(-)$	$\mathbf{M^0L^0T^0}$
Q	volumetric flow rate (m³/s)	$\mathbf{L^3T^{-1}}$
R	sphere radius (m)	$\mathbf{L}$
Re	particle Reynolds number $(-)$	$\mathbf{M^0L^0T^0}$
Re_B	particle Reynolds number based on Bingham plastic viscosity $(-)$	$\mathbf{M^0L^0T^0}$
Re', Re^*	modified Reynolds number for packed beds $(-)$, equations (5.75) and (5.66)	$\mathbf{M^0L^0T^0}$
Re_1	modified Reynolds number, equation (5.78) $(-)$	$\mathbf{M^0L^0T^0}$
r	coordinate (m)	$\mathbf{L}$
S	specific surface of a particle (m⁻¹)	$\mathbf{L^{-1}}$
S_B	specific surface of a bed of particles (m⁻¹)	$\mathbf{L^{-1}}$
T	tortuosity factor $(-)$	$\mathbf{M^0L^0T^0}$
V	terminal falling velocity of a sphere (m/s)	$\mathbf{LT^{-1}}$
V^+	dimensionless velocity equation (5.19) or (5.22)	$\mathbf{M^0L^0T^0}$
V^*	dimensionless velocity, equation (5.36) $(-)$	$\mathbf{M^0L^0T^0}$
V_i	interstitial or pore velocity (m/s)	$\mathbf{LT^{-1}}$
V_m	terminal falling velocity of a sphere in a bounded fluid medium (m/s)	$\mathbf{LT^{-1}}$
V_0	superficial velocity (m/s)	$\mathbf{LT^{-1}}$
X	drag correction factor $(-)$	$\mathbf{M^0L^0T^0}$
Y	yield parameter $(-)$	$\mathbf{M^0L^0T^0}$
Z	fluidisation/hindered settling index $(-)$	$\mathbf{M^0L^0T^0}$

Greek letters

$\dot{\gamma}_w$	wall shear rate (s⁻¹)	$\mathbf{T^{-1}}$
δ	ratio of viscosity of the fluid sphere and that of the continuous medium $(-)$	$\mathbf{M^0L^0T^0}$
ε	voidage $(-)$	$\mathbf{M^0L^0T^0}$
θ	coordinate $(-)$	$\mathbf{M^0L^0T^0}$
λ_f	fluid relaxation time (s)	$\mathbf{T}$
μ	viscosity (Pa s)	$\mathbf{ML^{-1}T^{-1}}$
ρ	density (kg/m³)	$\mathbf{ML^{-3}}$
σ	surface tension (N/m)	$\mathbf{ML^0T^{-2}}$
τ_0	yield stress (Pa)	$\mathbf{ML^{-1}T^{-2}}$
$\tau_{r\theta}$	$r\,\theta$-component of stress (Pa)	$\mathbf{ML^{-1}T^{-2}}$
ϕ	τ_0^B/τ_w	$\mathbf{M^0L^0T^0}$
χ	composite shape factor, equation (5.33)	$\mathbf{M^0L^0T^0}$
ψ	sphericity $(-)$	$\mathbf{M^0L^0T^0}$

Subscripts/superscripts

B	Bingham model parameter
b	bubble
c	continuous phase
Eff	effective
o	creeping flow
Pore	pore conditions
s	solid

Chapter 6

Heat transfer characteristics of non-Newtonian fluids in pipes

6.1 Introduction

In many chemical and processing applications, fluids need to be heated or cooled and a wide range of equipment may be utilized. Examples include double pipe and shell and tube heat exchangers, compact heat exchangers, scraped surface heat exchangers and stirred vessels fitted with cooling coils or jackets. Sometimes, heat is generated in the process, as in extrusion which is extensively carried out in the polymer and food industry. It may also be necessary to reduce the rate at which heat is lost from a vessel or to ensure that heat is removed at a sufficient rate in equipment such as screw conveyors. In most applications, it is the rate of heat transfer within process equipment which is of principal interest. However, with thermally sensitive materials (such as foodstuffs, fermentation froths, pharmaceutical formulations), the temperature profiles must be known and maximum permissible temperatures must not be exceeded.

Because of their high consistencies, non-Newtonian materials are most frequently processed under conditions of laminar flow. Furthermore, shear stresses are generally so high that viscous generation of heat can rarely be neglected, and the temperature dependence of the rheological properties adds to the complexity of the mass, momentum and energy balance equations. Numerical techniques are often needed to obtain solutions, even for highly idealized conditions of flow.

Much of the research activity in this area has related to heat transfer to inelastic non-Newtonian fluids in laminar flow in circular and non-circular ducts. In recent years, some consideration has also been given to heat transfer to/from non-Newtonian fluids in vessels fitted with coils and jackets, but little information is available on the operation of heat exchange equipment with non-Newtonian fluids. Consequently, this chapter is concerned mainly with the prediction of heat transfer rates for flow in circular tubes. Heat transfer in external (boundary layer) flows is discussed in Chapter 7, whereas the cooling/heating of non-Newtonian fluids in stirred vessels is dealt with in Chapter 8. First of all, however, the thermo-physical properties of the commonly used non-Newtonian materials will be described.

6.2 Thermo-physical properties

The most important thermo-physical properties of non-Newtonian fluids are thermal conductivity, density, specific heat, surface tension and coefficient of thermal expansion. While the first three characteristics enter into virtually all heat transfer calculations, surface tension often exerts a strong influence on boiling heat transfer and bubble dynamics in non-Newtonian fluids, as seen in Chapter 5. Likewise, the coefficient of thermal expansion is important in heat transfer by free convection.

Only a very limited range of measurements of physical properties have been made, and for dilute and moderately concentrated aqueous solutions of commonly used polymers including carboxymethyl cellulose, polyethylene oxide, carbopol, polyacrylamide, density, specific heat, thermal conductivity, coefficient of thermal expansion and surface tension differ from the values for water by no more than 5–10% (Porter, 1971; Cho and Hartnett, 1982; Irvine *et al.*, 1987; Chhabra, 1999, 2003). Thermal conductivity might be expected to be shear rate dependent, because both apparent viscosity and thermal conductivity are dependent on structure. Although limited measurements (Loulou *et al.*, 1992) on carbopol solutions and of Lin *et al.* (2003) on fruit juices confirm this assertion, the effect is small. For engineering design calculations, there will be little error in assuming that all the above physical properties of aqueous polymer solutions, except apparent viscosity, are equal to the values for water.

Some values of these properties of polymer melts are also available (Brandrup and Immergut, 1989; Domininghaus, 1993). Unfortunately though, no simple predictive expressions are available for their estimation. Besides, values seem to be strongly dependent on the method of preparation of the polymer, the molecular weight distribution, chain architecture, etc., and therefore extrapolation from one system to another, even under nominally identical conditions, can lead to significant errors. For industrially important slurries and pastes exhibiting strong non-Newtonian behaviour, the thermo-physical properties (density, specific heat and thermal conductivity) can deviate significantly from those of its constituents. Early measurements (Orr and Dallavalle, 1954) on aqueous suspensions of powdered copper, graphite, aluminium and glass beads suggest that both the density and the specific heat can be approximated by the weighted average of the individual constituents, i.e.,

$$\rho_{\text{sus}} = \phi \rho_s + (1 - \phi) \rho_L \tag{6.1}$$

$$C_{p_{\text{sus}}} = \phi C_{p_s} + (1 - \phi) C_{p_L} \tag{6.2}$$

where ϕ is the volume fraction of the solids, and the subscripts s and L refer to the values for the solid and the liquid medium, respectively.

The thermal conductivity of these systems, on the other hand, seems generally to be well correlated by the following expression (Tareef, 1940; Orr and Dallavalle, 1954; Skelland, 1967):

$$k_{\text{sus}} = k_L \left[\frac{1 + 0.5(k_s/k_L) - \phi(1 - k_s/k_L)}{1 + 0.5(k_s/k_L) + 0.5\phi(1 - k_s/k_L)} \right] \tag{6.3}$$

Thermal conductivities of suspensions up to 60% (by weight) in water and other suspending media are well represented by equation (6.3). It can readily be seen that for a suspension of highly conducting particles ($k_L/k_s \rightarrow 0$), the thermal conductivity of a suspension can be increased by several folds which in turn will promote heat transfer. However, the resulting increase in apparent viscosity from such addition would more than offset the effects of increase in thermal conductivity on the rate of heat transfer.

For suspensions of mixed particle size, the following expression due to Bruggemann (1935) is found to be satisfactory:

$$\frac{(k_{\text{sus}}/k_s) - 1}{(k_L/k_s) - 1} = \left(\frac{k_{\text{sus}}}{k_L} \right)^{1/3} (1 - \phi) \tag{6.4}$$

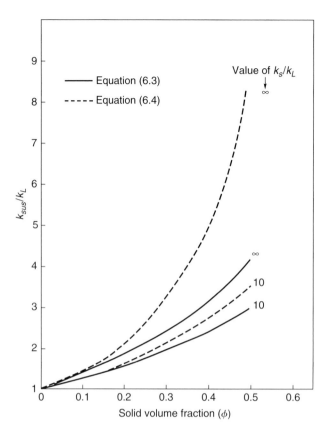

Figure 6.1   *Effect of concentration on thermal conductivity of suspensions*

The scant experimental data (Rajaiah *et al.*, 1992) for suspensions ($\phi \leq 0.3$) of alumina (0.5–0.8 μm) particles in a paraffin hydrocarbon are in line with the predictions of equation (6.4). An exhaustive review of the thermal conductivity of structured media including polymer solutions, filled and unfilled polymer melts, suspensions and food-stuffs has been published by Dutta and Mashelkar (1987). Figure 6.1 shows the predictions of equations (6.3) and (6.4) for a range of values of (k_L/k_s); the two predictions are fairly close, except for the limiting value of $k_L/k_s = 0$.

Example 6.1

Estimate the value of thermal conductivity at 20°C of 25% (by vol) for aqueous suspensions of (a) alumina, $k_s = 30\,\text{W/mK}$, (b) thorium oxide, $k_s = 14.2\,\text{W/mK}$ and (c) glass beads, $k_s = 1.20\,\text{W/mK}$.

Solution

Here $\phi = 0.25$; thermal conductivity of water, $k_L = 0.60\,\text{W/mK}$.
   The values of the thermal conductivity of various suspensions are calculated using equations (6.3) and (6.4) for the purposes of comparison.

Suspension	Value of k_{sus} (W/mK)	
	Equation (6.3)	Equation (6.4)
Alumina	1.92	2.2
Thorium oxide	1.85	2.06
Glass beads	1.29	1.20

The values obtained by the two methods are seen to be close and the difference, being 10%, is well within the limits of experimental error in such measurements. □

Of all the physico-chemical properties, it is the rheology which shows the strongest temperature dependence. For instance, the decrease in apparent viscosity at a fixed shear rate is well represented by the Arrhenius-type exponential expression; the pre-exponential factor and the activation energy are then both functions of shear rate. It is thus customary to denote the temperature dependence using rheological constants such as the power-law consistency coefficient and flow behaviour index. It is now reasonably well established that the flow behaviour index, *n*, of suspensions, polymer melts and solutions is nearly independent of temperature, at least over a range of 40–50°C, whereas the consistency coefficient exhibits an exponential dependence on temperature, i.e.,

$$m = m_0 \exp(E/\mathbf{R}T) \tag{6.5}$$

where the coefficient m_0 and E, the pre-exponential factor and the activation energy of viscous flow respectively, are evaluated using experimental results for the temperature range of interest. Similarly, in the case of Bingham plastic model, both the plastic viscosity and the yield stress decrease with temperature in a similar fashion, but each with different values of the pre-exponential factors and the activation energies. Temperature dependencies of the other rheological characteristics such as the primary and the secondary normal stress differences, extensional viscosity, etc., have been discussed in detail by Ferry (1980).

6.3 Laminar flow in circular tubes

Heat transfer through a liquid in streamline flow takes place by conduction in the radial direction. As mentioned earlier, the consistency of most non-Newtonian materials is high so that turbulent conditions do not usually occur and free convection also is seldom significant.

When a shear-thinning power-law fluid enters a pipe heated on the outside, the fluid near the wall will be both at a higher temperature and subject to higher shear rates than that at the centre, and therefore its viscosity will be lower. It thus follows that for a given volumetric flow rate the velocity of the fluid near the wall will be greater, and that near the centre correspondingly less, as compared with an unheated fluid (Figure 6.2). Thus the velocity profile is flattened when the fluid is heated and, conversely, sharpened when it is cooled. If the fluid has a high apparent viscosity, frictional heating may be sufficient to modify the temperature and velocity profiles and the analysis of the flow problem then becomes very complex (Section 6.8).

As in the case of Newtonian flow, it is necessary to differentiate between the conditions in the entry region and in the region of fully (thermally) developed flow.

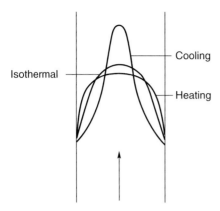

Figure 6.2   *Effect of heat transfer on velocity distribution in a tube*

Since the power-law and the Bingham plastic fluid models are usually adequate for modelling the shear dependence of viscosity in most engineering design calculations, the following discussion will therefore be restricted to cover just these two models; where appropriate, reference, however, will also be made to the applications of other rheological models. Theoretical and experimental results will be presented separately. For more detailed accounts of work on heat transfer in non-Newtonian fluids in both circular and non-circular ducts, reference should be made to one of the detailed surveys (Cho and Hartnett, 1982; Irvine and Karni, 1987; Shah and Joshi, 1987; Hartnett and Kostic, 1989; Hartnett and Cho, 1998; Chhabra, 1999; Irvine and Capobianchi, 2000).

6.4 Fully developed heat transfer to power-law fluids in laminar flow

The heating of a viscous fluid in laminar flow in a tube of radius R (diameter, D) will now be considered. Prior to the entry plane ($z < 0$), the fluid temperature is uniform at T_1; for $z > 0$, the temperature of the fluid will vary in both radial and axial directions as a result of heat transfer at the tube wall. A thermal energy balance will first be made on a differential fluid element to derive the basic governing equation for heat transfer. The solution of this equation for the power-law and the Bingham plastic models will then be presented.

Consider the differential control volume shown in Figure 6.3. The velocity profile is assumed to be fully developed in the direction of flow, i.e. $V_z(r)$. Furthermore, all physical properties including m and n for a power-law fluid and plastic viscosity and yield stress for a Bingham plastic fluid, are assumed to be independent of temperature. This simplifying assumption leads to the decoupling of the momentum and thermal energy equations, but at the same time it also restricts the applicability of these results to the situations where the temperature difference between the wall and fluid is small.

At steady state, the temperature of the fluid, T, will be a function of both r and z, and the rate of transfer of heat into the control volume is:

$$\underbrace{(2\pi r\,dz\cdot q_r)\big|_r}_{\substack{\text{(conduction in}\\ \text{r-direction)}}} \quad + \quad \underbrace{(2\pi r\,dr q_z)\big|_z}_{\substack{\text{(conduction in}\\ \text{z-direction)}}} \quad + \quad \underbrace{(2\pi r V_z \rho C_p (T - T_1)dr\big|_z}_{\substack{\text{(conduction in}\\ \text{z-direction)}}} \qquad (6.6)$$

Figure 6.3 *Schematics for heat balance in a tube*

Similarly, the rate of transfer of heat out of the control volume is given by:

$$(2\pi(r + dr)dz q_r)|_{(r+dr)} + (2\pi r\,dr q_z)|_{(z+dz)} + (2\pi r V_z \rho C_p(T - T_1)dr)|_{(z+dz)} \quad (6.7)$$

where q_r and q_z, respectively, are the heat fluxes due to conduction in the r- and z-directions.

Under steady state conditions, expressions (6.6) and (6.7) must be equal and slight re-arrangement yields:

$$\rho C_p V_z(r)\frac{\partial T}{\partial z} + \frac{\partial q_z}{\partial z} + \frac{1}{r}\frac{\partial}{\partial r}(r q_r) = 0 \quad (6.8)$$

Since heat transfer in r-direction is solely by conduction,

$$q_r = -k\frac{\partial T}{\partial r} \quad (6.9)$$

Similarly, the conduction heat flux in z-direction,

$$q_z = -k\frac{\partial T}{\partial z} \quad (6.10)$$

Combining equations (6.8)–(6.10),

$$\rho C_p V_z(r)\frac{\partial T}{\partial z} = k\frac{\partial^2 T}{\partial z^2} + \frac{k}{r}\frac{\partial}{\partial r}\left(r\frac{\partial T}{\partial r}\right) \quad (6.11)$$

It is implicit in equation (6.11) that thermal conductivities are isotropic. This is satisfactory for homogeneous systems (e.g. polymer solutions), but not always so for filled polymer melts (Dutta and Mashelkar, 1987).

Generally, conduction in the z-direction is negligible (a reasonable assumption for Peclet number >1000, as shown by Johnston (1991) in comparison with the convection, and the term $\partial^2 T/\partial z^2$ in equation (6.11) may thus be dropped to give:

$$\rho C_p V_z(r)\frac{\partial T}{\partial z} = \frac{k}{r}\frac{\partial}{\partial r}\left(r\frac{\partial T}{\partial r}\right) \quad (6.12)$$

Two important wall conditions will now be considered: (1) constant wall temperature (e.g. when steam is condensing on the outside of the tube) and (2) constant heat flux (e.g. when an electrical heating coil surrounds the pipe). The solutions for these two conditions are quite different, and the two cases will therefore be dealt with separately.

6.5 Isothermal tube wall

6.5.1 Theoretical analysis

Let the tube wall be at a uniform temperature T_0 ($z > 0$) and the entry temperature of the fluid be T_1 at $z = 0$. The solution of equation (6.12) is simplified by using a dimension-less temperature, θ, defined by:

$$\theta = \frac{T - T_0}{T_1 - T_0} \tag{6.13}$$

On substitution in equation (6.12):

$$V_z(r)\frac{\partial \theta}{\partial z} = \frac{\alpha}{r}\frac{\partial}{\partial r}\left(r\frac{\partial \theta}{\partial r}\right) \tag{6.14}$$

where $\alpha = k/\rho C_p$ is the thermal diffusivity of the fluid. This basic equation for temperature, $\theta(r, z)$, must be solved subject to the following boundary conditions:

At the tube wall, $r = R$, $\theta = 0$ for all values of $z > 0$ (6.15a)

At the centre of tube, $r = 0$, $\dfrac{\partial \theta}{\partial r} = 0$ for all values of z (6.15b)

and at the tube inlet, $z = 0$, $\theta = 1$ for all values of r (6.15c)

The solution depends on the form of the velocity distribution. Unfortunately, the closed form solutions are only possible for the following three forms of $V_z(r)$.

Piston or plug flow

This type of flow is characterized by the uniform velocity across the cross-section of the tube, i.e. $V_z(r) = V_0$, the constant value. This condition applies near the tube entrance, and is also the limiting condition of $n = 0$ with power-law model, i.e. infinite pseudo-plasticity. In view of its limited practical utility, though this case is not discussed here, but detailed solutions are given in several books, e.g. see Skelland (1967). However, Metzner *et al.* (1957) put forward the following expression for Nusselt number under these conditions (for $Gz > 100$):

$$Nu = \frac{h_m D}{k} = \frac{4}{\pi}(2 + Gz^{1/2}) \tag{6.16}$$

where h_m is the heat transfer coefficient averaged over the length L, of the tube, and Gz is the Graetz number defined by:

$$Gz = \frac{\dot{m}C_p}{kL} \tag{6.17}$$

where $\dot{m}$ is the mass flow rate of the fluid.

Fully developed power-law fluid flow

As seen in Chapter 3, the fully developed laminar velocity profile for power-law fluids in a tube is given by:

$$V_z(r) = V\left(\frac{3n+1}{n+1}\right)\left[1 - \left(\frac{r}{R}\right)^{(n+1)/n}\right] \tag{3.7}$$

which, upon substitution in equation (6.14), yields:

$$V\left(\frac{3n+1}{n+1}\right)\left[1 - \left(\frac{r}{R}\right)^{(n+1)/n}\right]\frac{\partial\theta}{\partial z} = \frac{\alpha}{r}\frac{\partial}{\partial r}\left(r\frac{\partial\theta}{\partial r}\right) \tag{6.18}$$

where V is the mean velocity of flow.

This differential equation can be solved by the *separation of variables* method by letting

$$\theta = F(r)Z(z) \tag{6.19}$$

where F is a function of r only and Z is a function of z only. Equation (6.18) then reduces to:

$$\frac{V}{\alpha}\left(\frac{3n+1}{n+1}\right)\frac{1}{Z}\frac{dZ}{dz} = \frac{1}{rF\{1 - (r/R)^{(n+1)/n}\}}\frac{d}{dr}\left(r\frac{dF}{dr}\right) \tag{6.20}$$

Both sides of equation (6.20) must be equal to a constant, say, $-\beta^2$, i.e.,

$$\frac{1}{Z}\frac{dZ}{dz} = -\beta^2\left(\frac{n+1}{3n+1}\right)\frac{\alpha}{V}$$

which can be integrated to obtain the Z component of θ as:

$$Z = \exp\left\{-\frac{\alpha(n+1)}{V(3n+1)}\beta^2 z\right\} \tag{6.21}$$

The variation of θ in the radial direction is given by the right hand side of equation (6.20), with slight rearrangement as:

$$\frac{d^2F}{dr^2} + \frac{1}{r}\frac{dF}{dr} + \beta^2\left[1 - \left(\frac{r}{R}\right)^{(n+1)/n}\right]F = 0 \tag{6.22}$$

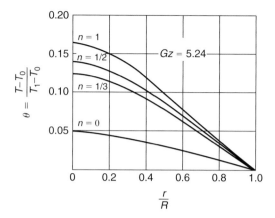

Figure 6.4   *Temperature distribution for power-law fluids in a tube (Bird et al., 1987)*

This equation is of the Sturm–Liouville type, and its complete solution for $n = 1$, $1/2$ and $1/3$ and for small values of the Graetz number is available (Lyche and Bird, 1956). Detailed tabulations of the eigen values of equation (6.22) are given in the original paper, and a representative temperature profile for $Gz = 5.24$ is shown in Figure 6.4, for a range of values of the power-law flow behaviour index. The temperature profile is seen to become progressively flatter as the flow index n is reduced (increasing degree of shear-thinning behaviour). However, difficulties in evaluating the series expansions at high Graetz numbers limited their predictions to the range $1.57 \leq Gz \leq 31.4$.

The resulting values of the mean Nusselt number, Nu_∞, under thermally fully developed conditions are 3.66, 3.95 and 4.18, respectively, for n values of 1, 1/2 and 1/3 (Bird *et al.*, 1987).

In addition, the so-called Leveque approximation (Leveque, 1928) has also been extended and applied to power-law fluids. The key assumption in this approach is that the temperature boundary layer is confined to a thin layer of the fluid adjacent to tube wall. This is a reasonable assumption for high flow rates and for short tubes, i.e., large values of Gz. A linear velocity gradient can then be assumed to exist within this thin layer:

$$V_z(r) = -\beta_v(R - r) \tag{6.23}$$

where β_v is the velocity gradient close to the tube wall. For power-law fluids, it can be evaluated using the velocity distribution given by equation (3.59a) as:

$$\beta_v \simeq \left.\frac{dV_z(r)}{dr}\right|_{r=R} = -\frac{V}{R}\left(\frac{3n+1}{n}\right) \tag{6.24}$$

and the velocity profile near the wall is therefore given approximately by:

$$V_z(r) = \frac{V}{R}\left(\frac{3n+1}{n}\right)(R - r) \tag{6.25}$$

When substituted in equation (6.14), this yields:

$$\frac{V}{R}\left(\frac{3n+1}{n}\right)(R-r)\frac{\partial\theta}{\partial z} = \frac{\alpha}{r}\frac{\partial}{\partial r}\left(r\frac{\partial\theta}{\partial r}\right)$$ (6.26)

Furthermore, since the heat transfer is confined to a thin layer near the wall, the effects of curvature can be neglected. Denoting the distance from the wall as y, i.e. $y = R - r$, the right hand side of equation (6.26) can be further simplified to give:

$$\left(\frac{3n+1}{n}\right)\xi\frac{\partial\theta}{\partial\zeta} = \frac{\partial^2\theta}{\partial\xi^2}$$ (6.27)

where $\xi = y/R$; $\zeta = \alpha z/VR^2 = \pi(z/L)(1/Gz)$. The boundary conditions (6.15a) and (6.15b) may be re-written as $\xi = 0$, $\theta = 0$ and $\xi \to \infty$, $\theta = 1$, respectively. Under these conditions, equation (6.27) can be solved by a *combination of variables* method as:

$$\theta = \theta(\chi)$$ (6.28)

with

$$\chi = \frac{\xi}{\left[9\zeta\left(\dfrac{n}{3n+1}\right)\right]^{1/3}}$$ (6.29)

When substituted in equation (6.27), this gives:

$$\frac{d^2\theta}{d\chi^2} + 3\chi^2\frac{d\theta}{d\chi} = 0$$ (6.30)

The first integration gives:

$$\frac{d\theta}{d\chi} = A_0 e^{-\chi^3}$$

where A_0 is a constant of integration. This equation can be again integrated:

$$\theta = A_0 \int e^{-\chi^3}d\chi + B_0$$

The two constants A_0 and B_0 are evaluated by applying the boundary conditions for dimensionless temperature θ. When $\chi = 0$, $\theta = 0$ and $\chi \to \infty$, $\theta = 1$.

Thus, substituting the value of $\theta = 0$, when $\chi = 0$, it gives $B_0 = 0$. Similarly, substituting $\chi \to \infty$ and $\theta = 1$:

$$A_0 = \frac{1}{\int_0^\infty e^{-\chi^3}d\chi}$$

The definite integral in the denominator is identified to be a gamma function, $\Gamma(4/3)$. Therefore, the dimensionless temperature θ is obtained as:

$$\theta = \frac{1}{\Gamma(4/3)} \int_0^\chi e^{-\chi^3} d\chi \tag{6.31}$$

The local value of the Nusselt number, in turn, is deduced by writing the rate of transfer of heat at the tube wall as:

$$h(T_1 - T_0) = -k \left.\frac{\partial T}{\partial r}\right|_{r=R} \tag{6.32}$$

which can be re-arranged in terms of dimensionless variables:

$$
\begin{aligned}
Nu_z &= \frac{h_z D}{k} = 2 \left.\frac{\partial \theta}{\partial \xi}\right|_{\xi=0} \\
&= 2 \left.\frac{d\theta}{d\chi}\right|_{\chi=0} \left(\frac{\partial \chi}{\partial \xi}\right)
\end{aligned}
\tag{6.33}
$$

Substitution of equations (6.29) and (6.31) in equation (6.33) gives:

$$
\begin{aligned}
Nu_z &= \frac{2}{\Gamma(4/3)} \left[\left(\frac{3n+1}{n}\right) \frac{VR^2}{9\alpha Z}\right]^{1/3} \\
&= 1.167 \left(\frac{3n+1}{4n}\right)^{1/3} Gz_z^{1/3}
\end{aligned}
\tag{6.34}
$$

where Gz_z relates to an arbitrary axial distance z instead of the tube length L. The average value of the Nusselt number is obtained as:

$$Nu = \frac{1}{L} \int_0^L Nu_z dz$$

which when evaluated yields:

$$Nu = 1.75 \Delta^{1/3} Gz^{1/3} \tag{6.35}$$

where

$$\Delta = \left(\frac{3n+1}{4n}\right) \tag{6.36}$$

When n is put equal to unity, equation (6.36) correctly reduces to its Newtonian form, $Nu = 1.75\, Gz^{1/3}$.

Hirai (1959) and Schechter and Wissler (1960) have extended this approach to include Bingham plastic fluids; the factor Δ is then given by:

$$\Delta = \frac{3}{(3 - \phi - \phi^2 - \phi^3)} \tag{6.37}$$

where $\phi = \tau^B_0/\tau_w$, the ratio of the yield stress to the wall shear stress. Indeed, Pigford (1955) has asserted that equation (6.35) is applicable to any type of time-independent fluid provided that n', the apparent flow behaviour index, replaces n in equation (6.35). For power-law shear-thinning fluids, equations (6.34) and (6.35) seem to be valid for $Gz > 100$. Blackwell (1985) improved upon the approximate analysis of Schechter and Wissler (1960) by including additional eigen values, especially in the entrance region of the pipe.

6.5.2 Experimental results and correlations

The experimental studies on heat transfer to/from purely viscous fluids in laminar flow in circular tubes have been critically reviewed in many publications (Porter, 1971; Cho and Hartnett, 1982; Irvine and Karni, 1987; Shah and Joshi, 1987; Hartnett and Kostic, 1989). Metzner *et al.* (1957) found it necessary to modify equation (6.35) to account for the temperature dependence of the consistency index as:

$$Nu = 1.75\Delta^{1/3}Gz^{1/3}\left(\frac{m'_b}{m'_w}\right)^{0.14} \tag{6.38}$$

where m'_b and m'_w, respectively, are the apparent consistency coefficients $\{m' = m((3n' + 1)/4n')^{n'}\}$ at the bulk temperature of the fluid and at the wall temperature. They asserted that this equation applies to all types of time-independent fluids provided that the local value of n' is used in evaluating Δ and in the viscosity correction term. Extensive comparisons between the predictions of equation (6.38) and experimental results suggest that it is satisfactory for $1 > n' > 0.1$ and $Gz > 20$. Subsequently (Metzner and Gluck, 1960), equation (6.38) has been modified to take into account free-convection effects:

$$Nu = 1.75\Delta^{1/3}\left[Gz + 12.6\left\{(PrGr)_w\left(\frac{D}{L}\right)\right\}^{0.4}\right]^{1/3}\left(\frac{m'_b}{m'_w}\right)^{0.14} \tag{6.39}$$

where the Grashof and Prandtl numbers are defined as:

$$Gr = \frac{\beta\Delta TD^3\rho^2 g}{\mu^2_{eff}} \tag{6.39a}$$

$$Pr = \frac{C_p\mu_{eff}}{k} \tag{6.39b}$$

The thermo-physical properties including the effective viscosity are evaluated at the wall conditions of shear rate and temperature. For a power-law fluid therefore the effective viscosity is evaluated at the shear rate of $\{(3n + 1)/4n\}(8V/D)$. However, Oliver and Jenson (1964) found that equation (6.39) underpredicted their results on heat transfer to carbopol solutions in 37 mm diameter tubes and that there was no effect of the (L/D) ratio. They correlated their results as $(0.24 \leq n' \leq 0.87)$:

$$Nu = 1.75[Gz + 0.0083(PrGr)^{0.75}_w]^{1/3}\left(\frac{m'_b}{m'_w}\right)^{0.14} \tag{6.40}$$

where the effective viscosity used in calculating the Prandtl and Grashof numbers is evaluated at the wall conditions.

The limited data on heat transfer to Bingham plastic suspensions of thoria (Thomas, 1960) in laminar flow seem to be well correlated by equations (6.35) and (6.37) except that a slightly different numerical constant must be used. Skelland (1967) has put it in a more convenient form as:

$$j_H = \left(\frac{h_m}{\rho C_p V}\right) Pr^{2/3} \left(\frac{\mu_w}{\mu_b}\right)^{0.4} \left(\frac{L}{D}\right)^{1/3} \Delta^{1/3} = 1.86 Re_B^{-1/3} \tag{6.41}$$

where the Reynolds number is based on the mean plastic viscosity of the suspension. The density and heat capacity were taken as weighted averages, and the thermal conductivity was estimated using equation (6.1).

Example 6.2

A 0.2% aqueous carbopol solution at 25°C is flowing at a mass flow rate of 200 kg/h through a 30 mm diameter copper tube prior to entering a 2 m long heated section. The initial unheated section is sufficiently long for the velocity profile to be fully established. The heated section is surrounded by a jacket in which steam condenses at a pressure of 70 kPa (saturation temperature 90°C). Estimate the mean heat transfer coefficient and the bulk temperature of the fluid leaving the heating section. Compare the predicted values given by the various equations/correlations presented in the preceding section.

Physical properties of carbopol solution:
   Density and specific heat as for water, 1000 kg/m^3 and 4.2 kJ/kg K, respectively.
   Thermal conductivity = 0.56 W/mK
   Power-law index, $n = 0.26$, applies in the range $15 \le T \le 85°C$; power-law consistency coefficient, m (Pa s^n) = 26 − 0.0566 T (288 $\le T < 363$ K) where T is in K.

Solution

The Graetz number for the given conditions will be calculated first.

$$Gz = \frac{\dot{m} C_p}{kL} = \frac{(200/3600) \times 4.2 \times 1000}{0.56 \times 2} = 208$$

Since this value is greater than 100, and it is not necessary to take account of the temperature dependence of viscosity, equation (6.35) may be used:

$$Nu = 1.75\Delta^{1/3}Gz^{1/3}$$
$$= 1.75\left(\frac{3 \times 0.26 + 1}{4 \times 0.26}\right)^{1/3} 208^{1/3}$$
$$= 12.40$$

and the heat transfer coefficient, $h = 12.40 \times \dfrac{k}{D}$

$$= \frac{12.40 \times 0.56}{0.03} = 232 \text{ W/m}^2\text{K}$$

The arithmetic mean of the temperature difference between the wall and the fluid may, as a first approximation, be used to calculate the rate of heat transfer:

At inlet, $\Delta T_1 = T_s - T_1 = 90 - 25 = 65°C$

At exit, $\Delta T_2 = T_s - T_2 = 90 - T_2$

$$\therefore \qquad \Delta T_{mean} = \frac{\Delta T_1 + \Delta T_2}{2} = \frac{65 + 90 - T_2}{2} = 77.5 - \frac{1}{2}T_2$$

Neglecting the resistance of the film on the steam side and of the copper tube wall (high thermal conductivity), the overall heat transfer coefficient, $U \approx h = 232\,\text{W/m}^2\text{K}$

From a heat balance on the fluid:

$$\dot{m}C_p(T_2 - 25) = UA\Delta T$$

Substituting values:

$$\frac{200}{3600} \times 4200 \times (T_2 - 25) = 232 \times 3.14 \times 0.03 \times 2 \times \left(77.5 - \frac{1}{2}T_2\right)$$

Solving: $T_2 = 36.1°C$.

The temperature of the fluid leaving the heating section is therefore 36.1°C.

It is important next to establish the influence of temperature dependence of the consistency index. Equation (6.38) will be used since the values of both n and the Graetz number lie within its range of validity. In this example, since the value of n is constant,

$$\frac{m_b'}{m_w'} = \left(\frac{m_b}{m_w}\right)$$

Strictly speaking, as the outlet temperature of the fluid is not known, a trial and error solution is required. Assuming the outlet fluid temperature to be 36.1°C, as calculated above, the mean fluid temperature is $(25 + 36.1)/2$, i.e. 30.55°C and the value of m is calculated as:

$$m_b = 26 - 0.0566T = 26 - 0.0566(273 + 30.55)$$
$$= 8.82 \text{ Pa } s^n$$

Similarly, $m_w = 26 - 0.0566(273 + 90)$

$$= 5.45 \text{ Pa } s^n$$

Substituting the values in equation (6.38):

$$Nu = 1.75\left(\frac{3 \times 0.26 + 1}{4 \times 0.26}\right)^{1/3} 208^{1/3}\left(\frac{8.82}{5.45}\right)^{0.14} = 13.26$$

and $h = 248\,\text{W/m}^2\text{K}$.

Under these conditions, the viscosity correction is small (only about 7%). The outlet temperature of the fluid in this case is found to be 36.8°C which is sufficiently close to the assumed value of 36.1°C for a second iteration not to be needed.

Finally, the contribution of free convection may be evaluated using equations (6.39) and (6.40) The wall shear rate for a power-law fluid is:

$$\dot{\gamma}_{wall} = \left(\frac{3n+1}{4n}\right)\left(\frac{8V}{D}\right)$$

The mean velocity of flow, $V = \dfrac{4\dot{m}}{\rho\pi D^2} = \dfrac{4 \times 200/3600}{1000 \times 3.14 \times (30 \times 10^{-3})^2}$

$$= 0.0785\,\text{m/s}$$

$$\therefore \dot{\gamma}_{\text{wall}} = \left(\frac{3 \times 0.26 + 1}{4 \times 0.26}\right)\left(\frac{8 \times 0.0785}{30 \times 10^{-3}}\right) = 35.8\,\text{s}^{-1}$$

The effective viscosity at the wall shear rate and temperature is:

$$\mu_{\text{eff}} = m(\dot{\gamma}_{\text{wall}})^{n-1}$$

The value of m at wall temperature (90°C) is 5.45 Pa s^n

$$\therefore \quad \mu_{\text{eff}} = 5.45(35.8)^{0.26-1} = 0.386\,\text{Pa s}$$

Grashof number, $Gr = (g\beta\Delta T D^3 \rho^2)/\mu_{\text{eff}}^2$

The coefficient of thermal expansion of the carbopol solution is assumed to be same as that for water, and the mean value in this temperature range is $0.000302\,\text{K}^{-1}$.
Substituting values:

$$Gr = \frac{0.000302 \times 9.81 \times (90 - 25) \times (30 \times 10^{-3})^3 \times 1000^2}{0.386^2}$$

$$= 31.7$$

and Prandtl number, $Pr = \dfrac{C_p\mu_{\text{eff}}}{k} = \dfrac{4200 \times 0.386}{0.56}$

$$= 2890$$

Now substituting the above values in equation (6.39):

$$Nu = 1.75 \times \left(\frac{0.26 \times 3 + 1}{4 \times 0.26}\right)^{1/3}$$

$$\times \left[208 + 12.6\left\{2890 \times 31.7 \times \frac{30 \times 10^{-3}}{2}\right\}^{0.4}\right]^{1/3}\left(\frac{8.82}{5.45}\right)^{0.14}$$

$$= 17$$

$$\therefore h = \frac{17 \times 0.56}{30 \times 10^{-3}} = 317\,\text{W/m}^2\text{K}$$

The resulting outlet temperature of the polymer solution in this case is 39.8°C which is significantly different from the assumed value of 36.1°C used in evaluating the viscosity correction term. However, the latter has an exponent of 0.14 which gives rise to a change in the value of h of less than 0.2%.
Similarly, equation (6.40) gives:

$$Nu = 13.85 \quad \text{and} \quad h = 258\,\text{W/m}^2\text{K}$$

The outlet fluid temperature is then about 36.2°C.

Using the minimum value of the consistency coefficient, which will apply in the wall region where the temperature is 90°C:

$$Re_{MR} = \frac{\rho V^{2-n} D^n}{8^{n-1} \left(\dfrac{3n+1}{4n} \right)^n m}$$

$$= \frac{1000 \times 0.0785^{2-0.26} \times (30 \times 10^{-3})^{0.26}}{8^{0.26-1} \left(\dfrac{0.26 \times 3 + 1}{4 \times 0.26} \right)^{0.26} \times 5.45}$$

$$= 3.57 \ll 2100$$

The flow is thus laminar.

Equations (6.35), (6.38) and (6.40) all give comparable results while equation (6.39) yields a much higher value of *h*. From design view point, it is desirable to establish upper and lower bounds on the value of the heat transfer coefficient and hence on the required heat transfer area. There does not appear to be much information available as to when natural convection effects become significant. Using the criterion that it should be so for $Gr/Re^2_{MR} > 1$, as for Newtonian fluids, natural convection effects will be important. □

6.6 Constant heat flux at tube wall

6.6.1 Theoretical treatments

In this case, the energy balance given by equation (6.12) is still applicable; however, the boundary conditions are amended as follows:

At the tube wall,

$$r = R, \quad -k\frac{\partial T}{\partial r} = q_0 \text{ (constant)} \quad \text{(for } z \geq 0) \tag{6.42a}$$

At the centre of the tube,

$$r = 0, \quad \frac{\partial T}{\partial r} = 0 \quad \text{(for } z \geq 0) \tag{6.42b}$$

and the inlet condition,

$$z = 0, T = T_1 \quad \text{(for all values of } r \leq R). \tag{6.42c}$$

The use of the Leveque approximation in this instance leads to the following expression for the mean Nusselt number over the entry length (Bird, 1959):

$$Nu = 2.11\Delta^{1/3}Gz^{1/3} \tag{6.43}$$

Similarly, the Nusselt number under fully developed thermal conditions is given by (Bird, 1959):

$$Nu_\infty = \frac{8(3n+1)(5n+1)}{(31n^2 + 12n + 1)} \tag{6.44}$$

On the other hand, Grigull (1956) estimated the value of Nu_∞ by modifying the value for Newtonian fluids by applying a factor of $\Delta^{1/3}$:

i.e. $$Nu_\infty = 4.36\Delta^{1/3} \qquad (6.45)$$

where $\Delta = [(3n + 1)/4n]$

For $n = 1$, both equations (6.44) and (6.45) reduce to the correct value $Nu_\infty = 48/11 = 4.36$. For $n = 0.1$, the two predictions differ by about 6.5%. Interestingly enough, the correction is smaller for shear-thickening fluids, e.g. for $n \to \infty$, $\Delta^{1/3} = 0.908$.

Similar results for square and triangular ducts are also available in the literature (Irvine and Karni, 1987; Chhabra, 1999; Irvine and Capobianchi, 2000; Park and Lee, 2001; Park and Lee, 2001, 2003).

6.6.2 Experimental results and correlations

For the constant heat flux condition at the tube wall, most investigators have expressed the local values of the Nusselt number as a function of the corresponding Graetz number. Cho and Hartnett (1982) concluded that for small values of temperature difference between the liquid and the tube wall, the radial variation in apparent viscosity is not significant, and most available experimental results are well correlated by the approximate isothermal solution, equation (6.43), due to Bird (1959). Indeed, even the data obtained with weakly visco-elastic solutions of polyacrylamide and polyethylene oxide are in line with equation (6.43). However, when an appreciable temperature difference exists, the temperature dependence of the viscosity and natural convection effects can no longer be neglected. For these conditions, Cho and Hartnett (1982) have recommended using the following equation due to Mahalingam *et al.* (1975):

$$Nu_z = 1.46\Delta^{1/3}[Gz + 0.0083(GrPr)_w^{0.75}]^{1/3}\left(\frac{m_b}{m_w}\right)^{0.14} \qquad (6.46)$$

where the subscripts, z, b, w refer, respectively, to an arbitrary value of z (length in the direction of flow), bulk and wall conditions; the Grashof and Prandtl numbers are defined using the apparent viscosity at the shear rate of $(8\Delta V/D)$ in equations (6.39a) and (6.39b). Equation (6.46) covers the following ranges of conditions: $0.24 \le n \le 1$; $2.5 \le q_0 \le 40\,\text{kW/m}^2$; $200 \le Gz \le 10000$ and $750 \le Gr \le 4000$. Furthermore, Mahalingam *et al.* (1975) have suggested that natural convection effects are negligible for $(Gr/Re_{MR}) < 1$.

Example 6.3

The polymer solution referred to in Example 6.2 is to be heated from 15°C to 25°C in a 25 mm diameter 1800 mm long steel tube at the rate of 100 kg/h. The steel tube is wrapped with an electrical heating wire so that a constant heat flux is maintained on the tube wall. Estimate the rate of heat transfer to the solution, the heat flux at the wall and the temperature of the tube wall at the exit end of the tube. The physical properties are given in Example 6.2.

Solution

Initially, the temperature dependence of the fluid consistency coefficient m will be neglected so that equation (6.43) can be used. For this case,

$$Gz = \frac{(100/3600) \times 4200}{0.56 \times 1.8} = 115.7$$

For $n = 0.26$ (applicable for T not exceeding 85°C),

$$\Delta = (3n + 1)/(4n) = (3 \times 0.26 + 1)/(4 \times 0.26) = 1.71$$

Using equation (6.43),

$$Nu = \frac{hD}{k} = 2.11\Delta^{1/3}Gz^{1/3}$$

or $\quad h = 2.11 \times 1.71^{1/3} \times 115.7^{1/3} \times \dfrac{0.56}{25 \times 10^{-3}} = 271 \, \text{W/m}^2\text{K}$

From a heat balance,

$$Q = \dot{m}C_p(T_{\text{out}} - T_{\text{in}})$$

$$= \frac{100}{3600} \times 4200 \times (25 - 15) = 1167 \, \text{W}$$

Also, $\quad q_0 \pi DL = Q$

$$\therefore q_0 = \frac{1167}{\pi \times 0.025 \times 1.8} = 8253 \, \text{W/m}^2 \text{ or } 8.25 \, \text{kW/m}^2$$

$$= h(T_{\text{w}} - T_{\text{b}})$$

For $T_{\text{b}} = 25°C$

$$T_{\text{w}} = \frac{8253}{271} + 25 = 55.5°C \tag{i}$$

The role of natural convection can now be ascertained by using equation (6.46). Initially, the above estimate of the wall temperature will be used to evaluate the power-law consistency coefficient. For the wall temperature of 55.5°C, $m = 26 - 0.0566(273 + 55.5) = 7.41 \, \text{Pa s}^n$ and for the bulk temperature of 25°C, $m = 9.13 \, \text{Pa s}^n$.

$$\dot{\gamma}_{\text{wall}} = \left(\frac{3n + 1}{4n}\right)\left(\frac{8V}{D}\right),$$

where $\quad V = \dfrac{100}{3600 \times 1000} \times \dfrac{4}{\pi \times 0.025^2} = 0.057 \, \text{m/s}$

$$\therefore \dot{\gamma}_{\text{wall}} = \left(\frac{3n + 1}{4n}\right)\left(\frac{8V}{D}\right)$$

$$= \left(\frac{3 \times 0.26 + 1}{4 \times 0.26}\right)\left(\frac{8 \times 0.057}{25 \times 10^{-3}}\right) = 15.92 \, \text{s}^{-1}$$

$$\therefore \mu_{\text{eff wall}} = m(\dot{\gamma}_{\text{wall}})^{n-1} = 7.41(15.92)^{0.26-1} = 0.96 \, \text{Pa s}$$

At the wall conditions,

$$Pr = \frac{C_p \mu_{eff}}{k} = \frac{4200 \times 0.96}{0.56} = 7170$$

$$Gr = \frac{\beta g \Delta T D^3 \rho^2}{\mu_{eff}^2}$$

$$= \frac{0.000302 \times 9.81 \times (55.5 - 25) \times (25 \times 10^{-3})^3 \times 1000^2}{0.96^2}$$

$$= 1.53$$

Note that this value is outside the range of the validity of equation (6.46), so its use here is only tentative.

$$Nu = \frac{hD}{k} = 1.46(1.71)^{1/3}[115.7 + 0.0083(7170 \times 1.53)^{0.75}]^{1/3}\left(\frac{9.13}{7.14}\right)^{0.14}$$

$$= 9$$

$$\therefore h = \frac{9 \times 0.56}{25 \times 10^{-3}} = 201 \, W/m^2 K$$

At the exit of the tube,

$$q_0 = h(T_w - T_b) = 201 \times (T_w - 25)$$

For $q_0 = 8.25 \, kW/m^2$; $T_w = 66°C$

This value is substantially different from the assumed value of 55.5°C (from equation (i)). Another iteration is carried out by assuming $T_w = 64°C$, at which $m = 6.93 \, Pa \, s^n$
Again, at $\dot{\gamma}_{wall} = 15.92 \, s^{-1}$

$$\mu_{eff} = m(\dot{\gamma}_{wall})^{n-1} = 6.93(15.92)^{0.26-1} = 0.894 \, Pa \, s$$

The new values of the Prandtl and Grashof numbers are

$$Pr = 6705 \quad \text{and} \quad Gr = 2.25$$

Using equation (6.46), $Nu = 9.12$

$$\therefore h = \frac{9.12 \times 0.56}{0.025} = 204 \, W/m^2 K$$

At the tube exit, $q_0 = h(T_w - T_b)$
Substituting values, $q_0 = 8250 \, W/m^2$; $h = 204 \, W/m^2 K$ and $T_b = 25°C$,

$$T_w = 65.4°C$$

Though this value is fairly close to the assumed value of 64°C, a second iteration yields $T_w = 65°C$. Thus, if the natural convection effects are ignored, the wall temperature at the end is found to be 55.5°C (see equation (i)) which is some 10°C lower than the value of 65°C. In a sense, this discrepancy also reflects the inherent inaccuracy of the predictive equations in this field.

The value of Re_{MR} (using the apparent viscosity near the wall) is calculated and this confirms that the flow is streamline. Thus,

$$Re_{MR} = \frac{\rho V^{2-n} D^n}{8^{n-1} \left(\dfrac{3n+1}{4n}\right)^n m} = \frac{1000 \times 0.057^{2-0.26} \times 0.025^{0.26}}{8^{0.26-1} \times 1.71^{0.26} \times 6.93}$$

$$= 1.53$$

Hence the flow in the tube is indeed laminar. □

6.7 Effect of temperature-dependent physical properties on heat transfer

The theoretical treatments considered so far have been based on the assumption that the thermo-physical properties are constant (i.e. independent of temperature and therefore the velocity profiles do not change over the heat transfer section of the tube). Christiansen and Craig (1962) investigated the effect of temperature-dependent power-law viscosity on the mean values of Nusselt number for streamline flow in tubes with constant wall temperature. They postulated that the flow behaviour index, n was constant and that the variation of the consistency coefficient, m, with temperature could be represented by equation (6.47) giving:

$$\tau_{rz} = m_0 \left[\left(-\frac{dV_z}{dr} \right) \exp\left(\frac{E}{\mathbf{R}T} \right) \right]^n \tag{6.47}$$

where m_0 and E are the pre-exponential factor and the activation energy of viscous flow; their values can be estimated by making rheological measurements at different temperatures covering the range of application. All other physical properties were assumed to be temperature independent. Figure 6.5 shows representative results for the mean values of Nusselt number, Nu, as a function of the Graetz number for $n = 1/3$.

The sign and magnitude of the dimensionless parameter, $\psi = (E/\mathbf{R})(1/T_1 - 1/T_0)$ depends upon the direction of heat transfer and the sensitivity of m to temperature. For a cold fluid entering the tube, ψ would be positive. This analysis predicts an enhancement in the rate of heat transfer which is qualitatively consistent with the limited data presented by Christiansen and Craig (1962). The thermal conductivity must be evaluated at the mean film temperature, i.e. $\frac{1}{2}\left[T_0 + \frac{1}{2}(T_1 + T_2)\right]$ where T_0, T_1, T_2 are the temperatures of the tube wall and of the fluid at inlet and outlet, respectively. Finally, the effect of the flow behaviour index is found to be small.

On the other hand, Forrest and Wilkinson (1973, 1974) modelled the temperature dependence of the power-law consistency coefficient as:

$$m = m_0 \{1 + \beta_w (T - T_1)\}^{-n} \tag{6.48}$$

where β_w is the temperature coefficient of viscosity. For the heating and cooling of power-law fluids in streamline flow in tubes with walls at a constant temperature, representative results from their study are shown in Figure 6.6, for a range of values β_w and $\phi = (T_0/T_1)$, which are qualitatively similar to those shown in Figure 6.5.

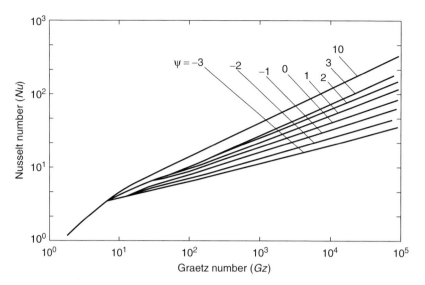

Figure 6.5   *Computed Nusselt number versus Graetz number for n = 1/3*

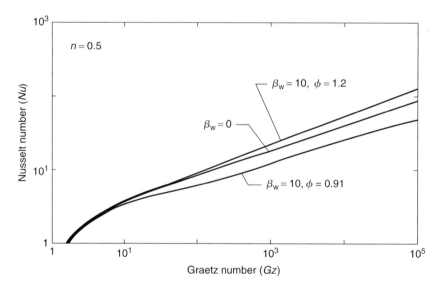

Figure 6.6   *Nusselt number versus Graetz number for heating and cooling of power-law fluids (n = 0.5)*

The enhancement of heat transfer at an isothermal tube wall due to temperature-dependent viscosity has been correlated by Kwant *et al.* (1973) as follows:

$$\frac{Nu_{VP}}{Nu} = 1 + 0.271 \ln \phi + 0.023 (\ln \phi)^2 \tag{6.49}$$

where Nu_{VP} and Nu are, respectively, the Nusselt numbers for the cases where consistencies are temperature dependent and independent, respectively. Physical properties are evaluated at the mean film temperature. A similar level of enhancement in heat transfer is obtained in viscoplastic media (Nouar *et al.*, 1994; Nouar, 2005).

To summarize, the heat transfer rate is enhanced as a result of the temperature dependence of the power-law consistency coefficient during heating ($\phi \geq 1$ or $\psi \geq 0$) while it is reduced during cooling of the fluid ($\phi < 1$, $\psi < 0$). This effect is much more pronounced for the case of constant wall temperature than that for the constant wall heat flux condition. The work of other investigators on the role of temperature-dependent viscosity has been reviewed by Lawal and Mujumdar (1987) and Chhabra (1999).

6.8 Effect of viscous energy dissipation

In the flow of all fluids, mechanical energy is degraded into heat and this process is called viscous dissipation. The effect may be incorporated into the thermal energy balance by adding a source term, S_V (per unit volume of fluid) to the right hand side of equation (6.12). Its magnitude depends upon the local velocity gradient and the apparent viscosity of the fluid. Although, in general, the viscous dissipation includes contributions from both shearing and normal stresses, but under most conditions the contribution of shearing components outweighs that of normal stress components. Thus, it can readily be seen that the viscous dissipation term, S_V, is equal to the product of the shear stress and the shear rate. Thus, for example, in the streamline flow of a power-law fluid in a circular tube, it is of the form:

$$S_V = \tau_{rz}\left(-\frac{dV_z}{dr}\right) = m\left(-\frac{dV_z}{dr}\right)^{n+1} \tag{6.50}$$

For fully developed isothermal flow, the velocity gradient is obtained by differentiating the expression for velocity (equation (3.10)) to give:

$$-\frac{dV_z}{dr} = \frac{V}{R}\left(\frac{3n+1}{n}\right)\left(\frac{r}{R}\right)^{1/n} \tag{6.51}$$

and thus the source term, S_V, becomes:

$$S_V = m\left\{\frac{V}{R}\left(\frac{3n+1}{n}\right)\right\}^{n+1}\left(\frac{r}{R}\right)^{(n+1)/n} \tag{6.52}$$

where V is the mean velocity of flow.

Evidently, for the limiting case of a shear-thinning fluid (Newtonian fluid $n = 1$), the velocity gradient is a maximum and hence, the viscous heat generation is also maximum. The effect of this process on the developing temperature profile can be illustrated by considering the situation depicted in Figure 6.7.

Here, a Newtonian fluid initially at a uniform temperature, T_0, enters a tube whose walls are also maintained at the same temperature T_0. As the fluid proceeds down the tube, heat is produced within the fluid at the rate given by equation (6.52). Near the entrance of the tube, the maximum temperature is obtained closed to the wall. Due to the conduction of heat in the radial direction (towards the centre), the temperature profile evolves

Figure 6.7   *Velocity (left) and temperature (right) profiles for the flow of a Newtonian fluid in a tube*

progressively, eventually the maximum temperature being reached at the centre of the tube (Figure 6.7). Obviously, viscous dissipation effects will be significant for the flow of high consistency materials (such as polymer melts) and/or where velocity gradients are high (such as in lubrication and rolling processes).

Viscous dissipation can be quantified in terms of the so-called Brinkman number, Br, which is defined as the ratio of the heat generated by viscous action to that dissipated by conduction. Thus for streamline flow in a circular tube (on the basis of per unit volume of fluid):

$$\text{heat generated by viscous action} \sim m(V/R)^{n+1}$$
$$\text{and the heat conducted in radial direction} \sim (k\Delta T)/(R^2)$$

In this case, the Brinkman number, Br is:

$$Br = \frac{m(V/R)^{n+1}}{k(\Delta T/R^2)} \qquad (6.53)$$

Clearly, the viscous dissipation effects will be significant whenever the Brinkman number is much greater than unity.

While the rigorous solutions of the thermal energy equation are quite complex (Min *et al.*, 1997; Min and Yoo, 1999), some useful insights can be gained by qualitative considerations of the results of Forrest and Wilkinson (1973, 1974) and Dinh and Armstrong (1982) amongst others or from the review papers (Winter, 1977; Lawal and Mujumdar, 1987).

6.9 Heat transfer in transitional and turbulent flow in pipes

Although turbulent flow gives higher values of the heat transfer coefficient, it frequently cannot be achieved in practice with non-Newtonian polymer solutions and particulate suspensions. However, the turbulent flow of the so-called drag reducing polymer solutions has been the subject of several studies and Cho and Hartnett (1982) have reviewed the literature on heat transfer to these dilute solutions. They have clearly shown that the results reported by different workers often do not agree and may differ by more than an order of magnitude under nominally identical conditions of flow. The rheological and heat transfer characteristics of these fluids are known to be extremely sensitive to the method of preparation, chemistry of solvent, etc. Furthermore these solutions show appreciable shear degradation and few reliable data are available on their heat transfer coefficients (Cho and Hartnett, 1982). Yoo (1974) has collated most of the available data on heat transfer to purely viscous fluids in turbulent region and has proposed the following correlation:

$$St = \frac{h}{\rho V C_p} = 0.0152(Re_{MR})^{-0.155} Pr^{-2/3} \tag{6.54}$$

where the effective viscosity used in the Reynolds and Prandtl numbers is evaluated at the wall shear rate of $[(3n + 1)/4n](8V/D)$ for a power-law fluid. Under fully turbulent flow conditions in a pipe, the yield stress does not seem to influence the rate of heat transfer and one can therefore use equation (6.54) for viscoplastic fluids also (Peixinho *et al.*, 2007).

Equation (6.54) correlates the data over wide ranges of conditions ($0.24 \leq n \leq 0.9$ and $3000 \leq Re_{MR} \leq 90000$) with an average error of $\pm 3\%$. On the other hand, the heat transfer data for particulate suspensions ($0.42 \leq n \leq 0.89$) seems to be well represented by:

$$j_{\mathrm{H}} = \frac{h}{\rho V C_p} Pr^{2/3} \left(\frac{\mu_{\mathrm{w}}}{\mu_{\mathrm{b}}} \right)^{0.14} = 0.027 Re_{\mathrm{B}}^{-0.2} \tag{6.55}$$

where $Re_{\mathrm{B}} = \rho VD/\mu_{\mathrm{B}}$.

Other similar correlations are available in the literature (Quader and Wilkinson, 1981), none of which, however, has been validated using independent data.

Thus, it is concluded that although adequate information is available on laminar heat transfer to purely viscous fluids in circular tubes, further work is needed in turbulent regime, particularly with visco-elastic fluids.

Further reading

Carreau, P.J., Dekee, D. and Chhabra, R.P., *Rheology of Polymeric Systems: Principles and Applications*. Hanser, Munich (1997).

Chhabra, R.P., *Advances in the Rheology and Flow of Non-Newtonian Fluids Part B* (edited by Siginer, D., Dekee, D. and Chhabra, R.P.), Elsevier, Amsterdam (1999), Chapter 39.

Cho, Y.I. and Hartnett, J.P., *Adv. Heat Transf.* **15** (1982) 59.

Hartnett, J.P. and Kostic, M., *Adv. Heat Transf.* **19** (1989) 247.

Kakac, S., Shah, R.K. and Aung, W., Eds, *Handbook of Single Phase Convective Heat Transfer*, Wiley, New York (1987).

Lawal, A. and Mujumdar, A.S., *Adv. Transport Proc.* **5** (1987) 352.

Rohsenow, W.M., Hartnett, J.P. and Cho, Y.I., Eds, *Handbook of Heat Transfer*, 3rd edn, McGraw-Hill, New York (1998).

References

Bird, R.B., *Chem. Ing. Tech.* **31** (1959) 569.

Bird, R.B., Armstrong, R.C. and Hassager, O., *Dynamics of Polymeric Liquids. Vol. 1 Fluid Dynamics*, 2nd edn, Wiley, New York (1987).

Blackwell, B.F., *J. Heat Transf. (ASME)* **107** (1985) 466.

Brandrup, J. and Immergut, E.H., Eds, *Polymer Handbook*, 3rd edn., Wiley, New York (1989).

Bruggemann, D.A.G., *Ann. Phys. (Leipzig)* **24** (1935) 636.

Chhabra, R.P., *Advances in the Flow and Rheology of Non-Newtonian Fluids* (edited by Siginer, D.A., DeKee, D. and Chhabra, R.P.), Elsevier, Amsterdam (1999), Chapter 39.
Chhabra, R.P., *Adv. Heat Transf.* **37** (2003) 77.
Cho, Y.I. and Hartnett, J.P., *Adv. Heat Transf.* **15** (1982) 59.
Christiansen, E.B. and Craig, S.E., *AIChE J.* **8** (1962) 154.
Dinh, S.M. and Armstrong, R.C., *AIChE J.* **28** (1982) 294.
Domininghaus, H., *Plastics for Engineers: Materials, Properties and Applications*, Hanser, Munich (1993).
Dutta, A. and Mashelkar, R.A., *Adv. Heat Transf.* **18** (1987) 161.
Ferry, J.D., *Visco-elastic Properties of Polymers*, 3rd edn., Wiley, New York (1980).
Forrest, G. and Wilkinson, W.L., *Trans. Inst. Chem. Engrs.* **51** (1973) 331.
Forrest, G. and Wilkinson, W.L., *Trans. Inst. Chem. Engrs.* **52** (1974) 10.
Grigull, U., *Chem. Ing. Tech.* 28 (1956) **553**, 655.
Hartnett, J.P. and Cho, Y.I., *Handbook of Heat Transfer* (edited by Rohsenow, W.M., Hartnett, J.P. and Cho, Y.I.), 3rd edn., McGraw Hill, New York (1998).
Hartnett, J.P. and Kostic, M., *Adv. Heat Transf.* **19** (1989) 247.
Hirai, E., *AIChE J.* **5** (1959) 130.
Irvine, T.F., Jr. and Karni, J., *Handbook of Single Phase Convective Heat Transfer* (edited by Kakac, S., Shah, R.K. and Aung, W.), Wiley, New York (1987), Chapter 5.
Irvine, T.F., Jr., Kim, I., Cho, K. and Gori, F., *Exp. Heat Transf.* **1** (1987) 155.
Irvine, T.F., Jr. and Capobianchi, M., *The CRC Handbook of Thermal Engineering*, CRC Press, Boca Raton, FL (2000), pp.3.56–3.64.
Johnston, P.R., *Int. J. Heat Mass Transf.* **34** (1991) 1209.
Kwant, P.B., Zwaneveld, A. and Dijkstra, F.C., *Chem. Eng. Sci.* **28** (1973) 1303.
Lawal, A. and Mujumdar, A.S., *Adv. Transport Proc.* **5** (1987) 352.
Leveque, J., *Ann. Mines* 13 (1928) **201**, 305, 381.
Lin, S.X.Q., Chen, X.D., Chen, Z.D. and Bandopadhyay, R., *J. Food Eng.* **57** (2003) 217.
Loulou, T., Peerhossaini, H. and Bardon, J.P., *Int. J. Heat Mass Transf.* **35** (1992) 2557.
Lyche, B.C. and Bird, R.B., *Chem. Eng. Sci.* **6** (1956) 34.
Mahalingam, R., Tilton, L.O. and Coulson, J.M., *Chem. Eng. Sci.* **30** (1975) 921.
Metzner, A.B. and Gluck, D.F., *Chem. Eng. Sci.* **12** (1960) 185.
Metzner, A.B., Vaughn, R.D. and Houghton, G.L., *AIChE J.* **3** (1957) 92.
Min, T. and Yoo, J.Y., *J. Heat Transf. (ASME)* **121** (1999) 556.
Min, T., Yoo, J.Y. and Choi, H., *Int. J. Heat Mass Transf.* **40** (1997) 3025.
Nouar, C., *Int. J. Heat Mass Transf.* **48** (2005) 5520.
Nouar, C., Devienne, R. and Lebouche, M., *Int. J. Heat Mass Transf.* **37** (1994) 1.
Oliver, D.R. and Jenson, V.G., *Chem. Eng. Sci.* **19** (1964) 115.
Orr, C. and Dallavalle, J.M., *Chem. Eng. Prog. Sym. Ser.* **50**(#9) (1954) 29.
Park, S. and Lee, D.R., *Heat Mass Transfer.* 38 (2003) 141, Also see *ibid.* **39** (2003) 645.
Peixinho, J., Desaubry, C. and Lebouche, M., *Int. J. Heat Mass Transf.* **51** (2007) 198.
Pigford, R.L., *Chem. Eng. Prog. Sym. Ser.* **51**(#17) (1955) 79.
Porter, J.E., *Trans. Inst. Chem. Engrs.* **49** (1971) 1.
Quader, A.K.M.A. and Wilkinson, W.L., *Int. J. Multiphase Flow* **7** (1981) 545.
Rajaiah, J., Andrews, G., Ruckenstein, E. and Gupta, R.K., *Chem. Eng. Sci.* **47** (1992) 3863.
Schechter, R.S. and Wissler, E.H., *AIChE J.* **6** (1960) 170.
Shah, R.K. and Joshi, S.D., *Handbook of Single Phase Convective Heat Transfer* (edited by Kakac, S., Shah, R.K. and Aung, W.), Wiley, New York (1987), Chapter 5.
Skelland, A.H.P., *Non-Newtonian Flow and Heat Transfer*, Wiley, New York (1967).
Tareef, B.M., *Colloid, J. (U.S.S.R.)* **6** (1940) 545.
Thomas, D.G., *AIChE J.* **6** (1960) 632.
Yoo, S.S., PhD Thesis, University of Illinois at Chicago Circle, Chicago, IL (1974).
Winter, H.H., *Adv. Heat Transf.* **13** (1977) 205.

Nomenclature

		Dimensions in $\mathbf{M,L,T,\theta}$
Br	Brinkman number, equation (6.53) $(-)$	$\mathbf{M^0 L^0 T^0}$
C_p	specific heat (J/kg K)	$\mathbf{L^2 T^{-2} \theta^{-1}}$
D	pipe diameter (m)	$\mathbf{L}$
E	activation energy (J/mole)	$\mathbf{M L^2 T^{-2}}$
Gr	Grashof number, equation (6.39a) $(-)$	$\mathbf{M^0 L^0 T^0}$
Gz	Graetz number, equation (6.17) $(-)$	$\mathbf{M^0 L^0 T^0}$
g	acceleration due to gravity (m/s^2)	$\mathbf{L T^{-2}}$
h	heat transfer coefficient (W/m^{2}K)	$\mathbf{M T^{-3} \theta^{-1}}$
j_H	heat transfer factor, equation (6.41) $(-)$	$\mathbf{M^0 L^0 T^0}$
k	thermal conductivity (W/mK)	$\mathbf{M L T^{-3} \theta^{-1}}$
L	length (m)	$\mathbf{L}$
m	power-law consistency coefficient (Pa s^n)	$\mathbf{M L^{-1} T^{n-2}}$
m'	apparent power-law consistency coefficient (Pa s$^{n'}$)	$\mathbf{M L^{-1} T^{n'-2}}$
m_0	pre-exponential factor, equation (6.47) (Pa s^n)	$\mathbf{M L^{-1} T^{n-2}}$
$\dot{m}$	mass flow rate (kg/s)	$\mathbf{M T^{-1}}$
n	power-law flow behaviour index $(-)$	$\mathbf{M^0 L^0 T^0}$
n'	apparent power-law flow behaviour index $(-)$	$\mathbf{M^0 L^0 T^0}$
Nu	Nusselt number $(=hD/k)$ $(-)$	$\mathbf{M^0 L^0 T^0}$
Pr	Prandtl number, equation (6.39b) $(-)$	$\mathbf{M^0 L^0 T^0}$
q	heat flux (W/m^2)	$\mathbf{M T^{-3}}$
R	pipe radius (m)	$\mathbf{L}$
$\mathbf{R}$	universal gas constant (J/kmol K)	$\mathbf{M L^2 T^{-2} \theta^{-1}}$
r	radial coordinate (m)	$\mathbf{L}$
Re_B	Reynolds number $(=\rho V D/\mu_B)$ $(-)$	$\mathbf{M^0 L^0 T^0}$
Re_{MR}	Metzner–Reed Reynolds number $(= \rho V^{2-n} D^n / m \Delta^n 8^{n-1})$ $(-)$	$\mathbf{M^0 L^0 T^0}$
St	Stanton number, equation (6.54) $(-)$	$\mathbf{M^0 L^0 T^0}$
S_V	energy produced by viscous dissipation per unit volume of fluid (J/m^3)	$\mathbf{M L^{-1} T^{-2}}$
$V_z(r)$	local velocity in z-direction (m/s)	$\mathbf{L T^{-1}}$
V	average velocity of flow (m/s)	$\mathbf{L T^{-1}}$
T	temperature (K)	θ
T_1	fluid temperature at inlet (K)	θ
T_0	constant temperature of tube wall (K)	θ
z	axial coordinate (m)	$\mathbf{L}$

Greek letters

α	thermal diffusivity (m²/s)	$\mathbf{L^2 T^{-1}}$
β	coefficient of expansion (K⁻¹)	$\mathbf{\theta^{-1}}$
Δ	correction factor ($=(3n + 1)/4n$) ($-$)	$\mathbf{M^0 L^0 T^0}$
θ	dimensionless temperature ($-$)	$\mathbf{M^0 L^0 T^0}$
μ	viscosity (Pa s)	$\mathbf{M L^{-1} T^{-1}}$
ρ	density (kg/m³)	$\mathbf{M L^{-3}}$
ϕ	volume fraction ($-$) or $= T_1/T_0$ ($-$)	$\mathbf{M^0 L^0 T^0}$
ψ	dimensionless factor $[=(E/\mathbf{R})(1/T_1 - 1/T_0)]$ ($-$)	$\mathbf{M^0 L^0 T^0}$

Subscripts

A	apparent
b	bulk condition
L	liquid
m	mean
O	outside film
S	solid
sus	suspension
VP	variable property
z	arbitrary z value
w	wall condition

Chapter 7

Momentum, heat and mass transfer in boundary layers

7.1 Introduction

When an incompressible liquid flows steadily over a solid surface, the liquid close to the surface experiences a significant retardation. The liquid velocity is zero at the surface (provided that the no-slip boundary condition holds) and gradually increases with distance from the surface, and a velocity profile is established. The velocity gradient is steepest adjacent to the surface and becomes progressively less with distance from it. Although theoretically, the velocity gradient is a continuous function becoming zero only at infinite distance from the surface, the flow may conveniently be divided into two regions:

(1) A boundary layer close to the surface where the fluid is retarded and a velocity gradient exists, and in which shear stresses are significant.
(2) The region outside the boundary layer where the liquid is all flowing at the free stream velocity.

It is thus evident that the definition of the boundary layer thickness is somewhat arbitrary. It is often defined as the distance (normal to flow) from the surface at which the fluid velocity reaches some proportion (e.g. 90%, 99%, 99.9%, etc.) of the free stream velocity; 99% is the most commonly used figure. As will be seen later, difficulties arise in comparing differently defined boundary layer thicknesses, because as the free stream velocity is approached, the velocity gradient becomes very low and a small difference in the velocity criterion will correspond to a very large difference in the resulting value of the boundary layer thickness. A thorough understanding of the flow in the boundary layer is of considerable importance in a range of chemical and processing applications because the nature of flow influences not only the drag at a surface or on an immersed object, but also the rates of heat and mass transfer when temperature or concentration gradients exist.

At the outset, it is convenient to consider an incompressible fluid flowing at a constant free stream velocity V_0 over a thin plate oriented parallel to the flow (Figure 7.1).

Although this is a simplified case, it facilitates the discussion of more complex two- and three-dimensional boundary layers. The plate is assumed to be sufficiently wide in the z-direction for the flow conditions to be uniform across any width W of the plate. In addition, the extent of the fluid in y-direction is assumed to be sufficiently large for the velocity of the fluid remote from the plate to be unaffected, and to remain constant at the value of V_0. From Bernoulli's equation, the consequence of this is that the pressure gradient is zero in the direction of flow.

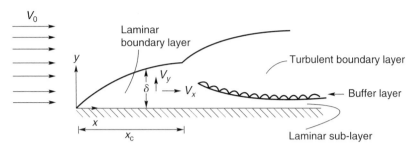

Figure 7.1   *Schematic development of boundary layer*

Although there is conflicting evidence in the literature concerning the validity of the no-slip boundary condition for non-Newtonian materials, it will be assumed here that it is satisfied, so that the fluid in contact with the surface ($y = 0$) can be considered to be at rest. At a distance x from the leading edge, the x-direction velocity V_x will increase from zero at the surface and approach the free stream velocity V_0 asymptotically. At the leading edge ($x = 0$, $y = 0$), the liquid will have been subjected to the retarding force exerted by the surface for only an infinitesimal time and the effective boundary layer thickness will therefore be zero. The liquid will then experience retardation for progressively longer periods of time as it flows over the surface. The retarding effects will affect fluid at greater depths, and the boundary layer thickness, δ, will therefore increase. The velocity gradient (dV_x/dy) at the surface ($y = 0$) decreases as the boundary layer thickens. Near the leading edge, the boundary layer thickness is small, the flow is laminar and the shear stresses arise solely from viscous shearing effects. However, when the boundary layer thickness exceeds a critical value, the flow in the boundary layer itself becomes turbulent. However, the transition from laminar to turbulent flow is not as sharply defined as in a conduit and is strongly influenced by small protuberances at the leading edge and by surface roughness and irregularities which may possibly give rise to an early transition. The flow parameter describing the nature of the flow is a Reynolds number, defined for power-law fluids as $\rho\delta^n V_0^{2-n}/m\,(=Re_\delta)$. Since the boundary layer thickness (δ) is a function of x, it is more convenient to define the local Reynolds number, Re_x as:

$$Re_x = \frac{\rho x^n V_0^{2-n}}{m} \qquad (7.1)$$

The relation between Re_x, and Re_δ will be developed later.

The transition for Newtonian fluids occurs at $Re_x \simeq 10^5$ (Schlichting, 1968).

Even when the flow in the body of the boundary layer is turbulent, flow remains laminar in the thin layer close to the solid surface, the so-called laminar sub-layer. Indeed, the bulk of the resistance to momentum, heat and mass transfer lies in this thin film and therefore interphase heat and mass transfer rates may be increased by decreasing its thickness. As in pipe flow, the laminar sub-layer and the turbulent regions are separated by a buffer layer in which viscous and inertial effects are of comparable magnitudes, as shown schematically in Figure 7.1.

The frictional drag force on a submerged object will depend on the flow conditions in the boundary layer. The analysis of boundary layer flow of non-Newtonian materials will now be based on a direct extension of that applicable to Newtonian fluids. Detailed accounts are given in the literature (Schlichting, 1968), but the salient features of this approach, which is based on Prandtl's analysis, will be re-capitulated. If the flow can be regarded as unidirectional (x-direction), it implies that the effects of velocity components normal to the surface may be neglected, except at very low Reynolds number where the boundary layer thickens rapidly. A further simplifying assumption is to neglect the existence of the buffer layer and to assume that there is a sharp interface between the laminar sub-layer and the turbulent region. If there is a negligible pressure gradient in the direction of flow, i.e. $(\partial p/\partial x) \to 0$, it follows from the application of Bernoulli equation to the free stream outside the boundary layer that its velocity must be constant. The pressure gradient may in practice be either positive or negative. For $(\partial p/\partial x) > 0$, the so-called adverse pressure gradient will tend to retard the flow and cause the boundary layer to thicken rapidly with the result that *separation* may occur. Conversely, the effect of a negative pressure gradient is to reduce the boundary layer thickness.

An integral form of the general momentum balance will be obtained by applying Newton's second law of motion to a control volume of fluid. The evaluation of the resulting integral necessitates a knowledge of the velocity profile and appropriate assumptions of its form must be made for both laminar and turbulent flow conditions.

7.2 Integral momentum equation

Schlichting (1968) points out that the differential equations for flow in boundary layers require numerical solutions even when the flow is laminar and fluid behaviour Newtonian. However, reasonably good estimates of drag on a plane surface can be obtained by using the integral momentum balance approach due to von Karman, as illustrated in this section.

Consider the steady flow of an incompressible liquid of density ρ over an immersed plane surface. Remote from the surface, the free stream velocity of liquid is V_0. A boundary layer of thickness δ develops near the surface, as shown schematically in Figure 7.1. Now consider the equilibrium of the control volume bounded by the planes AB and CD, as shown in Figure 7.2. The velocity component normal to the surface is assumed to be negligible ($V_y \ll V_x$). The rate at which the fluid is entering the control volume through the boundary AB is given by:

$$\dot{m}_{AB} = \int_0^H \rho V_x W \, dy \tag{7.2}$$

Similarly, the total rate of x-momentum transfer through plane AB

$$= W \int_0^H \rho V_x^2 \, dy \tag{7.3}$$

In passing from plane AB to CD, the mass flow changes by:

$$W \frac{d}{dx} \left[\int_0^H \rho V_x \, dy \right] dx \tag{7.4}$$

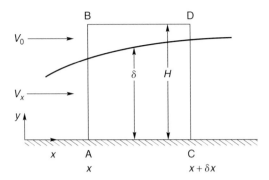

Figure 7.2   *Control volume for momentum balance on the fluid in boundary layer flow*

and the rate of momentum transfer changes by:

$$W \frac{\mathrm{d}}{\mathrm{d}x} \left[\int_0^H \rho V_x^2 \, \mathrm{d}y \right] \mathrm{d}x \qquad (7.5)$$

A mass flow of fluid equal to the difference between the flows at planes C–D and A–B must therefore occur through plane B–D. Since plane B–D lies outside the boundary layer, this entering fluid must have a velocity V_0 in the x-direction. Because the fluid in the boundary layer is being retarded, there will be a smaller flow at plane C–D than at A–B, and hence the amount of fluid entering through plane B–D is negative; that is to say fluid leaves the control volume through the plane B–D.

Thus, the rate of transfer of momentum through plane BD into the control volume

$$= WV_0 \frac{\mathrm{d}}{\mathrm{d}x} \left[\int_0^H \rho V_x \, \mathrm{d}y \right] \mathrm{d}x \qquad (7.6)$$

The net rate of change of momentum in the x-direction associated with the fluid in the control volume must be equal to the rate of addition of momentum from outside (through plane B–D) together with the net force acting on the control volume. With zero pressure gradient, there are no pressure forces and the only external force is that due to the shear stress, τ_{wx}, acting on the surface AC ($y = 0$). As this is a retarding force, τ_{wx} is negative.

Thus, the net force acting on the control volume is $\tau_{wx} W \, \mathrm{d}x$. Hence, the linear momentum balance on the control volume:

$$W \frac{\mathrm{d}}{\mathrm{d}x} \left[\int_0^H \rho V_x^2 \, \mathrm{d}y \right] \mathrm{d}x = WV_0 \frac{\mathrm{d}}{\mathrm{d}x} \left[\int_0^H \rho V_x \, \mathrm{d}y \right] \mathrm{d}x + \tau_{wx} W \, \mathrm{d}x$$

Since V_0 does not vary with x:

$$\frac{\mathrm{d}}{\mathrm{d}x} \int_0^\delta \rho(V_0 - V_x)V_x \, \mathrm{d}y = -\tau_{wx} \qquad (7.7)$$

Note that the upper limit of integration has been changed to δ because the integrand is zero (as $V_x = V_0$) in the range $\delta \leq y \leq H$.

This expression, (equation (7.7)), known as the integral momentum equation, is valid for both laminar and turbulent flow, and no assumption has also been made about the nature of the fluid or its behaviour. In order to integrate equation (7.7), the relation between V_x and y must be known.

Laminar flow will now be considered for both power-law and Bingham plastic model fluids, followed by a short discussion for turbulent flow.

7.3 Laminar boundary layer flow of power-law liquids over a plate

For the laminar flow of a power-law fluid, the only forces acting within the fluid are pure shearing forces, and no momentum transfer occurs by eddy motion. A third degree polynomial approximation may be used for the velocity distribution:

$$V_x = a + by + cy^2 + dy^3 \tag{7.8}$$

The constants a to d are evaluated by applying the following boundary conditions:

$$\text{At } y = 0, \quad V_x = 0 \text{ (no slip)} \tag{7.9a}$$

Since shear stress is constant at the wall (as explained below),

$$\frac{\partial}{\partial y}\left(\frac{\partial V_x}{\partial y}\right)^n = 0$$

i.e.,

$$n\left(\frac{\partial V_x}{\partial y}\right)^{n-1} \frac{\partial^2 V_x}{\partial y^2} = 0$$

As neither $n = 0$ nor $(\partial V_x/\partial y) = 0$ at $y = 0$, hence:

$$\therefore \frac{\partial^2 V_x}{\partial y^2} = 0 \tag{7.9b}$$

$$\text{At } y = \delta, \quad V_x = V_0 \tag{7.9c}$$

$$\frac{\partial V_x}{\partial y} = 0 \tag{7.9d}$$

Condition (7.9d) is necessary to ensure the continuity of the velocity at $y = \delta$. Condition (7.9b) can be explained by the following physical reasoning. The shearing stress in the fluid increases towards the plate, where the velocity is zero. Thus, the change in the momentum of the fluid in the x-direction must be very small near the plate and the shear stress in the fluid therefore must approach a constant value so that equation (7.9b) is applicable. Conversely, since the velocity gradient $(\partial V_x/\partial y)$ is maximum at $y = 0$, its second derivative, i.e., $\partial^2 V_x/\partial y^2 = 0$.

Evaluating the constants and incorporating them into equation (7.8):

$$\frac{V_x}{V_0} = \frac{3}{2}\left(\frac{y}{\delta}\right) - \frac{1}{2}\left(\frac{y}{\delta}\right)^3 \tag{7.10}$$

as for Newtonian fluids.

Thus, the shearing stress acting at any position on the plate can now be evaluated by substituting from equation (7.10) into equation (7.7):

$$-\tau_{wx} = \rho V_0^2 \frac{d}{dx}\int_0^\delta \left\{1 - \frac{3}{2}\left(\frac{y}{\delta}\right) - \frac{1}{2}\left(\frac{y}{\delta}\right)^3\right\}\left\{\frac{3}{2}\left(\frac{y}{\delta}\right) - \frac{1}{2}\left(\frac{y}{\delta}\right)^3\right\} dy \tag{7.11}$$

After integration and simplification, equation (7.11) gives:

$$-\tau_{wx} = \frac{39}{280}\rho V^2 \frac{d\delta}{dx} \tag{7.12}$$

For a power-law liquid,

$$\tau_{yx} = -m\left(\frac{dV_x}{dy}\right)^n$$

and

$$\tau_{yx}|_{y=0} = -m\left(\frac{dV_x}{dy}\bigg|_{y=0}\right)^n \tag{7.13}$$

From the velocity distribution given by equation (7.10),

$$\frac{dV_x}{dy}\bigg|_{y=0} = \frac{3V_0}{2\delta} \tag{7.14}$$

Now combining equations (7.12), (7.13) and (7.14) eliminating τ_{wx}:

$$\frac{39}{280}\rho V_0^2 \frac{d\delta}{dx} = m\left(\frac{3V_0}{2\delta}\right)^n \tag{7.15}$$

which upon further integration with respect to x, and noting that $\delta = 0$ at $x = 0$, gives:

$$\frac{\delta}{x} = F(n)Re_x^{-1/(n+1)} \tag{7.16}$$

where

$$F(n) = \left[\frac{280}{39} (n+1) \left(\frac{3}{2} \right)^n \right]^{1/(n+1)}$$ (7.17)

and the Reynolds number is given by equation (7.1):

$$Re_x = \frac{\rho V_0^{2-n} x^n}{m}$$ (7.1)

Putting $n = 1$ and $m = \mu$ for a Newtonian fluid, equation (7.16) reduces to the well-known result:

$$\frac{\delta}{x} = 4.64 Re_x^{-1/2}$$ (7.18)

The rate of thickening of the boundary layer for power-law fluids is given by:

$$\frac{d\delta}{dx} = \left(\frac{1}{n+1} \right) F(n) Re_x^{-1/(n+1)}$$

or

$$\frac{d\delta}{dx} \propto x^{-n/(n+1)}$$ (7.19)

which suggests that, for a pseudoplastic fluid ($n < 1$), the boundary layer thickens more rapidly with distance along the surface as compared with a Newtonian liquid, and conversely for a shear-thickening fluid ($n > 1$).

7.3.1 Shear stress and frictional drag on the plane immersed surface

As seen above, the shear stress in the fluid at the surface ($y = 0$) is given by:

$$\tau_{yx}|_{y=0} = -m \left(\frac{dV_x}{dy} \bigg|_{y=0} \right)^n$$

$$= -m \left(\frac{3V_0}{2\delta} \right)^n$$ (7.20)

Substitution for δ from equation (7.16) and with slight re-arrangement:

$$\tau_{yx}|_{y=0} = -\left(\frac{3}{2F(n)} \right)^n \rho V_0^2 Re_x^{-1/(n+1)}$$ (7.21a)

The shear stress acting on the plate will be equal and opposite to the shear stress in the fluid at the surface, i.e. $\tau_{wx} = -\tau_{yx}|_{y=0}$.

Thus,

$$\tau_{wx} = \left(\frac{3}{2F(n)}\right)^{n} \rho V_0^2 Re_x^{-1/(n+1)} \tag{7.21b}$$

Equation (7.21b) gives the local value of the shearing stress on the plate which can be averaged over the length of the plate to obtain a mean value, τ_w:

$$
\begin{aligned}
\tau_w &= \frac{1}{L} \int_0^L \tau_{wx}\, dx \\
&= \frac{\rho V_0^2}{L} \left\{\frac{3}{2F(n)}\right\}^{n} \int_0^L Re_x^{-1/(n+1)}\, dx \\
&= \rho V_0^2 (n+1) \left(\frac{3}{2F(n)}\right)^{n} Re_L^{-1/(n+1)}
\end{aligned}
\tag{7.22}
$$

where

$$Re_L = \rho V_0^{2-n} L^n / m$$

The total frictional drag force F_d exerted on one side of the plate of length L and width W is then obtained as:

$$F_d = \tau_w(LW) = \left(\frac{3}{2F(n)}\right)^{n} (n+1)\rho V_0^2 Re_L^{-1/(n+1)} WL \tag{7.23}$$

Introducing a dimensionless drag coefficient C_D defined as:

$$C_D = \frac{F_d}{\left(\frac{1}{2}\rho V_0^2\right)(WL)}$$

Equation (7.23) becomes:

$$
\begin{aligned}
C_D &= 2(n+1)\left(\frac{3}{2F(n)}\right)^{n} Re_L^{-1/(n+1)} \\
&= C(n) Re_L^{-1/(n+1)}
\end{aligned}
\tag{7.24}
$$

where

$$C(n) = 2(n+1)(3/2F(n))^n \tag{7.24a}$$

For the special case of Newtonian fluids ($n = 1$), equation (7.24) reduces to the well-known result of $C_D = 1.292 Re_L^{-1/2}$.

The values of the laminar boundary layer thickness and of the frictional drag are not very sensitive to the form of approximations used for the velocity distribution, as illustrated by Skelland (1967) for various choices of velocity profiles. The resulting values of $C(n)$ are compared in Table 7.1 with the more refined values obtained by Acrivos *et al.* (1960) who solved the differential momentum and mass balance equations numerically;

Table 7.1 *Values of C(n) in equation (7.24a)*

n	Value of $C(n)$	
	Numerical	Equation (7.24a)
0.1	2.132	1.892
0.2	2.094	1.794
0.3	1.905	1.703
0.5	1.727	1.554
1.0	1.328	1.292
1.5	1.095	1.128
2	0.967	1.014
3	0.776	0.872
4	0.678	0.79
5	0.613	0.732

the two values agree within 10% of each other. Schowalter (1978) has discussed the extension of the laminar boundary layer analysis for power-law fluids to the more complex geometries of two- and three-dimensional flows. Finally, it is useful to delineate the value of the Reynolds number below which the boundary layer analysis is not applicable. For power-law fluids, the critical values of the Reynolds number depend on the value of the power-law index (Wu and Thompson, 1996). Based on a 5% criterion (in the value of drag coefficients), the resulting critical values of the Reynolds number are 120, 170, 200, 100 and 45, for $n = 1, 0.8, 0.6, 0.3$ and 0.1. Thus, the foregoing analysis based on boundary layer flow approximation must only be used when the prevailing Reynolds number are greater than the above-noted critical values.

7.4 Laminar boundary layer flow of Bingham plastic fluids over a plate

Outside the boundary layer region, the velocity gradient is zero, and thus the shearing forces must also be zero in the case of Newtonian and power-law fluids, as seen in the preceding section. In contrast, in the case of Bingham plastic fluids, the shear stress approaches the yield stress of the fluid at the outer edge of the boundary layer which must eventually decay to zero over a relatively short distance and thus, once again, there is no shearing force present in the bulk of the fluid outside the boundary layer. It should be noted that in this case the position of the boundary layer is defined precisely, as a point where the shear stress in the fluid equals the yield stress of the fluid. Conversely, the fluid inside the boundary layer is sheared everywhere.

However, to take account of the yield stress acting at the edge of the boundary layer, the integral momentum balance equation (7.7) must be modified as:

$$\rho \frac{d}{dx} \left[\int_0^\delta (V_0 - V_x) V_x \, dy \right] = -(\tau_{wx} + \tau_0^B) \qquad (7.25)$$

Assuming that equation (7.10) still provides a satisfactory representation of the laminar boundary layer flow in the region $0 \le y \le \delta$, substituting equation (7.11) into (7.25) yields:

$$\frac{39}{280} \rho V_0^2 \frac{d\delta}{dx} = -(\tau_{wx} + \tau_0^B) \qquad (7.26)$$

For a Bingham plastic, the fluid shear stress at the surface, i.e., at $y = 0$ is given by:

$$\tau_{yx}|_{y=0} = \tau_0^B + \mu_B \frac{dV_x}{dy}\Big|_{y=0} \qquad (7.27)$$

The velocity gradient at the surface,

$$\frac{dV_x}{dy}\Big|_{y=0} = \frac{3V_0}{2\delta} \qquad (7.14)$$

Substituting in equation (7.27):

$$\tau_{wx} = -\tau_{yx}|_{y=0} = -\left(\tau_0^B + \frac{3}{2}\frac{\mu_B V_0}{\delta}\right) \qquad (7.28a)$$

or

$$-(\tau_{wx} + \tau_0^B) = \frac{3}{2}\frac{\mu V_0}{\delta} \qquad (7.28b)$$

Combining equations (7.26) and (7.28):

$$\delta \, d\delta = \left(\frac{140}{13}\right)\frac{\mu_B}{\rho V_0} dx$$

Integration with the initial condition that at $x = 0$, $\delta = 0$, yields:

$$\frac{\delta}{x} = \frac{4.64}{\sqrt{Re_x}} \qquad (7.29)$$

where Re_x is now defined as $\rho V_0 x / \mu_B$.

It is interesting to note that equation (7.29) is identical to that for a Newtonian fluid (see equation (7.18) with $n = 1$). In spite of this similarity with Newtonian fluids, the frictional drag on a plate submerged in a Bingham plastic fluid deviates from that in a Newtonian fluid owing to the existence of the fluid yield stress in equation (7.28), as will be seen in the next section.

7.4.1 Shear stress and drag force on an immersed plate

The shear stress in the fluid adjacent to the plate ($y = 0$) is given by equation (7.28a), i.e.,

$$\tau_{yx}|_{y=0} = -\left(\tau_0^B + \frac{3}{2}\frac{\mu_B V_0}{\delta}\right) \qquad (7.28a)$$

Substitution for δ from equation (7.29):

$$\tau_{yx}|_{y=0} = -\left(\tau_0^B + \frac{0.323}{x} V_0 \mu_B \sqrt{Re_x}\right) \tag{7.30}$$

The shear stress acting on the plate will be equal and opposite of this value, i.e. $\tau_{wx} = -\tau_{yx}|_{y=0}$, thus

$$\tau_{wx} = \left(\tau_0^B + \frac{0.323}{x} V_0 \mu_B \sqrt{Re_x}\right) \tag{7.31}$$

The average shear stress acting over the entire plate may be calculated as:

$$\tau_w = \frac{1}{L}\int_0^L \tau_{wx}\, dx = \frac{1}{L}\int_0^L (\tau_0^B + 0.323 V_0^{3/2} \rho^{1/2} \mu_B^{1/2} x^{-1/2})\, dx$$

or

$$\tau_w = \tau_0^B + 0.646\left(\frac{\rho V_0^2}{\sqrt{Re_L}}\right) \tag{7.32}$$

Introducing a drag coefficient C_D, as defined earlier, equation (7.32) can be re-written as:

$$C_D = Bi + \frac{1.292}{\sqrt{Re_L}} \tag{7.33}$$

where Bi, the Bingham number, is defined as $2\tau_0^B/\rho V_0^2$.

Again, for Newtonian fluids, $Bi = 0$ and equation (7.33) reduces to the expected form.

Example 7.1

A polymer solution (of density $1000\,\text{kg/m}^3$) is flowing parallel to a plate ($300\,\text{mm} \times 300\,\text{mm}$); the free stream velocity is $2\,\text{m/s}$. In the narrow shear rate range, the rheology of the polymer solution can be adequately approximated by both the power-law ($m = 0.3\,\text{Pa s}^n$ and $n = 0.5$) and the Bingham plastic model ($\tau_0 = 2.28\,\text{Pa}$ and $\mu_B = 7.22\,\text{mPa s}$). Using each of these models, estimate and compare the values of the shear stress and the boundary layer thickness $150\,\text{mm}$ away from the leading edge, and the total frictional force on each side of the plate.

Solution

(a) *Power-law model calculations*
First, check the value of the Reynolds number at $x = L = 0.3\,\text{m}$,

$$Re_L = \frac{\rho V_0^{2-n} L^n}{m}$$

$$= \frac{(1000)(2)^{2-0.5}(0.3)^{0.5}}{0.3} = 5164$$

The flow is likely to be streamline over the whole plate length (see Section 7.5).

The value of the local shear stress at $x = 0.15\,\mathrm{m}$ can be calculated using equation (7.21b):

$$\tau_{wx}\big|_{x=0.15\,m} = \left(\frac{3}{2F(n)}\right)^n \rho V_0^2 Re_x^{-1/(n+1)}$$

where

$$F(n) = \left[\frac{280}{39}(n+1)\left(\frac{3}{2}\right)^n\right]^{1/(n+1)}$$

For $n = 0.5$,

$$F(n) = \left[\frac{280}{39}(1.5)(3/2)^{0.5}\right]^{1/1.5} = 5.58$$

Substitution in $\tau_{wx}\big|_{x=0.15\,m}$ yields:

$$\tau_{wx}\big|_{x=0.15\,m} = \left(\frac{3}{2 \times 5.58}\right)^{0.5} \times 1000 \times 2^2 \left\{\frac{(1000)(2)^{2-0.5}(0.15)^{0.5}}{0.3}\right\}^{-1/1.5}$$

$$= 8.75\,\mathrm{Pa}$$

The boundary layer thickness δ at $x = 0.15\,\mathrm{m}$ is calculated using equation (7.16):

$$\frac{\delta}{x} = F(n)Re_x^{-1/(n+1)}$$

or

$$\delta\big|_{x=0.15\,m} = (0.15)(5.58)\left\{\frac{(1000)(2)^{2-0.5}(0.15)^{0.5}}{0.3}\right\}^{-1/1.5}$$

$$= 3.53 \times 10^{-3}\,\mathrm{m} \quad \text{or} \quad 3.53\,\mathrm{mm}$$

Finally, the frictional drag force on the plate is calculated using equation (7.24):

$$C_D = C(n)Re_L^{-1/(n+1)}$$

where

$$C(n) = 2(n+1)\left(\frac{3}{2F(n)}\right)^n$$

For $n = 0.5$,

$$C(n) = 2 \times 1.5 \times \left(\frac{3}{2 \times 5.58}\right)^{0.5} = 1.554$$

$$\therefore\ C_D = 1.554 \times (5164)^{-1/1.5} = 0.0052$$

Drag force on one side of plate $F_D = C_D \frac{1}{2} \rho V_0^2 WL$

$$F_D = 0.0052 \times \frac{1}{2} \times 1000 \times 2^2 \times 0.3 \times 0.3 = 0.94 \, \text{N}$$

(b) Bingham plastic model calculations

The maximum value of the Reynolds number occurs at the trailing edge of the plate, i.e.,

$$Re_L = \frac{\rho V_0 L}{\mu_B} = \frac{1000 \times 2 \times 0.3}{7.22 \times 10^{-3}} = 83100$$

Using the same transition criterion as for Newtonian fluids, namely, $Re_L < {\sim}10^5$, the flow in the whole of the boundary layer is likely to be laminar.

The value of the local shear stress at $x = 0.15 \, \text{m}$ is calculated using equation (7.31):

$$
\begin{aligned}
\tau_{wx}|_{x=0.15\,\text{m}} &= \tau_0^B + \frac{0.323}{x} V_0 \mu_B \sqrt{Re_x} \\
&= 2.28 + \frac{0.323}{0.15} (2)(7.22 \times 10^{-3}) \left(\frac{1000 \times 2 \times 0.15}{7.22 \times 10^{-3}} \right)^{1/2} \\
&= 8.62 \, \text{Pa}
\end{aligned}
$$

Note that, although about a quarter of the shear stress stems from the yield stress, the total value is comparable with that based on the use of the power-law model, as seen in part (a) above.

The boundary layer thickness δ is calculated using equation (7.29):

$$
\begin{aligned}
\delta &= 4.64 x Re_x^{-1/2} = 4.64 \times 0.15 \left(\frac{1000 \times 2 \times 0.15}{7.22 \times 10^{-3}} \right)^{-1/2} \\
&= 3.41 \times 10^{-3} \, \text{m} \quad \text{or} \quad 3.41 \, \text{mm}
\end{aligned}
$$

This value also closely corresponds with that calculated in part (a).

Finally, the drag coefficient is evaluated using equation (7.33):

$$
\begin{aligned}
C_D &= Bi + \frac{1.292}{\sqrt{Re_L}} = \frac{2\tau_0^B}{\rho V_0^2} + \frac{1.292}{\sqrt{Re_L}} \\
&= \frac{2 \times 2.28}{(1000)(2)^2} + \frac{1.292}{(83100)^{1/2}} = 0.00562
\end{aligned}
$$

and the drag force exerted on one side of the plate,

$$
\begin{aligned}
F_D &= C_D \frac{1}{2} \rho V_0^2 WL = 0.00562 \times \frac{1}{2} \times 1000 \times 2^2 \times 0.3 \times 0.3 \\
&= 1.015 \, \text{N}
\end{aligned}
$$

This value is slightly higher than that obtained in part (a) above. ☐

7.5 Transition criterion and turbulent boundary layer flow

7.5.1 Transition criterion

As mentioned in Section 7.1, the nature of the flow in the boundary layer is influenced by several variables including the degree of turbulence in the free stream, the roughness of the immersed surface and the value of the Reynolds number, defined by equation (7.1). While no experimental data are available on the limiting values of the Reynolds number for the flow of non-Newtonian materials in boundary layers, Skelland (1967) has suggested that the limiting value of the Reynolds number decreases with the increasing degree of shear-thinning behaviour. Based on heuristic considerations, he defined an effective viscosity for a power-law fluid in such a manner that the shear stress at the surface equals that in Newtonian fluids at the transition point, i.e. at $Re_x \sim 10^5$. Thus, equating the values of the shear stress at the surface for a Newtonian fluid of hypothetical viscosity to that for the power-law fluid (equation (7.21)), Skelland (1967) obtained the following transition criterion for power-law fluids:

$$\left(\frac{0.323}{B(n)}\right)^2 Re_c^{2/(n+1)} < 10^5 \qquad (7.34)$$

where $B(n) = (1.5/F(n))^n$, $F(n)$ is given by equation (7.17) and $Re_c = \rho V_0^{2-n} x_c^n / m$ where x_c is the distance (from the leading edge) at which the flow ceases to be streamline. For Newtonian fluids ($n = 1$), $B(n) = 0.323$ and $Re_c = 10^5$ which is the value at the transition point. For a liquid of flow behaviour index, $n = 0.5$, the limiting value of the Reynolds number is 1.14×10^4 which is an order of magnitude smaller than the value for a Newtonian fluid.

7.5.2 Turbulent boundary layer flow

As mentioned previously, even when the flow becomes turbulent in the boundary layer, there exists a thin sub-layer close to the surface in which the flow is laminar. This layer and the fully turbulent regions are separated by a buffer layer, as shown schematically in Figure 7.1. In the simplified treatments of flow within the turbulent boundary layer, however, the existence of the buffer layer is neglected. In the laminar sub-layer, momentum transfer occurs by molecular means, whereas in the turbulent region eddy transport dominates. In principle, the methods of calculating the local values of the boundary layer thickness and shear stress acting on an immersed surface are similar to those used above for laminar flow. However, the main difficulty stems from the fact that the viscosity models, such as equation (7.13) or (7.27), are applicable only for laminar flow and thus appropriate expressions for the shear stress at the surface must be known if they are to be inserted in equation (7.12). For Newtonian fluids, Blasius (1913) circumvented this difficulty by inferring the values of the shear stress at the wall from the turbulent friction factor–Reynolds number relationship for flow in circular tubes. Curvature effects are likely to be negligible because the laminar sub-layer thickness is generally much smaller than the diameter of the pipe. Schlichting (1968) and Coulson and Richardson (1999) have discussed turbulent boundary layer flow in depth. Skelland (1967) has extended this approach to power-law liquids by replacing the Blasius equation by a modified form for power-law fluids, equation (3.75). Furthermore, because non-Newtonian materials generally have high apparent viscosities, turbulent conditions are not often encountered in flow over immersed surfaces. Therefore, this topic will not be pursued further here because the turbulent flow of non-Newtonian fluids is not at all well understood and there are no well-established and validated equations for shear stress, even for flow in pipes. A thorough review of the pertinent literature is available (Chhabra, 2006).

7.6 Heat transfer in boundary layers

When the fluid and the immersed surface are at different temperatures, heat transfer will take place. If the heat transfer rate is small in relation to the thermal capacity of the flowing stream, its temperature will remain substantially constant. The surface may be maintained at a constant temperature, or the heat flux at the surface may be maintained constant; or surface conditions may be intermediate between these two limits. Because the temperature gradient will be highest in the vicinity of the surface and the temperature of the fluid stream will be approached asymptotically, a thermal boundary layer may therefore be postulated which covers the region close to the surface and in which the whole of the temperature gradient is assumed to lie.

Thus a momentum and a thermal boundary layer will develop simultaneously whenever the fluid stream and the immersed surface are at different temperatures (Figure 7.3). The momentum and energy equations are coupled, because the physical properties of non-Newtonian fluids are normally temperature and shear-rate dependent. The resulting governing equations for momentum and heat transfer require numerical solutions. However, if the physical properties of the fluid do not vary significantly over the relevant temperature interval, there is little interaction between the two boundary layers and they may both be assumed to develop independently of one another. As seen in Chapter 6, the physical properties other than apparent viscosity may be taken as constant for commonly encountered non-Newtonian fluids.

In general, the thermal and momentum boundary layers will not correspond. In the ensuing treatment, the simplest non-interacting case will be considered with physical properties taken as being temperature independent. The temperature of the bulk fluid will be taken to be T_0 (constant) and that of the immersed plate to be T_s (also constant). For convenience, the temperature scale will be chosen so that the surface temperature

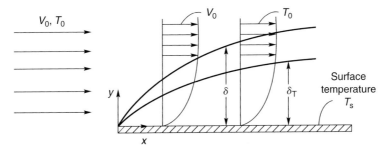

Figure 7.3   *Schematic representation of momentum and thermal boundary layers*

Figure 7.4   *Control volume for heat balance*

is zero, giving the boundary condition $T_s = 0$, corresponding to zero velocity in the momentum balance equation.

An integral procedure similar to that adopted previously for momentum boundary layers will be used here to obtain the expression for the rate of heat exchange between the fluid and the plane surface. A heat balance will be made over a control volume (Figure 7.4) which extends beyond the limits of both the momentum and thermal boundary layers.

At steady state and with no source or sink present in the control volume, the heat balance for the control volume ABCD can be stated as follows:

> The sum of heat convected in through planes AB and BD and
> that added by conduction at wall AC must be equal to the heat
> convected out at plane CD   (7.35)

The rate at which heat enters the control volume at plane AB is:

$$W \int_0^H \rho V_x C_p T \, \mathrm{d}y \tag{7.36a}$$

Similarly, the rate of transfer of heat at the plane CD:

$$W \int_0^H \rho V_x C_p T \, \mathrm{d}y + W \frac{\mathrm{d}}{\mathrm{d}x} \left[\int_0^H \rho V_x C_p T \, \mathrm{d}y \right] \mathrm{d}x \tag{7.36b}$$

It has already been seen that the mass rate of transfer of fluid across the plane BD is given by equation (7.4):

$$W \frac{\mathrm{d}}{\mathrm{d}x} \left(\int_0^H \rho V_x \, \mathrm{d}y \right) \mathrm{d}x \tag{7.4}$$

The enthalpy of this stream is:

$$C_p T_0 W \frac{d}{dx} \left(\int_0^H \rho V_x \, dy \right) dx \qquad (7.37)$$

Substituting for the terms in equation (7.35):

$$\rho W C_p \int_0^H V_x T \, dy + \rho C_p W T_0 \frac{d}{dx} \left(\int_0^H V_x \, dy \right) dx + q_w W \, dx$$

$$= \rho W C_p \int_0^H V_x T \, dy + \rho W C_p \frac{d}{dx} \left(\int_0^H V_x T \, dy \right) dx \qquad (7.38)$$

where q_w is the heat flux at the wall and is equal to $-k(dT/dy)|_{y=0}$ which upon substitution in equation (7.38) gives:

$$\frac{d}{dx} \int_0^{\delta_T} V_x (T_0 - T) dy = \alpha \left. \frac{dT}{dy} \right|_{y=0} \qquad (7.39)$$

where α is the thermal diffusivity ($=k/\rho C_p$) of the fluid. Since both V_x and V_y are zero at the surface, the only mechanism of heat transfer which is relevant is conduction. Also, the upper limit of the integration in equation (7.39) has been changed to δ_T, the thermal boundary layer thickness, because for $y \geq \delta_T$, $T = T_0$ and the integrand is identically zero.

The relation between V_x and y has already been obtained for the laminar flow in the boundary layer. A similar relation between the temperature of the fluid, T, and y for laminar flow of power-law fluids in the boundary layer will now be derived.

7.6.1 Heat transfer in laminar flow of a power-law fluid over an isothermal plane surface

Consideration is now given to the flow of an incompressible power-law liquid of temperature T_0 over a plane surface maintained at a higher temperature T_s. At any given distance from the leading edge, the temperature of the fluid progressively decreases with distance y from the surface reaching T_0 at the extremity of the thermal boundary layer, $y = \delta_T$. The temperature at a distance y from the surface can be approximated by a polynomial of the form:

$$T = a + by + cy^2 + dy^3 \qquad (7.40)$$

The boundary conditions for the temperature field are:

At $y = 0, \quad T = T_s$ \hfill (7.41a)

$$y = \delta_T, \ T = T_0 \quad \text{and} \quad \frac{dT}{dy} = 0 \qquad (7.41b)$$

The fourth boundary condition for temperature can be deduced as follows: the rate of heat transfer per unit area of the surface is given by:

$$q|_{y=0} = -k \frac{dT}{dy}\bigg|_{y=0} \tag{7.41c}$$

If the temperature of the fluid element adhering to the wall is to remain constant at $T = T_s$, the rate of heat transfer into and out of the element must be the same, i.e. the temperature gradient must be constant and therefore:

$$\frac{d}{dy}\left(\frac{dT}{dy}\bigg|_{y=0}\right) \quad \text{or} \quad \frac{d^2T}{dy^2}\bigg|_{y=0} = 0 \tag{7.41d}$$

One can also argue that the temperature gradient in the lateral direction (dT/dy) is maximum at the surface, i.e. $y = 0$, and therefore, its derivative (d^2T/dy^2) must be zero at $y = 0$.

The four unknown constants in the assumed temperature profile, equation (7.40), can now be evaluated by using equations (7.41a)–(7.41d) to give:

$$a = T_s; \quad b = \frac{3}{2}\left(\frac{T_0 - T_s}{\delta_T}\right)$$

$$c = 0; \quad d = -\frac{(T_0 - T_s)}{2\delta_T^3} \tag{7.42}$$

Substitution of these values into equation (7.40) yields

$$\frac{\theta}{\theta_0} = \frac{T - T_s}{T_0 - T_s} = \frac{3}{2}\left(\frac{y}{\delta_T}\right) - \frac{1}{2}\left(\frac{y}{\delta_T}\right)^3 \tag{7.43}$$

It should be noted that equation (7.10) for velocity distribution and equation (7.43) for temperature distribution in the laminar boundary layer are identical in form, because basically the same forms of equation and boundary conditions have been used for the velocity and temperature profiles.

It will now be assumed that the thermal boundary layer is everywhere thinner, i.e. $\delta_T < \delta$. The conditions under which this approximation is justified are examined later. Now substituting in equation (7.39) for V_x from equation (7.10), T and $(dT/dy)|_{y=0}$ from equation (7.43):

$$V_0\theta_0 \frac{d}{dx}\left[\int_0^{\delta_T}\left\{\frac{3}{2}\left(\frac{y}{\delta}\right) - \frac{1}{2}\left(\frac{y}{\delta}\right)^3\right\}\left\{1 - \frac{3}{2}\left(\frac{y}{\delta_T}\right) + \frac{1}{2}\left(\frac{y}{\delta_T}\right)^3\right\}dy\right] = \frac{3}{2}\alpha\frac{\theta_0}{\delta_T} \tag{7.44}$$

Integration with respect to y between the limits $y = 0$ to $y = \delta_T$ yields:

$$\theta_0 V_0 \frac{d}{dx}\left[\delta\left(\frac{3}{20}\varepsilon^2 - \frac{3}{280}\varepsilon^4\right)\right] = \frac{3}{2}\frac{\alpha\theta_0}{\delta\varepsilon} \tag{7.45}$$

where $\varepsilon = (\delta_T/\delta) < 1$. Neglecting the ε^4 term in equation (7.45) leads to

$$\frac{V_0}{10}\frac{d}{dx}(\delta\varepsilon^2) = \frac{\alpha}{\delta\varepsilon} \tag{7.46}$$

which may be expressed as:

$$\frac{V_0}{10}\left(2\varepsilon^2\delta^2\frac{d\varepsilon}{\delta x} + \delta\varepsilon^3\frac{d\delta}{dx}\right) = \alpha \tag{7.47}$$

Now introducing a new variable $z = \varepsilon^3$ and substituting for δ and $d\delta/dx$ from equations (7.16) and (7.19), respectively:

$$\frac{dz}{dx} + \frac{3}{2x(n+1)}z = \frac{15\alpha}{V_0}\frac{Re_x^{2/(n+1)}}{x^2(F(n))^2} \tag{7.48}$$

where $Re_x = (\rho V_0^{2-n}x^n)/m$.

The solution of the differential equation (7.48) is given by:

$$z = \varepsilon^3 = \left(\frac{\delta_T}{\delta}\right)^3 = \frac{15\alpha}{(F(n))^2 V_0}\cdot\frac{Re_x^{2/(n+1)}}{x}\cdot\frac{2(n+1)}{(2n+1)} + C_1 x^{-3/(n+1)} \tag{7.49}$$

Substituting for δ from equation (7.16):

$$\delta_T^3 = \frac{15\alpha F(n)}{V_0(2n+1)}2(n+1)x^2 Re_x^{-1/(n+1)} + C_1(F(n))^3 x^{3(2n+1)/2(n+1)}Re_x^{-3/(n+1)} \tag{7.50}$$

For the general case where the heating of the plate begins at $x = x_0$, the constant C_1 is evaluated using the condition $\delta_T = 0$ at $x = x_0$ yielding the expression for δ_T as:

$$\frac{\delta_T}{x} = \left[\frac{30(n+1)}{(2n+1)}F(n)\right]^{1/3}\left(\frac{\alpha}{xV_0}\right)^{1/3}Re_x^{-\frac{1}{3(n+1)}}\left[1 - \left(\frac{x_0}{x}\right)^{(2n+1)/2(n+1)}\right]^{1/3} \tag{7.51a}$$

For the special case where the whole plate is being heated, i.e. $x_0 = 0$, the thermal boundary layer thickness δ_T is given by:

$$\frac{\delta_T}{x} = \left[\frac{30(n+1)}{(2n+1)}F(n)\right]^{1/3}\left(\frac{\alpha}{xV_0}\right)^{1/3}Re_x^{-\frac{1}{3(n+1)}} \tag{7.51b}$$

A heat transfer coefficient, h, may be defined such that:

$$h(T_s - T_0) = -k\frac{dT}{dy}\bigg|_{y=0} \tag{7.52}$$

The temperature gradient is found by differentiating the temperature profile, equation (7.43):

$$\left.\frac{dT}{dy}\right|_{y=0} = \frac{3(T_0 - T_s)}{2\delta_T} \tag{7.53}$$

and the heat transfer coefficient is obtained by combining equations (7.52) and (7.53):

$$h = \left(\frac{3}{2}\right)\left(\frac{k}{\delta_T}\right) \tag{7.54}$$

Substitution for δ_T from equation (7.51) into (7.54) leads to

$$Nu_x = \frac{hx}{k} = \frac{3}{2}\left[\frac{30F(n)(n+1)}{(2n+1)}\right]^{-(1/3)} Pr_x^{1/3} Re_x^{(n+2)/(3(n+1))} \tag{7.55}$$

where the Prandtl number, Pr_x, is defined as:

$$Pr_x = \frac{C_p}{k} m \left(\frac{V_0}{x}\right)^{n-1} \tag{7.56}$$

The Prandtl number for a power-law fluid is not simply a combination of physical properties of the fluid but also depends on the point value of an apparent viscosity $m(V_0/x)^{n-1}$ which is a function of both x and y. The choice of linear dimension in this definition is somewhat arbitrary and no single definition may necessarily be appropriate for all applications. However, for $n = 1$, equation (7.55) reduces to the corresponding Newtonian expression, i.e.

$$Nu_x = 0.328 Pr_x^{1/3} Re_x^{1/2} \tag{7.57}$$

It is seen that the local value of the heat transfer coefficient varies with distance from the leading edge according to the relation:

$$h \propto x^{-(n+2)/3(n+1)} \tag{7.58}$$

The heat transfer coefficient h has a (theoretical) infinite value at the leading edge where the thickness of the thermal boundary layer is zero and decreases progressively as the boundary layer thickens. The mean value of the heat transfer coefficient over the plate length can be obtained by integrating it from $x = 0$ to $x = L$ as:

$$\begin{aligned} h_m &= \frac{1}{L}\int_0^L h\,dx \\ &= \frac{B_0(n)}{L}\,k\int_0^L \frac{Pr_x^{1/3} Re_x^{(n+2)/3(n+1)}}{x}\,dx \end{aligned} \tag{7.59}$$

where

$$B_0(n) = \frac{3}{2}\left[\frac{30F(n)(n+1)}{(2n+1)}\right]^{-(1/3)}$$

The evaluation of the integral in equation (7.59) followed by some simplification and re-arrangement finally yields:

$$Nu_m = \frac{h_m L}{k} = \frac{9(n+1)}{2(2n+1)} \left[\frac{30F(n)(n+1)}{(2n+1)} \right]^{-1/3} Pr_L^{1/3} Re_L^{(n+2)/3(n+1)} \qquad (7.60)$$

where $F(n)$, a function of n, is given by equation (7.17) and Pr_L and Re_L, respectively, are defined as:

$$Pr_L = \frac{C_p}{k} m \left(\frac{V_0}{L} \right)^{n-1} \qquad (7.61)$$

and

$$Re_L = \frac{\rho V_0^{2-n} L^n}{m} \qquad (7.62)$$

The corresponding analysis for the case when a constant heat flux is maintained on the surface of the plate is little more involved than this. This is due to the fact that the temperature of the fluid is now a function of both x and y coordinates, and therefore it is not possible to express the temperature distribution by a polynomial similar to equation (7.43). However, at steady state, each point on the plate is at a constant temperature (corresponding to the constant heat flux) which varies in the x-direction. For this case, an approximate expression for the local Nusselt number is given here (Chhabra, 1999a):

$$Nu_x = \frac{hx}{k} = \left(\frac{3}{2} \right) \sqrt{\frac{5}{3} \frac{\left| \frac{8n+7}{6n+3} \right|}{\left| \frac{2n+2}{2n+1} \right|}} B_0(n) Re_x^{(n+2)/3(n+1)} Pr_x^{1/3} \qquad (7.63)$$

For the limiting case of Newtonian fluid behaviour ($n = 1$), equation (7.63) reduces to the expected result:

$$Nu_x = 0.454 Re_x^{1/2} Pr^{1/3} \qquad (7.64)$$

More rigorous numerical solutions for the laminar boundary layers of power-law and other time-independent fluids flowing over plane surfaces and objects of two-dimensional axisymmetric shapes are given in the literature (Acrivos *et al.*, 1960; Huang and Chen, 1984; Agarwal *et al.*, 2002; Khan *et al.*, 2006) and these predictions are in good agreement (within $\pm 10\%$) with the approximate results given by equations (7.60) and (7.63). Most of the theoretical developments in the field of thermal boundary layers of time-independent fluids have been critically reviewed in references (Shenoy and Mashelkar, 1982; Nakayama, 1988; Chhabra, 1993, 1999b, 2006). The limited experimental results and empirical correlations are presented in a later section in this chapter.

Example 7.2

A dilute polymer solution at 25°C flows at 2 m/s over a 300 mm × 300 mm square plate which is maintained at a uniform temperature of 35°C. The average values of the power-law constants (over this temperature interval) may be taken as: $m = 0.3\,\mathrm{Pa\,s}^n$ and $n = 0.5$. Estimate the thickness of the thermal boundary layer 150 mm from the leading edge and the rate of heat transfer from one side of the plate only. The density, thermal conductivity and heat capacity of the polymer solution may be approximated as those of water at the same temperature.

Solution

As seen in Example 7.1, the flow conditions in the boundary layer appear to be laminar over the entire length of the plate under the given operating conditions. Therefore, the thickness of the thermal boundary layer can be estimated using equation (7.51). The value of Re_x is:

$$Re_x = \frac{\rho V_0^{2-n} x^n}{m} = \frac{1000 \times 2^{2-0.5} \times 0.15^{0.5}}{0.3} = 3650$$

For water at 25°C, the values of the thermo-physical properties are: $\rho = 1000\,\mathrm{kg/m^3}$; $k = 0.615\,\mathrm{W/m\,K}$; $C_p = 3800\,\mathrm{J/kg\,K}$.

$$\therefore \text{ thermal diffusivity } \alpha = \frac{k}{\rho C_p} = \frac{0.615}{1000 \times 3800}$$
$$= 1.62 \times 10^{-7}\,\mathrm{m^2/s}$$

We have already calculated in Example 7.1, $F(n) = 5.58$ when $n = 0.5$. Substituting these values in equation (7.51):

$$\delta_T = \left[\frac{30 \times 1.62 \times 10^{-7} \times 5.58 \times 0.15^2}{2} \times \left(\frac{1.5}{2}\right)\right]^{1/3} 3650^{-1/4.5}$$
$$= 9.88 \times 10^{-4}\,\mathrm{m}$$

or about 1 mm which is less than a third of the corresponding momentum boundary layer thickness.

The Prandtl number with L as characteristic linear dimension is evaluated:

$$Pr_L = \frac{C_p m}{k}\left(\frac{V_0}{L}\right)^{n-1} = \frac{3800 \times 0.3}{0.615}\left(\frac{2}{0.3}\right)^{0.5-1} = 718$$

and

$$Re_L = \frac{\rho V_0^{2-n} L^n}{m} = \frac{1000 \times 2^{2-0.5} \times 0.3^{0.5}}{0.3} = 5164$$

Substituting these values in equation (7.60):

$$Nu_m = \frac{h_m L}{k} = \frac{9 \times 1.5}{2 \times 2}\left[\frac{30 \times 5.58 \times 1.5}{2}\right]^{-1/3} (718)^{1/3}(5164)^{2.5/4.5}$$
$$= 697.5$$
$$\therefore h_m = \frac{697.5 \times 0.615}{0.3} = 1430\,\mathrm{W/m^2\,K}$$

The rate of heat loss from the plate is:

$$Q = h_m A \Delta T = 1430 \times (0.3 \times 0.3) \times (35 - 25) = 1287\,\mathrm{W}. \ \square$$

7.7 Mass transfer in laminar boundary layer flow of power-law fluids

The basic equation for mass transfer by molecular diffusion is Fick's law which may be expressed as:

$$N_A = -D_{AB} \frac{dC_A}{dy}$$

where N_A is the mass transfer rate per unit area (kmol/m²s), C_A is the molar concentration of the diffusing component and D_{AB} is the molecular diffusivity.

This equation is analogous to Fourier's law for heat transfer by conduction:

$$q = -k \frac{dT}{dy} = -\frac{k}{\rho C_p} \frac{d(\rho C_p T)}{dy} = -\alpha \frac{d(\rho C_p T)}{dy}$$

(assuming ρ and C_p to be constant).

It will be seen that there is a synergy between the two equations, $\rho C_p T$ the heat content per unit volume being equivalent to C_A, the molar concentration of the diffusing component.

The boundary layer equation for mass transfer may therefore be written by analogy with equation (7.39) as:

$$\frac{d}{dx} \int_0^H (C_{A_s} - C_A) V_x \, dy = D_{AB} \frac{dC_A}{dy} \bigg|_{y=0} \tag{7.65}$$

where C_{A_s} and C_A are respectively the concentration of the solute A in the free stream (outside the boundary layer) at a distance x from the leading edge and at a distance y from the surface. H is the thickness of the control volume normal to the surface as shown in Figure 7.4; it is chosen to be greater than the thickness of both the momentum and concentration boundary layers.

Again the form of concentration profile in the diffusion boundary layer depends on the boundary conditions at the surface and in the fluid stream. For the conditions analogous to those used in consideration of thermal boundary layer (constant concentrations both in the free stream outside the boundary layer and at the submerged surface), the concentration profile will be of similar form to that given by equation (7.43):

$$\frac{C_A - C_{A_0}}{C_{A_s} - C_{A_0}} = \frac{3}{2}\left(\frac{y}{\delta_m}\right) - \frac{1}{2}\left(\frac{y}{\delta_m}\right)^3 \tag{7.66}$$

and

$$\frac{1}{C_{A_s} - C_{A_0}} \frac{dC_A}{dy} = \frac{3}{2}\left[\frac{1}{\delta_m} - \frac{y^2}{\delta_m^3}\right] \tag{7.67}$$

where C_{A_0} is the solute concentration at the surface ($y = 0$) and δ_m is the thickness of the concentration boundary layer.

Substituting in equation (7.65) for the velocity V_x from equation (7.10), and for the concentration C_A and concentration gradient (dC_A/dy) from equations (7.66) and (7.67):

$$V_0 \frac{d}{dx}\left\{ \delta\left(\frac{\varepsilon_m^2}{10} - \frac{1}{140}\varepsilon_m^4 \right) \right\} = \frac{D_{AB}}{\delta\varepsilon_m} \tag{7.68}$$

where $\varepsilon_m = (\delta_m/\delta) < 1$. It should be noted that it has been implicitly assumed that the diffusion boundary layer is everywhere thinner than the momentum boundary layer. With this assumption, the fourth order term in ε_m in equation (7.68) can be neglected without incurring significant error. The resulting approximate solution is identical in form to equation (7.47) for heat transfer, except that the diffusivity D_{AB} replaces the thermal diffusivity α, and thus its solution for a power-law liquid, with the condition $\delta_m = 0$ at $x = 0$, is by analogy with equation (7.51b):

$$\frac{\delta_m}{x} = \left[\frac{30(n + 1)}{(2n + 1)} F(n) \right]^{1/3} \left[\frac{D_{AB}}{xV_0} \right]^{1/3} Re_x^{-1/3(n+1)} \tag{7.69}$$

One can now define a mass transfer coefficient, h_D, by the relation:

$$h_D(C_{A_0} - C_{A_s}) = -D_{AB} \left. \frac{\partial C_A}{\partial y} \right|_{y=0} = -\frac{3}{2}(C_{A_s} - C_{A_0})\frac{D_{AB}}{\delta_m} \tag{7.70}$$

Substituting for δ_m from equation (7.69) and re-arranging:

$$Sh_x = \frac{h_D x}{D_{AB}} = \frac{3}{2}\left[\frac{30(n + 1)F(n)}{(2n + 1)} \right]^{-1/3} Sc_x^{1/3} Re_x^{(n+2)/3(n+1)} \tag{7.71}$$

where the Schmidt number,

$$Sc_x = \frac{m}{\rho D_{AB}}\left(\frac{V_0}{x} \right)^{n-1} \tag{7.72}$$

Here, Sc_x is analogous to Pr_x for heat transfer (equation (7.56)). Finally, the average value of the transfer coefficient over the plate length L can be obtained in the same way as for heat transfer (equation (7.59)) by integration:

$$Sh_m = \frac{h_D L}{D_{AB}} = \frac{9(n + 1)}{2(2n + 1)}\left\{ \frac{30(n + 1)F(n)}{(2n + 1)} \right\}^{-1/3} Sc_L^{1/3} Re_L^{(n+2)/3(n+1)} \tag{7.73}$$

Both Mishra *et al.* (1976) and Ghosh *et al.* (1986) measured the rates of mass transfer from glass plates coated with benzoic acid to non-Newtonian solutions of carboxyme-thyl cellulose ($0.88 \leq n \leq 1$; $5 \leq Re_L \leq {\sim}200$), and they found a satisfactory ($\pm 25\%$) agreement between their data and the predictions of equation (7.73). In this narrow range of n values, there is a little difference between the predictions for $n = 1$ and $n = 0.88$, however. Thus, the predictive equations developed for Newtonian fluids can be applied without incurring appreciable errors.

For both heat and mass transfer in laminar boundary layers, it has been assumed that the momentum boundary layer is everywhere thicker than the thermal and diffusion boundary layers. For Newtonian fluids ($n = 1$), it can readily be seen that ε varies as

$Pr^{-1/3}$ and $\varepsilon_m \propto Sc^{-1/3}$. Most Newtonian liquids (other than molten metals) have the values of Prandtl number >1 and therefore the assumption of $\varepsilon < 1$ is justified. Likewise, one can justify this assumption for mass transfer provided $Sc > 0.6$. Most non-Newtonian polymer solutions used in heat and mass transfer studies to date seem to have large values of Prandtl and Schmidt numbers (Ghosh *et al.*, 1994), and therefore the assumptions of $\varepsilon \ll 1$ and $\varepsilon_m \ll 1$ are valid.

7.8 Boundary layers for visco-elastic fluids

For flow over a bluff body, the fluid elements are subjected to a rapid change in deformations near the frontal face; hence elastic effects are likely to be important in this region and the simple boundary layer approximations should not be applied to visco-elastic materials in this region. However, if elastic effects are negligibly small, the previous approach is reasonably satisfactory for visco-elastic fluids. For instance, the normal stresses developed in visco-elastic fluids will give rise to additional terms in the *x*-component of the momentum balance, equation (7.7) (Harris, 1977; Schowalter, 1978). This leads to the boundary layer having a finite thickness at the leading edge ($x = 0$) which is a function of the Deborah number. The effect of fluid visco-elasticity on the boundary layer appears to persist at considerable distances from the leading edge, resulting in thickening at all points (Harris, 1977; Ruckenstein, 1994). Although various treatments of visco-elastic boundary layers differ in detail, in all cases it is assumed that the fluid elasticity is small and the Deborah number is low in the flow regions of interest (Beard and Walters, 1964; Denn, 1967; Serth, 1973). Unfortunately, there is a lack of experimental results for boundary layer flows of visco-elastic fluids. However, Hermes and Fredrickson (1967), in a study of the flow of a series of carboxymethyl cellulose solutions over a flat plate, do show that the visco-elasticity has a significant effect on the velocity profile in the boundary layer. More quantitative comparisons between experiments and predictions are not possible, as some of the rheological parameters inherent in the theoretical treatments are not capable of evaluation from simple rheological measurements. On the other hand, Ruckenstein (1994) has used dimensional considerations to infer that the effect of fluid elasticity is to reduce the dependence of heat and mass transfer coefficients on the free stream velocity. This conclusion is qualitatively consistent with the only limited heat transfer experimental results which are available (James and Acosta, 1970; James and Gupta, 1971). It is thus not yet possible at present to suggest simple expressions for the thickness of visco-elastic boundary layers analogous to those developed for inelastic power-law fluids. Good accounts of development in the field have been presented by Schowalter (1978) and others (Shenoy and Mashelkar, 1982; Chhabra, 1999b, 2006).

7.9 Practical correlations for heat and mass transfer

In addition to the theoretical treatments just described, many workers have measured heat and mass transfer by forced convection from bodies such as plates, spheres and cylinders. The bulk of the literature in the field has been reviewed by numerous workers (Shenoy and Mashelkar, 1982; Irvine and Karni, 1987; Nakayama, 1988; Chhabra, 1993, 1999b, 2006; Ghosh *et al.*, 1994), but only a selection of reliable correlations for spheres and

Figure 7.5   *Overall correlation for heat and mass transfer from a single sphere immersed in power-law fluids*

cylinders is presented in this section which is based primarily on the reviews of Irvine and Karni (1987) and of Ghosh *et al.* (1994).

7.9.1 Spheres

For particle–liquid heat and mass transfer in non-Newtonian polymer solutions flowing over spheres fixed in tubes ($0.25 \leq d/D_t \leq 0.5$), Ghosh *et al.* (1992, 1994) invoked the usual heat and mass transfer analogy, i.e., $Sh \equiv Nu$ and $Sc \equiv Pr$ to correlate most of the literature data as follows:

$$Y = 1.428 Re_p^{1/3} \quad Re_p < 4 \tag{7.74a}$$

$$Y = Re_p^{1/2} \quad Re_p < 4 \tag{7.74b}$$

where $Y = (Nu - 2)(m_s/m_b)^{1/(3n+1)} Pr_p^{-1/3}$ for heat transfer and $Y = (Sh - 2)Sc_p^{-1/3}$ for mass transfer.

The predicted results are shown in Figure 7.5 for both heat and mass transfer over the range of conditions: $0.32 \leq n \leq 0.93$; $Re_p < 200$; $29\,000 \leq Sc_p \leq 4.9 \times 10^5$ and $9500 \leq Pr_p \leq 1.9 \times 10^6$. The effective viscosity used in the Reynolds, Prandtl and Schmidt numbers (denoted by suffix p) is $m(V/d)^{n-1}$ where V is the mean velocity of flow in the tube. The average deviation between predictions and data is 17%. Finally, the available limited numerical results for heat transfer from spheres (maintained at a constant temperature or subject to uniform heat flux) to power-law fluids are also consistent with these experimental results (Dhole *et al.*, 2006).

7.9.2 Cylinders in cross-flow

The limited work on heat and mass transfer between power-law fluids and cylinders with their axis normal to the flow has been summarized by Ghosh *et al.* (1994) who proposed the following correlation for convective heat and mass transfer.

Figure 7.6 *Overall correlation for heat and mass transfer from cylinders in cross-flow*

For heat transfer:

$$Nu = \frac{hd}{k} = 1.18Re_p^{1/3}Pr_p^{1/3} \quad \text{for } Re_p < 10 \qquad (7.75a)$$

$$= 0.76Re_p^{1/2}Pr_p^{1/3} \quad \text{for } Re_p > 10 \qquad (7.75b)$$

For mass transfer:

$$Sh = \frac{h_D d}{D} = 1.18Re_p^{1/3}Sc_p^{1/3} \quad \text{for } Re_p < 10 \qquad (7.76a)$$

$$= 0.76Re_p^{1/2}Sc_p^{1/3} \quad \text{for } Re_p > 10 \qquad (7.76b)$$

where the effective viscosity is evaluated, as for spheres, with the diameter of the cylinder as the characteristic length. The overall correlations for heat and mass transfer are given in Figure 7.6. It is appropriate to add here that these experimental results are also in line with the scant numerical results for heat transfer from long cylinders to power-law fluids (Soares *et al.*, 2005; Bharti *et al.*, 2007a, b).

Example 7.3

A polymer solution at 25°C flows at 1.8 m/s over a heated hollow copper sphere of diameter 30 mm, maintained at a constant temperature of 55°C (by steam condensing inside the sphere). Estimate the rate of heat loss from the sphere. The thermo-physical properties of the polymer solution may be approximated by those of water, the power-law constants in the temperature interval $25°C \le T \le 55°C$ are given below: $n = 0.26$ and $m = 26 - 0.0566T$ where T is in K. What will be the rate of heat loss from a cylinder 30 mm in diameter and 60 mm long, oriented normal to flow?

Solution

At the mean film temperature of $(25 + 55)/2 = 40°C$, the values of ρ, C_p and k for water:

$$\rho = 991\,\text{kg/m}^3;\ C_p = 4180\,\text{J/kg}°C;\ k = 0.634\,\text{W/mK}$$

The consistency index, m:

$$m = 26 - 0.0566(273 + 40) = 8.28\,\text{Pa s}^n$$

The effective viscosity,

$$\mu_{\text{eff}} = m\left(\frac{V}{d}\right)^{n-1}$$

$$= 8.28 \times \left(\frac{1.8}{30 \times 10^{-3}}\right)^{0.26-1} = 0.4\,\text{Pa s}$$

$$\therefore Re_p = \frac{\rho V d}{\mu_{\text{eff}}} = \frac{991 \times 1.8 \times 30 \times 10^{-3}}{0.4} = 134 > 4$$

Therefore, equation (7.74b) applies in this case:

$$(Nu - 2)\left(\frac{m_s}{m_b}\right)^{1/(3n+1)} = Re_p^{1/2} Pr_p^{1/3}$$

At $T_s = 328\,\text{K}: m_s = 7.44\,\text{Pa s}^n$
$\quad\ T_b = 313\,\text{K}: m_b = 8.28\,\text{Pa s}^n$

$$Pr_p = \frac{C_p \mu}{k} = \frac{4180 \times 0.4}{0.634} = 2637$$

$$\therefore Nu = \frac{hd}{k} = 2 + Re_p^{1/2} Pr_p^{1/3}\left(\frac{m_b}{m_s}\right)^{1/(3n+1)}$$

Substituting values,

$$\frac{hd}{k} = 2 + 134^{1/2} \times 2637^{1/3} \times \left(\frac{8.28}{7.44}\right)^{1/1.78}$$

$$= 172$$

$$\therefore h = \frac{172 \times 0.634}{30 \times 10^{-3}} = 3631\,\text{W/m}^2\,\text{K}$$

$$\therefore \text{rate of heat loss, } q = hA(T_s - T_f)$$

$$= 3631 \times \pi \times (30 \times 10^{-3})^2 \times (55 - 25)$$

$$= 308\,\text{W}$$

For a cylinder, the Reynolds number of the flow is still the same and therefore equation (7.75b) applies:

$$\therefore Nu = \frac{hd}{k} = 0.76\,Re_p^{1/2}\,Pr_p^{1/3}$$

$$= 0.76 \times (134)^{1/2} \times (2637)^{1/3} = 121.5$$

$$\therefore h = \frac{121.5 \times 0.634}{30 \times 10^{-3}} = 2570\,\text{W/m}^2\,\text{K}$$

$$\therefore \text{rate of heat loss, } q = hA(T_s - T_f)$$

$$= 2570 \times \pi \times 30 \times 10^{-3} \times 60 \times 10^{-3} \times (55 - 25)$$

$$= 436\,\text{W}$$

Note that the drop in the value of the heat transfer coefficient in this case has been compensated by the increase in surface area resulting in higher rate of heat loss. Also, the heat loss from the flat ends has been neglected. □

7.10 Heat and mass transfer by free convection

In free convection, there is a new complexity in that fluid motion arises from buoyancy forces due to the density differences, and the momentum, heat and mass balance equations are therefore coupled. The published analytical results for heat transfer from plates, cylinders and spheres involve significant approximations. This work has been reviewed by Shenoy and Mashelkar (1982), Irvine and Karni (1987) and Chhabra (2006); the simple expressions (which are also considered to be reliable) for heat transfer coefficients are given in the following sections.

7.10.1 Vertical plates

For a plate maintained at a constant temperature in a power-law fluid, the mean value of the Nusselt number based on the height of the plate, L, is given as:

$$Nu = \frac{hL}{k} = C_0(n)\, Gr^{1/(2(n+1))}\, Pr^{n/(3n+1)} \tag{7.77a}$$

where the Grashof number, $Gr = \dfrac{\rho^2 L^{n+2} (g\beta\Delta T)^{2-n}}{m^2}$

and the Prandtl number,

$$Pr = \frac{\rho C_\mathrm{p}}{k}\left(\frac{m}{\rho}\right)^{2/(n+1)} L^{(n-1)/2(n+1)} (g\beta\Delta T)^{3(n-1)/2(n+1)} \tag{7.77b}$$

All physical properties are evaluated at the mean film temperature, except the coefficient of expansion, β, which is evaluated at the bulk fluid temperature. The constant $C_0(n)$ has values of 0.60, 0.68 and 0.72 for n values of 0.5, 1 and 1.5, respectively; a mean value of 0.66 may be used for most design calculations. The available scant data on heat transfer are in line with the predictions of equation (7.77).

For constant heat flux at the plate surface, the point values of the Nusselt number at a distance, x, from the base of the plate has been correlated empirically by:

$$Nu_x = \frac{hx}{k} = C[Gr^{(3n+2)/(n+4)}\, Pr^n]^B \tag{7.78a}$$

where

$$Gr = \frac{\rho^2 x^4}{m^2}\left(\frac{g\beta q_\mathrm{w}}{k}\right)^{2-n}$$

$$Pr = \frac{\rho C_\mathrm{p}}{k}\left(\frac{m}{\rho}\right)^{5/(n+4)} x^{2(n-1)/(n+4)} \left(\frac{g\beta q_\mathrm{w}}{k}\right)^{3(n-1)/(n+4)} \tag{7.78b}$$

The recommended values (Irvine and Karni, 1987) of C and B are 0.6 and 0.21, respectively.

7.10.2 Isothermal spheres

Experimental results for heat transfer (Liew and Adelman, 1975; Amato and Tien, 1976), mass transfer (Lee and Donatelli, 1989) and the approximate analytical results (Acrivos, 1960; Stewart, 1971) are well represented by the following simple relationship:

$$Nu = \frac{hd}{k} = 2[Gr^{1/2(n+1)}Pr^{n/(3n+1)}]^{0.682} \tag{7.79a}$$

$$\text{for } Gr^{1/2(n+1)}Pr^{n/(3n+1)} < 10$$

and

$$Nu = \frac{hd}{k} = Gr^{1/2(n+1)}Pr^{n/(3n+1)} \tag{7.79b}$$

$$\text{for } 10 \leq Gr^{1/2(n+1)}Pr^{n/(3n+1)} \leq 40$$

where the Grashof and Prandtl numbers are given by equation (7.77b) with the height of the plate, L, replaced by the sphere radius, R. For mass transfer, the Nusselt and Prandtl numbers are replaced by the corresponding Sherwood and Schmidt numbers, respectively.

7.10.3 Horizontal cylinders

Gentry and Wollersheim (1974) and Kim and Wollersheim (1976) have measured the rates of heat transfer between isothermal horizontal cylinders and power-law polymer solutions. For pseudoplastic fluids, they correlated their results by the following expression:

$$Nu = \frac{hd}{k} = 1.19(GrPr)^{0.20} \tag{7.80}$$

where the Grashof and Prandtl numbers are given by equation (7.78b) with L replaced by the diameter of the cylinder.

Example 7.4

The heat flux at the surface of an electrically heated vertical plate (250 mm × 250 mm) immersed in a polymer solution at 20°C is constant at 250 W/m². Estimate the heat transfer coefficient and the average temperature of the plate. The physical properties of the polymer solution may be approximated by those of water, and the power-law constants may be taken as: $n = 0.5$ and $m = 20 - 0.05T$ (Pa s^n) in the range 288–360 K.

Solution

Since the temperature of the plate is not known, the physical properties of the polymer solution cannot immediately be evaluated. Assuming the average temperature of the plate to be 45°C, the mean film temperature, $T_f = (45 + 20)/2 = 32.5°C$ and the physical properties (of water) are:

$\rho = 993 \, \text{kg/m}^3$ $k = 0.625 \, \text{W/m K}$

$C_p = 4180 \, \text{J/kg K}$ $\beta \, (\text{at } 20°C) = 320.6 \times 10^{-6} \, (1/K)$

$m = 4.73 \, \text{Pa s}^n$ $n = 0.5$

$$\therefore Gr = \frac{\rho^2 x^4}{m^2} \left(\frac{g\beta q_w}{k} \right)^{2-n}$$

$$= \frac{993^3 \times x^4}{(4.73)^2} \times \left(\frac{9.81 \times 320.6 \times 10^{-6} \times 250}{0.625} \right)^{2-0.5}$$

$$= 62189 x^4$$

where x is in metres.

$$Pr = \frac{\rho C_p}{k} \left(\frac{m}{\rho} \right)^{5/(n+4)} x^{2(n-1)/(n+4)} \left(\frac{g\beta q_w}{k} \right)^{3(n-1)/(n+4)}$$

$$= \frac{993 \times 4180}{0.625} \left(\frac{4.73}{993} \right)^{5/(4.5)} \times x^{-(1/4.5)} \left(\frac{9.81 \times 320.6 \times 10^{-6} \times 250}{0.625} \right)^{-1/3}$$

$$= 16178 x^{-1/4.5}$$

$$Nu_x = \frac{h_x x}{k} = 0.6[Gr^{(3n+2)/(n+4)} Pr^n]^{0.21}$$

$$= 0.6[(62189 x^4)^{3.5/4.5} (16178 x^{-1/4.5})^{0.5}]^{0.21}$$

or $h_x = 6.29 x^{-0.37}$

The average value of heat transfer coefficient, $\overline{h}$:

$$\overline{h} = \frac{1}{L} \int_0^L h_x dx = \frac{1}{L} \int_0^L 6.29 x^{-0.37} dx$$

or

$$\overline{h} = \frac{6.29}{(250 \times 10^{-3})} \times \frac{(250 \times 10^{-3})^{0.63}}{0.63} = 16.7 \, W/m^2 \, K$$

From heat balance,

$$h_m(T_s - T_f) = 250$$

$$\therefore T_s = \frac{250}{16.7} + 20 = 35°C$$

Therefore, the assumed value of 45°C is too high and another iteration must be carried out. With the surface temperature of 35.5°C, the mean film temperature is $(35.5 + 20)/2 = 27.75°C$, at which $m = 4.96 \, Pa\,s^n$ and $\beta = 300 \times 10^{-6} \, K^{-1}$ and the other properties are substantially unaltered. The new values yield $\overline{h} = 16.4 \, W/m^2 \, K$. The plate temperature is then 35.3°C which is sufficiently close to the assumed value of 35.5°C. □

Further reading

Schlichting, H., *Boundary Layer Theory*, 6th edition, McGraw Hill, New York (1968).
Schowalter, W.R., *Mechanics of Non-Newtonian Fluids*, Pergamon, Oxford (1978).

References

Acrivos, A., *AIChE J.* **6** (1960) 584.

Acrivos, A., Shah, M.J. and Petersen, E.E., *AIChE J.* **6** (1960) 312.

Agarwal, M., Chhabra, R.P. and Eswaran, V., *Chem. Eng. Sci.* **57** (2002) 1331.

Amato, W.S. and Tien, C., *Int. J. Heat Mass Trans.* **19** (1976) 1257.

Beard, D.W. and Walters, K., *Proc. Camb. Phil. Soc.* **60** (1964) 667.

Bharti, R.P., Chhabra, R.P. and Eswaran, V., *Int. J. Heat Mass Trans.* **50** (2007) 977.

Blasius, H., *Forsch. Ver. Deut. Ing.* **131** (1913).

Chhabra, R.P., *Adv. Heat Trans.* **23** (1993) 187.

Chhabra, R.P., *J. Chem. Eng. Jpn.* **32** (1999a) 812.

Chhabra, R.P., *Advances in the Rhealogy and Flow of Non-Newtonian Fluids* (edited by Siginer, D., Dekec, D. and Chhabra, R.P.), Elsevier, Amsterdam (1999b), Chapter 39.

Chhabra, R.P., *Bubbles, Drops and Particles in Non-Newtonian Fluids*, 2nd edition CRC Press, Boca Raton, FL (2006).

Coulson, J.M. and Richardson, J.F., *Chemical Engineering*, 6th edition, Vol. 1, Butterworth-Heinemann, Oxford (1999).

Denn, M.M., *Chem. Eng. Sci.* **22** (1967) 395.

Dhole, S.D., Chhabra, R.P. and Eswaran, V., *AIChE J.* **52** (2006) 3658.

Gentry, C.C. and Wollersheim, D.E., *J. Heat Trans. (ASME)* **96** (1974) 3.

Ghosh, U.K., Kumar, S. and Upadhyay, S.N., *Chem. Eng. Commun.* **43** (1986) 335.

Ghosh, U.K., Kumar, S. and Upadhyay, S.N., *Polym.-Plast. Technol. Eng.* **31** (1992) 271.

Ghosh, U.K., Upadhyay, S.N. and Chhabra, R.P., *Adv. Heat Trans.* **25** (1994) 251.

Harris, J., *Rheology and Non-Newtonian Flow*, Longman, London (1977).

Hermes, R.A. and Fredrickson, A.G., *AIChE J.* **13** (1967) 253.

Huang, M.-J. and Chen, C.-K., *Int. Commun. Heat Mass Trans.* **11** (1984) 361.

Irvine, T.F., Jr and Karni, J., *Handbook of Single Phase Convective Heat Transfer* (edited by Kakac, S., Shah, R.K. and Aung, W.), Wiley, New York (1987). Chapter 5.

James, D.F. and Acosta, A.J., *J. Fluid Mech.* **42** (1970) 269.

James, D.F. and Gupta, O.P., *Chem. Eng. Prog. Sym. Ser.* **67**(111) (1971) 62.

Khan, W.R., Culham, J.R. and Yovanovich, M.M., *J. Heat Trans. (ASME)* **128** (2006) 870.

Kim, C.B. and Wollersheim, D.E., *J. Heat Trans. (ASME)* **98** (1976) 144.

Lee, T.-L. and Donatelli, A.A., *Ind. Eng. Chem. Res.* **28** (1989) 105.

Liew, K.S. and Adelman, M., *Can. J. Chem. Eng.* **53** (1975) 494.

Mishra, I.M., Singh, B. and Mishra, P., *Indian J. Tech.* **14** (1976) 322.

Nakayama, A., *Encyclopedia of Fluid Mechanics* (edited by Cheremisinoff, N.P.), Vol. 7, Gulf, Houston, TX, p. 305 (1988).

Ruckenstein, E., *Ind. Eng. Chem. Res.* **33** (1994) 2331.

Schlichting, H., *Boundary Layer Theory*, 6th edition, McGraw Hill, New York (1968).

Schowalter, W.R., *Mechanics of Non-Newtonian Fluids*, Pergamon, Oxford (1978).

Serth, R.W., *AIChE J.* **19** (1973) 1275.

Shenoy, A.V. and Mashelkar, R.A., *Adv. Heat Trans.* **15** (1982) 143.

Skelland, A.H.P., *Non-Newtonian Flow and Heat Transfer*, Wiley, New York (1967).

Soares, A.A., Ferreira, J.M. and Chhabra, R.P., *Ind. Eng. Chem. Res.* **44** (2005) 5815.

Stewart, W.E., *Int. J. Heat Mass Trans.* **14** (1971) 1013.

Wu, J. and Thompson, M.C., *J. Non-Newt. Fluid Mech.* **66** (1996) 127.

Nomenclature

		Dimensions in $\mathbf{M, N, L, T, \theta}$
a, b, c, d	unknown constants in equation (7.8)	$\mathbf{M^0 L^0 T^0}$
Bi	Bingham number $(-)$	$\mathbf{M^0 L^0 T^0}$
C_A	solute concentration (kmol/m^3)	$\mathbf{N}^{-3}$
C_{A_0}	solute concentration at the solid surface (kmol/m^3)	$\mathbf{N}^{-3}$
C_{A_s}	solute concentration in the free stream (kmol/m^3)	$\mathbf{N}^{-3}$

C_D	drag coefficient ($-$)	$M^0L^0T^0$
C_P	heat capacity (J/kg K)	$L^2T^{-2}\theta^{-1}$
d	sphere or cylinder diameter (m)	L
D_t	tube diameter (m)	L
D_{AB}	molecular diffusion coefficient (m²/s)	L^2T^{-1}
F_d	drag force (N)	MLT^{-2}
$F(n)$	function of n, equation (7.17) ($-$)	$M^0L^0T^0$
g	acceleration due to gravity (m/s²)	LT^{-2}
Gr	Grashof number, equation (7.77b) ($-$)	$M^0L^0T^0$
h	heat transfer coefficient (W/m² K)	$MT^{-3}\theta^{-1}$
h_m	average value of heat transfer coefficient (W/m² K)	$MT^{-3}\theta^{-1}$
h_D	mass transfer coefficient (m/s)	LT^{-1}
H	height of the control volume in y-direction (m)	L
k	thermal conductivity (W/m K)	$MLT^{-3}\theta^{-1}$
L	length or height of plate (m)	L
m	power-law consistency coefficient (Pa sⁿ)	$ML^{-1}T^{n-2}$
n	power-law flow behaviour index ($-$)	$M^0L^0T^0$
p	pressure (Pa)	$ML^{-1}T^{-2}$
Pr	Prandtl number ($-$)	$M^0L^0T^0$
q	heat flux (W/m²)	MT^{-3}
q_w	heat flux at wall or surface (W/m²)	MT^{-3}
R	sphere radius (m)	L
Re	Reynolds number ($-$)	$M^0L^0T^0$
Sc	Schmidt number ($-$)	$M^0L^0T^0$
Sh	Sherwood number ($-$)	$M^0L^0T^0$
T	temperature (K)	θ
T_0	free stream fluid temperature (K)	θ
T_s	temperature of surface (K)	θ
V_0	free stream fluid velocity (m/s)	LT^{-1}
V_x	velocity in x-direction (m/s)	LT^{-1}
W	width of plate (m)	L
x	coordinate in the direction of flow (m)	L
y	coordinate in the direction normal to flow (m)	L

Greek letters

α	thermal diffusivity (m²/s)	L^2T^{-1}
β	coefficient of expansion (1/K)	θ^{-1}
δ	momentum boundary layer thickness (m)	L
δ_m	diffusion boundary layer thickness (m)	L
δ_T	thermal boundary layer thickness (m)	L
ε	ratio of δ_T to δ ($-$)	$M^0L^0T^0$
ε_m	ratio of δ_m to δ ($-$)	$M^0L^0T^0$
θ	temperature difference $= T - T_s$ (K)	θ
θ_0	temperature difference $= T_0 - T_s$ (K)	θ
μ_B	Bingham plastic viscosity (Pa s)	$ML^{-1}T^{-1}$
ρ	fluid density (kg/m³)	ML^{-3}
τ_0^B	Bingham yield stress (Pa)	$ML^{-1}T^{-2}$
τ_w	wall shear stress (Pa)	$ML^{-1}T^{-2}$

Subscripts

p	Value using effective viscosity $m(V/d)^{n-1}$
m	Mean value
x	Value at an arbitrary point

Chapter 8
Liquid mixing

8.1 Introduction

Mixing is one of the most common operations in the chemical, biochemical, polymer processing and allied industries. Almost all manufacturing processes entail some sort of mixing, and the operation may constitute a considerable proportion of the total processing time. The term 'mixing' is applied to the processes used to reduce the degree of non-uniformity or gradient of a property such as concentration, viscosity, temperature, colour and so on. Mixing can be achieved by moving material from one region to another. It may be of interest simply as means of reaching a desired degree of homogeneity, but it may also be used to promote heat and mass transfer, often where a system is undergoing a chemical reaction.

At the outset, it is useful to consider some common examples of problems encountered in industrial mixing operations, since this will not only reveal the ubiquitous nature of the process, but will also provide an appreciation of some of the associated difficulties. One can classify mixing problems in many ways, such as the flowability of the final product in the mixing of powders, but it is probably most satisfactory to base this classification on the phases present: liquid–liquid, liquid–solid, gas–liquid, etc. This permits a unified approach to the mixing problems in a range of industries.

8.1.1 Single phase liquid mixing

In many instances, two or more miscible liquids must be mixed to give a product of a desired specification, as for example, in the blending of petroleum fractions of different viscosities. This is the simplest type of mixing as it does not involve either heat or mass transfer, or indeed a chemical reaction. Even such simple operations can, however, pose problems when the two liquids have vastly different viscosities, or if density differences are sufficient to lead to stratification. Another example is the use of mechanical agitation to enhance the rates of heat and mass transfer between a liquid and the wall of a vessel, or a coil. Additional complications arise in the case of highly viscous Newtonian and non-Newtonian materials.

8.1.2 Mixing of immiscible liquids

When two immiscible liquids are stirred together, one liquid becomes dispersed as droplets in the second liquid which forms a continuous phase. Liquid–liquid extraction, a process using successive mixing and settling stages, is one important example of this type of mixing. The two liquids are brought into contact with a solvent that selectively dissolves one of the components present in the mixture. Agitation causes one phase to disperse in the other and, if the droplet size is small, a high interfacial area is created for interphase mass transfer. When the agitation is stopped, phase separation may occur, but care must be taken to ensure that the droplets are not so small that a diffuse layer is formed instead of a well-defined interface; this can remain in a semi-stable state over a long period of time and prevent the completion of effective separation. The production of stable emulsions such as those encountered in food, brewing and pharmaceutical applications provides another important example of dispersion of two immiscible liquids. In these systems, the droplets are very small and are often stabilized by surface-active agents, so that the resulting emulsion is usually stable for considerable lengths of time.

8.1.3 Gas–liquid dispersion and mixing

Numerous processing operations involving chemical reactions, such as aerobic fermentation, wastewater treatment, oxidation, hydrogenation or chlorination of hydrocarbons and so on, require good contact between a gas and a liquid. The purpose of mixing here is to produce a large interfacial area by dispersing bubbles of the gas into the liquid. Generally, gas–liquid mixtures or dispersions are unstable and separate rapidly if agitation is stopped, provided that a foam is not formed. In some cases, a stable foam is needed (such as fire fighting foam); this can be formed by injecting gas into a liquid using intense agitation, and stability can be increased by the addition of a surface-active agent.

8.1.4 Liquid–solid mixing

Mechanical agitation may be used to suspend particles in a liquid in order to promote mass transfer or a chemical reaction. The liquids involved in such applications are usually of low viscosity, and the particles will settle out when agitation ceases. There is also an occasional requirement to achieve a relatively homogeneous suspension in a mixing vessel, particularly when this is being used to prepare materials for subsequent processes.

At the other extreme, in the formation of composite materials, especially filled polymers, fine particles must be dispersed into a highly viscous Newtonian or non-Newtonian liquid. The incorporation of carbon black powder into rubber is one such operation. Because of the large surface areas involved, surface phenomena play an important role in these applications.

8.1.5 Gas–liquid–solid mixing

In some applications such as catalytic hydrogenation of vegetable oils, slurry reactors, three phase fluidized beds, froth flotation, fermentation and so on, the success and efficiency of the process is directly influenced by the extent of mixing between the three phases. Despite its great industrial importance, this topic has received only scant attention and the mechanisms and consequences of interactions between the phases are almost unexplored.

8.1.6 Solid–solid mixing

Mixing together of particulate solids, sometimes referred to as blending, is a very complex process in that it is very dependent, not only on the character of the particles – density, size, size distribution, shape and surface properties – but also on the differences between these properties of the components. Mixing of sand, cement and aggregate to form concrete, and of the ingredients in gun powder preparation, are longstanding examples of the mixing of solids.

Other industrial sectors employing solids mixing include food, drugs, agriculture and the glass industries, for example. All these applications involve only physical contacting, although in recent years, there has been a recognition of the industrial importance of solid–solid reactions, and solid–solid heat exchangers. Unlike liquid mixing, fundamental research on solids mixing has been limited until recently. The phenomena involved are very different from those when a liquid phase is present, so solid–solid mixing will not be discussed further here. However, most of the literature on solid–solid mixing has been reviewed elsewhere (Lindley, 1991; Harnby *et al.*, 1992; van den Bergh, 1994; Xuereb *et al.*, 2006).

8.1.7 Miscellaneous mixing applications

Mixing equipment may be designed not only to achieve a predetermined level of homogenity but also to improve the rates of heat and mass transfer, and the yield of chemical reactions. For example, if the rotational speed of an impeller in a mixing vessel is selected so as to achieve a required rate of heat transfer, the agitation may then be more than sufficient for the mixing duty. Excessive or over-mixing should be avoided. For example, in biological applications, excessively high impeller speeds or power input are believed by many to give rise to shear rates which may damage micro-organisms. In a similar way, where the desirable rheological properties of some polymer solutions may be attributable to structured long-chain molecules, excessive impeller speeds or agitation over prolonged periods, may damage the structure particularly of molecular aggregates, thereby altering their properties. Similarly, the texture and structure of many products (such as food, pharmaceutical, house-hold products, fermentation broths, etc.) are extremely shear sensitive and obviously in such applications, the prediction and control of distribution of shear rates in the mixing vessel are much more important than the estimation of power consumption (Benz, 2003, 2004). Yet in certain other applications, the viscosity evolves with time thereby adding to the degree of difficulty to achieve the desired level of homogenization. For instance, lubricating greases are produced by carrying out *in situ* saponification of glyceride fatty acids, followed by dehydration, heating it to the phase transition temperature of the soap, cooling it down to achieve soap crystallization, milling and mixing various additives. At each stage, the contents of the reactor–mixer vessel possess different rheological characteristics. Obviously, no single impeller will be equally efficient under all conditions. Similarly, lignin-based slurry fuels offer a promising alternative to fossil fuels in kraft pulp processes. The preparation of the so-called lignogels entails effective mixing of lignin, fuel oil, water and surfactant, and indeed their non-Newtonian characteristics determine their atomization and burning characteristics. In turn, the non-Newtonian characteristics are influenced by the size of oil droplets produced during the mixing stage. As our final example, similarly, the quality of bread is strongly influenced by the gluten network present in the dough. When wheat flour and water are mixed, the protein in the dough forms gluten which is a highly elastic cohesive mass. About 20% of the wheat protein is soluble (albumen and globulin), and the remaining 80% is made up of gliadin and glutenin. The glutenin is responsible for the extensibility, strength and firmness of the dough while gliadin is soft, fluid-like and it imparts cohesiveness and elasticity to the dough. Dough consistency (measured in terms of its elasticity and mouldability) is therefore strongly influenced by the degree to which the various ingredients are mixed. It is therefore important to appreciate that 'over-mixing' may often be undesirable because it may result in both excessive energy consumption and impaired product quality. Equally, under-mixing also is obviously undesirable.

From the examples given here, it is abundantly clear that mixing cuts across the boundaries between industries, and there may be a need to mix virtually anything with anything – be it a gas or a solid or a liquid and indeed mixing operations constitute one of the key steps in most process engineering applications; it is clearly not possible to consider the whole range of mixing problems here. Instead attention will be given primarily to batch liquid mixing of viscous Newtonian and non-Newtonian materials, followed by short discussions of gas–liquid systems, and heat transfer in mechanically agitated systems. To a large extent, the rheology of the liquids concerned determines the equipment and procedures to be used in a mixing operation. However, when non-Newtonian fluids are involved, the process itself can have profound effects on the rheological properties of the product, as seen above in the case of bread making dough.

8.2 Liquid mixing

A considerable body of information is now available on batch liquid mixing and this forms the basis for the design and selection of mixing equipment. It also affords some physical insight into the nature of the mixing process itself. In mixing, there are two types of problems to be considered – how to design and select mixing equipment for a given duty, and how to assess whether an available mixer is suitable for a particular application. In both cases, the following aspects of the mixing process must be understood:

 (i) Mechanisms of mixing
 (ii) Scale-up or similarity criteria
 (iii) Power consumption
 (iv) Flow patterns
 (v) Rate of mixing and mixing time.

Each of these factors is now considered in detail.

8.2.1 Mixing mechanisms

If mixing is to be carried out in order to produce a uniform product, it is necessary to understand how mixtures of liquids move and approach uniformity of composition. For liquid mixing devices, it is necessary that two requirements are fulfilled. Firstly, there must be bulk or convective flow so that there are no dead or stagnant zones. Secondly, there must be a zone of intensive or high shear mixing in which the inhomogeneities are progressively broken down. Both of these processes are energy consuming and ultimately the mechanical energy is dissipated as heat; the proportion of energy attributable to each varies from one application to another. Depending upon the fluid properties, primarily viscosity, the flow in mixing vessels may be laminar or turbulent, with a substantial transition zone in between and frequently both types of flow occur simultaneously in different parts of the vessel. Laminar and turbulent flows arise from different mechanisms, and it is convenient to consider them separately.

(i) *Laminar mixing*

Large-scale laminar flow is usually associated with high viscosity liquids ($>{\sim}10$ Pa s) which may exhibit either Newtonian or non-Newtonian flow characteristics. Inertial forces therefore tend to die out quickly, and the impeller must sweep through a significant proportion of the cross-section of the vessel to impart sufficient bulk motion. Because the velocity gradients close to a moving impeller are high, the fluid elements in that region deform and stretch. They repeatedly elongate and become thinner each time they pass through the high shear zone. Figure 8.1 shows such a shearing sequence.

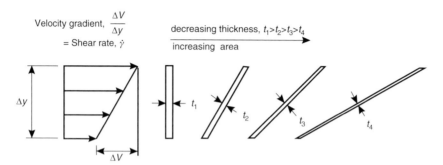

Figure 8.1   *Schematic representation of the thinning of fluid elements due to laminar shear flow*

In addition, extensional or elongational flow usually occurs simultaneously. As shown in Figure 8.2, this can be result of convergence of the streamlines and consequential increase of velocity in the direction of flow. Since for incompressible fluids, the volume remains constant, there must be a thinning or flattening of the fluid elements, as shown in Figure 8.2. Both of these mechanisms (shear and elongation) give rise to stresses in the liquid which then effect in a reduction in droplet size and an increase in interfacial area, by which means the desired degree of homogeneity is obtained.

In addition, molecular diffusion always acts in such a way as to reduce inhomogeneities, but its effect is not significant until the fluid elements have been sufficiently reduced in size for their specific areas to become large. It must be recognized, however, that the ultimate homogenization of miscible liquids can be brought about only by molecular diffusion. In the case of liquids of high viscosity, this is a slow process.

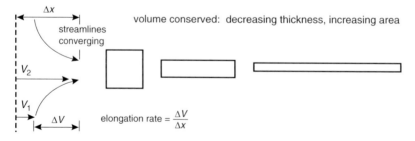

Figure 8.2   *Schematic representation of the thinning of fluid elements due to extensional flow*

In laminar flow, a similar mixing process occurs when a liquid is sheared between two rotating cylinders. During each revolution, the thickness of an initially radial fluid element is reduced, and molecular diffusion takes over when the fluid elements are sufficiently thin. This type of mixing is shown schematically in Figure 8.3 in which the tracer is pictured as being introduced perpendicular to the direction of motion. It will be realized that, if an annular fluid element had been chosen to begin with, then no obvious mixing would have occurred. This emphasizes the importance of the orientation of the fluid elements relative to the direction of shear produced by the mixer.

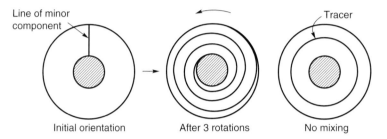

Figure 8.3   *Laminar shear mixing in a coaxial cylinder arrangement*

Finally, mixing can be induced by physically dividing the fluid into successively smaller units and then re-distributing them. In-line mixers for laminar flows rely primarily on this mechanism, as shown schematically in Figure 8.4.

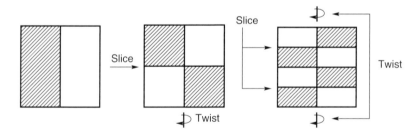

Figure 8.4   *Schematic representation of mixing by cutting and folding of fluid elements*

Thus, mixing in liquids is achieved by several mechanisms which gradually reduce the size or scale of the fluid elements and then re-distribute them. If, for example, there are initial differences in concentration of a soluble material, uniformity is gradually achieved, and molecular diffusion becomes progressively more important as the element size is reduced. Ottino (1989) has illustrated the various stages in mixing by means of colour photographs.

(ii) *Turbulent mixing*
In low viscosity liquids ($<\sim 10\,\text{mPa}\,\text{s}$), the flows generated in mixing vessels with rotating impellers are usually turbulent. The inertia imparted to the liquid by the impeller is sufficient to cause the liquid to circulate throughout the vessel and return to the impeller. Turbulence may occur throughout the vessel but will be greatest near the impeller. Mixing by eddy diffusion is much faster than mixing by molecular diffusion and, consequently, turbulent mixing occurs much more rapidly than laminar mixing.

Ultimately, homogenization at the molecular level depends on molecular diffusion, which in general takes place more rapidly in low viscosity liquids. Mixing is fastest near the impeller because of the high shear rates and associated Reynolds stresses in vortices formed at the tips of the impeller blades; furthermore, a high proportion of the energy is dissipated here.

Turbulent flow is inherently complex, and calculation of the flow fields prevailing in a mixing vessel are not amenable to rigorous theoretical treatment. If the Reynolds number of the main flow is sufficiently high, some insight into the mixing process can be gained by using the theory of local isotropic turbulence. Turbulent flow may be considered to contain a spectrum of velocity fluctuations in which eddies of different sizes are superimposed on an overall time-averaged mean flow. In a mixing vessel, it is reasonable to suppose that the large primary eddies, of a size corresponding approximately to the impeller diameter, would give rise to large velocity fluctuations but would have a low frequency. Such eddies are anisotropic and account for much of the kinetic energy present in the system. Interaction between these primary eddies and slow moving streams produces smaller eddies of higher frequency which undergo further disintegration until, finally, viscous forces cause dissipation of their energy as heat.

The description given here is a gross over-simplification, but it does give a qualitative representation of the salient features of turbulent mixing. This whole process is similar to that of the turbulent flow of a fluid close to a boundary surface. Although some quantitative results for the scale size of eddies have been obtained and some workers (van der Molen and van Maanen, 1978; Tatterson, 1991) have reported experimental measurements on the structure of turbulence in mixing vessels, these studies have little relevance to the mixing of non-Newtonian substances which are usually processed under laminar conditions.

8.2.2 Scale-up of stirred vessels

One of the problems confronting the designers of mixing equipment is that of deducing the most satisfactory arrangement for a large unit from experiments carried out with small units. An even more common problem is the specification of small scale experiments to optimize the use of existing large-scale equipment for new applications. In order to achieve similar flow patterns in two units, geometrical, kinematic and dynamic similarity and identical boundary conditions must be maintained. The problem of scale-up has been discussed in a number of books (Oldshue, 1983; Tatterson, 1991, 1994; Harnby *et al.*, 1992; Coulson and Richardson, 1999; Paul *et al.*, 2004; Xuereb *et al.*, 2006) and therefore only the salient features are re-capitulated here. It has been found convenient to relate the power used by the agitator to the geometrical and mechanical arrangement of the mixer, and thus to obtain a direct indication of the change in power arising from alteration of any of the factors relating to the mixer. A typical mixer arrangement is shown in Figure 8.5. The criteria for similarity between two systems is expressed in terms of dimensionless ratios of geometric dimensions and of forces occurring in the fluid in the mixing vessel. In the absence of heat and mass transfer and for geometrically similar systems, the resulting dimensionless groups are the Reynolds, Froude and Weber numbers for Newtonian fluids, defined respectively as:

$$\text{Reynolds number: } Re = \frac{\rho D^2 N}{\mu} \tag{8.1}$$

D	Diameter of agitator
D_T	Diameter of tank
C	Height of agitator from base of tank
H	Depth of liquid
N_b	Number of baffles
W_B	Width of baffles
N	Speed of agitator
p	Pitch of agitator
W	Width of blades of agitator or paddle

Figure 8.5   *Typical configuration and dimensions of an agitated vessel*

Froude number: $Fr = \dfrac{DN^2}{g}$ (8.2)

Weber number: $We = \dfrac{D^3 N^2 \rho}{\sigma}$ (8.3)

where D is the impeller diameter.

For heat and mass transfer in stirred vessels, additional dimensionless groups which are important include the Nusselt, Sherwood, Prandtl, Schmidt and Grashof numbers. Likewise, in the case of non-Newtonian fluids, an appropriate value of the apparent viscosity must be identified for use in equation (8.1). Furthermore, it may also be necessary to introduce dimensionless parameters indicative of non-Newtonian effects including a Bingham number (Bi) for viscoplastic materials and a Weissenberg number (We) for visco-elastic fluids. It is thus imperative that in order to ensure complete similarity between two systems, all the pertinent dimensionless numbers must be equal. In practice, however, this is not usually possible, owing to conflicting requirements. In general, it is necessary to specify the one or two key features that must be matched. Obviously, in the case of substances not exhibiting a yield stress, the Bingham number is redundant, as is the Weissenberg number for inelastic fluids. Similarly, the Froude number is usually important only when significant vortex formation occurs; this effect is almost non-existent for viscous materials.

Aside from these theoretical considerations, further difficulties can arise depending upon the choice of scale-up criteria, and these are strongly dependent on the type of mixing (e.g. gas–liquid or liquid–liquid) and on the ultimate goal of the mixing process. Thus, for geometrically similar systems, the size of the equipment is determined by the scale-up factor. In most practical situations, the equality of the Reynolds numbers ensures complete similarity of flow between two geometrically identical systems. Though the use of this criterion results in the smallest size of the impeller, it is not used often in practice. The use of constant tip speed (πND) is also common, especially when the product quality correlates well with the tip speed in laboratory scale equipment. However, the most common criterion used in industry is to maintain the power input per unit volume of fluid the same in two systems. With this criterion, the maximum shear rate in the impeller region increases with the scale of the equipment, but the average shear rate decreases, For instance for a scale up ratio of 10, the maximum shear rate may increase by a factor of 2, while the average shear rate may drop by a factor of 5. Similarly, processes involving heat transfer in stirred vessels are often scaled up, either on the basis of equal heat transfer per unit volume of liquid, or by maintaining the same value of the heat transfer coefficient. Other commonly used criteria are equal mixing times, the same specific interfacial area in gas–liquid systems or the same value of mass transfer coefficient.

8.2.3 Power consumption in stirred vessels

From a practical point of view, power consumption is perhaps the most important parameter in the design of stirred vessels. Because of the very different flow patterns and mixing mechanisms involved, it is convenient to consider power consumption in low and high viscosity systems separately.

(i) *Low viscosity systems*
Typical equipment for low viscosity liquids consists of a baffled vertical cylindrical tank, with a height to diameter ratio of 1.5–2, fitted with an agitator which generally operates in the turbulent flow regime. For low viscosity liquids, high speed impellers of diameter between one-third and one-half that of the vessel are suitable, running with tip speeds of around 1–3 m/s. Although work on single phase mixing of low viscosity liquids is of limited relevance to industrial applications involving non-Newtonian materials, it does, however, serve as a useful starting point for the subsequent treatment of high viscosity liquids.

For a stirred vessel of diameter D_T in which a Newtonian liquid of viscosity μ and density ρ is agitated by an impeller of diameter D rotating at a speed of N revolutions per unit time (the other dimensions are as shown in Figure 8.5); the power input P to the liquid depends on the independent variables (μ, ρ, D, D_T, N, g, other geometric dimensions) and may be expressed as:

$$P = f(\mu, \rho, D, D_T, N, g, \text{geometric dimensions}) \tag{8.4}$$

In equation (8.4), P is the impeller power, i.e. the energy per unit time dissipated within the liquid. Clearly, the electrical power required to drive the motor will be greater than P on account of transmission losses in the gear box, motor, bearings and so on.

It is readily acknowledged that the functional relationship expressed in equation (8.4) cannot be established from first principles. However, by using dimensional analysis, the number of variables can be reduced to give:

$$\frac{P}{\rho N^3 D^5} = f\left(\frac{\rho N D^2}{\mu}, \frac{N^2 D}{g}, \text{geometric ratios}\right) \tag{8.5}$$

where the dimensionless group on the left hand side is called the Power number, Po; ($\rho N D^2/\mu$) is the Reynolds number, Re and ($N^2 D/g$) is the Froude number, Fr. The geometric ratios relate to the specific impeller/vessel configuration. For geometrically similar systems, these ratios must be constant and the functional relationship between the Power number and the other groups therefore reduces to:

$$Po = f(Re, Fr) \tag{8.6}$$

In equation (8.6), the Froude number is generally important only when vortex formation occurs, and in single phase mixing, it can be neglected if the value of the Reynolds number is less than about 300. In view of the detrimental effect of vortex formation on the quality of mixing, tanks are usually fitted with baffles and hence in most situations involving low viscosity Newtonian fluids, the Power number is a function of the Reynolds number and geometry only. Likewise, as the liquid viscosity increases, the tendency for vortex formation decreases and thus so does the need of installing baffles in mixing tanks ($\mu > {\sim}5\,\text{Pa}\,\text{s}$). Figure 8.6 shows the functional relationship given by equation (8.6) for a range of impellers used to mix/agitate Newtonian liquids of relatively low viscosity (Bates *et al.*, 1963). For a fixed geometrical arrangement and a single phase liquid, the data can be represented by a unique power curve. Three distinct zones can be discerned in the power curve: at small values of the Reynolds number ($<{\sim}10$–30), laminar flow occurs and the slope of the power curve on log–log coordinates is -1. This region, which is characterized by slow mixing, is where the majority of highly viscous (Newtonian and non-Newtonian) liquids are processed. The limited experimental data suggest that the smaller the value of power-law index (<1), the larger is the value of the Reynolds number up to which the laminar flow conditions persist. Irrespective of the limiting value of the Reynolds number denoting the end of the laminar flow conditions, as the Power number (Po) is inversely proportional to the Reynolds number (Re) in the laminar regime, the product $Re\,Po = K_p$, where K_p is a constant dependent only on the tank–impeller geometrical arrangement. For instance, for the standard Rushton turbine, the value of K_p is 75 for $W/D = 0.2$ and it drops to 50 for $W/D = 0.125$. Similarly, the commonly reported values of K_p for an anchor and a helix are in the ranges 200–300 and 35–50, respectively. For a screw with or without a draft tube, $K_p = 200$–400. It needs to be emphasized that these values are provided here as a general guide to give the reader a feel for it, but it is always desirable to evaluate the value of K_p experimentally for each tank–impeller arrangement. Over the years, considerable effort has been made to develop predictive expressions for K_p as a function of the type of impeller and geometric

Figure 8.6 *Power number–Reynolds number correlation in Newtonian fluids for various turbine impeller designs (redrawn from Bates et al., 1963)*

ratios and these have been summarized by Delaplace and Guerin (2006). For helical ribbons used frequently for the mixing of viscous Newtonian and non-Newtonian fluids, Delaplace and Leuliet (2000) have collated most of the literature data relating to 148 tank–impeller configurations to develop the following expression for K_p:

$$K_p = 91(N_r)^{0.79} \left[0.5 \left(\left(\frac{D_T}{D} \right) - 1 \right) \right]^{-0.31} \left(\frac{s}{D} \right)^{-0.37} \left(\frac{W}{D} \right)^{0.16} \left(\frac{H}{D} \right) \qquad (8.7)$$

Equation (8.7) is based on experimental results covering the following ranges of conditions:

$$1.012 \le \left(\frac{D_T}{D} \right) \le 1.38; \quad 0.07 \le \left(\frac{W}{D} \right) \le 0.22; \quad 0.32 \le \left(\frac{s}{D} \right) \le 2.16;$$

$$0.88 \le \left(\frac{H}{D} \right) \le 1.30 \quad \text{and} \quad N_r = 1 \text{ or } 2.$$

where H is the height of the impeller immersed in liquid, W is the width of the ribbon, s is the pitch of the helical ribbon and N_r is the number of ribbons. Equation (8.7) was stated to reproduce most of the literature data on power consumption for viscous Newtonian liquids with a maximum error of $\pm 20\%$.

At very high values of the Reynolds number ($> \sim 10^4$), the flow is fully turbulent and inertia dominated, resulting in rapid mixing. In this region, the Power number is virtually constant and independent of Reynolds number, as shown in Figure 8.6 for Newtonian fluids and demonstrated recently for shear-thinning polymer solutions (Nouri and Hockey, 1998); however, it depends upon the impeller–vessel configuration. Often gas–liquid, liquid–solid and liquid–liquid contacting operations are carried out in this region. Though the mixing itself is quite rapid, the overall process may be mass transfer controlled.

In between the laminar and turbulent zones ($Re > \sim 10^4$), there exists a substantial transition zone in which the viscous and inertial forces are of comparable magnitudes. No simple mathematical relationship exists between Po and Re in this flow region and, at a given value of Re, the value of Po must be read off the appropriate power curve. Though it is generally accepted that laminar flow occurs for $Re < \sim 10$–30, the transition from transitional to the fully turbulent flow is strongly dependent on the impeller–vessel geometry. Grenville et al. (1995) correlated the critical Reynolds number for Newtonian fluids at the boundary between the transitional and turbulent regime as follows:

$$Re_{cr} = 6370 \, Po_t^{-1/3} \qquad (8.8)$$

where Po_t is the value of the Power number in fully turbulent conditions. The constant value of the Power number under fully turbulent conditions is only dependent upon the tank–impeller geometry. Typical values are shown in Table 8.1 for commonly used impellers (Paul et al., 2004).

Power curves for many different impeller geometries, baffle arrangements and so on are available in the literature (Skelland, 1983; Harnby et al., 1992; Tatterson, 1992, 1994; Ibrahim and Nienow, 1995), but it must always be remembered that though the power curve approach is applicable to any single phase Newtonian liquid, at any impeller speed, the curve will be valid for only one system geometry. While it is difficult to provide definitive information about the role of geometrical parameters on power consumption, some qualitative trends can be summarized here as a general guide. For instance, for a six-bladed Rushton turbine, the power number varies with the width of blades as

Table 8.1 *Typical values of Po_t (for vessels fitted with four standard baffles)*

Type of impeller	Value of Po_t
45° Pitched blade turbine*	
– 4 blades	1.27
– 6 blades	1.64
Marine propeller*	
– 1.0 pitch	0.34
– 1.5 pitch	0.62
Smith or concave – or hollow* blade with 6-blades	4.4
Hollow blade turbine	4.1
Ekato MIG-3 ($D/D_T = 0.7$)	0.55
Ekato Intermig – 2 ($D/D_T = 0.7$)	0.61
High-shear disk ($Re = 10^4$)	0.20
Lightnin A 310	0.30
Chemineer HE 3	0.30

*For $D/D_T = 1/3$; $(C/D_T) = 1/3$ and $(W/D) = 1/5$

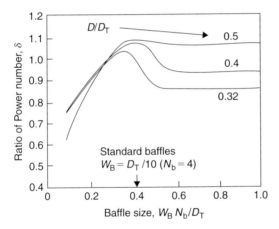

Figure 8.7 *Effect of the size and number of baffles on power number (replotted from Paul et al., 2004)*

$(W/D)^{1.45}$ whereas for a four-bladed 45° pitched blade turbine, it varies as $(W/D)^{0.65}$; similarly, all else being equal, the power number for three- to six-bladed turbines varies as $n_b^{0.8}$ and for six- to twelve-bladed turbines, this dependence weakens somewhat as the Power number varies as $n_b^{0.7}$, where n_b is the number of blades. Similarly, both the size (width W_B) and number (N_b) of baffles have a significant influence on the value of the Power number. As the value of the product (N_bW_B) increases, the Power number increases up to the value of the conventional configuration, namely, four equi-distant baffles with $W_B = (D_T/10)$. At higher values of ($W_BN_b/D_T > 0.6$), the Power number reaches constant values which are influenced only by the value of (D/D_T), as shown in Figure 8.7 (plotted here as the ratio of power numbers for the non-standard to standard

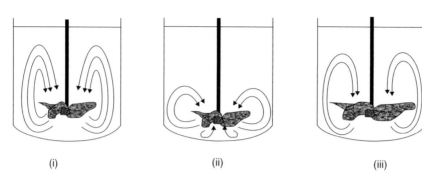

Figure 8.8   *Qualitative effect of bottom clearance ($Po_{ii} > Po_i$) and of impeller diameter ($Po_{iii} > Po_i$) for axial flow impellers (replotted from Xuereb et al., 2006)*

Figure 8.9   *Qualitative effect of bottom clearance ($Po_{ii} > Po_i$) and of impeller diameter ($Po_{iii} > Po_i$) for turbines (replotted from Xuereb et al., 2006)*

configurations). For instance, for a turbine impeller, the limiting values of the Power number range from ~0.85 to ~1.05 as the value of (D/D_T) is increased from 0.32 to 0.5. Figures 8.8 and 8.9 schematically show the influence of bottom clearance (C/D) and impeller diameter (D/D_T) on the performance of an axial flow impeller and a turbine, respectively. In both cases, as the bottom clearance is reduced below the generally rec-ommended value, the power consumption increases and their pumping efficiency dete-riorates by varying extents. Similarly, the use of larger than the usually recommended impeller size (D/D_T) leads to an increase in power as well as in poor pumping efficiency. Furthermore, the effect of the clearance between the impeller and the vessel bottom (C/D) is generally small for radial flow impellers (Figure 8.10). When multiple impel-lers are mounted on the same shaft, the overall power consumption may or may not be additive depending upon the spacing between the impellers and the type of the impeller. The most commonly used spacing between impellers is one impeller diameter (D). For spacing smaller than D, the overall power number is less than twice the single impel-ler power for axial flow impellers (like pitched blade) due to the interfering flow fields produced by the two impellers. On the other hand, for a flat-blade turbine, the power may exceed twice the single impeller power depending on the spacing. The use of mul-tiple impellers also leads to changes in the flow patterns as shown in Figures 8.11 and 8.12 for radial and axial impellers, respectively. Finally, in the case of multiple Rushton turbine impellers mounted on the same shaft and operating in the turbulent flow regime,

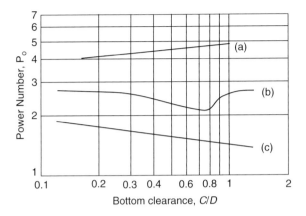

Figure 8.10   *Effect of bottom clearance (C/D) on Power number for turbines: (a) Rushton, (b) flat blade turbine and (c) pitched blade turbine for $(N_b W_B/D_T) = 0.33$ (replotted from Paul et al., 2004)*

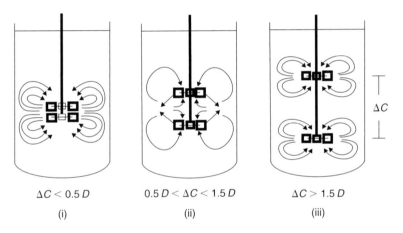

Figure 8.11   *Schematics of the effect of gap between two (radial flow) impellers on flow patterns (replotted from Xuereb et al., 2006)*

a sharp drop in the value of the Power number is observed if the baffle length (L_B) is in the range $(L_B/H) < 0.3$, whereas the Power number is little influenced in the range $0.3 \leq (L_B/H) \leq 0.5$. On the other hand, for $(L_B/H) > 0.5$, the Power number increases sharply (Markopoulos *et al.*, 2004). Overall, adequate information on low viscosity systems is now available for the estimation of power requirements for a given duty under most conditions of practical interest.

Example 8.1

On the assumption that the power required for mixing in a stirred tank is a function of the variables given in equation (8.4), obtain the dimensionless groups which are important in calculating power requirements for geometrically similar arrangements.

$$\Delta C < D$$ $$D < \Delta C < 1.5\,D$$ $$\Delta C > 1.5\,D$$

(i)   (ii)   (iii)

Figure 8.12   *Qualitative effect of gap between two (axial flow) impellers on flow patterns (replotted from Xuereb et al., 2006)*

Solution

The variables in this problem, together with their dimensions, are as follows:

$P : \mathbf{ML^2T^{-3}}$	$\mu : \mathbf{ML^{-1}T^{-1}}$
$\rho : \mathbf{ML^{-3}}$	$N : \mathbf{T^{-1}}$
$g : \mathbf{LT^{-2}}$	$D : \mathbf{L}$
$D_T : \mathbf{L}$	

There are seven variables and with three fundamental units ($\mathbf{M, L, T}$), there will be $7 - 3 = 4$ dimensionless groups.

Choosing as the recurring set ρ, N and D which themselves cannot be grouped together to form a dimensionless group, $\mathbf{M}$, $\mathbf{L}$ and $\mathbf{T}$ can now be expressed in terms of combinations of ρ, N and D. Thus:

$$\mathbf{L} \equiv D \quad \mathbf{T} \equiv N^{-1} \quad \mathbf{M} \equiv \rho D^3$$

Dimensionless group 1 $= P\mathbf{M^{-1}L^{-2}T^3} = P(\rho D^3)^{-1}(D)^{-2}(N^{-1})^3$
$$= P/\rho D^5 N^3$$

Dimensionless group 2 $= \mu\mathbf{M^{-1}LT} = \mu(\rho D^3)^{-1}(D)(N^{-1})$
$$= \mu/\rho D^2 N$$

Dimensionless group 3 $= g\mathbf{L^{-1}T^2} = g(D)^{-1}(N^{-1})^2$
$$= g/N^2 D$$

Dimensionless group 4 $= D_T\mathbf{L^{-1}} = D_T/D$

Hence,

$$\frac{P}{\rho N^3 D^5} = \mathrm{f}\left(\frac{\rho D^2 N}{\mu}, \frac{N^2 D}{g}, \frac{D_T}{D}, \dots\right)$$

which corresponds with equation (8.5). □

As discussed above, the Froude number exerts little influence on power requirement under most conditions of practical interest, and hence for geometrically similar systems:

$$Po = f(Re)$$

At low Reynolds numbers, $Po \propto (1/Re)$ and therefore power varies as $D^3 N^2/\mu$ and it is independent of the fluid density (negligible inertia). On the other hand, power varies as $\rho D^5 N^3$ under fully turbulent conditions when the fluid viscosity is of little importance.

(ii) *High viscosity Newtonian and inelastic non-Newtonian systems*
As noted previously, mixing in highly viscous liquids is slow both at the molecular scale, on account of the low values of molecular diffusivity and at the macroscopic scale, due to low levels of bulk flow. Whereas in low viscosity liquids momentum can be transferred from a rotating impeller through a relatively large body of fluid, in highly viscous liquids only the fluid in the immediate vicinity of the impeller is influenced by the agitator and the flow is normally laminar.

For the mixing of highly viscous and non-Newtonian fluids, it is usually necessary to use specially designed impellers with close clearances at the vessel walls (as discussed in a later section). High speed stirring with small impellers merely wastefully dissipates energy at the central portion of the vessel, particularly when the liquid is highly shear-thinning and/or possesses a yield stress. The power-curve approach is usually applicable. Although highly viscous Newtonian fluids include sugar syrups, glycerol and many lubricating oils, most of the highly viscous fluids of interest in the chemical and processing industries exhibit non-Newtonian flow characteristics.

A simple relationship has been shown to exist, however, between the power consumption for time-independent non-Newtonian liquids and for Newtonian liquids in the laminar region. This link, which was first established by Metzner and Otto (1957) for pseudoplastic fluids, depends on the fact that there appears to be a characteristic average shear rate $\dot{\gamma}_{avg}$ for a mixer (impeller–tank assembly) which characterizes power consumption, and which is directly proportional to the rotational speed of impeller:

$$\dot{\gamma}_{avg} = k_s N \qquad (8.9)$$

where k_s is a function of the type of impeller and the vessel configuration. If the apparent viscosity corresponding to the average shear rate defined above is used in the equation for a Newtonian liquid, the power consumption for laminar conditions is satisfactorily predicted for most inelastic non-Newtonian fluids.

The validity of the linear relationship given in equation (8.9) was subsequently confirmed by Metzner and Taylor (1960) and Wichterle and Wein (1981) by means of flow visualization experiments. The experimental evaluation of k_s for a given geometry proceeds as follows:

(i) The Power number (Po) is determined for a particular value of N.
(ii) The corresponding value of Re is obtained from the appropriate power curve (relating to an identical geometry) for a Newtonian liquid.
(iii) The 'equivalent' apparent viscosity is computed from the value of Re.
(iv) The value of the corresponding shear rate is obtained, either directly from a flow curve obtained by independent viscometric experiment, or by use of an appropriate fluid model such as the power-law model.
(v) The value of k_s is calculated for a particular impeller configuration using equation (8.9).

This procedure can be repeated for different values of N and an average value of k_s estimated. A compilation of the experimental values of k_s for a variety of impellers, turbines, propeller, paddle, anchor and so on, has been given by Skelland (1983), and an examination of Table 8.2 suggests that for pseudoplastic fluids, k_s lies approximately in the range 10–13 for most configurations of practical interest, while slightly larger values 25–30 have been reported for anchors and helical ribbons (Bakker and Gates, 1995). Skelland (1983) has also given a correlation for most of the data on the agitation of inelastic non-Newtonian fluids and this is shown in Figure 8.13.

The prediction of power consumption for the agitation of a given time-independent non-Newtonian fluid in a particular mixer, at a desired impeller speed, may be evaluated by the following procedure:

(i) The average shear rate is estimated from equation (8.9).
(ii) The corresponding apparent viscosity is evaluated, either from a flow curve, or by means of an appropriate viscosity model.
(iii) The value of the Reynolds number is calculated as $(\rho ND^2/\mu)$ and then the value of the Power number, and hence of P, is obtained from the appropriate curve in Figure 8.13.

Although this approach of Metzner and Otto (1957) has gained wide acceptance (Doraiswamy *et al.*, 1994), it has also come under some criticism. Equation (8.9) essentially reduces the complex three-dimensional flow field in a mixing tank to a single constant, namely, k_s. Since it is fair to postulate that the flow field produced by an impeller is influenced by the fluid rheology and tank–impeller geometry and therefore, in turn, it is not unreasonable to expect that the value of k_s should be a function of the fluid rheology and geometrical configuration. Consequently, it is not at all surprising that the literature is inundated with conflicting conclusions regarding the dependence of k_s on the rheological properties and the system geometry. For instance, Skelland (1967) and Mitsuishi and Hirai (1969) argued that a constant value of k_s does not always yield an unique power curve for a wide range of power-law flow behaviour index, thereby suggesting k_s to be a function of n. On the other hand, limited numerical (Aubin *et al.*, 2000; Murthy Shekhar and Jayanti, 2003) predict only a very weak dependence of k_s on the flow behaviour index, n. It is also appropriate to add here that care must be taken when comparing the values of k_s from different studies which are frequently based on geometrically dissimilar systems. While extensive compilations of experimental and analytical results regarding the dependence of k_s on power-law index and geometrical parameters are available in the literature (Chhabra, 2003; Delaplace and Guerin, 2006; Xuereb *et al.*, 2006), it is not yet established, however, how strongly the value of k_s depends on power-law index and on the geometrical arrangement of the system. For example, Calderbank and Moo-Young (1959), Beckner and Smith (1966), Yamamoto *et al.* (1998), Wang *et al.* (2000), Thakur *et al.* (2004) and Delaplace *et al.* (2006) and many others have all related k_s to the impeller/vessel configuration and the power-law index with varying degree of success. For moderately shear-thinning fluids ($n \geq \sim 0.4$), the dependence of k_s on power-law index is quite weak (± 25–30%) for a fixed geometrical arrangement. On the other hand, the value of k_s, for instance for helical ribbon impellers, can vary by up to 100% as the value of the power-law index changes from $n = 0.2$ to 0.4. Therefore, it is preferable to evaluate the value of k_s in each new application depending upon the range of power-law index likely to be encountered and in a geometrically similar laboratory scale arrangement. However, when this is not possible,

Table 8.2   *Values of k_s for various types of impellers and key to Figure 8.13*

Curve	Impeller	Baffles	D (m)	D_T/D	N (Hz)	k_s ($n < 1$)
A–A	Single turbine with 6 flat blades	4, $W_B/D_T = 0.1$	0.051–0.20	1.3–5.5	0.05–1.5	11.5 ± 1.5
A–A₁	Single turbine with 6 flat blades	None.	0.051–0.20	1.3–5.5	0.18–0.54	11.5 ± 1.4
B–B	Two turbines, each with 6 flat blades and $D_T/2$ apart	4, $W_B/D_T = 0.1$	–	3.5	0.14–0.72	11.5 ± 1.4
B–B₁	Two turbines, each with 6 flat blades and $D_T/2$ apart	4, $W_B/D_T = 0.1$ or none	–	1.02–1.18	0.14–0.72	11.5 ± 1.4
C–C	Fan turbine with 6 blades at 45°	4, $W_B/D_T = 0.1$ or none	0.10–0.20	1.33–3.0	0.21–0.26	13 ± 2
C–C₁	Fan turbine with 6 blades at 45°	4, $W_B/D_T = 0.1$ or none	0.10–0.30	1.33–3.0	1.0–1.42	13 ± 2
D–D	Square-pitch marine propellers with 3 blades (downthrusting)	None, (i) shaft vertical at vessel axis, (ii) shaft 10° from vertical, displaced $r/3$ from centre	0.13	2.2–4.8	0.16–0.40	10 ± 0.9
D–D₁	Same as for D–D but upthrusting	None, (i) shaft vertical at vessel axis, (ii) shaft 10° from vertical, displaced $r/3$ from centre	0.13	2.2–4.8	0.16–0.40	10 ± 0.9

(Continued)

Table 8.2 (*Continued*)

Curve	Impeller	Baffles	D (m)	D_T/D	N (Hz)	k_s ($n < 1$)
D–D$_2$	Same as for D–D	None, position (ii)	0.30	1.9–2.0	0.16–0.40	10 ± 0.9
D–D$_3$	Same as for D–D	None, position (i)	0.30	1.9–2.0	0.16–0.40	10 ± 0.9
E–E	Square-pitch marine propeller with 3 blades	4, $W_B/D_T = 0.1$	0.15	1.67	0.16–0.60	10
F–F	Double-pitch marine propeller with 3 blades (downthrusting)	None, position (ii)	–	1.4–3.0	0.16–0.40	10 ± 0.9
F–F$_1$	Double-pitch marine propeller with 3 blades (downthrusting)	None, position (i)	–	1.4–3.0	0.16–0.40	10 ± 0.9
G–G	Square-pitch marine propeller with 4 blades	4, $W_B/D_T = 0.1$	0.12	2.13	0.05–0.61	10
G–G$_1$	Square-pitch marine propeller with 4 blades	4, $W_B/D_T = 0.1$	0.12	2.13	1.28–1.68	–
H–H	Two-bladed paddle	4, $W_B/D_T = 0.1$	0.09–0.13	2–3	0.16–1.68	10
–	Anchor	None	0.28	1.02	0.34–1.0	11 ± 5
–	Cone impellers	0 or 4, $W_B/D_T = 0.08$	0.10–0.15	1.92–2.88	0.34–1.0	11 ± 5

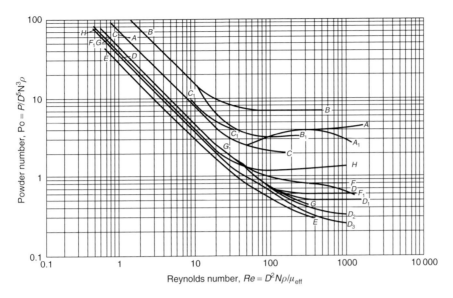

Figure 8.13   *Power curve for pseudoplastic liquids agitated by different types of impeller (from Skelland, 1983)*

one should use the value of k_s from the literature which relates to the fluid rheology and geometrical arrangement as closely as possible to the intended application and it is not uncommon for the value of k_s to have an uncertainty of ± 25–30%. Beyond the laminar flow conditions, the Metzner–Otto constant k_s shows additional dependence on the Reynolds number (Wichterle and Wein, 1981; Kelly and Gigas, 2003). Conversely, this implies that the value of k_s rises much more steeply with the rotational speed (N) of the impeller than the linear dependence seen in laminar flow conditions. On the other hand, Godfrey (1992) has asserted that the constant k_s is independent of equipment size, and thus, there are no scale-up problems.

   Data for power consumption in Bingham plastic and dilatant fluids have been reported and correlated in this manner (Metzner *et al.*, 1961; Nagata, 1975; Johma and Edwards, 1990; Delaplace *et al.*, 2004a, b), whereas Edwards *et al.* (1976) and Sestak *et al.* (1982, 1986) have studied the mixing of time-dependent thixotropic materials. In most instances, the approach of Metzner and Otto (1957), i.e. k_s, has been used successfully to correlate power consumption data. However, this approach poses special difficulties in the mixing of viscoplastic and shear-thickening fluids. For the mixing of viscoplastic systems, depending upon the torque applied to the shaft of the impeller, there might be regions in the tank where the fluid is unyielded and therefore, the effective shear rate in this case would vary from a finite value in the yielded region to zero in the unyielded region. This makes the averaging inherent in equation (8.9) even more questionable. Conversely, one can postulate that the value of k_s is again dependent on the rheology of the fluid, notably its yield stress. This is in line with the limited experimental results (Hirata and Aoshima, 1996; Curran *et al.*, 2000) and numerical simulations (Bertrand *et al.*, 1996; Tanguy *et al.*, 1996; Torrez and Andre, 1999; Marouche *et al.*, 2002; Anne-Archard *et al.*, 2006) for a range of impeller geometries. Strictly speaking, the value of k_s in viscoplastic fluids depends on the value of the Bingham number and on the tank–impeller geometry, and it is possible to assign a single value to k_s only when there are no unyielded regions present in the system of fixed geometry. Once again, it is probably best to evaluate the value of k_s over the range of conditions to be covered in an envisaged application. Obviously, the formation of the unyielded regions leads to poor mixing and must be avoided as far as possible. Similarly, during the mixing of shear-thickening fluids, an opposite effect is seen. Since the shear rate is maximum in the impeller region and therefore, the effective viscosity is maximum in this region and the impeller draws enormous power to rotate in this almost solidified material whereas the material shows more fluid-like characteristics far away from the impeller (Delaplace *et al.*, 2000).

This approach has also been used in the reverse sense, using impeller power data as a means of characterizing fluid rheology. Because of the indeterminate nature of the flow field produced by an impeller, the method is in principle suspect, though it can provide useful guidance in some circumstances, e.g. when working with suspensions. Even in these cases, it is imperative to limit agitation conditions to ensure laminar flow in the vessel, say, keeping $Re \leq 10$. Furthermore, it must be borne in mind that up to three- to four-fold variations in the value of the 'average' viscosity near the impeller and away from it are not uncommon in a mixing vessel. In spite of these uncertainties, this technique is used extensively in food and many other process engineering applications for routine quality control purposes (Kamiwano *et al.*, 1990; Steffe, 1996; Lai *et al.*, 2000; Chhabra, 2003; Arzate *et al.*, 2004; Franco *et al.*, 2005).

As the viscosity of the liquid increases, the performance of both high speed agitators and the close-clearance gate or anchor impellers deteriorates rapidly and therefore impellers with high pumping capacity are then preferred. Two such devices which have gained wide acceptance are the helical screw and helical ribbon impellers (see Figures 8.32 and 8.34). The approach of Metzner and Otto (1957) also provides a satisfactory correlation of power-consumption data with such impellers (e.g. see Ulbrecht and Carreau, 1985; Carreau *et al.*, 1993; Bakker and Gates, 1995); the value of k_s, however, is sensitive to the geometry of the impeller. In recent years, this approach has also been used to predict power consumption for kneaders (Nishi *et al.*, 2006, 2007), dual and hybrid designs of impellers (Wang *et al.*, 2000; Ascanio *et al.*, 2003; Foucault *et al.*, 2004; Farhat *et al.*, 2006; Rudolph *et al.*, 2006). However, in all such applications in which the two impellers (often of different type and size) rotate at different speeds, the choice of a single linear dimension and characteristic velocity is far from obvious (Delaplace *et al.*, 2005, 2007) which adds to the level of difficulty in interpreting and correlating experimental data in this field.

In conclusion, it is possible to estimate the power requirements for the agitation of single phase Newtonian and inelastic non-Newtonian fluids with reasonable accuracy under most conditions of interest. Many useful review articles on recent developments in this area are available in the literature (Chavan and Mashelkar, 1980; Harnby *et al.*, 1992; Chhabra, 2003; Paul *et al.*, 2004; Xuereb *et al.*, 2006).

(iii) *Effects of visco-elasticity*

Little is known about the effect of fluid visco-elasticity on power consumption, but early studies (Kelkar *et al.*, 1972; Chavan *et al.*, 1975; Carreau *et al.*, 1976; Yap *et al.*, 1979) seem to suggest that it is negligible under laminar flow conditions. Others have argued that the fluid visco-elasticity may increase and/or decrease the power consumption for turbine impellers as compared with that in Newtonian and inelastic non-Newtonian fluids under similar flow conditions (Nienow *et al.*, 1983; Ducla *et al.*, 1983; Seyssiecq *et al.*, 2003; Patterson, 2006). Since, most polymer solutions often show pseudoplastic as well as visco-elastic behaviour, it is not possible to delineate the contributions of these non-Newtonian characteristics on power consumption. However, the development of synthetic test fluids having a constant apparent viscosity but a range of high degrees of visco-elasticity (measured in terms of first normal stress difference) has helped resolve this difficulty. The experimental results to date suggest that the extent of visco-elastic effects is strongly dependent on the impeller geometry and operating conditions (Reynolds and Deborah numbers, etc.). For instance, for Rushton-type turbine impellers, the power consumption may be either greater or less than that for Newtonian fluids depending upon the values of the Reynolds and Deborah numbers (Oliver *et al.*, 1984; Prud'homme and Shaqfeh, 1984; Collias and Prud'homme, 1985). Recent work with helical ribbons and hydrofoil impellers suggests that visco-elasticity increases the power requirement for single phase agitation both in the laminar and in the transitional regimes (Carreau *et al.*, 1993; Özcan-Taskin and Nienow, 1995; Cavadas and Pinho, 2004). Thus caution must be used in applying results obtained from one geometry to another. For a given fluid and tank–impeller system, it is generally possible to correlate the power consumption data using the Metzner–Otto approach, but the constant k_s shows additional dependence on visco-elastic parameters like a suitably defined Deborah or Weissenberg number. Also, the fluid elasticity causes an early departure from the expected inverse relationship between the Power number and the Reynolds number, even when the Reynolds number is small. Therefore, extrapolation of results from one system to a different visco-elastic fluid and/or to a geometrical arrangement should be avoided as far as possible. This difficulty is further compounded due to the lack of reliable information about the dependence of k_s on visco-elasticity which seems to be much more sensitive to the geometrical details than that in the case of Newtonian and time-independent fluids (Seyssiecq *et al.*, 2003). No satisfactory correlations are thus available enabling the estimation of power consumption in visco-elastic fluids.

Finally, before concluding this section on power consumption, it should be noted that the calculation of the power draw requires a knowledge of the impeller speed which is necessary to blend the contents of a tank in a given time, or of the impeller speed needed to achieve a given mass transfer rate in a gas–liquid system. Since a full understanding of the mass transfer/mixing mechanism is not yet available, the selection of the optimum operating speed therefore remains primarily a matter of experience (Paul *et al.*, 2004).

Some guidelines for choosing an appropriate rotational speed for disc turbine blades are available for Newtonian fluids. Hicks *et al.* (1976) introduced a scale of agitation, S_A, which ranges from 1 to 10 with 1 being *mildly mixed* and 10 being *intensely mixed*. The scale of agitation is defined as (Fasano *et al.*, 1994; Bakker and Gates, 1995):

$$S_A = 32.8 \frac{N_q ND^3}{(\pi/4) D_{T_{eff}}^2} \qquad (8.10)$$

where N_q is the pumping number ($=Q/ND^3$), Q being the bulk flow induced by the impeller and $D_{T_{eff}}$ is the effective tank diameter evaluated as $(4V_l/\pi)^{1/3}$; V_l is the volume of liquid batch. Most chemical and processing applications are characterized by scales of agitation in the range 3–6, while values of 7–10 are typical of applications requiring high fluid velocities such as in chemical reactors, fermentors and in mixing of highly viscous systems. Some guidelines for choosing a suitable value of S_A for specific

Figure 8.14   *Effect of impeller size (D/D$_T$) on pumping number for a pitched blade turbine (replotted from Paul et al., 2004)*

Table 8.3   *Typical values of pumping number (N_q) under turbulent conditions*

Type of impeller	Value of N_q
Propeller	0.4–0.6
Pitched blade turbine	~0.8
Hydrofoil	0.55–0.75
Retreat curve blades	0.3
Flat blade turbine	0.7
Rushton turbine	0.72
Hollow-blade turbine	0.76

applications employing turbine agitators are also available in the literature (Gates *et al.*, 1976). Typical values of the pumping number, N_q, as a function of the Reynolds number and geometrical arrangement, are often provided by the manufacturers of mixing equipment. Broadly speaking, the pumping number (N_q) increases with the impeller Reynolds number and it attains a constant value under fully turbulent conditions. The constant value of N_q in turbulent conditions is strongly dependent on the type of the impeller and the geometrical arrangement, especially the value of (D/D$_T$), as shown in Figure 8.14 for a pitched blade turbine. Table 8.3 lists representative values of the pumping number, N_q, for a variety of impellers operating under fully turbulent conditions.

Typical power consumptions for common processing applications (kW/m^3) are shown in Table 8.4.

Table 8.4   *Typical power consumptions in common processes*

Duty	Power (kW/m^3)
Low power	
Suspending light solids, blending of low viscosity liquids	0.2
Moderate power	
Gas dispersion, liquid–liquid contacting, heat transfer, etc.	0.6
High power	
Suspending heavy solids, emulsification, gas dispersion, etc.	2
Very high power	
Blending pastes, doughs	4

Example 8.2

A thickened lubricating oil exhibiting power-law behaviour is to be agitated using the configuration shown in Figure 8.5. Using dimensional analysis, obtain the relevant dimensionless parameters for calculating power consumption in geometrically similar equipment.

Solution

The variables in this problem, together with their dimensions, are as follows:

$P : \mathbf{ML^2T^{-3}}$	$m : \mathbf{ML^{-1}T^{n-2}}$
$n : \mathbf{M^0L^0T^0}$	$N : \mathbf{T^{-1}}$
$g : \mathbf{LT^{-2}}$	$D : \mathbf{L}$
$D_T : \mathbf{L}$	$\rho : \mathbf{ML^{-3}}$

By Buckingham's π theorem, there will be $8 - 3 = 5$ dimensionless groups. Since n is already a dimensionless parameter, there will be four more π-groups.

Choosing ρ, N and D as the recurring set as in Example 8.1; then expressing $\mathbf{M}$, $\mathbf{L}$ and $\mathbf{T}$ in terms of these variables:

$$\mathbf{L} \equiv D; \quad \mathbf{M} \equiv \rho D^3 \quad \text{and} \quad \mathbf{T} \equiv N^{-1}$$

$$\pi_1 = PM^{-1}L^{-2}T^3 = P(\rho D^3)^{-1}(D)^{-2}(N^{-1})^3$$
$$= P/\rho D^5 N^3$$

$$\pi_2 = mM^{-1}LT^{2-n} = m(\rho D^3)^{-1}(D)(N^{-1})^{2-n}$$
$$= m/\rho D^2 N^{2-n}$$

$$\pi_3 = gL^{-1}T^2 = g(D)^{-1}(N^{-1})^2 = g/DN^2$$

$$\pi_4 = D_T L^{-1} = D_T/D$$

Thus, the functional relationship is given:

$$\frac{P}{\rho D^5 N^3} = \mathrm{Po} = f\left(\frac{\rho D^2 N^{2-n}}{m}, \frac{N^2 D}{g}, \frac{D_T}{D}, n \right)$$

where $\rho D^2 N^{2-n}/m$ is the Reynolds number which can be further re-arranged as:

$$Re = \frac{\rho D^2 N}{m(N)^{n-1}}$$

On comparing this expression with the corresponding definition for Newtonian fluids, the effective viscosity of a power-law fluid is given by:

$$\mu_{eff} = m(N)^{n-1}$$

This, in turn, suggests that the average shear rate, $\dot{\gamma}_{avg}$ for a mixing tank can be defined as being proportional to N, which is consistent with equation (8.8). $\square$

Example 8.3

In a polymerization reactor, a monomer/polymer solution is to be agitated in a baffled mixing vessel using a double turbine (six flat blades) impeller, with the configuration B–B in Table 8.1, at a rotational speed of 2 Hz. The solution exhibits power-law behaviour with $n = 0.6$ and $m = 12$ Pa s$^{0.6}$. Estimate the power required for a 300 mm diameter impeller. The density of the solution is 950 kg/m^3.

Solution

Since the mixing tank is fitted with baffles, one can assume that no vortex formation will occur and the Power number is a function only of the Reynolds number.
From Table 8.1 for configuration A–A,

$$k_s = 11.5 \quad \text{and} \quad \dot{\gamma}_{avg} = k_s N = 11.5 \times 2 = 23\,s^{-1}$$

$\therefore$ the corresponding effective viscosity of the solution.

$$\mu_{eff} = m(\dot{\gamma}_{avg})^{n-1} = 12 \times (23)^{0.6-1} = 3.42\,Pa\,s$$

The Reynolds number, $Re = \dfrac{\rho D^2 N}{\mu_{eff}} = \dfrac{950 \times (300 \times 10^{-3})^2 \times 2}{3.42}$

$$= 50$$

From Figure 8.13 at $Re = 50$
$$Po = \sim 7.4$$
and $P = Po\,\rho N^3 D^5$
$$= 7.4 \times 950 \times 2^3 \times (300 \times 10^{-3})^5$$
$$= 137\,W\ \square$$

Example 8.4

It is desired to scale-up a mixing tank for the agitation of a power-law liquid. The same fluid is used in both model and large-scale equipment. Deduce the functional dependence of power consumption per unit volume of fluid on the size of the impeller and the speed of rotation.

Solution

For geometrically similar systems and in the absence of vortex formation,

$$Po = f(Re)$$

Regardless of the nature of the function, f, in order to ensure similarity between the two systems Re_1 and Re_2 in the model and large-scale equipment, respectively must be equal, and Po_1 would then be equal to Po_2. The equality of the two Reynolds numbers yields:

$$\frac{\rho_1 N_1 D_1^2}{\mu_{eff1}} = \frac{\rho_2 N_2 D_2^2}{\mu_{eff2}}$$

For a power-law fluid, $\mu_{eff} = m(k_s N)^{n-1}$. Thus,

$$\mu_{eff1} = m_1 (k_s N_1)^{n_1-1} \quad \text{and} \quad \mu_{eff2} = m_2 (k_s N_2)^{n_2-1}$$

Since the same fluid is to be used in the two cases (and assuming that the power-law constants are independent of the shear rate over the ranges encountered), $m_1 = m_2 = m$; $n_1 = n_2 = n$; and $\rho_1 = \rho_2 = \rho$, therefore the equality of Reynolds number gives:

$$\frac{N_1 D_1^2}{(N_1)^{n-1}} = \frac{N_2 D_2^2}{(N_2)^{n-1}}$$

which can be further simplified as:

$$\frac{N_1}{N_2} = \left(\frac{D_2}{D_1} \right)^{2/(2-n)}$$

From the equality of the Power numbers:

$$\frac{P_1}{\rho_1 N_1^3 D_1^5} = \frac{P_2}{\rho_2 N_2^3 D_2^5}$$

Noting that $\rho_1 = \rho_2$, the ratio of the powers per unit volume is given by:

$$\frac{(P_1/D_1^3)}{(P_2/D_2^3)} = \left(\frac{N_1}{N_2} \right)^3 \left(\frac{D_1}{D_2} \right)^2$$

Substituting for $\dfrac{N_1}{N_2}$

$$\frac{(P_1/D_1^3)}{(P_2/D_2^3)} = \left(\frac{D_1}{D_2} \right)^{\frac{2n+2}{n-2}}$$

Similarly, expressing the final result in terms of the rotational speed:

$$\frac{(P_1/D_1^3)}{(P_2/D_2^3)} = \left(\frac{N_1}{N_2} \right)^3 \left(\frac{N_1}{N_2} \right)^{n-2} = \left(\frac{N_1}{N_2} \right)^{n+1}$$

Thus, power per unit volume ($\propto P/D^3$) increases less rapidly with both stirrer speed and impeller diameter for a shear-thinning fluid than for a Newtonian fluid (as would be expected). The converse applies to a shear-thickening fluid ($n > 1$).

Other scale-up criteria such as equal mixing times, or equality of interfacial areas per unit volume, etc., would in general yield different relations between variables. Most significantly, complete similarity is not achievable as it would lead to conflicting requirements. In such cases, it is necessary to decide the most important similarity feature for the particular problem and then to equate the appropriate dimensionless groups. Furthermore, almost without exception, scale-up of small-scale data gives non-standard agitator speeds and power requirements for the large-scale equipment, and thus one must choose the unit nearest to that calculated from scale-up considerations. □

8.2.4 Flow patterns in stirred tanks

A qualitative picture of the flow field created by an impeller in a mixing vessel in a single phase liquid is useful in establishing whether there are stagnant or dead regions in the vessel, and whether or not particles are likely to be maintained in suspension. In addition, the efficiency of mixing equipment, as well as product quality, are influenced by the flow patterns occurring in the mixing vessel.

Flow patterns produced in a mixing vessel are very much dependent upon the geometry of the impeller. It is thus convenient to classify the agitators used in non-Newtonian applications into three types:

(i) Those which operate at relatively high speeds, producing high shear rates in the vicinity of the impeller as well as giving good momentum transport rates throughout the whole of the liquid; typical examples include turbine impellers and propellers;

(ii) The second type is characterized by close-clearance impellers (such as gates and anchors) which extend over the whole diameter of the vessel and rely on shearing the fluid in the small gaps at the walls;

(iii) Finally, there are slowly rotating impellers which do not produce high shear rates but rely on their effective pumping action to ensure that an adequate velocity is imparted to the fluid in all parts of the vessel; typical examples include helical screw and helical ribbon impellers (see Figures 8.32 and 8.34).

(i) *Class I impellers*
The flow patterns for single phase Newtonian and non-Newtonian fluids in tanks agitated by class 1 impellers have been reported in the literature by, amongst others, Metzner and Taylor (1960), Norwood and Metzner (1960), Godleski and Smith (1962), Wichterle and Wein (1981), Adams and Barigou (2006), Saeed *et al.* (2007) and Ein-Mozaffari *et al.* (2007). The experimental methods used have included the introduction of tracer liquids, neutrally buoyant particles or hydrogen bubbles; and measurement of local velocities by means of pitot tubes, laser doppler velocimeters, ultrasonic velocimetry, particle image velocimetry (PIV), magnetic resonance imaging (MRI) and so on. The salient features of the flow patterns produced by propellers and disc turbines are shown in Figures 8.15a and b, respectively. Essentially, the propeller creates an axial flow through the impeller, which may be upwards or downwards depending upon its direction of rotation. Though the flow field is three-dimensional and unsteady, circulation patterns such as those shown in Figure 8.15a are useful in avoiding the formation of dead regions.

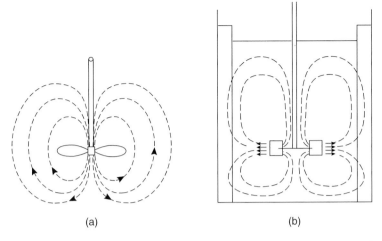

Figure 8.15   *(a) Flow pattern from propeller mixer and (b) radial flow pattern for disc turbine*

If the propeller is mounted centrally and there are no baffles in the tank, there is a tendency for the lighter liquid to be drawn in to form a vortex, and for the degree of mixing to be reduced. These difficulties are usually circumvented by fitting baffles to the walls of the tank. The power requirement is then increased, but an improved flow pattern is obtained as seen in Figure 8.16a. Another way of minimizing vortex formation is to mount the agitator off-centre to give a flow pattern similar to that depicted in Figure 8.16b. This, of course, adds to the complexity of the flow and leads to an increase in power consumption (Montante *et al.*, 2006).

The flat-bladed turbine impeller produces a strong radial flow outwards from the impeller (Figure 8.15b), thereby creating circulation zones in the top and bottom of the tank. The flow pattern can be altered by changing the impeller geometry and, for example if the turbine blades are angled to the vertical, a stronger axial flow component is produced. This can be useful in applications where it is necessary to suspend solids. Furthermore, as the Reynolds number decreases (by lowering the speed of rotation and/ or due to the increase in the liquid viscosity), the flow is mainly in the radial direction. The liquid velocities are very weak further away from the impeller and the quality of mixing deteriorates. A flat paddle produces a flow field with large tangential components of velocity, and this does not promote good mixing. Propellers, turbines and paddles are the principal types of impellers used for low viscosity Newtonian and pseudoplastic liquids operating in the transitional and turbulent regimes.

Generally, shear rates are highest in the region of the impeller and taper off towards the walls of the vessel. Thus, for a pseudoplastic fluid, the apparent viscosity is lowest in the impeller region and the fluid motion decreases much more rapidly for a pseudoplastic than for a Newtonian fluid as the walls are approached. Dilatant fluids display exactly the opposite behaviour. Using square cross-section vessels, Wichterle and Wein (1981) marked out zones with motion and those with no motion for pseudoplastic fluids agitated by turbine and by propeller impellers. Typical results showing stagnant zones are illustrated in Figure 8.17 in which it is clearly seen that the diameter (D_c) of the well-mixed region corresponds with the diameter of impeller D at low Reynolds numbers,

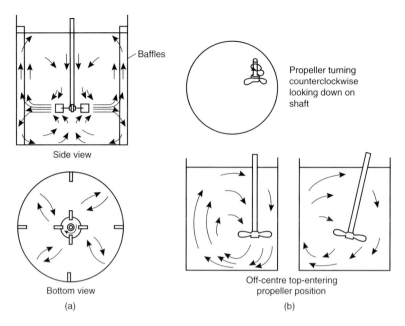

Side view

Bottom view

(a)

Baffles

Propeller turning
counterclockwise
looking down on
shaft

Off-centre top-entering
propeller position

(b)

Figure 8.16   *(a) Flow pattern in vessel with vertical baffles and (b) flow patterns with agitator offset from centre*

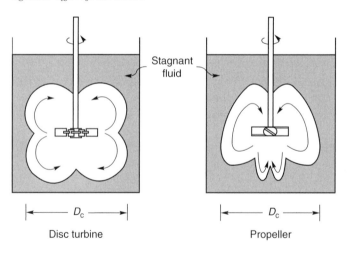

Stagnant
fluid

D_c

Disc turbine

D_c

Propeller

Figure 8.17   *Shape of the mixing cavity in a shear-thinning suspension. (Wichterle and Wein, 1981)*

but covers an increasing volume of the liquid with increasing Reynolds number or rotational speed. The authors presented the following expressions for D_c:

$$\frac{D_c}{D} = 1 \qquad Re < \frac{1}{a_0^2}$$

(8.11a)

$$\frac{D_c}{D} = a_0 \sqrt{Re} \qquad Re > \frac{1}{a_0^2}$$

(8.11b)

The constant a_0 was found to be 0.3 for propellers, 0.6 for turbines and $\sim 0.375\ (Po_t)^{1/3}$ for other types where Po_t is the constant value of the Power number under fully turbulent conditions (see Table 8.1). The Reynolds number here is defined by assuming $k_s = 1$, i.e. $Re = \rho N^{2-n} D^n / m$.

A direct link between the flow pattern and the corresponding power consumption is well illustrated by the study of Nagata *et al.* (1970) which relates to the agitation of viscoplastic media. These workers reported a cyclic increase and decrease in power consumption which can be explained as follows: Initially, the power consumption is high due to the high viscosity of the solid-like structure; however, once the yield stress is exceeded and the material begins to behave like a fluid, the power consumption decreases. The structure then becomes re-established and the solid-like zones re-form leading to an increase in the power consumption; then the cycle repeats. There was a tendency for a vortex to form at the liquid surface during this cyclic behaviour. This tendency was either considerably reduced or eliminated with class II impellers. More quantitative information on flow patterns in viscoplastic materials agitated by Rushton disc turbine impellers has been obtained using X-rays and hot wire anemometry (Solomon *et al.*, 1981; Elson *et al.*, 1986; Elson, 1990; Amanullah *et al.*, 1998). When the stress levels induced by the impeller rotation drop below the yield stress, relative motion within the fluid ceases in viscoplastic fluids. Solomon *et al.* (1981) attempted to predict the size of the well-mixed cavern, shown in Figure 8.17. While the height of such caverns is typically 0.4 D_c, they suggested the following relation for D_c:

$$\frac{D_c}{D} = \left[\left(\frac{1.36 Po}{\pi^2} \right) \left(\frac{\rho N^2 D^2}{\tau_0^B} \right) \right]^{1/3} \tag{8.12}$$

Equation (8.12) was stated to be applicable in the ranges of conditions as: $(\rho N^2 D^2 / \tau_0^B) \leq 4/\pi^3 Po \leq (\rho N^2 D^2 / \tau_0^B)(D/D_T)^3$, i.e. when $D \leq D_c \leq D_T$, though more recent experimental and numerical studies suggest that equation (8.12) is not completely satisfactory (Adams and Barigou, 2006; Ein-Mozaffari *et al.*, 2007). In spite of these deficiencies, equation (8.12) is used widely in industry (Etchells *et al.*, 1987). Evidently, in order to avoid the occurrence of unsheared material anywhere in the mixing vessel, the diameter of the cavern D_c must be equal to the vessel diameter (D_T). Once the cavern diameter coincides with the vessel diameter, it slowly rises up along the wall of the vessel. Therefore, enormous energy input is needed to achieve shearing everywhere in the vessel. The most common and also perhaps effective approach to achieve this is to use multiple impellers mounted on the same shaft; the separation between two impellers must be such that two intersecting caverns are formed, thereby inducing shearing everywhere. Typical values of the geometrical parameters include $(D/D_T) = \sim 0.5$–0.6 in a vessel with working depth of (H/D_T) and with the two impellers being situated at distances of $\sim 0.25 H$ and $\sim 0.75 H$ from the bottom of the tank. Under these conditions, the two impellers operate in almost a non-interactive fashion, and the power requirement is simply twice of that required for a single impeller under otherwise identical conditions (Paul *et al.*, 2004).

In contrast, the influence of fluid visco-elasticity is both more striking and difficult to assess. Photographs of rotating turbine and propeller-type impellers in visco-elastic fluids suggest two distinct flow patterns (Giesekus, 1965). In a small region near the impeller the flow is outwards, whereas elsewhere the liquid is flowing inwards towards

Figure 8.18   *Shear rate profiles for an anchor impeller rotating in a visco-elastic liquid. (Peters and Smith, 1967)*

the impeller in the equatorial plane and outwards from the rotating impeller along the axis of rotation. The two regions are separated by a streamline and thus there is no convective transport between them. A more quantitative study made by Kelkar *et al.* (1973) suggests that, irrespective of the nature of the secondary flow pattern, the primary flow pattern (i.e. tangential velocity) around a rotating body is virtually unaffected by the visco-elasticity of the fluid. Indeed, various types of flow patterns may be observed depending upon the relative magnitudes of the elastic, inertial and viscous forces, i.e. the values of the Reynolds and Deborah numbers.

(ii) *Class II impellers*
While anchors and gate-type impellers are known to produce poor axial circulation of the liquid in a vessel, in one study (Peters and Smith, 1967) it seems that the fluid viscoelasticity promotes axial flow. For instance, axial flow is reported to be almost 15 times greater in a visco-elastic solution as compared with a Newtonian medium. The shear rate profiles reported by these authors shown in Figure 8.18 clearly indicate that the liquid in the tank is virtually unaffected by the blade passage. In recent years, there has been a growing trend to use dual impellers, typically consisting of an anchor (to sweep large volume of fluid) and a turbine (to create a high shear zone), both of which rotate at different speeds in order to minimize the formation of stagnant regions or caverns in highly shear-thinning and viscoplastic fluids (Thibault and Tanguy, 2002; Foucault *et al.*, 2006). By appropriate selection of their rotational velocities and the direction of rotation (cocurrent or counter current), it is possible to improve the quality of mixing

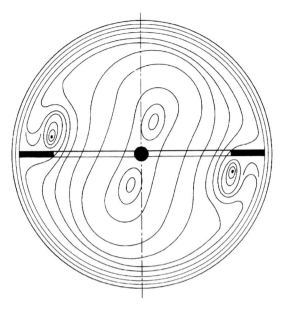

Figure 8.19 *Streamlines for a visco-elastic liquid in a tank with a gate agitator, drawn relative to the arm of the stirrer*

without significantly increasing the power consumption of such systems. Sometimes, a four-blade impeller is split into two, two-blade units mounted on the same shaft, but with adjacent pairs oriented at right angles to each other.

Broadly speaking, both gate and anchor agitators promote fluid motion close to the wall but leave the region near the shaft relatively stagnant, as can be seen in the typical pattern of streamlines in Figure 8.19. Furthermore, due to the only modest top to bottom turnover, vertical concentration gradients usually exist, but these may be minimized by using a helical ribbon or a screw twisted in the opposite sense, pumping the fluid downward near the shaft. Typical flow patterns for an anchor impeller are shown schematically in Figure 8.20. In such systems, the flow pattern changes with the impeller speed and the average shear rate cannot be described adequately by a linear equation such as equation (8.9). Furthermore, any rotational motion induced within the tank will produce a secondary flow in the vertical direction; the liquid in contact with the tank bottom is essentially stationary while that at higher levels is rotating and will experience centrifugal forces. Consequently, the prevailing unbalanced forces within the fluid lead to the formation of a toroidal vortex. Depending upon the viscosity and type of fluid, the secondary flow pattern may be single celled as in Figure 8.20 or double-celled, as shown schematically in Figure 8.21. Indeed, these schematic flow patterns have been substantiated by recent experimental and numerical studies for pseudoplastic fluids (Abid *et al.*, 1992; Pedrosa and Nunhez, 2000).

(iii) Class III impellers
Apart from the qualitative results for a composite impeller (anchor fitted with a ribbon or screw) referred to in the preceding section, little is known about the flow patterns induced by helical ribbon and screw impellers. The salient features of the flow pattern

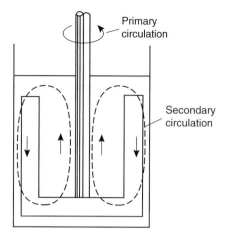

Figure 8.20   *Secondary circulation in an anchor agitated tank. (Peters and Smith, 1967)*

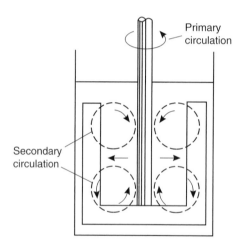

Figure 8.21   *Schematic double-celled secondary flow pattern*

produced by a helical ribbon impeller are shown in Figure 8.22 (Nagata *et al.*, 1956). The primary top-to-bottom circulation, mainly responsible for mixing, is due only to the axial pumping action of the ribbons. The shear produced by the helical ribbon is localized in the regions inside and outside the blade, whereas the shear between the wall and the bulk liquid is cyclic in nature. Notwithstanding the considerable scatter of results of Bourne and Butler (1969), the velocity data shown in Figure 8.23 seem to be independent of the scale of equipment, and the nature of fluid, e.g. pseudoplastic or visco-elastic. There was virtually no radial flow except in the top and bottom regions of the tank, and the vertical velocity inside the ribbon helix varied from 4% to 18% of the ribbon speed.

In addition to the primary flow pattern referred to above, secondary flow cells develop with the increasing impeller speed and these are similar to those observed by Peters and Smith (1967) for class II agitators as can be seen in Figure 8.20. Carreau *et al.* (1976) have also recorded flow patterns for a helical ribbon impeller. Visco-elasticity seems to

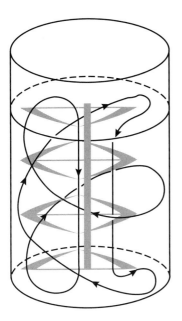

Figure 8.22   *Flow pattern produced by a helical ribbon impeller. (Nagata et al., 1956)*

cause a considerable reduction in the axial circulation, as can be seen in Figure 8.24 where the dimensionless axial velocity is plotted for an inelastic (2% carboxymethyl cellulose) and a visco-elastic (1% Separan) solutions. The axial velocities in the inelastic solution were found to be of the same order as in a Newtonian fluid.

On the other hand, the tangential velocities were so increased in visco-elastic fluids that the whole contents of the vessel, except for a thin layer adhering to the wall, rotated as a solid body with an angular velocity equal to that of the impeller. More definitive conclusions regarding the role of non-Newtonian rheology, especially visco-elasticity, must await additional work in this area.

Virtually nothing is known about the flow patterns produced by helical screw impellers. In a preliminary study, Chapman and Holland (1965) presented pictures of dye flow patterns for an off-centre helical screw impeller, pumping upwards with no draft tube. There seems to exist a dispersive flow between the flights of the screw, the dispersion being completed at the top of the screw. The flow into the screw impeller was from the other side of the tank, whereas the fluid in the remaining parts of the tank appeared to be virtually stagnant. Limited numerical predictions based on the assumption of three-dimensional flow induced by a helical ribbon-screw impeller show good agreement with experimental data for power consumption, mixing time and flow fields in Newtonian (Tanguy *et al.*, 1992), in shear-thinning (Murthy Shekhar and Jayanthi, 2003) and in viscoplastic fluids (Marouche *et al.*, 2002; Ihejirika and Ein-Mozaffari, 2007).

The limited results available on the flow patterns encountered in mixing devices used for thick pastes with complex rheology have been discussed by Hall and Godfrey (1968) and Kappel (1979). One common geometry used in mixing thick pastes is that of the sigma-blade mixer (Figure 8.32). Such devices have thick S- or Z-shaped blades which look like high pitch helical ribbon impellers. Generally, two units are placed

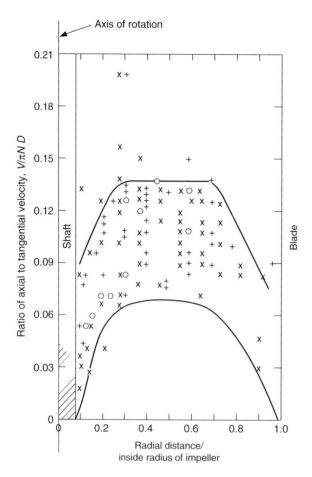

Figure 8.23   *Variation of axial fluid velocity in the core region of helical ribbon impellers pumping downwards in 27 and 730 L tanks. The curves indicate the upper and lower bounds of data (Bourne and Butler, 1969); +: $D/D_T = 0.889$; ×: $D/D_T = 0.952$ (small tank); ○: $D/D_T = 0.954$ (large tank)*

horizontally in separate troughs inside a mixing chamber and the blades rotate in opposite directions at different speeds. Preliminary results obtained using a positive displacement mixer suggest it has advantages over helical ribbon and sigma mixers for thick pastes and extremely viscous materials (Cheng *et al.*, 1974).

Thus, the flow patterns established in a mixing tank depend critically on the vessel–impeller configuration, the rheology of the liquid batch and the operating conditions. In selecting the appropriate combination of equipment, care must be taken to ensure that the resulting flow pattern is suitable for the required application.

8.2.5 *Rate and time of mixing*

Before considering the question of the rate and time of mixing, it is necessary to have some means of assessing the quality of the product mixture. This difficulty stems from

Figure 8.24 *Axial velocity distribution in a vessel agitated by a helical ribbon impeller (Carreau et al., 1976)*

two aspects of mixing, namely, the scale and the intensity of segregation. The scale of segregation is a measure of the large-scale breakup process (governed by bulk and eddy diffusivity) without the action of molecular diffusion, as shown schematically in Figure 8.25 (size of inhomogeneity is increasing from left to right in each row). In this case, the diminishing size of black squares indicates reduction in the scale of segregation, i.e. boxes on the left are more mixed than the ones on the right. On the other hand, the intensity of segregation is a measure of the difference in concentration between the 'black' and 'white' (representing two liquids) squares, shown schematically in Figure 8.25 (intensity is increasing from bottom to top in each column) and hence the degree of mixedness increases from top to bottom. Intuitively, one would expect the final mixture to become grey in colour somewhere in between black and white. Since the size of even the smallest eddies is much larger than the molecular scale, this occurs by molecular diffusion only. In practice, however, both processes occur simultaneously, i.e. as one moves along the diagonal from the top right corner to the bottom left corner. Furthermore, the wide scope and spectrum of mixing problems make it impossible to develop a single criterion for all applications. One intuitive and convenient, but perhaps unscientific, criterion is whether or not the mixture meets the required specification. Whatever the criteria used, mixing time is defined as the time needed to produce a mixture or a product of predetermined quality, and the rate of mixing is the rate at which the mixing progresses towards the final state. When a tracer is added to a single phase liquid in a stirred tank, the mixing time is measured as the interval between the introduction of tracer and the time when the contents of the vessel have reached the required degree of uniformity or mixedness. If the tracer is completely miscible and has the same viscosity and density as the liquid in the tank, the tracer concentration may be measured as a function of time at any point in the vessel by means of a suitable detector, such as a colour meter, or by electrical conductivity. For a given amount of tracer, the equilibrium

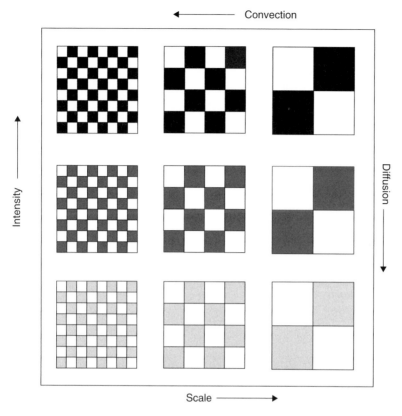

Figure 8.25   *Qualitative representation of the relationship between the intensity and scale, and of the role of convection and molecular diffusion in mixing*

concentration C_α may be calculated; this value will be approached asymptotically at any point as shown in Figure 8.26.

In practice, the mixing time will be that required for the mixture composition to come within a specified (95% or 99%) deviation from the equilibrium value C_∞, and this will be dependent upon the way in which the tracer is added and the location of the detector. It may therefore be desirable to record the tracer concentration at several locations, and to define the variance of concentration σ^2 about the equilibrium value as:

$$\sigma^2 = \frac{1}{p-1} \sum_{i=0}^{i=p} (C_i - C_\infty)^2 \tag{8.13}$$

where C_i is the tracer concentration at time t recorded by the ith detector. A typical variance curve is shown in Figure 8.27.

Several experimental techniques may be used, such as acid/base titration, electrical conductivity or temperature measurement, measurement of refractive index, light absorption and so on. In each case, it is necessary to specify the manner of tracer addition, the position and number of recording points, the sample volume of the detection system, and the criterion used in locating the end point (such as a predetermined cut-off point).

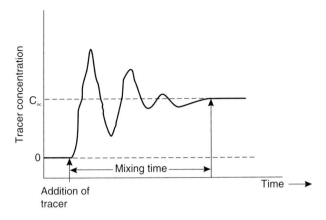

Figure 8.26   *Mixing-time measurement curve*

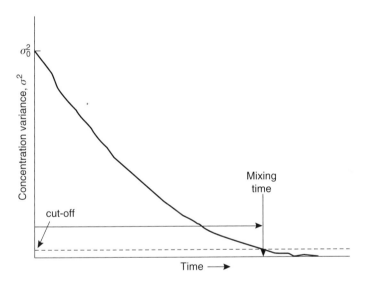

Figure 8.27   *Reduction in variance of concentration of tracer with time*

Each of these factors will influence the measured value of mixing time, and therefore, care must be exercised in comparing results from different investigations (Manna, 1997; Delaplace *et al.*, 2000, 2004a, b). Irrespective of the technique used to measure the mixing time, the response curve may show periodic behaviour. This may most probably be due to the repeated passage of a fluid element with a locally high concentration of tracer. The time interval between any two successive peaks is known as the circulation time.

For a given experiment and configuration, the mixing time, t_m, will depend upon the process and operating variables as follows:

$$t_\text{m} = \text{f}(\rho, \mu, N, D, g, \text{geometrical dimensions of the system})　　　　(8.14)$$

Using dimensional analysis, the functional relationship may be re-arranged as:

$$Nt_m = \theta_m = f\ (Re, Fr, \text{geometrical ratios}) \tag{8.15}$$

For geometrically similar systems and assuming that the Froude number, $Fr\ (=DN^2/g)$ is not important,

$$\theta_m = f(Re) \tag{8.16}$$

Broadly speaking, the dimensionless mixing times in both the laminar and fully turbulent regions are independent of the Reynolds number; a substantial transition zone exists between these two asymptotic values. Undoubtedly, the functional relationship between θ_m and Re is strongly dependent on the mixer geometry and the flow patterns produced. Another difficulty in making comparisons between results from different sources arises from the differences in criteria used in locating the end points. For instance, there may be applications (such as in agricultural, construction materials related settings) where it is sufficient to reach the 90% level of homogenization, while in food and pharmaceutical product formulations it maybe necessary to attain >99% homogenization. Therefore, the need to extrapolate existing results based on one criterion to a slightly different level of homogenization frequently arises. By treating the blending as a first order process, Grenville and Nienow (2004) put forward the following relationship linking two different levels of homogenization (i.e. x and y are fractional values indicating the different levels of homogenization):

$$\theta_x = \theta_y\ \frac{\ln(1-x)}{\ln(1-y)} \tag{8.17}$$

Thus, for instance, equation (8.17) can be used to estimate the value of the mixing time for $x = 0.98$ if the value of θ is known for $y = 0.95$. Thus,

$$\theta_{98} = \theta_{95}\ \frac{\ln(1-0.98)}{\ln(1-0.95)} = \frac{\ln(0.02)}{\ln(0.05)} = 1.31\theta_{95}$$

The highly non-linear relationship embodied in equation (8.17) needs to be emphasized here when extrapolating results based on one mixing criterion to another.

For class I impellers, the limited available work (Norwood and Metzner, 1960) confirms this dependence of θ_m on Re for turbine impellers in baffled tanks, albeit these results are believed to be rather unreliable. Nonetheless, Norwood and Metzner (1960) suggested that the correlations developed for Newtonian liquids can also be used for purely inelastic shear-thinning fluids, simply by using a generalized Reynolds number based on the effective viscosity evaluated at the average shear rate given by equation (8.9). Although, it is widely reported that θ_m is quite sensitive to impeller–vessel geometry for low viscosity liquids, there have been few studies of the effect of physical properties on θ_m. In a numerical study, Montante et al. (2005) have reported good agreement between the experimental and predicted profiles of tracer concentration in moderately viscous Newtonian and shear-thinning fluids being agitated by a down-pumping 45° pitched blade turbine impeller. Early work of Bourne and Butler (1969) suggests that the rates of mixing, and hence the mixing times, are not very sensitive to the fluid rheology for both Newtonian and inelastic non-Newtonian materials. On the other hand, Godleski and Smith (1962) reported mixing times for pseudoplastic fluids (agitated by turbines) up to 50 times greater than those expected from the corresponding results for materials exhibiting Newtonian behaviour. This emphasizes the care which must be exercised in applying any generalized conclusions to a particular system. Intuitively, one might expect a similar deterioration in mixing for visco-elastic liquids, especially when phenomena such as secondary flow and flow reversal occur.

The only study relating to the use of class II impellers for non-Newtonian media is that of Peters and Smith (1967) who reported a reduction in mixing and circulation times for visco-elastic polymer solutions agitated by an anchor impeller. The decrease in mixing time can be ascribed to the enhanced axial circulation.

In contrast to this, class III impellers have generated much more interest. Circulation times with helical impellers are known not to be affected by shear-thinning behaviour (Chavan and Ulbrecht, 1972, 1973; Carreau *et al.*, 1976). Thus, the circulation time is constant in the laminar regime ($Re < {\sim}10$); it decreases in the transition zone both with increasing Reynolds number and with increasingly shear-thinning behaviour. A subsequent more detailed study (Guerin *et al.*, 1984) shows that, even though the average circulation times are not influenced significantly by shear-thinning characteristics, their distribution becomes progressively narrower.

Mixing times in inelastic systems follow a similar pattern; namely, θ_m is independent of Reynolds number in the laminar region ($Re < {\sim}10$). For helical screw impellers in the intermediate zone ($10 \leq Re \leq 1000$), θ_m decreases with Reynolds number. With shear-thinning behaviour, the apparent viscosity should be evaluated at $\dot{\gamma}_{avg}$ estimated using equation (8.9). On the other hand, Carreau *et al.* (1976) have reported that non-Newtonian fluids required considerably longer mixing times as compared with Newtonian fluids at comparable circulation rates.

Grenville and Nienow (2004) have collated much of the literature data on the blending of Newtonian and shear-thinning fluids with turbine, helical ribbon and hydrofoil impellers and of visco-plastic fluids obtained in mixing vessels up to 3 m in diameter and fitted with four standard size baffles. For shear-thinning fluids, helical ribbon data relates to laminar flow conditions whereas the turbine and hydrofoil data falls in the transitional and turbulent regimes. Using dimensional analysis together with experimental results in turbulent flow, they put forward the following correlation to estimate the mixing time for the homogenization of Newtonian fluids (based on the 95% homogenization criterion):

$$P_0^{1/3}\theta_m \left(\frac{D}{D_T}\right)^2 \left(\frac{D_T}{H}\right)^{0.5} = 5.2 \tag{8.18}$$

For the commonly used configuration of $D_T = H$, it is possible to re-arrange this equation in the following dimensionless form:

$$P_0^{1/3}Re = \frac{5.2\rho D_T^2 N}{\mu\theta_m} = 5.2 \frac{Re_T}{\theta_m} \tag{8.19}$$

where Re_T is the so-called vessel Reynolds number and the empirical constant 5.2 has an uncertainty of $\pm 10\%$.

The mixing time data in the transitional regime is given by:

$$P_0^{1/3}Re = \frac{183}{\sqrt{Re_T/\theta_m}} \tag{8.20}$$

The standard deviation on the constant is $\pm 17.5\%$.

The critical Reynolds number, Re_{cr}, denoting the onset of the turbulent flow conditions in given by the intersection of equations (8.19) and (8.20) as:

$$P_0^{1/3} Re_{cr} = 6370 \quad \text{and} \quad \frac{\theta_m}{Re_T} = 1225 \tag{8.21}$$

The above-noted correlations lead to the following qualitative trends. In the turbulent regime (i) mixing time is almost independent of physical properties, (ii) all impellers of the same diameter require the same mixing time at the same power input per unit mass of fluid and finally (iii) an increase in the impeller diameter will reduce the mixing time at a constant power input per unit mass of liquid. In the transition regime, while mixing time is proportional to the kinematic viscosity of the liquid, the role of impeller size is similar to that in turbulent conditions, as noted above.

Aside from the development of these correlations, Grenville (1992) noted that there was a very little difference between the local and global values of the mixing time for Newtonian fluids in the turbulent regime. However, as the viscosity of the liquid is gradually increased (and therefore the Reynolds number decreases), the mixing time measured in regions with poor or no circulation (such as behind the baffles) increased significantly while those measured below the impeller and in the middle of the tank increased only slightly as compared with the turbulent values. Thus, the value of the mixing time close to the wall of the vessel determines the overall mixing time. Using this idea, Grenville (1992) proposed to use an effective viscosity of shear-thinning fluids in the wall region in equations (8.19) and (8.20). This necessitates the knowledge of the shear rate or the shear stress in the fluid close to the wall. Neglecting the contribution of pressure forces acting on the baffles, Grenville (1992) obtained the following expression for the effective shear stress at the wall of the vessel:

$$\tau_W = 0.62 \frac{\Lambda}{D_T^3} \tag{8.22}$$

where Λ is the torque acting on the shaft of the impeller. Using the rheological data, one can thus estimate the effective viscosity in the wall region which, in turn, can be used to evaluate the two Reynolds numbers appearing in equations (8.19) and (8.20) to estimate the mixing time for a shear-thinning fluid in the transitional and turbulent regions.

Finally, in the laminar regime, it is readily recognized that turbine impellers are grossly inadequate for the mixing of viscous Newtonian and shear-thinning fluids and helical ribbons are generally preferred. Based on the literature data, Grenville *et al.* (2001) proposed the following equation for mixing time:

$$\theta_m = 896 \times 10^3 K_P^{-1.69} \tag{8.23}$$

This equation can also be used for shear-thinning liquids. Clearly, larger is the value of K_P, lower is the mixing time. The critical Reynolds number, Re_{cr}, denoting the end of the laminar flow is given by:

$$Re_{cr} = \frac{183}{P_0^{1/3}} \tag{8.24}$$

The utility of some of these developments is illustrated in Example 8.5.

The results of the scant work with visco-elastic fluids are conflicting but suggest that relationship between mixing and circulation times is strongly dependent on the geo-metrical arrangement and that it is not yet possible to account quantitatively for the effects of visco-elasticity (e.g. see Chhabra *et al.*, 2007). Visco-elastic fluids appear to be much more difficult to mix than inelastic fluids. Reference should be made to review papers for further details of studies in this area (Takahashi, 1988; Delaplace *et al.*, 2000; Chhabra, 2003). Finally, the dimensionless mixing time seems to be relatively insensi-tive to the impeller size (D/D_T), at least for axial flow impellers (Wilkens *et al.*, 2003).

Example 8.5

A helical ribbon impeller–tank assembly with the following specifications is available in a plant:

Tank diameter, $D_T = 2.2$ m; Tank height $= 3.5$ m
Number of ribbons in the impeller, $N_r = 2$;
Clearance at the wall/impeller diameter $((D_T - D)/D) = 0.05$
Width to diameter ratio (W/D) $= 0.08$
Pitch to diameter ratio (s/D) $= 1$
Liquid height to impeller diameter ratio (H/D) $= 1.5$

This facility is to be used to disperse a liquid additive (small quantity) into a viscous (inelas-tic) polymer melt (density 980 kg/m^3) whose rheological behaviour can be approximated by the power-law fluid model ($m = 200$ Pa s^n and $n = 0.50$). Estimate the power requirements and the mixing time if the operating tip velocity of the impeller is 1 m/s. The value of the Metzner–Otto constant k_s is given to be 30 for this geometry.

Solution

In order to calculate the power consumption, we need to estimate the value of the Reynolds number, which, in turn necessitates knowledge of the value of the mean shear rate, given by equa-tion (8.9).

For a tip velocity of 1 m/s, the rotational speed of the impeller is calculated first as:

$$\pi DN = 1$$

For a tank diameter, $D_T = 2.2$ m and $(D_T - D)/D = 0.05$, the impeller diameter is calculated as:

$$\frac{D_T - D}{D} = 0.05$$

Therefore, $D = 2.095$ m and therefore, the rotational speed, N,

$$N = \frac{1}{\pi D} = \frac{1}{3.14 \times 2.095} = 0.152 \text{ Hz, i.e. 9.1 rpm.}$$

Thus, the average shear rate in the tank, $\dot{\gamma}_{avg}$, is calculated using equation (8.9) as:

$$\dot{\gamma}_{avg} = k_s N = 30 \times 0.152 = 4.56 \text{ s}^{-1}$$

Now substituting this value in the power-law equation, the average viscosity is calculated as:

$$\mu_{avg} = m(\dot{\gamma}_{avg})^{n-1} = 200 \times (4.56)^{0.5-1} = 93.7 \text{ Pa s}$$

The value of the Reynolds number, Re, is calculated as:

$$Re = \frac{\rho N D^2}{\mu_{avg}} = \frac{980 \times 0.152 \times (2.095)^2}{93.7} = 6.98$$

Thus, the flow is likely to be laminar and therefore, the product $(Po.Re)$ must be a constant equal to K_P, i.e.,

$$PoRe = K_P$$

The value of K_P is calculated using equation (8.7) for the geometrical details as follows:

$$K_P = 91 N_r^{0.79} \left\{ \left(\frac{0.5(D_T - D)}{D} \right)^{-0.37} \left(\frac{s}{D} \right)^{-0.37} (W/D)^{0.16} (H/D) \right\}$$

Substituting values to evaluate K_P as:

$$K_P = 91 \times (2)^{0.79} \times \{ (0.5 \times 0.05)^{-0.31} \times (1)^{-0.37} \times (0.08)^{0.16} \times (1.5) \} = 494$$

Now substituting for $Po = \dfrac{P}{\rho N^3 D^5}$ to calculate the power for the application as:

$$\left(\frac{P}{\rho N^3 D^5} \right) Re = K_P = 494$$

or $$P = \frac{494 \times \rho N^3 D^5}{Re}$$

Substituting numerical values as:

$$P = \frac{494 \times 980 \times 0.152^3 \times (2.095)^5}{6.98} = 9838W$$

Thus, the power drawn by the impeller is 9.84 kW.

In order to calculate, the mixing time for this application, equation (8.23) is used as:

$$\theta = N t_m = 896 \times 10^3 K_P^{-1.69}$$

Substituting values to get the mixing time t_m as:

$$t_m = \frac{896 \times 10^3 K_P^{-1.69}}{N}$$
$$= \frac{896 \times 10^3 \times 494^{-1.69}}{0.152}$$

or $t_m = 165\,s$

Therefore, the mixing time is estimated to be 2 min and 45 s.

Now checking that the flow is indeed laminar, the critical Reynolds number is calculated using equation (8.24).

$$Re_{Cr} = \frac{183}{P_0^{1/3}} = \frac{183}{\left(P \Big/ \rho N^3 D^5 \right)^{1/3}} = \frac{183}{\left\{ \dfrac{9838}{980 \times 0.152^3 \times (2.0945)^5} \right\}^{1/3}} = 44$$

Since the actual value of 6.98 is much smaller than this critical value, thus the flow in the tank is confirmed to be laminar. □

8.3 Gas–liquid mixing

Many gas–liquid reactions of industrial significance are carried out in agitated tank reactors, and the design requirements vary from one application to another. For instance, in effluent aeration and in some fermentation reactions, the systems are dilute and reactions are slow so that mass transfer is not likely to be a limiting factor. Energy efficiency is then the most important consideration, and large tanks giving long hold-up times are used. Chlorination and sulphonation reactions, on the other hand, are fast and the gases have high solubilities; and it is then desirable to have high rates of heat and mass transfer and short contact times. In the food industry, the non-Newtonian properties of batters and creams are of dominant importance and the flow field and temperature must be closely controlled. Similarly, highly viscous and shear-thinning (and with yield stress) fermentation broths are encountered in the manufacturing of commercially important antibiotics and polysaccharides. In such applications, the productivity, and product quality are strongly dependent on bulk mixing and oxygen mass transfer, which, in turn are governed by impeller design and rheological properties. On the other hand, in fungal fermentation, extreme agitation conditions can significantly alter the morphology of the product, and hence, the quality of mixing will be compromised. Thus, a delicate balance exists between the product quality on one hand and the hydrodynamic conditions needed for adequate mixing on the other. Similarly, the process design of mixing equipment for use with animal and insect cell cultures is even more challenging due to their extreme shear sensitivity. Irrespective of the application, a rational understanding of the fluid mechanical aspects of gas dispersion into liquids is a necessary precursor to the modelling of heat and mass transfer, and reactions in such systems.

Gas dispersion in agitated tanks may be described in terms of bubble size, gas hold-up, interfacial area and mass transfer coefficient. While gas dispersion in low viscosity systems (Smith, 1985; Tatterson, 1991; Harnby *et al.*, 1992) has been extensively studied, little is known about the analogous process in highly viscous Newtonian and non-Newtonian media, such as those encountered in polymer processing, pulp and paper manufacturing and fermentation applications (Paul *et al.*, 2004; Xuereb *et al.*, 2006).

Good dispersion of a gas into a liquid can only be achieved by using high speed (class I) agitators which unfortunately are not very effective for mixing high viscosity liquids. Hence, for gas dispersion into highly viscous media, there are two inherently conflicting requirements which may be met in practice by using a combination of two impellers mounted on a single shaft. For instance, the Rushton disc turbine with a 45° pitched blade impeller combines the advantages of the low flow and high shear of a disc turbine with the high flow and low shear produced by the second component of the assembly. Such composite agitators provide a good compromise between those agitators which cause mixing as a result of local turbulence generated by their shape, and those which give large-scale convective flows. Figure 8.28a shows a system which combines turbulence generation at the blade tips with an induced large-scale flow from the angled-blade arms; two or more agitators can be mounted on the same shaft at 90° to each other. If suitably designed, such assemblies can be rotated at moderately high speeds without excessive power consumption.

In some applications, two independent agitators are employed. One is of the positive displacement type and rotates slowly to give good mixing, whereas the other operates at a high rotational speed to facilitate gas dispersion. Another more widely used impeller for dispersing gases into liquids is the so-called scaba SRGT impeller (shown in Figure 8.28b)

Figure 8.28   *Composite impellers used for gas–liquid dispersions: (a) the Intermig and (b) the Scaba SRGT impeller*

in which half-cut pipes are used instead of plane blades in a four- or six-bladed Rushton turbine. The gas is invariably introduced through a sparger placed beneath the impeller.

Basically, the process of gas dispersion involves two competing processes: break-down of the gas into bubbles which occurs predominantly in the high shear region near the impeller, and coalescence which takes place in quiescent regions away from the impeller. Depending upon the impeller speed and the physical properties of the liq-uid phase (mainly viscosity and surface tension), the gas bubbles may either re-circu-late through the impeller region or escape from the system through the free surface, or they may coalesce. In addition, the dispersion of gases into highly viscous liquids differs from that in low viscosity systems because the gas bubbles have a greater tendency to follow the liquid motion when the viscosity is high. Admittedly, the exact mechanism of gas dispersion is not fully understood even for the extensively studied low viscos-ity systems; however, the mechanisms of dispersion in highly viscous Newtonian and non-Newtonian systems appear to be qualitatively similar. Essentially, the sparged gas gets sucked into the low pressure regions ('cavities') behind the impeller. The shape and stability of these cavities are strongly influenced by the liquid rheology, and the Froude and the Reynolds numbers for the agitator. There is, however, a minimum impeller speed required for these cavities to form. For instance, van't Reit (1975) suggested that the Froude number should be greater than 0.1 in liquids agitated by a disc turbine. As remarked earlier, high speed agitators are, however, not effective in dispersing a gas in highly viscous liquids ($Re < {\sim}10$) (Nienow *et al.*, 1983). For $Re > 1000$, the minimum impeller speed required for gas dispersion rises slowly with increasing liquid viscos-ity, e.g. a 50% increase in impeller speed is required for a 10-fold increase in viscos-ity. In the intermediate zone, $10 \le Re \le 1000$, the formation and stability of cavities is determined by complex interactions between Reynolds and Froude numbers, and the gas flow rate. Preliminary results seem to suggest that the effects of shear-thinning viscosity are negligible. Yield stress also does not appear to play any role in this process except that some gas may be entrapped in cavities, even after the impeller has been stopped (Nienow *et al.*, 1983). On the other hand, cavities which are much bigger and of different shapes have been observed in visco-elastic liquids (Solomon *et al.*, 1981). Furthermore, dispersion of a gas in visco-elastic media is more difficult because the gas entrapped in a cavity does not disperse until the gas flow is stopped. The effect of visco-elasticity is even more pronounced with the composite Intermig impeller (Figure 8.28a). In this case, the gas-filled cavities may extend behind the outer twin-split blades, almost all the way back to the following blade. Qualitatively, these differences in behaviour have been attributed to large extensional viscosities of highly visco-elastic media (Nienow and Ulbrecht, 1985).

8.3.1 Power consumption

The limited work on Rushton turbines suggests that at low Reynolds numbers (<10), the Power number is almost unaltered by the introduction of gas, possibly due to the fact that no gas-filled cavities are formed at such low impeller speeds. In the intermediate region ($10 < Re < 1000$), gas cavities begin to form, and the Power number decreases with increasing Reynolds number, goes through a minimum value *ca. Re* ~ 300–500 and begins to increase again, as shown in Figure 8.29 for a range of polymer solutions. However, the gas flow rate does not seem to influence the value of the Power number at a fixed impeller speed. The decrease in Power number arises from the combined effect of streamlining, the reduced pumping capacity of the impeller and the increased pressure in the gas-filled cavities. When the Power number is a minimum, the cavities are of maximum size and the impeller rotates in a pocket of gas without causing any dispersion. Undoubtedly, the reduction in power consumption, as well as its minimum value, are manifestations of the complex interplay between the size and structure of the cavities, the rheological characteristics of the liquid and the kinematic conditions, but the nature of these interactions is far from clear (Nienow and Ulbrecht, 1985; Tatterson, 1991; Harnby *et al.*, 1992).

Subsequent work with a helical ribbon impeller (Carreau *et al.*, 1992; Cheng and Carreau, 1994) suggests that the power consumption in the presence of a gas may either increase or decrease, depending upon the non-Newtonian flow properties of the liquid. For instance, Carreau *et al.* (1992) report reduced levels of power consumption in highly pseudoplastic liquids when gas is present (presumably due to the enhanced levels of shearing) whereas, in aerated visco-elastic liquids, the power always increases. The currently available correlations for the estimation of power under aerated conditions are too tentative to be included here (Cheng and Carreau, 1994).

Some information is also available on the performance of more widely used composite or modified impellers, for instance, Intermig and six-bladed Scaba SRGT impellers shown in Figure 8.28 (Galindo and Nienow, 1993). When a gas is present, the power is reduced considerably in non-Newtonian liquids agitated by a Scaba impeller. Furthermore, a point is reached where the Power number becomes independent of the gas flow rate for a fixed impeller speed. As far as scale-up is concerned, a preliminary study suggests that the power requirements for complete mixing decrease rapidly with increasing size of impeller irrespective of whether or not gas is present (Solomon *et al.*, 1981). Similarly, scant results indicate better performance of composite impellers

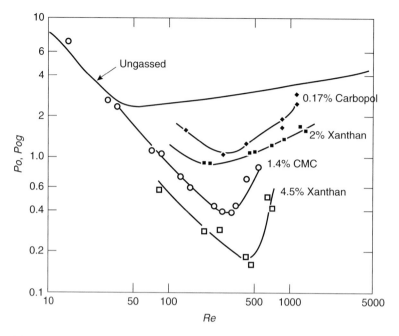

Figure 8.29 *Power number with gas-filled cavities on impeller versus Reynolds number at $Q_G = 0.5$ and 1 vvm (volumetric gas flow rate per minute per unit volume of liquid): $10 \leq Re \leq 1100$ (0, 1.4% CMC; $\square$, 4.5% Xanthan gum; $\blacksquare$, 2% Xanthan gum; $\blacklozenge$, 0.17% Carbopol)*

(for instance, a helical ribbon and a turbine) than that of a single impeller in gas–liquid mixing when the liquid phase is highly viscous and shear-thinning (Espinosa-Solares *et al.*, 2002). Finally, it is of interest to note here that owing to low viscosity of organic (Newtonian) liquids, most gas–liquid mixing applications are carried out under turbulent conditions. Naturally, the ratio of the power consumption in gassed and ungassed conditions varies with the gas flow number (Q_g/ND^3) and from one impeller to another. At the gas flow number value of 0.1, this ratio ranges from 0.3 to 0.4 for turbines to about 0.9 for impellers of special design like Chemineer CD 6 (shown in Figure 8.34). Unfortunately, no such information is available for viscous and/or non-Newtonian liquids.

8.3.2 Bubble size and hold-up

Bubble size (distribution) and hold-up, together with the specific interfacial area and volumetric mass transfer coefficients, may be used to characterize the effectiveness of gas dispersion into liquids. It is important to emphasize that these parameters, and bubble coalescence, are extremely sensitive to the presence of surface-active agents. Although all these variables show spatial variation, only globally averaged values are usually reported and these are frequently found to be adequate for the engineering design calculations. For a given liquid, the mean bubble size does not show a strong dependence on the level of agitation. Ranade and Ulbrecht (1978) studied the influence of polymer addition on hold-up and gas–liquid mass transfer in agitated vessels. Even small amounts of dissolved polymer were shown to give rise to substantial reductions in

Figure 8.30   *Typical relationship between hold-up and power input when dispersing gas in a high viscosity, shear-thinning fluid: (a) a single disc turbine and (b) a combination of 45° downward pumping impeller and disc turbine*

both hold-up and mass transfer, albeit the degree of reduction showing strong dependence on the type and concentration of the polymer. Qualitatively similar results have been reported for markedly shear-thinning fluids (Machon *et al.*, 1980). This reduction has been generally ascribed to the formation of large bubbles which have a shorter residence time. The average gas hold-up was found to vary with power P_g and gas flow rate, Q_g, as:

$$\phi \propto P_g^{0.3} Q_g^{0.7} \tag{8.25}$$

or with power consumption per unit mass, ε, and the superficial gas velocity, V_g, as:

$$\phi \propto \varepsilon^{0.3} V_g^{0.7} \tag{8.26}$$

For a fixed rate of feed of gas, both hold-up and power consumption decrease with the increasing polymer concentration, i.e. with increasing apparent viscosity. In these systems, bubbles have generally been found to be predominantly of two sizes, with a large population of small bubbles and a very few large bubbles. Figure 8.30 shows typical results for gas hold-up in highly viscous shear-thinning polymer solutions, obtained with a single disc turbine and a composite impeller (45° downward pumping impeller and disc turbine) for a range of (D/D_T) ratios. Finally, the point of minimum power consumption (largest amount of gas present in cavities), as expected, corresponds to the point of maximum gas hold-up (Hickman, 1988).

8.3.3 Mass transfer coefficient

In view of what has been said so far regarding the mixing of highly viscous Newtonian and non-Newtonian systems, a reduction in the mass transfer coefficient, $k_L a$ is inevitable. This is indeed borne out well by the few studies available on this subject (Perez and Sandall, 1974; Yagi and Yoshida, 1975; Ranade and Ulbrecht, 1978; Nishikawa *et al.*, 1981).

Dimensionless empirical correlations relating $k_L a$ to the system geometry, and kinematic and physical variables are available in the literature (Tatterson, 1991). For example, the following correlation due to Kawase and Moo-Young (1988) is one which embraces a very wide range of power-law parameters:

$$k_L a = \frac{0.675 \rho^{0.6} (P/V)^{(9+4n)/(10(1+n))}}{(m/\rho)^{0.5(1+n)} \sigma^{0.6}} D_L^{0.5} \left(\frac{V_g}{V_t}\right)^{0.5} \left(\frac{\mu_{\text{eff}}}{\mu_w}\right)^{-0.25} \tag{8.27}$$

Equation (8.27) is not dimensionally consistent and all quantities must be expressed in S.I. units; that is, $k_L a$, the volumetric mass transfer coefficient (s^{-1}); ρ, the density of liquid (kg/m^3); (P/V), the power input per unit volume of dispersion (W/m^3); σ, the interfacial tension (N/m); V_g, the superficial velocity of gas (m/s); V_t, the terminal rising velocity of a single bubble in a quiescent medium (m/s), (Kawase and Moo-Young recommended a constant value of 0.25 m/s); μ_{eff}, the effective viscosity estimated using equation (8.9) (Pa s); μ_w, the viscosity of water (Pa s); D_L, the diffusivity of gas into liquid (m^2/s) and m is the power-law consistency coefficient (Pa s^n). Equation (8.27) applies over the following ranges of conditions: $0.59 \leq n \leq 0.95$; $0.0036 \leq m \leq 10.8$ Pa s^n and $0.15 \leq D_T \leq 0.6$ m.

A comprehensive discussion of the other contemporary works in this field is available elsewhere (Nienow and Ulbrecht, 1985; Harnby *et al.*, 1992; Herbst *et al.*, 1992 Paul *et al.*, 2004; Xuereb *et al.*, 2006).

8.4 Heat transfer

The rate of heat transfer to process materials may be enhanced by externally applied motion both within the bulk of the material and in the proximity of heat transfer surfaces. In most applications, fluid motion is promoted either by pumping through tubes (Chapter 6) or by mechanical agitation in stirred vessels. A simple jacketed vessel is very commonly used in chemical, food, agro, biotechnological and pharmaceutical processing applications to carry out a range of operations. In many cases, heat has to be added or removed from the contents of vessel, either to control the rate of reaction, or to bring it to completion. The removal/addition of heat is customarily accomplished by using water or steam in a jacket fitted on the outside of the vessel or in an immersed cooling coil in the tank contents. In either case, an agitator is used both to achieve uniform temperature distribution in the vessel and to improve the heat transfer rate. As is the case with power requirement and mixing time, the rate of heat transfer in stirred vessels is strongly dependent on the tank–impeller configuration, type and number of baffles, the liquid rheology, the rotational speed and the type of heat transfer surface, e.g., jacket or coil. In the following sections, two methods (jacket and coil) of heat removal or addition for non-Newtonian materials are discussed separately.

8.4.1 Helical cooling coils

A vessel fitted with a cooling coil and an agitator is shown schematically in Figure 8.31. In this case the thermal resistances to heat transfer arise from the fluid film on the inside of the cooling coil, the wall of the tube (usually negligible), the fluid film on the outside

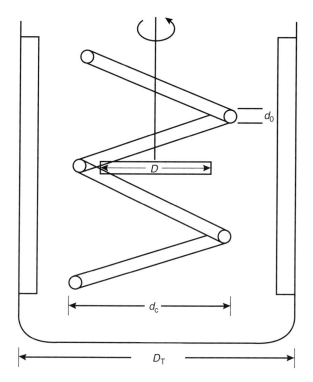

Figure 8.31 *Schematics of a vessel fitted with a cooling coil and an agitator*

of the coil, and the scale that may form on either surface. The overall heat transfer coefficient, U, can thus be expressed as:

$$\frac{1}{UA} = \frac{1}{h_i A_i} + \frac{x_w}{k_w h_w} + \frac{1}{h_o A_o} + \frac{R_o}{A_o} + \frac{R_i}{A_i} \tag{8.28}$$

where the subscripts 'i', 'o' and 'w' refer to the inside, outside and the tube wall conditions, respectively, and R is the scale resistance.

(i) *Inside film coefficients*
The inside film coefficient, h_i, may be estimated using the correlations available in the literature such as the well known Dittus–Boelter equation for low viscosity coolants, as recommended in the literature (Coulson and Richardson, 1999):

$$Nu = \frac{h_i d}{k} = 0.023 \left(\frac{\rho V d}{\mu} \right)^{0.8} \left(\frac{C_p \mu}{k} \right)^{0.4} \tag{8.29}$$

For more viscous liquids such as concentrated brine solutions, the Sieder–Tate equation is preferable:

$$\frac{h_i d}{k} = 0.027 Re^{0.8} Pr^{0.33} \left(\frac{\mu_b}{\mu_w} \right)^{0.14} \tag{8.30}$$

where the subscripts '*b*' and '*w*' relate to the bulk and wall conditions, respectively. Both of these equations were based on data for straight tubes and it is necessary to apply a small correction factor for the coil configuration (Jeschke, 1925) as:

$$h_i(\text{coil}) = h_i(\text{tube})\left[1 + 3.5\frac{d}{d_c}\right] \tag{8.31}$$

where *d* is the inside diameter of the tube and d_c that of the helix, as shown in Figure 8.31.

For steam condensing inside the tube, the heat transfer coefficient is usually sufficiently high for its contribution to the overall heat transfer resistance to be neglected.

(ii) *Outside film coefficients*

The value of h_o is much more difficult to determine, particularly in view of the strong interplay between the non-Newtonian rheology of the liquid and the geometry of the tank–impeller combination. Other internal fixtures such as baffles affect the flow pattern and heat transfer, as well as power consumption and mass transfer. It is thus dangerous to make detailed cross-comparisons between different studies unless the two systems are completely similar.

The results of heat transfer studies are usually expressed in terms of the relevant dimensionless groups as (Edwards and Wilkinson, 1972; Poggermann *et al.*, 1980; Chhabra, 2003; Paul *et al.*, 2004):

$$Nu = \text{f} \ (Re, Pr, Gr, \text{geometric ratios}) \tag{8.32}$$

The effect of geometric ratios on the Nusselt number is difficult to quantify, although some investigators have incorporated the principal length parameters as ratios in their heat transfer correlations. This factor alone precludes the possibility comparisons of the results of some workers!

For geometrically similar systems, equation (8.32) simplifies to:

$$Nu = \text{f} \ (Re, Gr, Pr) \tag{8.33}$$

The Grashof number, $Gr \ (= g\beta\Delta T L_c^3 \rho^2/\mu^2)$, is indicative of free convection effects which are generally significant in low viscosity liquids, but become increasingly less important with rising viscosity and increasing impeller speed (Carreau *et al.*, 1994). For a power-law fluid, the apparent viscosity estimated using the average shear rate given by equation (8.9) is used in evaluating the Grashof number.

Owing to the inherently different flow fields produced by each class of agitators, it is convenient to consider separately the application of equation (8.33) to equipment of particular forms. Since most of the work related to heat transfer to pseudoplastic materials has been critically reviewed elsewhere (Gluz and Pavlushenko, 1966; Edwards and Wilkinson, 1972; Desplanches *et al.*, 1980; Poggermann *et al.*, 1980), only a selection of widely used correlations is given here.

(a) *Class I impellers*

As mentioned earlier, these impellers operate at relatively high speeds and are effective only in low to medium viscosity liquids. In most cases, the main flow in the vessel tends to be transitional and/or turbulent. For shear-thinning polymer solutions and particulate suspensions agitated by paddle, turbine and propeller type impellers, many correlations of varying complexity and form are available for the estimation of the outside film coefficient. One such correlation, based on wide ranges of conditions ($400 \leq Re \leq 10^6$; $4 \leq Pr \leq 1900$; $0.65 \leq \mu_{eff} \leq 283\,\text{mPa s}$), is due to Edney and Edwards (1976):

$$Nu = \frac{h_o d_o}{k} = 0.036 \left(\frac{\rho ND^2}{\mu_{eff}} \right)^{0.64} \left(\frac{C_p \mu_{eff}}{k} \right)^{0.35} \left(\frac{\mu_{eff,b}}{\mu_{eff,w}} \right)^{0.2} \left(\frac{d_c}{D_T} \right)^{-0.375} \tag{8.34}$$

where d_o is the outside (coil) tube diameter, d_c is the coil helix and D_T is the vessel diameter, respectively. The effective viscosity is evaluated using equation (8.9) with $k_s = 11.6$.

(b) *Class II impellers*
These impellers, including gates and anchors, reach the far corners of the vessels directly instead of relying on momentum transport by bulk motion. Among others, Pollard and Kantyka (1969) carried out an extensive study on heat transfer from a coil to aqueous chalk slurries ($0.3 \leq n \leq 1$) in vessels up to 1.1 m in diameter fitted with anchor agitators; they proposed the following equation for Nusselt number:

$$Nu = \frac{hD_T}{k} = 0.077 Re^{2/3} Pr^{1/3} \left(\frac{\mu_{eff,b}}{\mu_{eff,w}} \right)^{0.14} \left(\frac{D_T}{d_o} \right)^{0.48} \left(\frac{D_T}{d_c} \right)^{0.27} \tag{8.35}$$

Equation (8.35) applies over the range $200 \leq Re \leq 6 \times 10^5$ and the effective viscosity is estimated using equation (8.9) and the value of k_s given by:

$$k_s = \left(9.5 + \frac{9}{\{1 - (D/D_T)^2\}} \right) \left(\frac{3n + 1}{4n} \right)^{n/(n-1)} \tag{8.36}$$

The indeterminate form of equation (8.36) should be noted for $n = 1$, but then k_s is redundant for Newtonian fluids.

(c) *Class III impellers*
This class of impellers has generated much more interest than the gates and anchors. Consequently, numerous workers (Coyle *et al.*, 1970; Nagata *et al.*, 1972; Heim, 1980; Blasinski and Kuncewicz, 1981; Kuriyama *et al.*, 1983; Ayazi Shamlou and Edwards, 1986; Kai and Shengyao, 1989; Carreau *et al.*, 1994; Delaplace *et al.*, 2001) have investigated the rates of heat transfer to viscous non-Newtonian materials in vessels fitted with helical ribbons and screw-type impellers. For instance, Carreau *et al.* (1994) studied heat transfer between a coil (acting as a draft tube) and viscous Newtonian, inelastic shear-thinning and visco-elastic polymer solutions agitated by a screw impeller. The rate of heat transfer was measured for both cooling and heating of solutions. The flow rate of water inside the coil was sufficiently high for the inside film resistance to be negligible (i.e. large values of h_i). The effective viscosity of shear-thinning and visco-elastic polymer solutions was evaluated using equation (8.9) with $k_s = 16$, the value deduced from their results on power consumption. These workers proposed a single correlation for Newtonian, inelastic power-law and visco-elastic fluids as:

$$Nu = \frac{h_o d_o}{k} = 0.39 Re^{0.51} Pr^{1/3} \left(\frac{d_o}{D} \right)^{0.594} \tag{8.37}$$

All physical properties are evaluated at the mean film temperature, i.e. $(T_w + T_b)/2$, and equation (8.37) encompasses the range of conditions: $3 \leq Re \leq 1300$ and $500 \leq Pr \leq 30\,000$. The absence of a visco-elastic parameter in equation (8.37) suggests the rate of heat transfer to be governed primarily by viscous properties of liquid.

8.4.2 Jacketed vessels

In many applications, it is not practicable to install cooling coils inside a tank, and heating of the contents of the vessel is achieved using condensing steam or coolant in a jacket. Indeed, this is a preferred arrangement for moderately viscous systems. The numerous studies of heat transfer to non-Newtonian fluids agitated by a range of impellers have been reviewed by Edwards and Wilkinson (1972), and more recently by Dream (1999); most of the work to date has been carried out in relatively small vessels (<650 mm diameter). Furthermore, investigators have used diverse methods of estimating the apparent viscosity which should be used in the evaluation of the Reynolds and Prandtl numbers and it is therefore not possible to discriminate between predictive expressions. In the following sections, a selection of the more reliable correlations is presented for each class of agitators.

For a steam jacketed tank (360 mm in diameter) fitted with baffles, Hagedorn and Salamone (1967) measured the rates of heat transfer to water, glycerol and aqueous carbopol solutions over wide ranges of conditions ($0.36 \leq n \leq 1$; $35 \leq Re \leq 6.8 \times 10^5$; $Pr \leq 2.4 \times 10^4$). They measured temperatures at various locations in the vessel and suggested the following general form of heat transfer correlation:

$$Nu = \frac{hD}{k} = C\ Re^{a/((n+1)+b)}\ Pr^d \left(\frac{m_b}{m_w}\right)^e \left(\frac{D_T}{D}\right)^f \left(\frac{W}{D}\right)^g n^i \qquad (8.38)$$

where the apparent viscosity is evaluated at the shear rate, $\dot{\gamma}_{eff} = 11\ N$. The values of the constants in equation (8.38) for various impellers are listed in Table 8.5.

Hagedorn and Salamone (1967) reported the mean error resulting from the use of equation (8.38) to be 14% for moderately shear-thinning materials ($n > 0.69$), increasing to 20% for highly shear-thinning fluids ($n < 0.69$); both are well within the limits of experimental errors generally associated with this type of work. Similarly, Sandall and Patel (1970) and Martone and Sandall (1971) have developed correlations for the heating of pseudoplastic (carbopol solutions) and Bingham plastic slurries (chalk in water)

Table 8.5   *Values of constants in equation (8.38)*

Type of impeller	a	b	C	d	e	f	g	i	$\dfrac{D_T}{D}$
Anchor	1.43	0	0.56	0.30	0.34	–	–	0.54	1.56
Paddle	0.96	0.15	2.51	0.26	0.31	−0.46	0.46	0.56	1.75–3.5
Propeller	1.28	0	0.55	0.30	0.32	−0.40	–	1.32	2.33–3.41
Turbine	1.25	0	3.57	0.24	0.30	0	0	0.78	2–3.50

in a steam-jacketed vessel fitted with a turbine impeller and baffles or with an anchor agitator. In their study which was limited to only one size of vessel, they were able to correlate their results in the following simple form:

$$Nu = \frac{hD_T}{k} = C\left(\frac{\rho ND^2}{\mu_{\text{eff}}}\right)^a \left(\frac{C_p \mu_{\text{eff}}}{k}\right)^b \left(\frac{\mu_{\text{eff,b}}}{\mu_{\text{eff,w}}}\right)^d \tag{8.39}$$

where the apparent viscosity was evaluated at $\dot{\gamma}_{\text{eff}} = k_s N$ and k_s is given by equation (8.36). Based on their data ($0.35 \le n \le 1$; $80 \le Re \le 10^5$ and $2 \le Pr \le 700$), they proposed $a = 2/3$ $b = 1/3$ (the values also applicable for Newtonian liquids) and $d = 0.12$. The remaining constant C showed some dependence on the type of the impeller, having values of 0.482 for turbines and 0.315 for anchors.

Limited numerical studies (Kaminoyama *et al.*, 1999; Delaplace *et al.*, 2001; Niedzielska and Kuncewicz, 2005) lend support to the above-noted results for the prediction of heat transfer in viscous Newtonian and shear-thinning fluids in mixing vessels employing anchor and ribbon impellers. Broadly speaking, the heat transfer coefficients for ribbon impellers are typically 20–30% larger than that for anchor agitators of the same diameter and at the same power consumption. As expected, even greater difference is observed in the performance of ribbon and turbine impellers, i.e. to obtain the same value of the heat transfer coefficient as that of a ribbon (at speed N), a turbine requires speeds up to $8N$, which, of course leads to enormous power consumption.

In conclusion, it should be emphasized that most of the currently available information on heat transfer to non-Newtonian fluids in stirred vessels relates to specific geometrical arrangements. Few experimental data are available for the independent verification of the individual correlations presented here which, therefore, must be regarded somewhat tentative. Reference should also be made to the extensive compilations (Edwards and Wilkinson, 1972; Poggermann *et al.*, 1980; Dream, 1999) of other correlations available in the literature. Although the methods used for the estimation of the apparent viscosity vary from one correlation to another, especially in terms of the value of k_s, this appears to exert only a moderate influence on the value of h, at least for shear-thinning fluids. For instance, for $n = 0.3$ (typical of suspensions and polymer solutions), a twofold variation in the value of k_s will give rise to a 40% reduction in viscosity, and the effects on the heat transfer coefficient will be further diminished because $Nu \propto \mu_{\text{eff}}^{0.33 \text{ to } 0.7}$. Thus, an error of 100% in the estimation of μ_{eff} will result in an error of only 25–60% in the value of h which is a reasonable engineering estimate.

Example 8.6

A polymer solution is to be heated from 18°C to 27°C before use as a thinner in a wall paint. The heating is to be carried out in a stainless steel vessel (1 m diameter) fitted with an anchor agitator of diameter equal to 0.9 m which is rotated at 100 RPM. The tank which is filled up to 0.8 m depth is fitted with a helical coil (helix diameter 0.8 m) made of 25 mm od and 22 mm id copper tube (total external heat transfer area of 2 m^2). Hot water at a mean temperature of 45°C (assumed to be approximately constant) is fed to the coil at a rate of 30 kg/min.

The thermal conductivity, heat capacity and density of the polymer solution can be taken as the same as for water. The values of the power-law constants are: $n = 0.36$ and $m = 26 - 0.0566$ $T(\text{Pa s}^n)$ in the range $288 \le T \le 323$ K. Estimate the overall heat transfer coefficient and the time needed to heat one batch of liquid.

Solution

The inside film coefficient will be evaluated first using equations (8.29) and (8.31):
For water, $k = 0.59\,\text{W/mK}$; $\mu = 0.85\,\text{mPa s}$; $C_p = 4180\,\text{J/kgK}$.

$$\text{mean velocity of water, } V = \frac{30}{60 \times 1000} \times \frac{4}{\pi(0.022)^2} = 1.315\,\text{m/s}.$$

$$\therefore \text{ the Reynolds number, } Re = \frac{\rho Vd}{\mu} = \frac{1000 \times 1.315 \times 0.022}{0.85 \times 10^{-3}}$$
$$= 34\,035$$

$$\text{Prandtl number, } Pr = \frac{C_p \mu}{k} = \frac{4180 \times 0.85 \times 10^{-3}}{0.59} = 6$$

From equation (8.29),

$$Nu = 0.023Re^{0.8}Pr^{0.4} = 0.023 \times (34035)^{0.8} \times (6)^{0.4}$$
$$Nu = 200$$
$$\text{or } h = \frac{200 \times 0.59}{0.022} = 5360\,\text{W/m}^2\text{K}$$

This value is for straight tubes and the corresponding value for the coil is given by equation (8.31):

$$h_i = 5360\left(1 + 3.5 \times \frac{0.022}{0.80}\right) = 5876\,\text{W/m}^2\text{K. (based on the inside tube area)}$$

$$\text{or } 5170\,\text{Wm}^2\text{K (based on the outer area)}.$$

Now, the outside coefficient, h_o, will be evaluated using equation (8.35).
For $D = 0.9\,\text{m}$ and $D_T = 1\,\text{m}$, the value of k_s is obtained from equation (8.36):

$$k_s = \left[9.5 + \frac{9}{1 - \left(\frac{0.9}{1}\right)^2}\right]\left[\frac{3 \times 0.36 + 1}{4 \times 0.36}\right]^{0.36/(0.36-1)} = 46.2$$

$\therefore$ the effective shear rate, $\dot{\gamma}_{\text{eff}} = k_s N = 46.2 \times (100)/(60) = 77\,\text{s}^{-1}$

The mean temperature of the liquid is $(18 + 27)/2 = 22.5°C$.
Therefore, $m = 26 - 0.0566 \times (273 + 22.5) = 9.27\,\text{Pa s}^n$

$$\therefore \mu_{\text{eff,b}} = 9.27 \times (77)^{0.36-1} = 0.575\,\text{Pa s}$$

$$Re = \frac{\rho ND^2}{\mu_{\text{eff}}} = \frac{1000 \times \frac{100}{60} \times 0.9^2}{0.575} = 2346$$

This value is within the range of applicability of equation (8.35).

$$Pr = \frac{C_p \mu_{\text{eff}}}{k} = \frac{4180 \times 0.575}{0.59} = 4074$$

Neglecting the thermal resistance of the coil, the outside coil surface would be at a mean temperature of 45°C, at which,

$$m = 26 - 0.0566 \times (273 + 45) = 8\,\text{Pa s}^n$$
$$\therefore \mu_{\text{eff,w}} = 8 \times (77)^{0.36-1} = 0.5\,\text{Pa s}$$

Now substituting values in equation (8.35):

$$Nu = \frac{h_o \times 1}{0.59} = 0.077 \times (2346)^{2/3} \times (4074)^{1/3} \times \left(\frac{0.575}{0.5}\right)^{0.14}$$

$$\times \left(\frac{1}{0.25}\right)^{0.48} \times \left(\frac{1}{0.8}\right)^{0.27}$$

or $\quad h_o = 816 \text{ W/m}^2\text{K}.$

Neglecting the thermal resistance of the coil wall, the overall heat transfer coefficient based on the external area of the coil, U is given by:

$$U = \frac{816 \times 5170}{816 + 5170} = 705 \text{ W/m}^2\text{K}.$$

Note that it has been assumed here that the inside and outside areas of the coil are nearly the same.

$$\text{The mass of liquid in the tank} = \rho \frac{\pi D_T^2}{4} L$$

$$= \frac{1000 \times \pi \times 1^2 \times 0.8}{4} = 628 \text{ kg}$$

Let the temperature of the solution be T_f at any time t, a heat balance on the solution gives:

$$mC_p \frac{dT_f}{dt} = UA(T_s - T_f)$$

where T_s is the mean surface temperature of the coil, assumed here to be 45°C. Substituting values and integrating from $T_f = 18°C$ to $T_f = 27°C$:

$$t = \frac{mC_p}{UA} \ln \frac{T_s - 18}{T_s - 27} = \frac{628 \times 4180}{705 \times 2} \ln \frac{45 - 18}{45 - 27}$$

$$= 755 \text{ s, i.e. } 12.6 \text{ min}$$

Heat losses to the surroundings have been neglected. □

8.5 Mixing equipment and its selection

The wide range of mixing equipment available commercially reflects the enormous variety of mixing duties encountered in the processing industries. It is reasonable to expect therefore that no single item of mixing equipment will be suitable for carrying out such a range of duties. Few manufacturers have taken into account non-Newtonian characteristics of the fluid, but in general those mixers which are suitable for high viscosity Newtonian fluids are also likely to be appropriate for shear-thinning fluids. This normally means the mixer should have small clearances between the moving and fixed parts, and be designed to operate at low speeds. This has led to the development of a number of distinct types of mixers over the years. Very little has been done, however, by way of standardization of equipment, and no design codes are available. The choice of a mixer type and its design is therefore primarily governed by experience. In the following sections, the main mechanical features of commonly used types of equipment, together with their range of applications, are described briefly. Detailed descriptions of design and selection of various types of mixers have been presented by Oldshue (1983), Harnby *et al.* (1992), Tatterson (1994) and Paul *et al.* (2004). Most equipment manufacturers also provide performance charts of their products and offer some guidelines for the selection of most appropriate configuration for a specific application.

8.5.1 Mechanical agitation

This is perhaps the most commonly used method of mixing liquids, and essentially there are three elements in such devices.

(i) *Vessels*
These are often vertically mounted cylindrical tanks, up to 10 m in diameter, and height to diameter ratios of at least 1.5, and typically filled to a depth equal to about one diameter. In some gas–liquid applications, tall vessels are used and the liquid depth is then up to three tank diameters; multiple impellers fitted on a single shaft are frequently used, e.g. see Kuboi and Nienow (1982). The base of the tanks may be flat, dished, conical, or specially contoured, depending upon factors such as ease of emptying, or the need to suspend solids, etc. Usually, large tanks are constructed with flat bottoms or with a shallow cone or inverted cone, as in crude oil storage tanks. Deep cone bottoms are used in tanks when inventory must be minimized. Such a geometry may require a special impeller to conform to the cone geometry.

For the batch mixing of thick pastes and doughs using ribbon impellers and Z- or sigma blade mixers, the tanks may be mounted horizontally. In such units, the working volume of pastes and doughs is often relatively small, and the mixing blades are massive in construction.

(ii) *Baffles*
To prevent gross vortexing, which is detrimental to mixing, particularly in low viscosity systems, baffles are often fitted to the walls of the vessel. These take the form of thin strips about one-tenth of the tank diameter in width and typically four equi-spaced baffles may be used. Smaller width baffles ($\sim 0.02\,D_T$) are sometimes used for medium to high viscosity fluids, or for the entrainment of neutrally buoyant particles, or when a small vortex is desired. In some cases, the baffles are mounted flush with the wall, although more usually a small clearance is left between the wall and the baffle to facilitate fluid motion in the wall region. Baffles are, however, generally not required for high viscosity fluids ($>\sim 5$ Pa s) because the viscous stresses are sufficiently large to damp out the rotary motion. Some times, the problem of vortexing is circumvented by mounting impellers off-centre or horizontally.

(iii) *Impellers*
Figure 8.32 shows some of the impellers which are frequently used. Propellers, turbines, paddles, anchors, helical ribbons, gate and screws are usually mounted on a central vertical shaft in a cylindrical tank, and they are selected for a particular duty, largely on the basis of liquid viscosity. It is generally necessary to move from a propeller to a turbine and then, in order, to a paddle, to an anchor or a gate, and then to a helical ribbon, and finally to a screw, as the viscosity of the liquid increases. The speed of rotation or agitation is reduced as the viscosity increases.

Propellers, turbines and paddles are generally used with relatively low viscosity liquids and operate at high rotational speeds (in the transitional–turbulent region). A typical velocity for the tip of the blades of a turbine is of the order of 3 m/s, with a propeller being a little faster and a paddle little slower. These are classed as remote-clearance impellers having diameters in the range 13–67% of the tank diameter. For instance, Figure 8.32b shows a standard six flat-bladed Rushton turbine, and possible variations are shown elsewhere in Figure 8.33. Thus, it is possible to have retreating blades, angled-blades, four- to twenty-bladed, hollow bladed turbines, wide blade hydrofoils (see Figure 8.34) and so on.

(a) Three-bladed propeller

(b) Six-bladed disc turbine (Rushton turbine)

Plan view

(c) Simple paddle

(d) Anchor impeller

(e) Helical ribbon

(f) Helical screw with draft tube

Draft tube

(g) Gate impeller

Shaft

(h) Banbury mixer

Ram

Figure 8.32 *A selection of commonly used impellers*

(a) Flat blade

(b) Disc flat blade

(c) Pitched vane

(d) Curved blade

(e) Tilted blade

(f) Shrouded

or

(g) Pitched blade

(h) Pitched curved blade

(i) Arrowhead

Figure 8.33 *Variation in turbine impeller designs*

(a)

(b)   (c)   (d)

Figure 8.34   *Specially designed impellers: (a) HE-3, (b) CD-6, (c) helix and (d) Maxflo 'T' impeller (courtesy Chemineer, Inc, Dayton, Ohio)*

For dispersion of gases in liquids, turbines and modified turbines are usually employed (Figure 8.28). Commonly two or more disc turbine impellers ($D_T/2$ distance apart) are mounted on the same shaft to ensure mixing over the whole depth of the tank.

Anchors (several design are shown in Figure 8.35), gate, helical ribbons and screws (a selection of possible variations shown in Figure 8.36), are generally used for high viscosity and non-Newtonian liquids. The anchor and ribbon types are arranged with a close clearance at the vessel wall, whereas the helical screw has a smaller diameter and is often used inside a draft tube to promote liquid motion through out the vessel. However, anchor and gate impellers induce very little vertical fluid motion and thus are generally ineffective unless used in conjunction with another impeller. Helical ribbons or interrupted ribbons are often used in horizontally mounted cylindrical vessels. A variation of the simple helix mixer is the helicone (Figure 8.37), which has the additional advantage that the clearance between the blade and the vessel wall is easily adjusted by a small axial shift of the impeller. The direction of pumping by a helical ribbon impeller depends on whether the helix is left handed or the right handed and the direction of the rotation

Figure 8.35   *Variation in anchor impeller designs*

of the impeller (clockwise or anti-clockwise). For viscous liquids, downward pumping at the wall is very effective, since the bottom of the tank imposes a direct return of fluid at the centre of the mixing tank. On the other hand, this mode is not desirable, especially with dry solids as down-pumping may cause crushing or compaction. For some applications involving anchor-stirrers, the shear stresses generated are not adequate for the

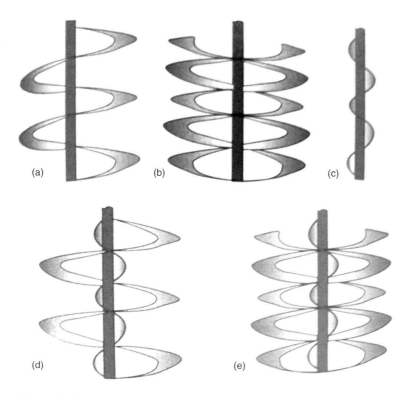

Figure 8.36   *Variation in ribbon/screw impeller designs. Some designs use single helix (a and c) while other designs (b, d, e) employ two helices (replotted from D.S. Dickey, Chemical Engineering, p. 73, May, 2000)*

Figure 8.37   *A double helicone impeller*

Figure 8.38 *Mastermix HVS/TS high speed/low-speed impeller combination (courtesy, Mastermix)*

breakup and dispersion of agglomerated particles. In such cases, it may be necessary to use an anchor to promote general flow in the vessel in conjunction with a high shear mixing device mounted on a separate eccentric shaft and operating at high speed. Similarly, Figures 8.38 and 8.39 show composite impellers involving a modified paddle and a small high speed dispenser and a helical blade–turbine combinations, respectively.

Kneaders, Z- and sigma-blade (Figure 8.40), plough (Figure 8.41) and Banbury mixers (sketched in Figure 8.32) are generally used for the mixing of high viscosity liquids, pastes, rubbers, doughs and so on, many of which have non-Newtonian flow characteristics. The tanks are usually mounted horizontally and two impellers are often used. The impellers are massive and the clearances between blades, as well as between the wall and the blade, are very small thereby ensuring that the entire mass of liquid is sheared. While mixing heavy pastes and doughs using a sigma blade mixer, it is usual for the two blades to rotate at different speeds – a ratio of 3:2 is common. Various blade profiles of different helical pitch are used. The blade design differs from that of the helical ribbons considered above in that the much higher effective viscosities (of the order of 10^4 Pa s) require a more solid construction; the blades consequently tend to sweep a greater quantity of fluid in front of them, and the main small-scale mixing process takes place by extrusion between the blade and the wall. Partly for this reason, mixers of this type are usually operated only partially full, though the Banbury mixer (Figure 8.32) used in the

Figure 8.39   *Composite impeller consisting of two turbines and an anchor or a helical blade*

Figure 8.40   *A sigma blade mixer*

Figure 8.41   *A typical plough mixer*

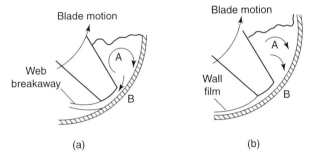

Figure 8.42   *Fluid motion in a sigma blade mixer: (a) non-wetting paste and (b) surface-wetting fluid*

rubber industry is filled completely and pressurized as well. The pitch of the blades produces the necessary motion along the channel, and this gives the larger scale blending that is required in order to limit batch blending times to reasonable periods. The Banbury mixer is a very energy intensive device consuming up to 5–6 kW/kg. In contrast, power inputs for sigma blade mixers range from 20 to 500 W/kg and these are available in sizes ranging from 1 L to 50 m^3 capacity. Sigma blade mixers are extensively used in food industry to make thick gravy base, pet food, cookie dough, etc. The plough mixers are used for free-flowing pasty materials and the power requirement is not too high.

The motion of the material in the sigma blade mixer can be considered in three stages, as illustrated in Figure 8.42. Material builds up in front of the blade in region A where it will undergo deformation and flow – the relative extent of these processes depending on the material properties. Materials will tend to be rolled and deformed until some is trapped in the gap B.

The difference between the solid and liquid behaviour will be evident in the regions where the shear stresses are least, i.e. less than the yield stress. Whether or not there is a radially inward flow in front of the blade will depend on the magnitude of this yield stress. Unfortunately, the dynamics of this region are so complex that it is not possible to quantify this generalization.

Once the material has entered region B it will be subject to an unsteady state developing flow situation; in the fully developed form the velocity gradient is linear, changing uniformly from the blade velocity at the blade surface to zero at the vessel wall.

The general situation is the same as that existing between moving planes; normally the radius of the vessel is large and the shear stress is sensibly uniform throughout the gap. Under these circumstances fully developed flow will be established providing that the *channel* length (i.e. the thickness of the edge of the blade facing the wall) is of the order of two or three times the clearance between the blade and the wall. For this situation the mean velocity of the material in the gap is just half the velocity of the blade relative to the wall. A solid material may well not achieve this condition, however, since the rate at which it is drawn into the gap is largely controlled by the deformation in zone A.

On discharge from the gap, the material will either break away from the wall (as in *a*) or remain adhering to it (as in *b*). In the former case, a web may be formed which will eventually break off to be incorporated with the material coming in front of the blade on its next revolution. If the extruded material remains adhering to the wall, the amount of material left on the wall is determined from mass balance considerations, at a thickness equal to about half the clearance.

(iv) *Side-entering mixers*

A side-entering mixer consists of a propeller or a hydrofoil impeller, which enters the tank from the side below the level of the liquid, typically near the bottom. This kind of mixer generates a horizontal spiralling jet that creates the fluid motion to keep the contents well mixed, or prevents the deposition of sludge such as that encountered in crude oil storage tanks. The mixer shaft is positioned with a tilt angle of about ~10–30°. Since crude oil storage tanks can be very large (up to 90 m in diameter), it is not uncommon to use multiple mixers to provide adequate mixing. For instance, one mixer is adequate in tanks up to 30 m in diameter whereas 4 to 5 mixers may be required in tanks with diameters larger than 60 m. Multiple mixers generally are of the same size and can be installed in a distributed (Figure 8.43a) or in a clustered fashion (Figure 8.43b). The major disadvantage of side-entering mixers is a submerged shaft seal, which is exposed to the process fluid. Process fluids may be abrasive as in paper pulp and slurries, or lubricants such as in petroleum products. The abrasive slurries cause wear and lubricants require a positive seal. The key advantages of this class of mixers include low capital and simple installation as no mounting support on the top of the tank is needed. Similarly, in certain situations, bottom entering mixers are used when the process requirements or tank geometry do not permit the use of the top entering and side-entering mixers. Figure 8.44 shows a typical side-entering mixer.

(v) *High shear mixers*

This class of impellers (Figure 8.45) has many features, which are opposite to that of high viscosity impellers like an anchor or a helical ribbon. Typical high shear impellers have small diameter ($0.1–0.2\ D_T$) and operate at high rotational speeds (16.7–60 Hz). The effect of high speed on power consumption is somewhat offset by their small diameter. Furthermore, high shear impellers have small blades which may appear as teeth on the edge of a disc, or slots and holes in a rotating cylinder (Figure 8.45). The slotted-cylinder design is generally used for both the rotating and the stationary component also known as rotor-stator design (Figure 8.46). This kind of impeller is often used to disperse

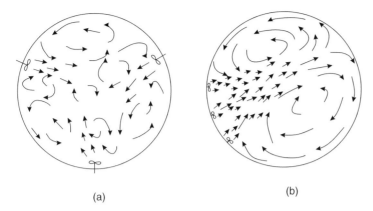

(a)   (b)

Figure 8.43   *Schematics of the flow patterns produced by multiple side-entering impellers. (a) uniformly distributed and (b) cluster configuration*

Figure 8.44   *A typical side entering mixer (courtesy, Chemineer)*

Figure 8.45   *High shear disk impeller with teeth (courtesy, INDCO)*

Figure 8.46   *Typical rotor-stator (slotted cylinder) impeller (courtesy, IKA)*

a second phase (liquid, solid, powder) in grinding, dispersing pigments and emulsification applications.

8.5.2 Rolling operations

Several blending operations involving highly viscous or non-Newtonian fluids have to be carried out in a flow regime which is essentially laminar. Although mixing is less efficient in a laminar flow system than under turbulent conditions, the natural extension of elements of fluid in a shear field reduces the effective path length over which the final dispersive stage of diffusion has to take place. With materials possessing effective viscosities of the order of 100 Pa s, one of the more usual ways of imposing a suitable velocity gradient for a significant period is to feed the material through the gap between two counter-rotating rollers. This operation is often termed *calendering*. In principle, the rolls need not be rotating at the same speed, nor even need they be the same diameter. The simplest case to consider, however, is that where the diameters and speeds are the same. The difficulties in the analysis lie in the fact that it is not possible to solve unambiguously for the position at which the breakaway from the roll surface will occur. Figure 8.47 shows the situation in the 'nip' region between two rolls.

Rolling machines are commonly used for batch blending very viscous materials, and the method of operation is shown in Figure 8.48. The preferential adherence of the extruded film to one of the rollers can be obtained with a small speed differential. The viscosity of the fluid is high enough to prevent centrifugal forces throwing it clear of the roller, and it returns to the feed side of the machine where it is accumulated in the highly sheared circulating fillet of excess material. This flow is, of course, simply circumferential around the roller and through the gap, and it is necessary to ensure adequate lateral mixing along the length of the machine by separate mechanical or manual means from time to time.

Figure 8.47   *Flow between rolls*

Figure 8.48   *Mixing with a rolling machine*

In addition to the basic impeller designs shown in Figures 8.32 and 8.33, many spe-
ciality impellers developed by the manufacturers of mixing equipment are also available
which supposedly give better performance than the basic configurations. For instance, the
so-called HE-3 high efficiency impeller shown in Figure 8.34 requires a much smaller
motor than the standard turbine to achieve comparable fluid velocities in the vessel.
Similarly, the concave blade disk impeller (CD-6), also shown in Figure 8.34, results in
up to 100% larger values of liquid phase mass transfer coefficients in gas–liquid systems
than those obtained with the six flat-bladed turbine impeller. Both these designs have
been developed and/or marketed by Chemineer, Inc, Dayton, Ohio. Obviously, it is not
possible to provide a complete list of designs offered by different manufacturers but it
is always desirable to be aware of such developments prior to selecting equipment for a
given application. Some guidelines for equipment design and selection are also available
in the literature (Bakker *et al.*, 1994; Paul *et al.*, 2004).

8.5.3 Portable mixers

For a wide range of applications, a portable mixer which can be clamped on the top
or side of the vessel is often used. This is commonly fitted with two propeller blades
so that the bottom rotor forces the liquid upwards and the top rotor forces the liquid

downwards. The power supplied is up to about 2 kW, though the size of the motor becomes too great at higher powers. To avoid excessive strain on the armature, some form of flexible coupling should be fitted between the motor and the main propeller shaft. Although details of impeller types vary from one design to another, axial flow impellers such as marine propellers or hydrofoils are used most often. Units of this kind are usually driven at a fairly high speed (up to 30 Hz) and, for use with high viscosity and non-Newtonian materials, a reduction gear can be fairly easily fitted to the unit for low speed operation, although this increases the mass of the unit.

8.6 Mixing in continuous systems

The mixing problems considered so far have related to batch systems in which two or more materials are mixed together and uniformity is maintained by continued operation of the agitator. Consideration will now be given to some of the equipment used for continuous mixing duties.

8.6.1 Extruders

Mixing duties in the plastics industry (and to a lesser extent in food industry) are often carried out in either single or twin screw extruders. The feed to such units usually contains the base polymer in either granular or powder form, together with additives such as stabilizers, pigments, plasticizers, fire retardants and so on. During processing in the extruder, the polymer is melted and the additives thoroughly mixed. The extrudate is delivered at high pressure and at a controlled rate from the extruder for shaping by means of either a die or a mould. Considerable progress has been made in the design of extruders in recent years, particularly by the application of finite-element methods. One of the problems is that an enormous amount of heat is generated (by viscous action), and the fluid properties may change by several orders of magnitude as a result of temperature changes. It is therefore always essential to solve the coupled equations of flow and heat transfer.

   In the typical single screw shown in Figure 8.49, the shearing which occurs in the helical channel between the barrel and the screw is not intense, and therefore this equipment does not give good mixing. Twin-screw extruders, shown in Figure 8.50, may be co- or counter-rotating and, as there are regions where the rotors are in close proximity,

Figure 8.49   *Single screw extruder*

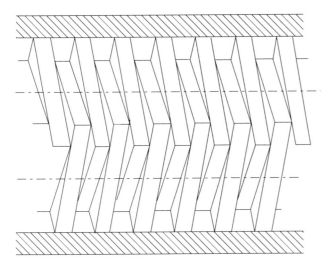

Figure 8.50   *Counter-rotating twin-screw extruder*

Figure 8.51   *Twisted-blade type of static mixer elements*

extremely high shear stresses may be generated. Clearly, twin-screw units can yield a product of better mixture quality than a single screw unit. Detailed accounts of the design and performance of extruders are available in the literature (Schenkel, 1966; Janssen, 1978; Rauendaal, 1992).

8.6.2 Static mixers

All the mixers so far described have been of the dynamic type in the sense that moving blades are used to impart motion to the fluid and produce the mixing effect. In static mixers, sometimes called 'motionless' or in-line mixers, the materials to be mixed are pumped through a pipe containing a series of specially shaped stationary blades. Static mixers can be used with liquids of a wide range of viscosities in either the laminar or turbulent regimes, but their features are perhaps best appreciated in relation to laminar flow mixing which is the normal condition for high viscosity and non-Newtonian fluids.

Figure 8.51 shows a particular type of static mixer in which a series of stationary helical blades mounted in a circular pipe is used to divide, split, and twist the flowing streams (Figure 8.52). In streamline flow, the material divides at the leading edge of each of these elements and follows the channels of complex shape created by the element.

Figure 8.52   *Twisted-blade type of static mixer operating in the laminar flow regime: (a) distributive mixing mechanism showing, in principle, the reduction in striation thickness produced and (b) radial mixing contribution from laminar shear mechanism*

At each succeeding element, the two channels are further divided, and mixing proceeds by a distributive process similar to the cutting and folding mechanism shown schematically in Figure 8.4. In principle, if each element divided the streams neatly into two, feeding two dissimilar streams to the mixer would give a striated material in which the thickness of each striation would be of the order $D_t/2^n$ where D_t is the diameter of the tube and n is the number of elements. However, the helical elements shown in Figure 8.51 also induce further mixing by a laminar shear mechanism (illustrated in Figures 8.1 and 8.2). This, combined with the twisting action of the element, helps to promote radial mixing which is important in eliminating any radial gradients of composition, velocity and possibly temperature that might exist in the material. Figure 8.53 shows how these mixing mechanisms together produce after only 10–12 elements, a well-mixed material.

Figure 8.54 shows a Sulzer-type SMX static mixer where the mixing element consists of a lattice of intermeshing and interconnecting bars contained in an 80-mm diameter pipe. It is recommended for viscous Newtonian and non-Newtonian materials in laminar flow. The mixer shown is used in food processing, for example for mixing fresh cheese with whipped cream, and in polymer processing, as in dispersing TiO_2 into polymer melts for de-lustering. While initially developed primarily for viscous fluids, over the years, the ever expanding range of applications (including liquid–liquid blending, dispersion of solid ingredients in to liquids, gas–liquid system, even to improve heat transfer, etc (Paul *et al.*, 2004)) has led to the development of a wide variety of designs of static mixers. A selection is shown here in Figure 8.55.

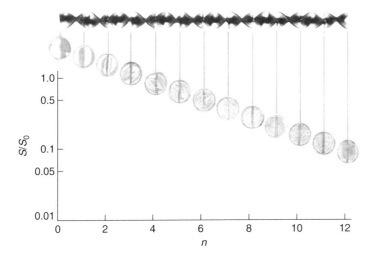

Figure 8.53   *Static mixer in laminar flow: reduction in relative standard deviation of samples indicating improvement in mixture quality with increasing number n of elements traversed*

Figure 8.54   *Static mixer for viscous Newtonian and non-Newtonian materials*

Quantitatively, a variety of methods (Heywood *et al.*, 1984) have been proposed to describe the degree or quality of mixing produced in a static mixer. One such measure of mixing quality is the relative standard deviation s/s_0, where s is the standard deviation in composition of a set of samples taken at a certain stage of the mixing operation, and s_0 is the standard deviation for a similar set of samples taken at the mixer inlet. Figure 8.53 shows schematically how the relative standard deviation decreases as the number of elements, n, through which the material has passed increases, a perfectly mixed product having a zero relative standard deviation. One of the problems in using relative standard deviation or a similar index as a measure of mixing is that this depends on sample size which therefore must be taken into account in any assessment. Some general observations are made here.

One of the most important considerations in comparing or selecting static mixers is the power consumed by the mixer to achieve a desired mixture quality. The pressure

SMV
 · SMX

SMXL
SMF

HEV
HEV (viscous liquids)

KM (laminar mixer)
KM (turbulent mixer)

KM (gas–liquid)

KM (liquid–liquid)

Lightnin static mixer

Figure 8.55 *Selection of commercially available static mixer designs. Sulzer-Koch type (first two rows); Chemineer-Kenic type (middle three rows) and Lightnin design (bottom)*

drop characteristics of a mixer are most conveniently described by the ratio of mixer pressure drop to empty tube pressure drop for the same flow rate and diameter. Thus, different static mixers designs can be compared on a basis of mixing quality, pressure drop ratio, initial cost and convenience of installation. It should be noted that pressure drop ratio may be dependent on the rheology of the material, which may well change significantly during the course of its passage through the static mixer. Great care should therefore be taken to ensure that the material used to determine the pressure drop criterion has a rheology which closely matches that of the material to be used in the static mixer.

Static mixers are widely used for highly viscous and non-Newtonian fluids in processes in which polymers, fibres and plastics of all kinds are manufactured, and where the material is often hot and at high pressures. However, static mixers have also achieved widespread use for mixing and blending of low viscosity liquids, liquid–liquid dispersions, and even gas–liquid dispersions. In some cases, the designs used for high viscosity liquids have also proved effective in the turbulent mixing regime for low viscosity fluids. In other cases, equipment manufacturers have developed special designs for turbulent mixing, and a wide variety of static mixer devices in now available. It is useful to offer some guidelines in selecting and/or assessing the performance of a static mixer. In particular, consideration will now be given to estimate the degree of mixing in static mixers.

At the end of the last mixing element, the cross-section can be divided into a number of sampling locations, say p. The sampling areas should be sized such that the area multiplied by the local velocity (normal to the area), i.e., volumetric flow rate is the same for each sampling part. Naturally, this approach will necessitate relatively large sampling areas in the wall region where the fluid velocity is relatively low. For the mixing of two liquids (1, 2), one can thus define the flow-rate averaged concentration of component 1 as:

$$\bar{C}_1 = \frac{1}{p}\sum_{i=1}^{i=p} C_{1i} \tag{8.40}$$

where C_1 and C_2 are the normalized concentrations and obviously $C_1 + C_2 = 1$. One can easily show that the average concentration $\bar{C}_1$ is also equal to the inlet volume fraction of component 1, i.e.,

$$\bar{C}_1 = \frac{Q_1}{Q_1 + Q_2} \tag{8.41}$$

where Q_1 and Q_2 are the volumetric flow rates of the two components. The standard deviation of concentration is now given as:

$$S_1 = \sqrt{\frac{1}{(p-1)}\sum_{i=1}^{i=p}(C_{1i} - \bar{C}_1)^2} \tag{8.42}$$

Two commonly used criteria for assessing the quality (degree of homogeneity) of a mixture are the so-called relative standard deviation (RSD), defined as $S_1/\sqrt{\bar{C}_1(1 - \bar{C}_1)}$

and the coefficient of variation (COV), defined as $S_1/\bar{C}_1$. For a two-component system, the two are related as:

$$\text{COV} = \left(\frac{Q_2}{Q_1}\right)^{1/2} \text{RSD} \tag{8.43}$$

For a completely segregated mixture (such as at the inlet of the mixer), both RSD and COV = 1. On the other hand, both COV and RSD will tend to zero for a perfectly homogenized (mixed) system. Naturally, the desired value of COV must be as small as possible. Grosz-Röll (1980) suggested a value of COV <0.05 to be sufficient for a product to be regarded homogeneous, but this definition is not only arbitrary and is also inadequate for applications such as blending of colours, precision extrusion, food and pharmaceutical processes which require the value of COV to be as small as 0.01 or even smaller to meet the desired product specifications.

For a given design of the static mixer, the value of COV is influenced by the Reynolds number, Schmidt number, ratio of the viscosities and densities of two liquids, the ratio of the flow rates and the size of the mixer (L/D). In laminar flow conditions and neglecting the role of molecular diffusion, it is possible to fit most of the available experimental data as follows (Thakur *et al.*, 2003):

$$\text{COV} = a \exp\left(\frac{-bL}{D}\right) \tag{8.44}$$

where a and b are two fitted constants for a fixed design of mixer. Broadly, the value of a depends on the value of $\bar{C}_1$ or on the ratio of the two flow rates Q_1/Q_2 for a two-component system, at least in the laminar flow conditions whereas the value of b is determined primarily by the design of the mixer. Table 8.6 summarizes the representative values of a and b for some of the commonly used designs of static mixers (Pahl and Muschelknautz, 1982; Thakur *et al.*, 2003). Alternately, the reduction in the coefficient of variation COV$_r$ (defined as the ratio of target value and the existing value) correlates rather well with the value of (L/D) for a specific mixer design. This relationship is given as:

$$\text{COV}_r = K_i^{L/D} \tag{8.45}$$

where the values of K_i for a few designs are also listed in Table 8.6 for laminar flow conditions.

The results shown in Table 8.6 do confirm the assertion that the value of constant b is more or less determined by the mixer geometry whereas the value of a is influenced by the desired value of COV; broadly, the smaller the value of COV (i.e. high degree of homogenization), larger is the value of constant a. Included in this table are also the values of the required (L/D) of different mixer designs to achieve the same degree of mixing, i.e. COV = 0.05.

Another consideration in the design and selection of a static mixer is the pressure drop. This information is conveniently presented in terms of a suitably defined friction factor and Reynolds number, though the choice of a characteristic linear dimension is far from obvious. Furthermore, the functional relationship between the friction factor and the Reynolds number is also strongly influenced by the detailed geometry of the mixer. However, some general observations can be made here. As a guide, liquids with viscosities higher than 100 mPa s are processed in laminar flow conditions. Unlike in an empty tube where the laminar flow occurs up to about $Re_{MR} \sim 2100$ for both Newtonian and inelastic non-Newtonian

Table 8.6   *Values of a and b for use in equation (8.44)*

Type of mixer	$\bar{C}_1$	a	b	(L/D) for COV = 0.05	K_i
Kenics	0.1	3.04	0.171	24	0.87
Komax	0.1	4.14	0.151	29.2	
Hi-Mixer	0.1	3.53	0.412	10.3	
Lightnin	0.1	3.30	0.059	71	
SMX	0.5	1.51–1.64	0.51	6.7	0.63
	0.1	4.52–5.54	0.51	9	
	0.01	~15	0.50	11.4	
	0.001	48.7–53.3	0.51	13.8	
SMV	0.5	0.17	0.74	1.7	
	0.1	0.51	0.74	3.1	
	0.01	1.70	0.75	4.7	
	0.001	5.17	0.75	6.2	
SMXL	0.5	1.25	0.163	19.7	0.85
	0.001	40	0.163	41	
Etoflow HV	–	5.58	0.138	34.2	

fluids, streamline flow conditions in static mixers cease to exist at much lower values of the Reynolds number. Depending upon the mixer design, values ranging from 15 to 55 have been reported in the literature for the end of the laminar flow conditions (Thakur *et al.*, 2003), albeit most of these results are based on data for Newtonian liquids only.

Irrespective of the type of mixer, the friction factor is inversely proportional to the Reynolds number in the laminar flow conditions, and therefore the product $(f\,Re)$ is constant, say A. The value of A is a function of the mixer design only and it is often provided by the manufacturers of equipment. By using the analogy with the streamline flow in packed beds, Streiff *et al.* (1999) proposed the following method to estimate pressure drop in static mixers. For the laminar flow of a Newtonian fluid in a circular pipe,

$$-\Delta p = \frac{32\mu VL}{D^2} \tag{8.46}$$

and the Reynolds number is defined as:

$$Re_D = \frac{\rho VD}{\mu} \tag{8.47}$$

For static mixers, Streiff *et al.* (1999) suggested the use of an effective velocity (V/ε), and of the hydraulic diameter D_h instead of the superficial velocity and pipe diameter together with a tortuosity factor, τ, as follows:

$$Re = \frac{\rho(V/\varepsilon)D_h}{\mu} = Re_D\left(\frac{D_h}{\varepsilon D}\right) \tag{8.48}$$

where the hydraulic diameter, D_h, is defined as four times the area of flow divided by the wetted perimeter. Equation (8.46) is modified as:

$$(-\Delta p) = \frac{32\mu(V/\varepsilon)L\tau^2}{D_h^2} \tag{8.49}$$

Finally, the Fanning friction factor, f, defined as:

$$f = \frac{2(-\Delta p)D_h}{L\rho(V/\varepsilon)^2} \tag{8.50}$$

is related to the pipe Reynolds number by the following general form of equation:

$$f = \frac{K_L}{Re_D} + K_T \tag{8.51}$$

Typical values of the friction factor f for static mixers range from 0.4 to 3 depending upon the element geometry and the value of the Reynolds number.

For non-Newtonian fluids, the effective viscosity is estimated at an effective shear rate $\dot{\gamma}_{eff}$ given as:

$$\dot{\gamma}_{eff} = \frac{K_G V}{D} \tag{8.52}$$

where

$$K_G = \frac{8\tau}{\varepsilon}\left(\frac{D}{D_h}\right)$$

Streiff *et al.* (1999) mention values of the tortuosity factor τ in the range of 1.3 to 1.8, which are similar to those encountered in packed beds. Most manufacturers provide the values of the overall constants K_L, K_T, K_G, etc. rather than that of the individual parameters like voidage, ε, tortuosity, τ and D_h. Table 8.7 gives a representative summary of such values for Koch-type mixers, as compiled by Streiff *et al.* (1999).

For the purpose of scaling up (experimental data obtained in a laboratory equipment) to design an industrial scale unit, the most commonly used criterion is to maintain geometrical similarity between the two scales, i.e. all linear dimensions (like length, diameter) bear a constant ratio. Thus, if the desired throughput ratio in the two units is S, the linear dimensions scale as $S^{1/3}$. This suggests that a twofold increase in length and diameter yields an eightfold increase in the throughput of the mixer. In this approach, while the residence time is unchanged, the Reynolds number scales as $S^{2/3}$ and the pressure drop remains unchanged in the laminar flow conditions. Similarly, the surface area of the mixer also scales as $S^{2/3}$ whereas the heat duty scales as S and therefore this scale-up scheme is not always satisfactory for heat transfer applications. Merits and de-merits of the other scale-up criteria have been discussed by Thakur *et al.* (2003) and by Etchells and Meyer (2004).

Finally, this chapter is concluded by providing quantitative guidelines for choosing an appropriate impeller for a new application. Bearing in mind that as the liquid viscosity increases, laminar flow conditions are generally encountered and therefore, one moves

Table 8.7 *Values of K_L, K_T, and K_G*

Type of mixer	K_L	K_T	K_G
SMX	600	2.5	64
SMV	715	~0.5–1	50
SMXL	125	~0.5	30
Kenics (KMS)	110	~0.75	28

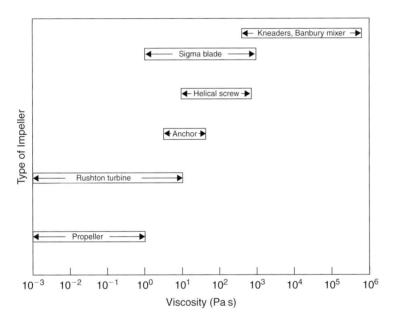

Figure 8.56 *Suggested ranges of operations of various impellers used in batch mixing operations depending upon viscosity*

away from high speed impellers to slow moving and close-clearance class of impellers. Figure 8.56 depicts the useful ranges for a selection of widely used geometries in terms of the liquid viscosity. Also, one is often confronted with the dilemma of batch versus continuous systems. Figure 8.57 offers some help in this regard. Typically, the batch systems have large working volumes. Of course, the final choice of an impeller design is influenced strongly by the fluid properties, degree of homogenization required, product specifications in terms of drop or bubble size, etc.

Example 8.7

A product ($\rho = 1200 \, \text{kg/m}^3$) is processed in five different mixing vessels (of different sizes and designs) and therefore the streams leaving these vessels contain varying concentrations of a pigment as 96, 48, 73, 64, 69 ppm, each at a flow rate of 1 ton per hour. It is proposed to design a static mixer (of SMX or SMXL type) of 75 mm diameter to mix these five streams to reduce

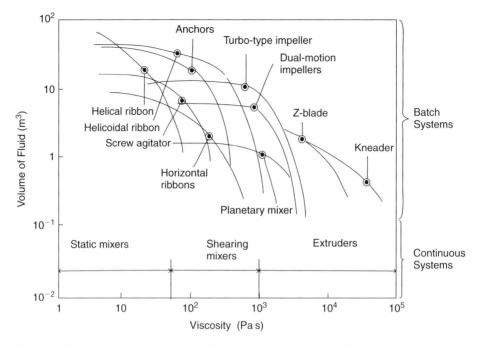

Figure 8.57   *Suggested range of applications of commonly used mixer configurations*

the variability in the pigment concentration to $\pm 0.75\,\text{ppm}$. The small variation is pigment content does not seem to influence the rheological behaviour of the product, and the mean values of the power-law constants are: $m = 15\,\text{Pa s}^n$ and $n = 0.4$. Calculate the length of SMX and SMXL-type static mixers and the corresponding pressure drops for the two designs.

Solution

In the feed, the pigment concentration ranges from 48 to 96 ppm, i.e. a spread of $96 - 48 = 48\,\text{ppm}$ and it is desired to reduce it to $\pm 0.75\,\text{ppm}$, i.e. a spread of $2 \times 0.75 = 1.5\,\text{ppm}$. Therefore, the desired reduction in the value of COV_r is:

$$\text{COV}_r = \frac{1.5}{48} = 0.03125$$

Using this value of COV_r together with the values of K_i from Table 8.6, for SMX and SMXL-type mixers, the required values of (L/D) are calculated using equation (8.45). For instance, for SMX mixer, $K_i = 0.63$, from equation (8.45).

$$\therefore \text{COV}_r = K_i^{(L/D)}$$

Now solving for (L/D) and substituting values:

$$\frac{L}{D} = \frac{\log \text{COV}_r}{\log K_i} = \frac{\log(0.03125)}{\log(0.63)} = 7.5$$

Similarly, for SMXL-type mixer using $K_i = 0.85$, the value of (L/D) is found to be 21.33.

The total flow rate is 5 ton per hour and therefore, the mean velocity V, in a 75 mm diameter pipe is estimated as:

$$V = \frac{5 \times 10^3}{3600 \times 1200 \times \dfrac{\pi}{4} \times (0.075)^2} = 0.262 \, \text{m/s}$$

Next, in order to calculate the value of the friction factor to estimate the pressure loss for this duty, one must evaluate the value of the Reynolds number for these conditions. The effective viscosity value to be used in the Reynolds number is calculated at shear rates of $(K_G V/D)$ where the values of K_G are given in Table 8.7.

For a SMX mixer, $K_G = 64$, and hence the average shear rate is:

$$\dot{\gamma}_{avg} = \frac{64 \times 0.262}{0.075} = 223.6 \, \text{s}^{-1}$$

Using this value in the power-law model as:

$$\mu_{avg} = m(\dot{\gamma}_{avg})^{n-1} = 15 \times (223.6)^{0.4-1}$$

i.e.

$$\mu_{avg} = 0.584 \, \text{Pa s}$$

The Reynolds number, Re_D is given as:

$$Re_D = \frac{\rho V D}{\mu_{avg}} = \frac{1200 \times 0.262 \times 0.075}{0.584} = 40.4$$

The friction factor, f, can now be calculated using equation (8.51):

$$f = \frac{K_L}{Re_D} + K_T$$

From Table 8.7, for a SMX mixer, $K_L = 600$; $K_T = 2.5$

$$\therefore f = \frac{600}{40.4} + 2.5 = 17.36$$

and finally, the pressure drop $(-\Delta p)$ is estimated as:

$$-\Delta p = \frac{2 f \rho V^2 L}{D}$$

Here, $L/D = 7.5$, and substituting values:

$$-\Delta p = 2 \times 17.36 \times 1200 \times 0.262^2 \times 7.5 = 2.145 \times 10^4 \, \text{Pa, i.e. } 21.45 \, \text{kPa}$$

One can similarly repeat these calculations for the SMXL design and a summary is provided in the table below:

Mixer	$\dot{\gamma}_{avg}$ (s^{-1})	μ_{avg} (Pa s)	Re_D	L/D	f	$-\Delta p$ (kPa)
SMX	223	0.584	40.3	7.5	17.36	21.45
SMXL	105	0.919	25.65	21.33	5.37	18.90

For almost identical values of the pressure drop, the SMX mixer is much shorter (less volume) than the SMXL design. Therefore, the final choice will depend upon the secondary criterion like shear sensitivity of the material, space economy, etc. □

Further reading

Harnby, N., Edwards, M.F. and Nienow, A.W., *Mixing in the Process Industries*, 2nd edition, Butterworth-Heinemann, Oxford (1992).
McDonough, R.J., *Mixing for the Process Industries*, Van Nostrand Reinhold, New York (1992).
Nagata, S., *Mixing: Principles and Applications*, Wiley, New York (1975).
Oldshue, J.Y., *Fluid Mixing Technology*, McGraw Hill, New York (1983).
Ottino, J.M., *The Kinematics of Mixing*, Cambridge University Press, London (1990).
Paul, E.L., Atiemo-Obeng, V.A. and Kresta, S.M., Eds, *Handbook of Industrial Mixing*, Wiley, New York (2004).
Tatterson, G.B., *Fluid Mixing and Gas Dispersion in Agitated Tanks*, McGraw Hill, New York (1991).
Tatterson, G.B., *Scaleup and Design of Industrial Mixing Processes*, McGraw Hill, New York (1994).
Ulbrecht, J.J. and Patterson, G.K., Eds, *Mixing of Liquids by Mechanical Agitation*, Gordon and Breach, New York (1985).
Xuereb, C., Poux, M. and Bertrand, J., Eds, *Agitation et Melange*, Dunod, Paris (2006).
Zlokarnik, M., *Stirring: Theory and Practice*, Wiley-VCH, New York (2001).
Zlokarnik, M., *Scale-up in Chemical Engineering, 2nd Edition*, Wiley-VCH, New York (2006).

References

Abid, M., Xuereb, C. and Bertrand, J., *Chem. Eng. Res. Des.* **70** (1992) 377.
Adams, L.W. and Barigou, M., *Proc. 12th European Conf. Mixing*, Bologna (Italy) p. 65 (2006).
Amanullah, A., Hjorth, S.A. and Nienow, A.W., *Chem. Eng. Sci.* **53** (1998) 455.
Anne-Archard, D., Marouche, M. and Boisson, H.C., *Chem. Eng. J.* **125** (2006) 15.
Arzate, A., Reglat, O. and Tanguy, P.A., *Flow Meas. Instrum.* **15** (2004) 77.
Ascanio, G., Foucault, S. and Tanguy, P.A., *Chem. Eng. Technol.* **26** (2003) 908.
Aubin, J., Naude, I., Bertrand, J. and Xuereb, C., *Chem. Eng. Res. Des.* **78A** (2000) 1105.
Ayazi Shamlou, P. and Edwards, M.F., *Chem. Eng. Sci.* **41** (1986) 1957.
Bakker, A. and Gates, L.E., *Chem. Eng. Prog.* **91** (Dec, 1995) 25.
Bakker, A., Morton, J.R. and Berg, G.M., *Chem. Eng.* **101** (Mar, 1994) 120.
Bates, R.L., Fondy, P.L. and Corpstein, R.R., *Ind. Eng. Chem. Proc. Des. Dev.* **2** (1963) 310.
Beckner, J.L. and Smith, J.M., *Trans. Inst. Chem. Engrs.* **44** (1966) 224.
Benz, G.T., *Chem. Eng. Prog.* **99** (May 2003) 32.
Benz, G.T., *Chem. Eng. Prog.* **100** (Feb, 2004) 18S.
Bertrand, F., Tanguy, P.A. and Brito De la Fuente, E., *J. Chem. Eng. Jpn.* **29** (1996) 51.
Blasinski, H. and Kuncewicz, C., *Int. Chem. Eng.* **21** (1981) 679.
Bourne, J.R. and Butler, H., *Trans. Inst. Chem. Engrs.* **47** (1969) 11.
Calderbank, P.H. and Moo-Young, M., *Trans. Inst. Chem. Engrs.* **37** (1959) 26.
Carreau, P.J., Patterson, I. and Yap, C.Y., *Can. J. Chem. Eng.* **54** (1976) 135.
Carreau, P.J., Paris, J. and Guerin, P., *Can. J. Chem. Eng.* **70** (1992) 1071.
Carreau, P.J., Chhabra, R.P. and Cheng, J., *AIChE J* **39** (1993) 1421.
Carreau, P.J., Paris, J. and Guerin, P., *Can. J. Chem. Eng.* **72** (1994) 966.
Cavadas, A.S. and Pinho, F.T., *Can. J. Chem. Eng.* **82** (2004) 289.
Chhabra, R.P., *Adv. Heat Trans.* **37** (2003) 77.
Chhabra, R.P., Cuvelier, G., Domenek, S., Andre, C. and Delaplace, G., *Chem. Eng. Technol.* **30** (2007) 1686.
Chapman, F.S. and Holland, F.A., *Trans. Inst. Chem. Engrs.* **43** (1965) 131.
Chavan, V.V. and Mashelkar, R.A., *Adv. Transp. Process.* **1** (1980) 210.
Chavan, V.V. and Ulbrecht, J.J., *Trans. Inst. Chem. Engrs.* **50** (1972) 147.
Chavan, V.V. and Ulbrecht, J.J., *Ind. Eng. Chem. Proc. Des. Dev.* **12** (1973) 472, Corrigenda ibid **13** (1974) 309.
Chavan, V.V., Arumugam, M., and Ulbrecht, J. *AIChE J*, **21** (1975) 613. Also see *Can. J. Chem. Eng.* **53** (1975) 62.
Cheng, J. and Carreau, P.J., *Chem. Eng. Sci.* **49** (1994) 1965.
Cheng, D.C.-H., Schofield C. and Jane, R.J. *Proc. First Eng. Conf. on Mixing & Centrifugal Sep., BHRA Fluid Eng.* Paper #C2–15 (1974).
Collias, D.J. and Prud'homme, R.K., *Chem. Eng. Sci.* **40** (1985) 1495.
Coulson, J.M. and Richardson, J.F., *Chemical Engineering, 6th edn*, Vol. 1, Butterworth-Heinemann, Oxford (1999).

Coyle, C.K., Hirschland, H.E., Michel, B.J. and Oldshue, J.Y., *Can. J. Chem. Eng.* **48** (1970) 275.

Curran, S.J., Hayes, R.E., Afacan, A., Williams, M.C. and Tanguy, P.A., *Ind. Eng. Chem. Res.* **39** (2000) 195.

Delaplace, G., Guerin, R., *Techniques de l'Ingenieur*, Paris, F3350 (2006).

Delaplace, G. and Leuliet, J.-C., *Entropie* **227** (2000) 10.

Delaplace, G., Leuliet, J.-C. and Relandeau, V., *Exp. Fluids* **28** (2000) 170.

Delaplace, G., Torrez, C., Leuliet, J.-C., Belaubre, N. and Andre, C., *Chem. Eng. Res. Des.* 79A (2001) 927.

Delaplace, G., Bouvier, L., Moreau, A., Guerin, R. and Leuliet, J.-C., *Expts. Fluids* **36** (2004a) 437.

Delaplace, G., Leuliet, J.-C. and Ronse, G., *Chem. Eng. Technol.* **23** (2004b) 329.

Delaplace, G., Guerin, R. and Leuliet, J.-C., *AIChE J* **51** (2005) 3094.

Delaplace, G., Guerin, R., Leuliet, J.-C. and Chhabra, R.P., *Chem. Eng. Sci.* **61** (2006) 3250.

Delaplace, G., Thakur, R.K., Bouvier, L., Andre, C. and Torrez, C., *Chem. Eng. Sci.* **62** (2007) 1442.

Desplanches, H., Llinas, J.R. and Chevalier, J.L., *Can. J. Chem. Engrs.* **58** (1980) 160.

Doraiswamy, D., Grenville, R.K. and Etchells, III, A.W., *Ind. Eng. Chem. Res.* **33** (1994) 2253.

Dream, R.F., *Chem. Eng.* **106** (Jan. 1999) 90.

Ducla, J.M., Desplanches, H. and Chevalier, J.J., *Chem. Eng. Commun.* **21** (1983) 29.

Edney, H.G.S. and Edwards, M.F., *Trans. Inst. Chem. Engrs.* **54** (1976) 160.

Edwards, M.F., Godfrey, J.C. and Kashim, M.M., *J. Non-Newt. Fluid Mech.* **1** (1976) 309.

Edwards, M.F. and Wilkinson, W.L., *The Chem. Engr.* **257** (1972) 310, 328.

Ein-Mozaffari, F., Bennington, C.P.J., Dumont, G.A. and Buckingham, D., *Chem. Eng. Res. Des.* **85A** (2007) 591.

Elson, T.P., *Chem. Eng. Commun.* **96** (1990) 303.

Elson, T.P., Cheesman, D.J. and Nienow, A.W., *Chem. Eng. Sci.* **41** (1986) 2555.

Espinosa-Solares, T., Brito De la Fuente, E., Tecante, A. and Tanguy, P.A., *Chem. Eng. Technol.* **25** (2002) 723.

Etchells, III, A.W. and Meyer, C.F., *Handbook of Industrial Mixing* (editors E.L. Paul, V.A. Atiemo-Obeng, and S.M. Kresta), Wiley, New York (2004), Chapter 7.

Etchells, A. W., Ford, W. N. and Short, D. G. R., *Fluid Mixing-3,* I Chem E symposium Ser., **108** (1987) 1.

Farhat, M., Rivera, C., Fradette, L., Heniche, M. and Tanguy, P.A., *Proc. 12th European Conf. Mixing*, Bologna (Italy) p. 89 (2006).

Fasano, J.B., Bakker, A. and Penney, W.R., *Chem. Eng.* **101** (Aug, 1994) 110.

Foucault, S., Ascanio, G. and Tanguy, P.A., *Chem. Eng. Technol.* **27** (2004) 324.

Foucault, S., Ascanio, G. and Tanguy, P.A., *Ind. Eng. Chem. Res.* **45** (2006) 352.

Franco, J.M., Delgado, M.A., Valencia, C., Sanchez, M.C. and Gallegos, C., *Chem. Eng. Sci.* **60** (2005) 2409.

Galindo, E. and Nienow, A.W., *Chem. Eng. Technol.* **16** (1993) 102.

Gates, L.E., Hicks, R.W. and Dickey, D.S., *Chem. Eng.* **83** (Dec 6, 1976) 165.

Giesekus, H., *Rheol. Acta.* 4 (1965) 85.

Gluz, M. and Pavlushenko, I.S., *J. App. Chem. (USSR)* **39** (1966) 2223.

Godfrey, J.C., *Mixing in the Process Industries* (edited by N. Harnby, M.F. Edwards, and A.W. Nienow), 2nd edition, Butterworth-Heinemann, Oxford, p. 185 (1992).

Godleski, E.S. and Smith, J.C., *AIChE J* **8** (1962) 617.

Grenville, R.K., *et al.* (1995). Paper presented at *Mixing XV*, 15th Biennial North American Mixing Conf., Banff, Alberta, Canada, June 18–23, 1995.

Grenville, R.K., PhD Thesis, Cranfield Institute of Technology, Cranfield, U.K. (1992).

Grenville, R.K., Hutchinson, T.M. and Higbee, R.K., Paper presented at *Mixing-18*, North American Forum on Mixing (2001).

Grenville, R.K. and Nienow, A.W., *Handbook of Industrial Mixing* (editors E.L. Paul, V.A. Atiemo-Obeng, and S.M. Kresta), Wiley, New York (2004), Chapter 9.

Grosz-Röll, F., *Int. Chem. Eng.* **20** (1980) 542.

Guerin, P., Carreau, P.J., Patterson, I. and Paris, J., *Can. J. Chem. Eng.* **62** (1984) 301.

Godleski, E.S. and Smith, J.C., *AIChE J* **8** (1962) 617.

Hagedorn, D. and Salamone, J.J., *Ind. Eng. Chem. Proc. Des. Dev.* **6** (1967) 469.

Hall, K.R. and Godfrey, J.C., *Trans. Inst. Chem. Engrs.* **46** (1968) 205.

Harnby, N., Edwards, M.F. and Nienow, A.W., Eds. *Mixing in the Process Industries*, 2nd edition, Butterworth-Heinemann, Oxford (1992).

Heim, A., *Intl. Chem. Eng.* **20** (1980) 271.

Herbst, H., Schumpe, A. and Deckwer, W.-D., *Chem. Eng. Technol.* **15** (1992) 425.

Heywood, N.I., Viney, L.J. and Stewart, I.W., *I. Chem. E. Sym. Ser. No. 89 Fluid Mixing II* (1984) p. 147.

Hickman, D. *Proc. 6th European Conf. Mixing, Pavia (Italy), BHRA Fluid Eng.*, Cranfield, p. 369 (1988).

Hicks, R.W., Morton, J.R. and Fenic, J.G., *Chem. Eng.* **83** (Apr 26, 1976) 102.

Hirata, Y. and Aoshima, Y., *Chem. Eng. Res. Des.,* **74A** (1996) 438.

Ibrahim, S. and Nienow, A.W., *Trans. Inst. Chem. Engrs.* **73A** (1995) 485.

Ihejirika, I. and Ein-Mozaffari, F., *Chem. Eng. Technol.* **30** (2007) 606.

Janssen, L.P.B.M., *Twin Screw Extrusion,* Elsevier, Amsterdam (1978).

Jeschke, D., *Z. Ver. Deut. Ing.* **69** (1925) 1526.

Johma, A.I. and Edwards, M.F., *Chem. Eng. Sci.* **45** (1990) 1389.

Kai, W. and Shengyao, Y., *Chem. Eng. Sci.* **44** (1989) 33.

Kaminoyama, M., Watanabe, M., Nishi, K. and Kamiwano, M., *J. Chem. Eng. Jpn.* **32** (1999) 23.

Kamiwano, M., Saito, F. and Kaminoyama, M., *Int. Chem. Eng.* **30** (1990) 274.

Kappel, M., *Int. Chem. Eng.* **19** (1979) 571.

Kawase, Y. and Moo-Young, M., *Chem. Eng. Res. Des.* **66** (1988) 284.

Kelkar, J.V., Mashelkar, R.A. and Ulbrecht, J., *Trans. Inst. Chem. Engrs.* **50** (1972) 343.

Kelkar, J.V., Mashelkar, R.A. and Ulbrecht, J., *Chem. Eng. Sci.* **17** (1973) 3069.

Kelly, W. and Gigas, G., *Chem. Eng. Sci.* **58** (2003) 2141.

Kuboi, R. and Nienow, A.W. *Proc. Fourth European Conf. Mixing, BHRA Fluid Eng.,* Cranfield, p. 247 (1982).

Kuriyama, M., Arai, K. and Saito, S., *J. Chem. Eng. Japan.* **16** (1983) 489.

Lai, K.P., Steffe, J.F. and Ng, P.K.W., *Cereal Chemistry* **77** (2000) 713.

Lindley, J.A., *J. Agr. Eng. Res.* **49** (1991) 1.

Machon, V., Vlcek, J., Nienow, A.W. and Solomon, J., *Chem. Eng. J.* **14** (1980) 67.

Manna, L., *Chem. Eng. Jl.* **67** (1997) 167.

Markopoulos, J., Babalona, E. and Tsiliopoulou, E., *Chem. Eng. Technol.* **27** (2004) 1212.

Marouche, M., Anne-Archard, D. and Boisson, H.C., *Applied Rheology* **12** (2002) 182.

Martone, J.A. and Sandall, O.C., *Ind. Eng. Chem. Proc. Des. Dev.* **10** (1971) 86.

Metzner, A.B., Feehs, R.H., Romos, H.L., Otto, R.E. and Tuthill, J.P., *AIChE J* **7** (1961) 3.

Metzner, A.B. and Otto, R.E., *AIChE J* **3** (1957) 3.

Metzner, A.B. and Taylor, J.S., *AIChE J* **6** (1960) 109.

Mitsuishi, N. and Hirai, N.J., *J. Chem. Eng. Jpn.* **2** (1969) 217.

Montante, G., Mostek, M., Jahoda, M. and Magelli, F., *Chem. Eng. Sci.* **60** (2005) 2427.

Montante, G., Bakker, A., Paglionti, A. and Magelli, F., *Chem. Eng. Sci.* **61** (2006) 2807.

Murthy Shekhar, S. and Jayanti, S., *AIChE J* **49** (2003) 2768.

Nagata, S., *Mixing; Principles and Applications,* Wiley, New York (1975).

Nagata, S., Yanagimoto, M. and Yokoyama, T., Kyoto University (Japan). *Memoirs of Fac. of Eng.* **18** (1956) 444.

Nagata, S., Nishikawa, M., Tada, H., Hirabayashi, H. and Gotoh, S., *J. Chem. Eng. Jpn.* **3** (1970) 237.

Nagata, S., Nishikawa, M., Kayama, T. and Nakajima, M., *J. Chem. Eng. Jpn.* **5** (1972) 187.

Niedzielska, A. and Kuncewicz, C., *Chem. Eng. Sci.* **60** (2005) 2439.

Nienow, A.W. and Ulbrecht, J., *Mixing of Liquids by Mechanical Agitation* (edited by J.J. Ulbrecht, and G.K. Patterson), Gordon & Breach, New York (1985), Chapter 6.

Nienow, A.W., Wisdom, D.J., Solomon, J., Machon, V. and Vlacek, J., *Chem. Eng. Commun.* 19 (1983) 273.

Nishi, K., Matsuda, K., Suzuku, Y., Misumi, R., Kamiwano, M. and Kaminoyama, M., *J. Chem. Eng. Jpn.* 39 (2006) 1041, Also see *ibid* **40** (2007) 658.

Nishikawa, M., Nakamura, M., Yagi, H. and Hashimoto, K., *J. Chem. Eng. Jpn.* **14** (1981) 219, 227.

Norwood, K.W. and Metzner, A.B., *AIChE J* **6** (1960) 432.

Nouri, J.M. and Hockey, R.M., *J. Chem. Eng. Jpn.* **31** (1998) 848.

Oldshue, J.Y., *Fluid Mixing Technology,* McGraw Hill, New York (1983).

Oliver, D.R., Nienow, A.W., Mitson, R.J. and Terry, K., *Chem. Eng. Res. Des.* **62** (1984) 123.

Ottino, J.M., *Sci. Am.* 260 (1989) 56.

Özcan-Taskin, N.G. and Nienow, A.W., *Trans. Inst. Chem. Engrs.* **73C** (1995) 49.

Pahl, M.H. and Muschelknautz, E., *Int. Chem. Eng.* **22** (1982) 197.

Patterson, G.K., *Proc. 12th European Conf. Mixing,* Bologna (Italy) p. 113 (2006).

Pedrosa, S.M.C.P. and Nunhez, J.R., *Comp. Chem. Eng.* **24** (2000) 1745.

Perez, J.F. and Sandall, O.C., *AlChEJ* **20** (1974) 770.

Peters, D.C. and Smith, J.M., *Trans. Inst. Chem. Engrs.* **45** (1967) 360.

Poggermann, R., Steiff, A. and Weinspach, P.M., *Ger. Chem. Eng.* **3** (1980) 163.

Pollard, J. and Kantyka, T.A., *Trans. Inst. Chem. Engrs.* **47** (1969) 21.

Prud'homme, R.K. and Shaqfeh, E., *AlChEJ* **30** (1984) 485.

Ranade, V.R. and Ulbrecht, J., *AlChEJ* **24** (1978) 796.

Rauendaal, R.C., *Mixing in Polymer Processing*, Hanser, Munich (1992).

Rudolph, L., Schaefer, M., Atiemo-Obeng, V. and Kraume, M., *Proc. 12th European Conf. Mixing*, Bologna (Italy) p. 73 (2006).

Saeed, S., Ein-Mozaffari, F. and Upreti, S.R., *Ind. Eng. Chem. Res.* **46** (2007) 2172.

Sandall, O.C. and Patel, K.G., *Ind. Eng. Chem. Proc. Des. Dev.* **9** (1970) 139.

Schenkel, G., *Plastics Extrusion Technology*, Cliffe Books, London (1966).

Sestak, J., Houska, M. and Zitny, R., *J. Rheol.* **26** (1982) 459.

Sestak, J., Zitny, R. and Houska, M., *AIChE J* **32** (1986) 155.

Seyssiecq, I., Tolofoudye, A., Desplanches, H. and Gaston-Bonhomme, Y., *Chem. Eng. Technol.* **26** (2003) 1155.

Skelland, A.H.P., *Non-Newtonian Flow and Heat Transfer*, Wiley, New York (1967).

Skelland, A.H.P., *Handbook of Fluids in Motion* (edited by N.P. Cheremisinoff, and R. Gupta), Ann Arbor Sci, Ann Arbor (1983), Chapter 7.

Smith, J.M., *Mixing of Liquids by Mechanical Agitation* (edited by J.J. Ulbrecht, and G.K. Patterson), Gordon and Breach, New York (1985), Chapter 5.

Solomon, J., Nienow, A.W. and Pace, G.W., *Fluid Mixing I, I. Chem. E. Sym. Ser.* #64 (1981), Paper No. A1.

Steffe, J.F., *Rheological Methods in Food Process Engineering*, 2nd edition, Freeman Press, East Lansing, MI (1996).

Streiff, F.A., Jaffer, S. and Schneider, G., *Proc. 3rd Int. Sym. on Mixing in Industrial Processes*, Japan p. 107 (1999).

Takahashi, K., *Encyclopedia of Fluid Mechanics* (edited by N.P. Cheremisinoff), Volume 7, Gulf, Houston, p. 869 (1988).

Tanguy, P., Lacorix, A., Bertrand, F., Choplin, L. and DeLa Fuente, E.B., *AlChEJ* **38** (1992) 939.

Tanguy, P.A., Bertrand, F., Labrie, R. and Brito De la Fuente, E., *Chem. Eng. Res. Des.* **74** (1996) 499.

Tatterson, G.B., *Fluid Mixing and Gas Dispersion in Agitated Tanks*, McGraw Hill, New York (1991).

Tatterson, G.B., *Scaleup and Design of Industrial Mixing Processes*, McGraw Hill, New York (1994).

Thakur, R.K., Vial, C., Nigam, K.D.P., Nauman, E.B. and Djelveh, G., *Chem. Eng. Res. Des.* **81** (2003) 787.

Thakur, R.K., Vial, C., Djelveh, G. and Labbafi, M., *Chem. Eng. Process.* **43** (2004) 1211.

Thibault, F. and Tanguy, P.A., *Chem. Eng. Sci.* **57** (2002) 3861.

Torrez, C. and Andre, C., *Chem. Eng. Technol.* **22** (1999) 701.

Ulbrecht, J.J. and Carreau, P.J., *Mixing of Liquids by Mechanical Agitation* (edited by J.J. Ulbrecht, and G.K. Patterson), Gordon and Breach, New York (1985), Chapter 4.

van den Bergh, W., *Chem. Eng.* **101** (Dec. 1994) 70.

van der Molen, K. and van Maanen, H.R.E., *Chem. Eng. Sci.* **33** (1978) 1161.

van't Reit, K. (1975). *PhD Thesis*, Delft. The Netherlands.

Wang, J.-J., Feng, L.-F., Gu, X.-P., Wang, K. and Hu, C.-H., *Chem. Eng. Sci.* **55** (2000) 2339.

Wichterle, K. and Wein, O., *Int. Chem. Eng.* **21** (1981) 116.

Wilkens, R.J., Henry, C. and Gates, L.E., *Chem. Eng. Prog.* **99** (May 2003) 44.

Yagi, H. and Yoshida, F., *Ind. Eng. Chem. Proc. Des. Dev.* **14** (1975) 488.

Yamamoto, K., Abe, K., Tarumoto, A., Nishi, K., Kaminoyama, M. and Kamiwano, M., *J. Chem. Eng. Jpn.* **31** (1998) 355.

Yap, C.Y., Patterson, W.I. and Carreau, P.J., *AlChEJ* **25** (1979) 516.

Nomenclature

		Dimensions in M, N, L, T, θ
A	area for heat transfer (m²)	L^2
a	constant, equation (8.44) ($-$)	$-$
b	constant, equation (8.44) ($-$)	$-$
Bi	Bingham number ($-$)	$-$
C	clearance between impeller and tank bottom (m)	L
C_i	tracer concentration recorded by ith detector (kmol/m³)	NL^{-3}

		Dimensions in M, N, L, T, θ
C_∞	equilibrium concentration of tracer (kmol/m³)	NL^{-3}
C_p	heat capacity (J/kg K)	$L^2T^{-2}\theta^{-1}$
D	impeller or pipe diameter (m)	L
D_c	size of cavity in a shear-thinning fluid, equation (8.11) (m)	L
D_h	hydraulic diameter (m)	L
D_L	molecular diffusivity (m²/s)	L^2T^{-1}
D_T	tank diameter (m)	L
d_c	mean helix diameter (m)	L
d_0	coil tube diameter (m)	L
Fr	Froude number $(-)$	$-$
f	friction factor, equation (8.50) $(-)$	$-$
Gr	Grashof number $(-)$	$-$
g	acceleration due to gravity (m/s²)	LT^{-2}
h	heat transfer coefficient (W/m²K)	$MT^{-3}\theta^{-1}$
K_G	constant, equation (8.52) $(-)$	$-$
K_i	constant, equation (8.45) $(-)$	$-$
K_L	constant, equation (8.51) $(-)$	$-$
K_P	product (Po Re) in laminar regime $(-)$	$-$
K_T	constant, equation (8.51) $(-)$	$-$
k	thermal conductivity (W/m K)	$MLT^{-3}\theta^{-1}$
k_s	constant, equation (8.9)	$-$
m	power-law consistency coefficient (Pa s^n)	$ML^{-1}T^{n-2}$
N	speed of rotation of agitation (s^{-1})	T^{-1}
N_b	number of baffles $(-)$	$-$
N_q	pumping number $(-)$	$-$
N_r	number of ribbons in a impeller $(-)$	$-$
Nu	Nusselt number $(-)$	$-$
n	power-law index $(-)$	$-$
P	Power (W)	ML^2T^{-3}
Po	Power number $(-)$	$-$
Pr	Prandtl number $(-)$	$-$
Q_g	volumetric flow rate of gas (m³/s)	L^3T^{-1}
r	radial distance from centre of tank (m)	L
R	thermal resistance due to scale formation (m²K/W)	$M^{-1}T^3\theta$
Re	Reynolds number $(-)$	$-$
Re_{cr}	critical Reynolds number $(-)$	$-$
Re_T	vessel Reynolds number $(-)$	$-$
S	pitch of screw (m)	$L\neg$
S_A	scale of agitation $(-)$	$-$
t_r	residence time (s)	T
T	temperature (K)	θ
$D_{T_{eff}}$	effective diameter of vessel $(4V_l/\pi)^{1/3}$ (m)	L
ΔT	temperature difference (K)	θ
t_m	mixing time (s)	T
U	overall heat transfer coefficient (W/m²K)	$MT^{-3}\theta^{-1}$
V	volume of gas–liquid dispersion in the vessel (m³)	L^3
V_g	gas superficial velocity (m/s)	LT^{-1}
V_l	volume of the liquid batch (m³)	L^3
V_t	terminal rise velocity of a gas bubble (m/s)	LT^{-1}
W	width of ribbon (m)	L
W_B	width of baffle (m)	L
We	Weber number $(-)$	$-$
$\dot{x}_w$	wall thickness (m)	L

Greek letters

α	thermal diffusivity (m^2/s)	L^2T^{-1}
β	coefficient of thermal expansion (K^{-1})	θ^{-1}
$\dot{\gamma}_{avg}$	average shear rate, equation (8.9) s^{-1}	T^{-1}
Λ	torque (N m)	ML^2T^{-2}
μ	Newtonian viscosity (Pa s)	$ML^{-1}T^{-1}$
μ_{eff}	effective shear viscosity evaluated at $\dot{\gamma} = \dot{\gamma}_{avg}$ (Pa s)	$ML^{-1}T^{-1}$
ϕ	gas hold-up ($-$)	$-$
θ_m	dimensionless mixing time ($-$)	$-$
ρ	liquid density (kg/m^3)	ML^{-3}
σ	surface tension (N/m)	MT^{-2}
τ	tortuosity factor ($-$)	$-$
τ_0^B	Bingham yield stress (Pa)	$ML^{-1}T^{-2}$
τ_w	shear stress at the wall of vessel (Pa)	$ML^{-1}T^{-2}$

Subscripts

b	bulk
cr	critical (to indicate laminar–turbulent transition)
g	with gas present in vessel
i	inside
o	outside
t	fully turbulent
w	wall
x, y	fractional degree of homogenization

Further exercises

The level of difficulty of problems has been graded: (a) straightforward, (b) somewhat more complex and (c) most difficult. In any given chapter the readers are recommended to tackle problems in increasing order of difficulty.

LEVEL
(a)

1.1 The following rheological data have been obtained for a liquid at 295.5 K.

Shear rate (s^{-1})	Shear stress (Pa)	Shear rate (s^{-1})	Shear stress (Pa)
2.22	1.32	35.16	20.92
4.43	2.63	44.26	26.33
7.02	4.17	70.15	41.74
8.83	5.25	88.31	52.54
11.12	6.62	111.17	66.15
14	8.33	139.96	83.27
17.62	10.48	176.2	104.84
22.18	13.20	221.82	132.0
27.93	16.62	279.25	166.15

By plotting these data on linear and logarithmic scales, ascertain the type of fluid behaviour, e.g. Newtonian, or shear-thinning, or shear-thickening, etc. Also, if the liquid is taken to have power-law rheology, calculate the consistency and flow-behaviour indices respectively for this liquid.

1.2 The following rheological data have been reported for a 0.6% (by weight) carbopol solution in a 1.5% (by weight) NaOH aqueous solution at 292 K. (a)

$\dot{\gamma}(s^{-1})$	τ(Pa)	$\dot{\gamma}(s^{-1})$	τ(Pa)
0.356	1.43	7.12	4.86
0.449	1.54	8.96	5.40
0.564	1.65	11.25	6.03
0.712	1.79	14.17	6.62
0.896	1.97	17.82	7.45
1.13	2.17	22.47	8.40
1.42	2.44	28.30	9.46
1.78	2.61	35.57	10.41
2.25	2.84	44.89	12.06
2.83	3.27	56.43	13.60
3.56	3.60	71.15	15.14
4.49	3.98	89.55	17.03
5.64	4.38	112.5	19.16

Plot these data in the form of $\tau - \dot{\gamma}$ and $\mu - \dot{\gamma}$ on logarithmic coordinates. Evaluate the power-law parameters for this fluid. Does the use of the Ellis fluid (equation (1.16)) or of the truncated Carreau fluid (equation (1.14)) model offer any improvement over the power-law model in representing these data? What are the mean and maximum % deviations from the data for these three models?

1.3 The following data for shear stress (τ) and first normal stress difference (N_1) have been reported for a 2% (by weight) Separan AP-30 solution in water measured at 289.5 K using a cone and plate rheometer. (a)

$\dot{\gamma}(s^{-1})$	$\tau(Pa)$	$N_1(Pa)$	$\dot{\gamma}(s^{-1})$	$\tau(Pa)$	$N_1(Pa)$
0.004 49	0.26	–	4.49	19.85	57.8
0.005 64	0.33	–	7.12	22.96	72.3
0.007 12	0.42	–	11.25	26.13	90.6
0.008 96	0.53	–	17.83	29.93	125.30
0.0113	0.66	–	28.30	34.44	154.20
0.0142	0.75	–	44.9	40.38	221.70
0.0178	0.96	–	71.2	46.32	318.60
0.0225	1.14	–	112.5	53.44	424.10
0.0356	1.65	–	89.6	49.88	366.30
0.0283	1.39	–	56.4	43.00	269.90
0.0449	1.99	–	35.6	37.5	192.8
0.0564	2.30	–	22.5	32.1	163.8
0.0712	2.85	–	14.2	28	120.5
0.0896	3.33	–	8.96	25.2	96.4
0.113	3.83	4.82	5.64	21.1	84.8
0.178	5.50	7.23	3.56	18.2	62.7
0.283	6.94	11.60	2.25	15.6	43.4
0.449	8.37	19.80	1.42	13.3	38.6
0.712	10.16	22.20	0.896	11	25.1
1.13	12.32	27.95	0.564	9.09	20.3
1.78	14.60	34.70	0.356	7.41	10.6

(i) Plot these shear stress, apparent viscosity and first normal stress difference data against shear rate on log–log scales. Does the shear stress data extend to the lower Newtonian region? What is the value of the zero-shear viscosity for this solution?

(ii) Suggest and fit suitable viscosity models covering the entire range. What is their maximum deviation?

(iii) Calculate and plot the Maxwellian relaxation time (equation (1.37)) as a function of shear rate for this polymer solution.

(iv) Is it possible to characterize the visco-elastic behaviour of this solution using a single characteristic time (using equation (1.32)) in the higher shear rate region? At what shear rate does it coincide with the Maxwellian relaxation time calculated in part (iii)?

1.4 The following rheological data for milk chocolate at 313 K are available. Determine the Bingham plastic (equation (1.17)) and Casson model (equation (1.19)) (a) parameters for this material. What are the mean and maximum deviations for both these models?

$\dot{\gamma}(s^{-1})$	$\tau(Pa)$	$\dot{\gamma}(s^{-1})$	$\tau(Pa)$
0.099	28.6	6.4	123.8
0.14	35.7	7.9	133.3
0.20	42.8	11.5	164.2
0.39	52.4	13.1	178.5
0.79	61.9	15.9	201.1
1.60	71.4	17.9	221.3
2.40	80.9	19.9	236
3.9	100		

It is likely that the model parameters are strongly dependent on the shear rate range covered by the rheological data. Compare the values of the model parameters by considering the following shear rate intervals:

(a) $\dot{\gamma} \leq 20\,s^{-1}$
(b) $1.6 \leq \dot{\gamma} \leq 20\,s^{-1}$
(c) $\dot{\gamma} \leq 1.6\,s^{-1}$

1.5 The following shear stress–shear rate data demonstrate the effect of temperature on the power-law constants for a concentrated orange juice containing 5.7% fruit pulp.

T = 254.3 K		T = 267.7 K		T = 282.6 K		T = 302.3 K	
$\dot{\gamma}(s^{-1})$	$\tau(Pa)$	$\dot{\gamma}(s^{-1})$	$\tau(Pa)$	$\dot{\gamma}(s^{-1})$	$\tau(Pa)$	$\dot{\gamma}(s^{-1})$	$\tau(Pa)$
0.5	14.4	0.6	4.3	1.1	2.6	8	3.6
1	24.3	1	6.5	8	10.3	20	7.6
10	142	10	38.4	15	17	40	13.1
20	240.4	20	65.4	30	29.5	60	17.5
30	327	30	89	60	50.3	120	31.2
40	408	40	111	90	69	240	54.5
50	484	50	132	150	103	480	94.4
60	556	60	152	250	154	800	142
70	635	70	171.3	350	200	1000	170
80	693	80	189.4	450	243	1100	183
150	1120	150	309				
		300	527				

How do the values of the power-law flow behaviour and consistency indices depend upon temperature? Estimate the activation energy of viscous flow (E) from these data by fitting them to the equation $m = m_0\,\exp(E/\mathbf{R}T)$ where m_0 and E are constants and $\mathbf{R}$ is the universal gas constant.

1.6 The following shear stress–shear rate data are available for an aqueous carbopol   (b)
solution at 293 K.

$\dot{\gamma}(s^{-1})$	τ(Pa)	$\dot{\gamma}(s^{-1})$	τ(Pa)
0.171	53.14	1.382	78.18
0.316	57.86	1.92	84.37
0.421	61.59	2.63	90.23
0.603	66	3.67	98.26
0.812	70	5.07	106.76
1.124	75.47		

By plotting these data on linear and logarithmic scales, ascertain the type of fluid
behaviour exhibited by this solution. Suggest a suitable viscosity model and evaluate the parameters for this solution. Does the fluid appear to have a yield stress?
Using the vane method (Nguyen, Q.D. and Boger, D.V., *Annu. Rev. Fluid Mech.*
24 (1992) 47), the yield stress was found to be 46.5 Pa. How does this value compare with that obtained by the extrapolation of data to $\dot{\gamma} = 0$ and that obtained by
fitting Bingham, Casson and Herschel–Bulkley fluid models?

1.7 The following rheological data have been reported for a 100-ppm polyacrylamide   (b)
solution in 96% (by weight) aqueous wheat syrup solution at 294 K.

$\dot{\gamma}(s^{-1})$	τ(Pa)	N_1(Pa)	$\dot{\gamma}(s^{-1})$	τ(Pa)	N_1(Pa)
0.025	0.70	–	0.315	8.79	18.1
0.0315	0.89	–	0.396	11.01	27.7
0.0396	1.12	–	0.500	13.87	42.0
0.050	1.42	0.0892	0.628	17.46	65.4
0.0628	1.78	0.158	0.790	21.80	97.5
0.0791	2.25	0.890	0.995	27.16	141.0
0.0995	2.83	1.26	1.25	33.75	196.0
0.125	3.56	2.47	1.58	42.34	283.0
0.158	4.49	3.64	1.99	53.53	440.0
0.199	5.65	6.71	2.50	67.03	661.0
0.250	7.08	11.60	3.15	84.11	863.0

(i)   Is this solution shear-thinning?
(ii) Can this solution be treated as a Newtonian fluid? If not, why?
(iii) Is it a visco-elastic fluid? Estimate the value of the Maxwellian relaxation
time for this solution.

1.8 The following shear stress–shear rate values have been obtained for aqueous silica   (b)
(bulk density $= 800\,kg/m^3$) suspensions to elucidate the effect of concentration on
the rheological behaviour of suspensions:

880 kg/m³		905 kg/m³		937 kg/m³		965 kg/m³		995 kg/m³	
$\dot{\gamma}(s^{-1})$	τ(Pa)	$\dot{\gamma}(s^{-1})$	τ(Pa)	$\dot{\gamma}(s^{-1})$	τ(Pa)	$\dot{\gamma}(s^{-1})$	τ(Pa)	$\dot{\gamma}(s^{-1})$	τ(Pa)
3.9	3.46	1.9	3.79	1.1	5.20	1.9	6.25	0.9	7.55
5.3	3.54	2.6	3.90	2.1	5.40	3.4	6.46	1.8	8.01
5.9	3.61	3.8	4.02	3.3	5.62	4.7	6.67	2.8	8.33
7.0	3.68	4.9	4.12	5.6	5.90	7.2	6.92	4.9	8.75
8.2	3.74	5.9	4.19	7.9	6.12	9.1	7.16	8.0	9.25
9.2	3.80	7.0	4.30	10.8	6.32	10.7	7.27	9.9	9.43
10.3	3.86	8.0	4.38	12.4	6.48	12.0	7.38	11.6	9.64
11.3	3.91	9.0	4.44	13.9	6.58	13.1	7.48	13.0	9.88
12.2	3.96	12.2	4.64	15.2	6.68	14.2	7.55	14.3	9.95
14.9	4.10	13.5	4.70	16.4	6.78	15.1	7.70	15.5	10.10
16.2	4.17	14.9	4.79	17.6	6.88	15.9	7.74	16.6	10.27
19.3	4.30	18.2	5.00	23.1	7.34	17.7	7.98	19.4	10.53
22.5	4.43	21.4	5.15	25.7	7.41	19.4	8.10	22.0	10.80
25.5	4.48	24.7	5.30	28.3	7.59	24.2	8.47	24.5	11.10
28.6	4.70	28.0	5.40	33.2	7.91	27.0	8.62	26.9	11.25
34.6	4.98	34.4	5.82			30.0	8.87	31.8	11.60
40.5	5.15	40.7	5.97			37.1	9.30	36.5	11.91

(i) Plot these data on linear and logarithmic scales and fit the Herschel–Bulkley
viscosity model to represent the behaviour of these suspensions.
(ii) How do the model parameters depend upon the concentration?

1.9 The following rheological data have been obtained for a 0.244% Polyiso-   (b)
butylene/92.78% Hyvis Polybutene/6.98% kerosene (by weight) solution at
293 K.

$\dot{\gamma}(s^{-1})$	τ(Pa)	N_1(Pa)	$\dot{\gamma}(s^{-1})$	τ(Pa)	N_1(Pa)
0.05	0.165	–	0.792	2.55	–
0.0629	0.202	–	0.998	3.18	0.30
0.0792	0.260	–	1.26	3.98	0.57
0.0998	0.330	–	1.58	4.58	1.20
0.126	0.413	–	1.99	6.24	2.21
0.158	0.518	–	2.51	7.72	3.29
0.199	0.663	–	3.15	9.61	5.09
0.251	0.823	–	3.97	12	8.10
0.315	1.03	–	5.00	15	12.10
0.397	1.29	–	6.29	18.6	18.20
0.50	1.61	–	7.92	23.2	28.00
0.629	2.03	–	9.97	29.2	46.90

(i) Does this fluid exhibit Newtonian, or shear-thinning, fluid behaviour?

(ii) Is the liquid visco-elastic? Show the variation of the Maxwellian relaxation time with shear rate.

2.1 The following volumetric flow rate–pressure gradient data have been obtained   (c)
using a capillary viscometer ($D = 10$ mm and $L = 0.5$ m) for a viscous material. Obtain the true shear stress — shear rate data for this substance and suggest a suitable viscosity fluid model.

Q(mm^3/s)	1.3	15	140	1450	14 500	1.5×10^5	1.4×10^6
$-\Delta p$(Pa)	0.5	1.5	5	15	50	160	500

2.2 A polymer solution was tested in a cone-and-plate viscometer (cone angle 0.1 rad   (b)
and cone radius 25 mm) at various rotational speeds. Use the following torque–speed data to infer shear stress–shear rate behaviour and suggest an appropriate fluid model to describe the fluid behaviour.

Ω(rad/s)	10^{-4}	10^{-3}	0.01	0.1	1	10	100
T(N m)	0.003	0.033	0.26	1	2.2	3.3	6.6

2.3 The following data have been obtained with a capillary viscometer for an aqueous   (c)
polymer solution, density 1000 kg/m^3 at 293 K.

Capillary data	$-\Delta p$(kPa)	Mass flow rate (kg/s)
$L = 200$ mm	224.3	1.15×10^{-3}
$D = 2.11$ mm	431	2.07×10^{-3}
	596.3	2.75×10^{-3}
	720.3	3.88×10^{-3}
$L = 50$ mm	87.5	1.01×10^{-3}
$D = 2.11$ mm	148.1	1.71×10^{-3}
	361.7	4.29×10^{-3}
	609.8	6.69×10^{-3}
$L = 50.2$ mm	43.40	3.95×10^{-3}
$D = 4.14$ mm	79.22	8.18×10^{-3}
	117.12	1.092×10^{-2}
	160.53	1.48×10^{-2}

Obtain the true shear stress–shear rate data for this polymer solution. Are end effects significant in this case?

2.4 The following capillary data are available for a suspension with capillaries of dif-   (c)
ferent diameters and L/D ratios. Obtain the true shear stress–shear rate data for this suspension and fit an appropriate viscosity model (Data from Lam, Y.C., Wang, Z.Y., Chen, X. and Joshi, S.C., *Powder Technol.* **177** (2007) 162).

D = 0.5 mm				D = 0.75 mm				D = 1 mm				D = 1.25 mm			
Q (mm³/s)	$(-\Delta p)$ (bar)			Q (mm³/s)	$(-\Delta p)$ (bar)			Q (mm³/s)	$(-\Delta p)$ (bar)			Q (mm³/s)	$(-\Delta p)$ (bar)		
	$L/D = 5$	$L/D = 10$	$L/D = 20$		$L/D = 5$	$L/D = 10$	$L/D = 20$		$L/D = 5$	$L/D = 10$	$L/D = 20$		$L/D = 5$	$L/D = 10$	$L/D = 20$
0.136	15.95	26.57	47.87	0.452	15.76	26.95	49.34	0.984	13.92	24.81	48.85	1.7	14.22	25.52	48.12
0.271	21.82	36.20	64.97	0.882	22.02	37.72	69.13	1.96	19.98	35.17	68.80	3.39	10.86	35.63	67.17
0.656	33.6	55.78	100.20	2.15	33.87	58.02	106.3	4.90	31.90	55.11	106.8	8.48	30.74	55.1	103.8
1.36	47.3	78.4	140.5	4.18	45.9	78.31	143.1	9.82	44.26	76.31	146.9	17.8	43.2	77.24	145.3
1.92	55.13	90.98	160.7	6.33	55.4	94.35	172.2	14.70	52.71	91.26	174.9	26.6	51.77	92.4	173.7
2.60	63.3	104.2	186.1	8.37	62.84	106.8	194.7	19.60	59.45	102.40	196.6	35.1	58.35	104	195.2
3.00	67.1	110.3	196.6	14.70	79.53	134.2	243.5	34.40	74.65	127.9	246.6	63.1	74.35	131.7	246.5
4.64	80.7	131.8	234	20.90	91.7	153.9	278.2	49	84.71	146.6	282	90.5	85.46	150.7	281.1
7.46	98.2	159	280.7	31.70	107.7	179.3	322.4	73.6	98.29	168.6	326.2	136	97.87	171.1	317.5
9.95	109.9	176.9	310.9	32.8	109.3	181.8	326.8	–	–	–	–	139	100.5	176.1	327.3

2.5 The following data are available (based on Cohen, Y. and Metzner, A.B., *J. Rheol.*   (c)
29 (1985) 67) for a polymer solution (at 298 K) which have been obtained with capillaries of different diameters and lengths to ascertain the significance of end effects
and wall slip. Obtain the true shear stress–shear rate data for this polymer solution
after making the necessary corrections.

$Q \times 10^{10} (\text{m}^3/\text{s})$	$(-\Delta p)(\text{kPa})$		
	$L = 531\,\text{mm}$	$L = 1094\,\text{mm}$	$L = 1195\,\text{mm}$
	$D = 0.191\,\text{mm}$		
0.366	60.25	121.7	132.7
0.760	85.27	172.8	188.6
1.03	99.15	201.3	219.6
1.96	129.8	264.0	288.1
2.48	142.3	289.7	316.2
3.33	163.2	332.6	362.9
3.97	177.2	361.2	394.1
4.53	189.7	386.9	422.2
5.18	206.4	421.0	459.5
6.01	216.2	441.3	481.6
	$D = 0.266\,\text{mm}$		
0.998	48.38	97.55	106.4
1.85	63.37	128.1	139.8
3.09	79.35	160.8	175.4
4.22	88.39	179.2	195.5
5.63	100.4	203.8	222.3
7.55	115.3	234.4	255.8
8.80	126.3	256.8	280.2
10.3	135.3	275.3	300.4
11.2	143.2	291.6	318.2
14.0	163.1	332.4	362.7
	$D = 0.626\,\text{mm}$		
3.41	14.87	29.14	31.70
15.9	26.13	51.89	56.51
34.3	35.57	71.08	77.44
46.8	40.71	81.53	88.85
53.4	43.27	86.75	94.55
65.2	47.53	95.45	104.0
78.4	51.38	103.3	112.6
96.7	56.08	112.9	123.1
107	58.64	118.1	128.8
122	62.06	125.1	136.4
155	69.31	139.9	152.6
169	71.46	144.3	157.4

$D = 1.10$ mm			
6.49	8.81	16.96	18.42
121	18.46	36.19	39.37
173	21.89	43.15	46.96
232	24.36	48.15	52.41
315	27.79	55.12	60.02
390	30.48	60.59	66.00
473	33.17	66.07	71.97
620	37.09	74.05	80.67
778	41.23	82.51	89.91
904	43.92	88.00	95.90
1040	47.09	94.46	103.0

2.6 The following data have been obtained for an apple puree with several degrees   (c)
(% water) of dilution using a 25-mm internal diameter pipe. The pressure drop
(in bars) is measured over a 7-m long straight section for a range of flow rates Q
(m^3/hr). Obtain the true shear stress–shear rate data for this system. Fit an appropriate viscosity model to these results. Also, develop the relationships between the
model parameters and the concentration of the puree. (Perona, P., *J. Food Eng.* **60**
(2003) 137).

% Dilution

0%		3%		10%		20%		40%		50%		60%	
Q	$(-\Delta p)$	Q	$(-\Delta p)$	Q	$(-\Delta p)$	Q	$(-\Delta p)$	Q	$(-\Delta p)$	Q	$(-\Delta p)$	Q	$(-\Delta p)$
0.14	0.114	0.314	0.124	0.31	0.0557	0.484	0.0498	0.508	0.0133	0.512	0.0106	0.404	0.0024
0.181	0.123	0.505	0.159	0.442	0.0695	0.766	0.0665	0.723	0.0191	0.612	0.0131	0.501	0.0034
0.312	0.162	0.712	0.189	0.718	0.094	1.083	0.0841	0.913	0.024	0.708	0.015	0.594	0.0046
0.42	0.187	1.02	0.234	0.932	0.112	1.255	0.0929	1.208	0.0319	0.815	0.0181	0.680	0.0059
5.46	0.719	1.25	0.268	1.27	0.138	1.496	0.104	1.44	0.038	0.905	0.0191	0.789	0.0077
6.04	0.761	1.77	0.319	1.61	0.161	2.014	0.131	1.659	0.0442	1.005	0.0207	0.912	0.0123
0.621	0.226	2.25	0.364	2.02	0.185	2.34	0.150	1.803	0.0491	1.11	0.023	1.031	0.0176
0.827	0.263	2.50	0.385	2.41	0.212	2.66	0.166	1.961	0.0585	1.308	0.0272	1.202	0.0241
3.01	0.295	3.02	0.426	2.75	0.237	3.11	0.187	2.210	0.0812	1.407	0.0302	1.352	0.0301
3.25	0.337	3.24	0.448	3.24	0.268	3.75	0.226	2.118	0.0735	1.51	0.0336	1.472	0.0350
4.76	0.401	3.75	0.488	3.76	0.301	4.10	0.252	2.421	0.0978	1.604	0.0382	1.603	0.0406
4.01	0.428	4.00	0.506	4.32	0.334	4.30	0.282	2.612	0.112	1.803	0.0556	1.709	0.0451
3.51	0.483	4.30	0.527	4.94	0.376	4.61	0.328	2.801	0.126	2.15	0.0774	1.810	0.0496
3.02	0.533	4.60	0.547	5.30	0.406	4.90	0.367	3.215	0.156	2.46	0.0969	1.926	0.055
3.70	0.588	4.90	0.568	5.80	0.490	5.19	0.399	3.502	0.179	2.75	9.116	2.15	0.0656
3.98	0.611	5.29	0.593	5.50	0.434	5.41	0.424	3.729	0.197	3.106	0.141	2.32	0.0744
4.38	6.642	5.64	0.619	6.035	0.525	5.60	0.446	4.005	0.220	3.691	0.183	2.70	0.095
4.97	0.687	6.04	0.649	4.01	0.316	6.00	0.492	4.252	0.241	4.009	0.207	2.93	0.108
—	—	—	—	—	—	—	—	—	—	—	—	3.16	0.122

2.7 The following flow rate–pressure drop data have been reported (Yeow *et al.*, *J.*   (b)
Food Sci, **67** (2001) 1407) for fruit purees in different diameter pipes. Obtain the
true shear stress–shear rate data and fit appropriate viscosity models.

Nectarine Purée			
D = 36.5 mm		D = 25 mm	
Q(m³/s) × 10⁴	$(-\Delta p/L)$(Pa/m)	Q(m³/s) × 10⁴	$(-\Delta p/L)$(Pa/m)
0.325	750	0.344	1500
0.667	867	0.728	1914
0.986	933	1.103	2186
1.35	1050	1.43	2314
1.71	1133	1.80	2457
2.06	1183	2.14	2614
2.54	1233	2.53	2771
2.93	1300	2.83	2929
3.54	1383	3.48	3214
4.10	1500	4.094	3514
4.78	1567	4.88	3786
5.41	1617	5.61	4029
6.06	1700	6.34	4200
6.66	1767	7.94	4671
7.31	1850	9.46	5228
8.06	1917	10.96	5686
8.86	1967	12.64	6143
9.73	2033	14.35	6600
10.54	2100	16.04	7071
12.03	2283	15.18	6800
13.70	2433	13.52	6357
16.24	2634	–	–

2.8 Kim and Dealy (*J. Rheol.*, **45** (2001) 1413) have reported the following data for a   (c)
linear metallocene polyethylene at 150°C using a capillary rheometer (capillary
diameter = 0.762 mm).

	Measured pressure drop $(-\Delta p)$ (MPa)		
Q (mm³/min)	(L/D) = 5	(L/D) = 10	(L/D) = 20
21.36	1.95	3.58	7.0
35.69	3.00	5.60	11.0
49.75	3.85	7.00	14.0
71.11	4.93	9.00	17.1
106.80	6.00	11.1	21.1
142.50	7.00	12.9	24.1
213.60	8.00	14.6	27.5

Obtain true shear stress–shear rate data for the behaviour (In view of the rather small values of (*L/D*) used, it is necessary to correct these results for end effects).

2.9 A four-bladed vane (*D* = 26.15 mm; *H/D* = 1.92) is rotated at the rotational (a) speed of 0.1 rpm in a 68.2% (by wt) red mud suspension. Estimate the yield stress of this suspension using the following torque–time data (based on Nguyen, Q.D. and Boger, D.V., *J. Rheol.*, **29** (1985) 335).

t (s)	8.3	16.6	25	33.3	41.7	50	66.7	83.3	100	116.7	133.3	150	167
T (mNm)	12.5	15.7	17.5	18.50	18.40	18.20	17.5	17.33	16.55	16.00	15.55	14.80	14.3
t (s)	183	200	217										
T (mNm)	13.9	13.4	13.2										

2.10 The following data have been reported for a star-branched polybutadiene at (b) 652 K. The end and slip effects are known to be negligible. Obtain the apparent viscosity-shear rate (true) data for this substance. Also, evaluate the power law constants, *m* and *n*, for this polymer (From Kraus, G. and Gruver, J.T., *J. Polym. Sci.*, **3A** (1965) 105).

$\frac{8V}{D}$ (s^{-1})	τ_w(MPa)	$\frac{8V}{D}$ (s^{-1})	τ_w(MPa)
0.021	44	1.15	163
0.051	57	1.56	180
0.15	66	1.77	192
0.21	84	2.56	217
0.48	112	3.13	236
0.64	128	3.68	254
1.01	151	–	–

2.11 The following shear data have been obtained for a high density polyethylene (b) (HDPE) using four capillaries of different values of *L/D* ratio. Obtain the true shear stress–shear rate data for this material after applying the end correction.

	Values of τ_w(MPa)			
$\frac{8V}{D}$ (s^{-1})	*L/D* = 5	*L/D* = 10	*L/D* = 15	*L/D* = 20
12	11	17.3	23.6	29.9
60	7.03	11.36	15.69	20.02
120	6	9.5	13	16.5
600	4.6	7.5	10.4	13.3

2.12   The following steady shear data has been obtained for a fire fighting foam at   (b)
340 kPa.

D = 6.95 mm		D = 9.9 mm		D = 15.8 mm	
τ_w (Pa)	$\dfrac{8V}{D}$ (s^{-1})	τ_w (Pa)	$\dfrac{8V}{D}$ (s^{-1})	τ_w (Pa)	$\dfrac{8V}{D}$ (s^{-1})
35.62	742.8	25.57	254.6	14.55	63.63
47.96	1016	29.29	348.2	18.92	87.01
58.03	1304	38.73	446.9	24.84	111.7
72.98	1617	47.91	554.3	31.54	138.5
78.88	1925	51.14	660	37.85	168
102.2	2648	60.57	907.6	35.48	165
115.7	3340	65.29	1145	42.18	226.8
		70.25	1367	44.95	286.1
		77.95	1583	48.50	341.5
				12.18	55.58
				17.54	83.42
				22.79	111.7
				29.53	139.4

The length of each capillary is 1400 mm. Obtain the true shear stress – shear rate
data after making all the corrections for this foam. (Data courtesy Dr. B.S. Gardiner,
University of Melbourne, Melbourne, Australia).

3.1   A low molecular weight polymer melt, which can be modelled as a power-law   (a)
fluid with $m = 5\,\text{k Pa s}^n$ and $n = 0.25$, is pumped through a 13-mm inside diameter
tube over a distance of 10 m under laminar flow conditions. Another pipe is needed
to pump the same material over a distance of 20 m at the same flow rate and with
the same frictional pressure loss. Calculate the required diameter of the new pipe.

3.2   The flow behaviour of a tomato sauce follows the power-law model, with   (b)
$n = 0.50$ and $m = 12\,\text{Pa s}^n$. Calculate the pressure drop per metre length of pipe
if it is pumped at the rate of 1000 kg/h through a 25-mm diameter pipe. The sauce
has a density of 1130 kg/m^3. For a pump efficiency of 50%, estimate the required
power for a 50-m long pipe.
How will the pressure gradient change if

(a)   the flow rate is increased by 50%,
(b)   the flow behaviour consistency coefficient increases to 14.75 Pa s^n without
altering the value of n, due to changes in the composition of the sauce,
(c)   the pipe diameter is doubled,
(d)   the pipe diameter is halved.
Is the flow still streamline in this pipe?

3.3   A vertical tube whose lower end is sealed by a moveable plate is filled with a visco-   (b)
plastic material having a yield stress of 20 Pa and density 1100 kg/m^3. Estimate
the minimum tube diameter for this material to flow under its own weight when the
plate is removed. Does the depth of the material in the tube have any influence on
the initiation of flow?

3.4 A power-law fluid ($m = 5\,\mathrm{Pa\,s}^n$ and $n = 0.5$) of density $1200\,\mathrm{kg/m}^3$ flows down an   (b)
inclined plane at $30°$ to the horizontal. Calculate the volumetric flow rate per unit
width if the fluid film is 6-mm thick. Assume laminar flow conditions.

3.5 A Bingham plastic material is flowing under streamline conditions in a circu-   (c)
lar pipe. What are the conditions for one-third of the total flow to be within the
central plug region across which the velocity profile is flat? The shear stress acting within the fluid τ varies with the velocity gradient dV_x/dr according to the
relation:

$$\tau = \tau_0^B + \mu_B\left(-\frac{dV_x}{dr}\right)$$

where τ_0^B and μ_B are, respectively, the Bingham yield stress and the plastic viscosity of the material.

3.6 Tomato purée of density $1100\,\mathrm{kg/m}^3$ is pumped through a 50-mm diameter pipe-   (c)
line at a flow rate of $1\,\mathrm{m}^3/\mathrm{h}$. It is suggested that, in order to double the rate of
production,

(a) a similar line with pump should be put in parallel with the existing one, or
(b) a larger pump should be used to force the material through the present line, or
(c) the cross-sectional area of the pipe should be doubled.

The flow behaviour of the tomato puree can be described by the Casson equation
(1.19), i.e.,

$$(+\tau_{rz})^{1/2} = (+\tau_0^C)^{1/2} + \left[\mu_C\left(-\frac{dV_z}{dr}\right)\right]^{1/2}$$

where τ_0^C, the Casson yield stress, is equal to $20\,\mathrm{Pa}$ and μ_C, the Casson plastic
viscosity, has a value of $5\,\mathrm{Pa\,s}$.
    Evaluate the pressure gradient for the three cases. Also, evaluate the viscosity
of a hypothetical Newtonian fluid for which the pressure gradient would be the
same. Assume streamline flow under all conditions.

3.7 A polymer solution is to be pumped at a rate of $11\,\mathrm{kg/min}$ through a 25-mm   (b)
inside diameter pipe. The solution behaves as a power-law fluid with $n = 0.5$ and
has an apparent viscosity of $63\,\mathrm{m\,Pa\,s}$ at a shear rate of $10\,\mathrm{s}^{-1}$, and a density of
$950\,\mathrm{kg/m}^3$.

(a) What is the pressure gradient in the pipe line?
(b) Estimate the shear rate and the apparent viscosity of the solution at the
    pipe wall?
(c) If the fluid were Newtonian, with a viscosity equal to the apparent viscosity
    at the wall as calculated in (b), what would be the pressure gradient?
(d) Calculate the Reynolds numbers for the polymer solution and for the
    hypothetical Newtonian fluid.

3.8 A concentrated coal slurry (density $1043\,\mathrm{kg/m}^3$) is to be pumped through a 25-mm   (b)
inside diameter pipe over a distance of $50\,\mathrm{m}$. The flow characteristics of this slurry
are not fully known, but the following preliminary information is available on
its flow through a smaller tube, 4mm in diameter and 1-m long. At a flow rate
of $0.0018\,\mathrm{m}^3/\mathrm{h}$, the pressure drop across the tube is $6.9\,\mathrm{kPa}$, and at a flow rate
of $0.018\,\mathrm{m}^3/\mathrm{h}$, the pressure drop is $10.35\,\mathrm{kPa}$. Evaluate the power-law constants

from the data for the small diameter tube. Estimate the pressure drop in the 25-mm diameter pipe for a flow rate of $0.45\,\mathrm{m^3/h}$.

3.9 A straight vertical tube of diameter D and length L is attached to the bottom of a large cylindrical vessel of diameter D_T $(\gg D)$. Derive an expression for the time required for the liquid height in the large vessel to decrease from its initial value of $H_0(\ll L)$ to $H(\ll L)$ as shown in the following sketch.   (c)

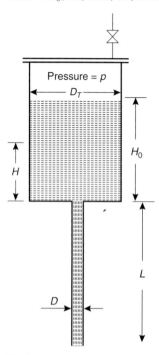

Neglect the entrance and exit effects in the tube as well as the changes in kinetic energy. Assume laminar flow in the tube and (i) power-law behaviour and (ii) Bingham plastic behaviour.

3.10 Estimate the time needed to empty a cylindrical vessel ($D_T = 101\,\mathrm{mm}$) (open to atmosphere) filled with a power-law liquid ($m = 4\,\mathrm{Pa\,s^n}$ and $n = 0.6$, density $= 1010\,\mathrm{kg/m^3}$). A 6-mm ID capillary tube 1.5-m long is fitted to the base of the vessel as shown in the diagram for Problem 3.9. The initial height of liquid in the vessel, $H_0 = 230\,\mathrm{mm}$. Estimate the viscosity of a hypothetical Newtonian fluid of the same density which would empty in the same time.   (c)

3.11 Using the same equipment as in Problem 3.10, the following height–time data have been obtained for a coal slurry of density $1135\,\mathrm{kg/m^3}$. Evaluate the power-law model parameters for this slurry (Assume streamline flow under all conditions).   (b)

t (s)	0	380	808
H (m)	0.25	0.20	0.15

3.12 A pharmaceutical formulation having a consistency coefficient of $2.5\,\mathrm{Pa\,s^n}$ and a flow behaviour index of 0.65 must be pumped through a stainless steel pipe   (a) of 40-mm inside diameter. If the shear rate at the pipe wall must not exceed $140\,s^{-1}$ or fall below $50\,s^{-1}$, estimate the minimum and maximum acceptable volumetric flow rates.

3.13 Two storage tanks, A and B, containing a coal slurry ($m = 2.7\,\text{Pa s}^n$; $n = 0.5$; (c) density $= 1040\,\text{kg/m}^3$) discharge through pipes each 0.3 m in diameter and 1.5 km long to a junction at D. From D, the slurry flows through a 0.5-m diameter pipe to a third storage tank C, 0.75 km away as shown in the following sketch. The free surface of slurry in tank A is 10 m above that in tank C, and the free surface in tank B is 6 m higher than that in tank A. Calculate the initial rates of discharge of slurry into tank C. Because the pipes are long, the minor losses (in fittings) and the kinetic energy of the liquid can be neglected. Assume streamline flow conditions in all pipes.

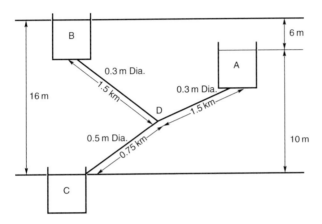

3.14 An engineer having carefully calculated the optimum internal diameter required (c) of a proposed long pipeline for transporting china clay slurries (approximating to power-law behaviour) at a given flow rate, discovers that a full range of pipe sizes is not available at the remote location. The diameters that are closest 25% too large and 15% too small. Not wanting to over design, and fearing to exceed the available pressure gradient, the engineer decided to use a composite pipeline made up of the two available sizes in series. If the cost per unit length of pipe is proportional to its inside diameter, what would be the %saving as a function of n (0.2–1.2) in using the composite line instead of a uniform line of the larger diameter when the flow in all pipes is (i) laminar, and (ii) turbulent in which case the friction factor is given by $f \propto Re_{MR}^{-1/(3n+1)}$. Assume that in the range of conditions encountered, the values of the power-law constants (m and n) do not vary appreciably.

3.15 A power-law liquid is flowing under streamline conditions through a horizontal (a) tube of 8 mm diameter. If the mean velocity of flow is 1 m/s and the maximum velocity at the centre is 1.2 m/s, what is the value of the flow behaviour index?

For a Newtonian, organic liquid of viscosity 0.8 m Pa s, flowing through the tube at the same mean velocity, the pressure drop is 10 kPa compared with 100 kPa for the power-law liquid. What is the power-law consistency coefficient, m, of the non-Newtonian liquid?

3.16 The rheology of a polymer solution can be approximated reasonably well by (b) either a power-law or a Casson model over the shear rate range of 20–100 s^{-1}.

If the power law consistency coefficient, m, is $10\,Pa\,s^n$ and the flow behaviour index, n, is 0.2, what will be the approximate values of the yield stress and the plastic viscosity in the Casson model?

Calculate the pressure drop using the power-law model when this polymer solution is in laminar flow in a pipe 200-m long and 40-mm inside diameter for a centreline velocity of 1 m/s. What will be the calculated centreline velocity at this pressure drop if the Casson model is used?

3.17 Using dimensional analysis, show that the frictional pressure drop for the fully developed flow of a Bingham fluid in circular tubes is given by: (a)

$$f = \phi(Re, He)$$

where f and Re are the usual fanning friction factor and pipe Reynolds number $(\rho VD/\mu_B)$ respectively; He, the Hedström number, is defined as $He = \rho\tau_0^B D^2/\mu_B^2$.

Use these dimensionless groups to rearrange equation (3.28) as:

$$f = \frac{16}{Re}\left[1 + \frac{1}{6}\left(\frac{He}{Re}\right) - \frac{1}{3}\frac{He^4}{f^3 Re^7}\right]$$

Give typical plots of f versus Re for a range of values of the Hedström number.

A carbopol solution ($\rho = 1000\,kg/m^3$), Bingham plastic rheology ($\tau_0^B = 8\,Pa$ and plastic viscosity of $50\,mPa\,s$), is flowing through a 25-mm diameter pipeline. Estimate the pressure drop per metre of pipe when the mean velocity is 1 m/s. Also, estimate the radius of the unsheared plug in the core region.

During maintenance, it is necessary to use a standby pump and the maximum attainable pressure gradient is 30% less than that necessary to maintain the original flow rate. By what percentage will the flow rate fall? Assume streamline flow conditions in the pipe.

3.18 A viscous plastic fluid of density $1400\,kg/m^3$ and containing 60% (by weight) (b) of a pigment is to be pumped in streamline flow through 75-, 100- and 125-mm diameter pipes within the plant area. The corresponding flow rates of the liquid in these pipes are 6×10^{-3}, 0.011 and $0.017\,m^3/s$, respectively. The following data were obtained on an extrusion rheometer using two different size capillaries:

Gas pressure in reservoir, $-\Delta p$ (kPa)	Time to collect 5 g liquid from the tube (s)	
	Tube A	Tube B
66.2	913	112
132.4	303	39
323.7	91	11.6
827.4	31	3.9

Tube A: diameter = 1.588 mm, length = 153 mm.
Tube B: diameter = 3.175 mm, length = 305 mm.
Estimate the pressure gradient in each of the pipes to be used to carry this liquid.

3.19 It is necessary to pump a non-Newtonian slurry (density 1000 kg/m³) over a dis-   (a)
tance of 100 m through a 200-mm diameter smooth-walled pipe. What is the
maximum possible flow rate at which the flow will be laminar and what then is
the pressure drop across the pipe?

   The following laboratory results are available for the laminar flow of this
slurry in small diameter tubes.

$(8V/D)$, s^{-1}	35	104	284	677	896	1750	4300	8000	10 900	23 100	51 500
τ_w (Pa)	4.7	9.3	17.2	29	37.9	51.4	90.3	153	160	291	429

Can these data be scaled up directly, without assuming a rheological model?
Compare the values of the pressure gradient obtained by such a method with
that obtained by assuming the power-law model behaviour. Assume streamline
flow conditions in all cases.

3.20 A coal-in-oil slurry (density 1640 kg/m³) containing 35% (by weight) of coal   (b)
(particle size <50 μm) is pumped at a rate of 3.5 m³/h from a storage tank,
through a 50-m long 12.5-mm diameter pipe to a boiler where it is burnt to raise
steam. The pressure in the gas space in the tank is atmospheric and the fuel
slurry must be delivered to the burner at an absolute pressure of 240 kPa.

(a) Estimate the pumping power required to deliver the slurry to the boiler
if the slurry is assumed to be a Newtonian fluid having a viscosity of
200 mPa s.
(b) Subsequently, preliminary rheological tests suggest the slurry to exhibit
Bingham plastic behaviour with a yield stress of 80 Pa and a plastic
viscosity of 200 mPa s. How will this knowledge affect the predicted value
of the pressure drop.

Comment on the difference between the two values of the pump power.

3.21 In a horizontal pipe network, a 150-mm diameter and 100-m long pipe branches   (a)
out into two pipes, one 100-mm diameter and 20-m long and the other 75-mm
diameter and 12-m long; the branches re-join into a 175-mm diameter and 50-m
long pipe. The volumetric flow rate of a liquid (density 1020 kg/m³) is 3.4 m³/
min and the pressure at the inlet to 150-mm diameter is 265 kPa. Calculate the
flow rate in each of the two smaller diameter branches and the pressure at the
beginning and end of the 175-mm diameter pipe for:

(a) a liquid exhibiting power-law behaviour, $n = 0.4$; $m = 1.4 \, \text{Pa} \, \text{s}^n$.
(b) a liquid exhibiting Bingham plastic behaviour, (yield stress 4.3 Pa and
plastic viscosity 43 m Pa s).

Due to process modifications, the flow rate must be increased by 20%; Calculate
the pressure now required at the inlet to the pipe network.

3.22 The fluid whose rheological data are given in Problem 3.19 is to be pumped at   (b)
0.34 m³/s through a 380-mm diameter pipe over a distance of 175 m. What will
be the required inlet pressure, and what pump power will be needed?

3.23 The same fluid as that used in Problem 3.19 is to be pumped through a 400-mm (b) diameter pipe over a distance of 500 m and the pressure drop must not exceed 53 kPa. Estimate the maximum possible flow rate achievable in the pipe. (Hint: By analogy with the method used for Newtonian fluids, prepare a chart of $\sqrt{Re}$ versus $\sqrt{fRe_{MR}^{2/(2-n)}}$ to avoid trial and error procedure).

3.24 Determine the diameter of a pipe, 175-m long, required to carry the liquid (b) whose rheology is given in Problem 3.19 at 0.5 m³/s with a pressure drop of 50 kPa. (Hint: Prepare a plot of $\sqrt{Re}$ versus $(\sqrt{fRe_{MR}^{5/(4-3n)}})$ to avoid the necessity of using a trial and error procedure; express average velocity in terms of the volumetric flow rate).

3.25 The following data have been obtained for the flow of the slurry specified in (a) Problem 3.19 in small tubes of three different diameters:

D(m)	L(m)	$-\Delta p$(kPa)	V(m/s)
0.016	4.27	136	5.76
		173	6.39
		205	7.14
		236	7.94
		284	9.01
		341	10.1
		395	11
		438	11.8
		500	12.8
		575	14
0.027	4.34	52.2	4.6
		64	5.26
		79.6	6.0
		98.7	6.91
		122	7.91
		152	9.03
		184	10.3
		235	11.7
		281	13.1
		352	15
0.053	2.64	8.41	3.33
		9.6	3.61
		11.4	4.03
		13.6	4.54
		17.6	5.37

Using these data, evaluate A, b, c in Bowen's relation and calculate the frictional pressure drop for flow rate of 0.34 m³/s in a 380-mm diameter pipe of length of 175 m. How does this value compare with that obtained in Problem 3.22?

3.26 Wilhelm *et al.* (*Ind. Eng. Chem.* **3** (1939) 622) presented the following pressure   (b)
drop–flow rate data for a 54.3% (by weight) rock slurry flowing through tubes
of two different diameters.

D(m)	L(m)	$-\Delta p$(kPa)	V(m/s)
0.019	30.5	338	3.48
		305	3.23
		257	2.97
		165	2.26
		107	1.81
		62	1.38
		48.4	1.20
		43	0.89
		38.6	0.44
		33.4	0.36
0.038	30.5	78.3	2.41
		50.3	1.86
		34.6	1.52
		19.1	1.09
		15.2	0.698
		17.3	0.512
		17.7	0.375

Using Bowen's method, develop a general scale-up procedure for predicting
the pressure gradients in turbulent flow of this rock slurry. Estimate the pump
power required for a flow rate of 0.45 m³/s in a 400-mm diameter pipe, 500-
m long. The pump has an efficiency of 60%. Take the density of slurry
1250 kg/m³.

3.27 What is the maximum film thickness of an emulsion paint that can be applied to   (a)
an inclined surface (15° from vertical) without the paint running off the surface?
The paint has a yield stress of 12 Pa and a density of 1040 kg/m³.

3.28 When a non-Newtonian liquid flows through a 7.5-mm diameter and 300-mm   (a)
long straight tube at 0.25 m³/h, the pressure drop is 1 kPa.

(i) Calculate the viscosity of a Newtonian fluid for which the pressure drop
would be the same at that flow rate?

(ii) For the same non-Newtonian liquid, flowing at the rate of 0.36 m³/h
through a 200-mm long tube of 7.5 mm in diameter, the pressure drop
is 0.8 kPa. If the liquid exhibits power-law behaviour, calculate its flow
behaviour index and consistency coefficient.

(iii) What would be the wall shear rates in the tube at flow rates of 0.18 and
0.36 m³/h. Assume streamline flow.

3.29 Two liquids of equal densities, the one Newtonian and the other a power-law   (c)
liquid, flow at equal volumetric rates down two wide inclined surfaces (30°
from horizontal) of the same widths. At a shear rate of 0.01 s⁻¹, the non-
Newtonian fluid, with a power-law index of 0.5, has the same apparent viscosity

as the Newtonian fluid. What is the ratio of the film thicknesses if the surface velocities of the two liquids are equal?

3.30 For the laminar flow of a time-independent fluid between two parallel plates (c) (Figure 3.18), derive a Rabinowitsch–Mooney type relation giving:

$$\left(-\frac{dV_z}{dy}\right)_{wall} = \left(\frac{3V}{h}\right)\left[\frac{2}{3} + \frac{1}{3}\frac{d\,ln(3V/h)}{d\,ln\,\tau_w}\right]$$

where $2h$ is the separation between two plates of width W ($\gg 2h$), τ_w is the wall shear stress. What is the corresponding shear rate at the wall for a Newtonian fluid?

3.31 A drilling fluid consisting of a china clay suspension of density $1090\,kg/m^3$ (a) flows at $0.001\,m^3/s$ through the annular cross-section between two concentric cylinders of radii 50 and 25 mm, respectively. Estimate the pressure gradient if the suspension behaves as:

(i) a power-law liquid: $n = 0.3$ and $m = 9.6\,Pa\,s^n$.
(ii) a Bingham plastic fluid: $\mu_B = 0.212\,Pa\,s$ and $\tau_0^B = 17\,Pa$.

Use the rigorous methods described in Section 3.6 as well as the approximate method presented in Section 3.7. Assume streamline flow.

    Due to pump malfunctioning, the available pressure gradient is only 75% of that calculated above. What will be the corresponding flow rates on the basis of the power-law and Bingham plastic models?

3.32 A power-law fluid (density $1000\,kg/m^3$) whose rheological parameters are $m =$ (a) $0.4\,Pa\,s^n$ and $n = 0.68$ is flowing at a mean velocity of 1.2 m/s in ducts of several different cross-sections:

(i)   a circular pipe
(ii) a square pipe
(iii) a concentric annulus with outer and inner radii of 26.3 and 10.7 mm, respectively
(iv) a rectangular pipe with aspect ratio of 0.5.
(v)   an elliptic pipe with minor-to major-axis ratio of 0.5.
(vi) an isosceles triangular with apex angle of 40°.

Using the geometric parameter method, estimate the pressure gradient required to sustain the flow in each of these conduits, all of which have the same hydraulic radius as the concentric annulus referred to in (iii). Also, calculate the Reynolds number for each case to check whether the flow is streamline.

    How will the pressure gradient in each pipe change if they were all to have same flow area as opposed to the same hydraulic diameter as the annulus?

3.33 The relation between cost per unit length C of a pipeline installation and its (c) diameter is given by

$$C = a + bD$$

where a and b are constants and are independent of pipe size. Annual charges are a fraction β of the capital cost C. Obtain an expression for the optimum pipe diameter on a minimum cost basis for a power-law fluid of density, ρ, flowing at a mass flow rate $\dot{m}$. Assume the flow to be (i) streamline and (ii) turbulent with friction factor $f \propto Re_{MR}^{-1/(3n+1)}$.

Discuss the nature of the dependence of the optimum pipe diameter on the flow behaviour index.

3.34 For streamline flow conditions, calculate the power needed to pump a power-law fluid through a circular tube when the flow rate is subject to sinusoidal variation of the form: (b)

$$\dot{Q} = Q_m \sin \omega t$$

where Q_m is the maximum flow rate at $t = \pi/2\omega$.

By what factor will the power requirement increase due to the sinusoidal variation in flow rate as compared with flow with uniform velocity?

Repeat this calculation for turbulent flow using the friction factor given by equation (3.75).

3.35 An aqueous bentonite suspension of density $1300 \, \text{kg/m}^3$, used to model an oil-based drilling mud, is to be pumped through the annular passage between the two concentric cylinders of radii 101.6 and 152.4 mm, respectively. The suspension, which behaves as a Bingham plastic fluid with $\mu_B = 20 \, \text{mPa s}$ and $\tau_0^B = 7.2 \, \text{Pa}$, is to be pumped at the rate of $0.13 \, \text{m}^3/\text{s}$ over a distance of 300 m. Estimate the pressure drop. Over what fraction of the area of the annulus, the material is in plug flow?   (a)

Plot the velocity distribution in the annular region and compare it with that for a Newtonian liquid having a viscosity of 20 mPa s under otherwise similar conditions.

3.36 The following data (Slatter, P., PhD thesis, University of Cape Town, Cape Town, 1994) have been obtained for the turbulent flow of a kaolin-in-water suspension (of density $1071 \, \text{kg/m}^3$) in pipes of different sizes: (a)

$D = 207$ mm

V (m/s)	2.93	2.67	2.46	2.27	2.04	1.71	1.55	1.35	1.15
τ_w (Pa)	19.2	17.11	13	11.6	7.87	6.14	5.1	4.16	2.89

$D = 140.5$ mm

V (m/s)	6.34	5.74	5.39	4.93	4.13	3.49	3	2.56	2	1.4
τ_w (Pa)	71.8	57.3	53.4	43.4	32.4	23.3	17.6	13.37	8.3	4.1

$D = 79$ mm

V (m/s)	6.48	5.44	4.5	3.63	2.69	1.92	2.62	2.1	1.71	1.18
τ_w (Pa)	83.5	59.1	41.23	29.5	16.9	12.3	10.7	8.94	7.31	3.93

$D = 21.6\,\text{mm}$

V (m/s)	7.09	6.80	6.48	6.08	5.62	5.1	4.77	4.53	4.1	3.7	2.91	1.47	1.3
τ_w (Pa)	123	114.3	105.6	93.58	82.1	69.3	62.1	55.4	46.2	39	25	7.6	5.9

$D = 13.2\,\text{mm}$

V (m/s)	6.84	6.41	5.82	5.4	4.76	4.1	3.77	3.46	3.18	2.7	2.41	1.91	1.49
τ_w (Pa)	136	120.8	102	89	72	55	47.5	40.6	35	26	21.2	13.74	7.45

$D = 5.6\,\text{mm}$

V (m/s)	4.55	3.89	3.50	3.22	3.08	2.84	2.65	2.08	2.22	1.7	1.51	1.61
τ_w (Pa)	64.6	58.5	48.5	40.5	33.5	27.4	22.9	20.6	18.6	14	13	10.7

(i) Use Bowen's method in conjunction with the second three sets of data for small diameter tubes to predict the pressure gradient in the remaining three large diameter pipes and compare them with the experimental values. Why does the discrepancy increase as the mean velocity of flow is decreased?

(ii) Slatter also fitted rheological data to the power-law and Bingham plastic models and the best values of the parameters were: $m = 0.56\,\text{Pa}\,\text{s}^n$ and $n = 0.31$; $\tau_0^B = 2.04\,\text{Pa}$ and $\mu_B = 3.56\,\text{mPa}\,\text{s}$. Compare the experimental and the calculated values of pressure gradient for all sizes of pipes.

3.37 Measurements are made of the yield stress of two carbopol solutions (density $1000\,\text{kg/m}^3$) and of a 52.9% (by weight) silica-in-water suspension (density $1491\,\text{kg/m}^3$) by observing their behaviour in an inclined tray which can be tilted to the horizontal. The values of the angle of inclination to the horizontal, θ, at which flow commences for a range of liquid depths, H, are given below. Determine the value of yield stress for each of these liquids. (b)

0.08% carbopol solution		0.09% carbopol solution		52.9% silica-in-water suspension	
H (mm)	θ (degrees)	H (mm)	θ (degrees)	H (mm)	θ (degrees)
2.0	6.8	6.4	5.5	13.7	3.55
2.6	6.1	7.0	4.6	17.3	2.85
3.2	4.9	12.0	2.8	22.5	2.20
3.9	3.8	12.1	2.6	24.1	1.90
5.2	3.0	15.3	2.1		
8.4	2.0	19.9	1.55		
14.0	1.0	24.1	1.35		
		30.0	1.10		
		32.8	0.95		

3.38   Viscometric measurements suggest that an aqueous carbopol solution behaves   (b)
as a Bingham plastic fluid with yield stress of 1.96 Pa and plastic viscosity 3.80 Pa s. The liquid flows down a plate inclined at an angle θ to the horizontal. Derive an expression for the volumetric flow rate per unit width
of the plate as a function of the system variables. Then, show that the following experimental results for $\theta = 5°$ are consistent with the theoretical
predictions.

Q (mm²/s)	H (mm)	Free surface velocity (mm/s)
4.8	7.19	0.8
10.8	8.62	1.54
18.0	9.70	2.20
26.2	10.51	3.14
35.9	11.50	3.94

3.39   The following rheological data for a 3% CMC solution (density of 1000 kg/m³)   (b)
and a 33% lime-in-water slurry (density of 1450 kg/m³) have been obtained
using a concentric cylinder viscometer (Christiansen, E.B., Ryan, N.W. and
Stevens, W.E., *AIChE J.*, **1** (1955) 544):

3% CMC solution		33% lime slurry	
$\dot{\gamma}(s^{-1})$	τ (Pa)	$\dot{\gamma}(s^{-1})$	τ (Pa)
20	40.7	40	11.01
37	69.9	95	12.21
78	105.33	130	13.31
88	123.53	190	13.41
118	137.41	370	15.56
145	153.22	560	17.00
185	173.3		
230	188.17		
265	206.36		
300	220.25		
345	233.65		
385	249.00		
415	262.86		
448	270.52		
500	280.10		
550	297.30		

Obtain the power-law constants for these two systems and compare the predicted values of the pressure drop gradients for their flow in pipes of different radius (R) with the experimental values listed below:

CMC solution			33% lime slurry		
R (mm)	Q (m³/s) × 10⁵	$(-\Delta p/L)$ (kPa/m)	R (mm)	Q (m³/s) × 10⁴	$(-\Delta p/L)$ (kPa/m)
9.81	1.64	11.64	26.37	7.6	0.98
	3.65	17.78		9.66	1.08
	5.69	23.72		12.9	1.17
				17.95	1.22
19.72	25.2	8.03		25.20	1.24
	45	11.17		31.70	1.28
	77.9	15.33			
	86	16.31	19.72	7.22	1.78
	136	20.81		7.39	1.71
	142	21.14		9.30	1.81
				10.90	1.84
				13.90	1.92
				17.40	2.06
			9.94	3.85	3.85
				5.47	4.27
				6.80	4.56

Also, calculate the value of the Reynolds number to confirm the occurrence of streamline flow over these ranges of conditions.

3.40 It is proposed to design a flume of rectangular cross-section (inclined to horizontal by an angle β) to transport a mineral slurry (which behaves like a Herschel–Bulkley model fluid). The flow is known to be laminar and nearly one-dimensional in the form of a sheet of small thickness which is much smaller than the other two dimensions of the flume. Derive the necessary relationship between the volumetric flow rate and the rheological and physical properties of the medium. Using this expression, obtain the corresponding relationships for a power-law fluid and a Bingham plastic fluid. (c)

For a kaolin slurry ($n = 0.3$, $\tau_0^H = 8.8$ Pa and $m = 1.7$ Pa s^n, density of 1200 kg/m³) flowing in a 300-mm wide flume, obtain the volumetric flow rate if the film thickness is 3.5 mm.

3.41 A stable coal slurry (approximated as a power law fluid) is stored under pressure (P_0) in a tank. The tank is fitted with two drain pipes of equal length, but of diameters D and $2D$, respectively which discharge the slurry into another open tank. What is the value of the flow behaviour index if the flow rate (Q) of the slurry in the large diameter pipe is thirty two times of that in the smaller tube? Neglect the gravity effects and assume the flow to be laminar. (b)

3.42 A bentonite slurry (of density 1400 kg/m³) behaves like a Herschel–Bulkley fluid ($m = 0.68$ Pa s^n; $n = 0.6$; $\tau_0^H = 20$ Pa) and it is to be heated in a 4-m long double pipe heat exchanger ($R_i = 16.5$ mm; $R_o = 25.5$ mm); Calculate the pressure drop when this slurry flows inside the tube and in the annular region when the flow rate is 3.41 m³/h. Assume the flow to be laminar in both cases. (b)

3.43   A coal suspension which behaves like a power-law model fluid ($m = 0.3\,\text{Pa s}^n$;   (c)
n = 0.64; $\rho = 1080\,\text{kg/m}^3$) is to be pumped through ducts of various cross-
sections of equal flow area, as detailed below. Estimate the pressure gradient
required to maintain an average velocity of 1 m/s in each of these ducts. The
values of ($f.Re$) for the laminar flow of Newtonian fluids in each duct are also
listed below along with the geometry of ducts.

   (i)   Circular tube of 50-mm inside diameter ($f.Re = 16$)
   (ii) Semi-circular cross-section ($f.Re = 15.768$)
   (iii) Sector (subtending an angle at the centre, $\alpha = 60°$) of a circle ($f\cdot Re = $
          14.17).
   (iv) Segment (subtending an angle at the centre, $\alpha = 60°$) of a circle ($f\cdot Re = $
          15.6).
   (v)   Concentric annulus ($\sigma = R_i/R_o = 0.7$).

3.44   A clay suspension ($m = 0.35\,\text{Pa s}^n$; n = 0.5; $\rho = 1050\,\text{kg/m}^3$) is being trans-   (b)
ported from a storage tank (point A located at an elevation of 6 m) to a process-
ing plant (located at point B at an elevation of 15 m) in two pipes arranged in
parallel (as shown below). The total flow rate of the suspension is 0.4 m³/s.

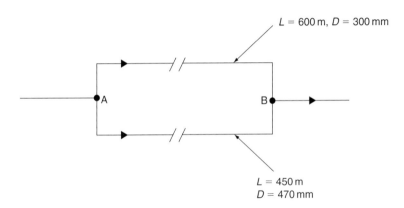

L = 600 m, D = 300 mm

L = 450 m
D = 470 mm

   Calculate the individual volumetric flow rates in the two pipes. Also estimate
the absolute pressure at point B if the gauge pressure at A is 690 kPa. Neglect
the minor losses in fittings, bends, etc. Also, the prevailing levels of the wall
shear stress are within the range of the rheological characterization of the
suspension.

3.45   In a kaolin clay processing plant area, a pipe network (as shown below) is to   (c)
be designed to pump kaolin slurries via different routes within the plant. The
concentration of the slurry may vary from day to day in the range 5–15% by
volume. The rheological behavior was measured for the extreme concentrations, and the resulting values of the power-law constants and density are as
follows:

Concentration	$n\,(-)$	$m\,(\text{Pa s}^n)$	$\rho\,(\text{kg/m}^3)$
5%	0.65	0.23	1075
15%	0.41	2.6	1225

Estimate the minimum and maximum values of the pressure drop between A and B if this facility is to operate at a total flow rate (entering at point A) of $700\,\mathrm{m^3/hr}$. The point B is located at a height of $100\,\mathrm{m}$ above point A. Neglect minor losses in fittings. All pipes are of $300\,\mathrm{mm}$ internal diameter which may be assumed smooth.

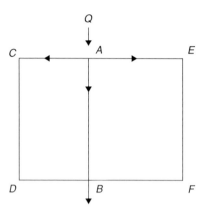

How will your answer change if (i) branch CD is not available due to maintenance schedule? (ii) the total flow rate is to be doubled and all three branches are available for flow. The lengths of the various sections are as follows:

$$DC = EF = AB = 500\,\mathrm{m};\ AC = BD = 300\,\mathrm{m};\ AE = BF = 400\,\mathrm{m}$$

3.46   Re-do problem 3.45 if the diameters of pipes are as follows: $D_{ACDB} = 200\,\mathrm{mm};$ (c)
$D_{AB} = 270\,\mathrm{mm};\ D_{AEFB} = 350\,\mathrm{mm}$.

3.47   For a power-law fluid flowing in a circular tube under laminar flow conditions,   (c)
how does the shear rate vary with radial position from being zero at the axis of
the tube to its maximum value at the wall of the tube? In the range $0 \leqslant r \leqslant R$,
does it ever become equal to the true shear rate for a Newtonian value at the
same value of $r_c(x^* = r_c/R)$?
   Plot x^* versus power law index in the range $0.1 \leqslant n \leqslant 1.2$. What will be
the % error in the value of viscosity if one were to use a constant value of
$x^* = 0.83, 0.84$ and 0.85 for all values of n in the range $0.1 \leqslant n \leqslant 1.2$?

3.48   An aqueous fire fighting foam is stored under pressure ($35\,\mathrm{kPa}$ gauge) in a tank   (c)
as shown below. It is pumped through a $200\,\mathrm{mm}$ internal diameter pipe (which
may be assumed smooth) as shown and issues out of a nozzle to form a free
jet directed into the broken window of a house on fire. Though the rheology
of foam is strongly dependent upon the existing pressure (expansion ratio), the
mean values of power-law constants are $n = 0.3$ and $m = 0.5\,\mathrm{Pa\,s^n}$. The density
of foam is $1030\,\mathrm{kg/m^3}$.
   The total equivalent length of pipe including bends, fittings and joints from
the tank to the nozzle tip is $100\,\mathrm{m}$ of $200\,\mathrm{mm}$ internal diameter pipe. Estimate the
power supplied by the pump.

3.49 In order to study the influence of the eccentricity on the frictional pressure drop for the laminar flow of a power-law drilling mud ($m = 0.35\,\text{Pa s}^n$; $n = 0.72$; $\rho = 1200\,\text{kg/m}^3$) in an eccentric annulus, a laboratory scale equipment is to be designed. The outer tube is of 50 mm internal diameter and the inner tube is of 15 mm OD. The centre-to-centre distance, C, between the two tubes is to vary from 0 mm (concentric annulus) to 12 mm. Estimate the power requirements if the flow rate is to vary in the range $0.5-2 \times 10^{-3}\,\text{m}^3/\text{s}$.

For the laminar flow of Newtonian fluids in an eccentric annulus, the following information is available:

$$f\,\text{Re} = \frac{16(1 - \sigma)^2(1 - \sigma^2)}{\phi}$$

where ϕ is a shape factor:

$$\phi = \left[1 - \sigma^4 - \frac{4C^2M^2}{R_0^4(\beta - \alpha)} - \frac{8C^2M^2}{R_0^4}\sum_{j=1}^{\infty}\frac{j\exp\{-j(\alpha + \beta)\}}{\sinh j(\beta - \alpha)}\right]^{-1}$$

The various symbols have the following meanings:
C = centre-to-centre distance between the two tubes

$$M = (F^2 - R_0^2)^{1/2}$$

$$F = \frac{(R_0^2 - R_i^2 + C^2)}{2C}$$

R_i & R_0 = Radii of inner and outer tubes respectively & $\sigma = R_i/R_0$.

$$\alpha = \frac{1}{2}\ln\left\{\frac{(F + M)}{(F - M)}\right\}$$

$$\beta = \frac{1}{2}\ln\left\{\frac{(F - C + M)}{(F - C - M)}\right\}$$

4.1   A 19.5% (by volume) kaolin-in-water suspension is flowing under laminar con-   (a)
ditions through a horizontal pipe, 42 mm diameter and 200 m long, at a volumetric flow rate of 1.25 m³/h. The suspension behaves as a power-law liquid with
$n = 0.16$ and $m = 9.6\,Pa\,s^n$, and has a density of 1320 kg/m³. Estimate the
pressure drop across the pipe. Air is 298 K is now introduced at an upstream
point at the rate of 5 m³/h (measured at the pressure at the mid-point of
the pipe length). What will be the two-phase pressure drop over the pipe
according to:

(i) the simple plug flow model
(ii) the generalized method of Dziubinski and Chhabra, equations (4.19) and
(4.24)
(iii) the method of Dziubinski, equation (4.26).

The experimental value of $-\Delta p$ is 105 kPa. Suggest the possible reasons for the
discrepancy between the calculated and actual values of $-\Delta p$.
    Calculate the average liquid holdup in the pipe.

4.2   A 25% aqueous suspension of kaolin is to be pumped under laminar conditions   (b)
through a 50-mm diameter and 50-m long pipe at the rate of 2 m³/h. The suspension behaves as a power-law fluid with $n = 0.14$, $m = 28.6\,Pa\,s^n$ and has a
density of 1400 kg/m³. Calculate the power needed for this duty when using a
pump of 50% efficiency.
    It is proposed to reduce the two-phase pressure drop by 50% by introduc-
ing air into the pipeline at an upstream point. Calculate the superficial velocity
of air required to achieve this if the air at 293 K enters the pipeline at a pressure
of 0.35 MPa. Assume isothermal expansion of gas. Use all three methods mentioned in Problem 4.1 to obtain the superficial velocity of the air. Using an
appropriate model, determine the maximum reduction in two-phase pressure
drop achievable for this slurry. What is the air velocity under these conditions?
What proportion of the volume of the pipe is filled with liquid and what flow
pattern is likely to occur in the pipe under these flow conditions? Neglect the
effect of air expansion.

4.3   A 50.4% (by weight) coal-in-water suspension of density 1070 kg/m³ is to be   (a)
transported at the flow rate of 5.5 m³/h through a pipeline of 75-mm diameter
and 30-m long. The suspension behaves as a Bingham plastic fluid with a yield
stress of 51.4 Pa and plastic viscosity of 48.3 mPa s. It is decided to introduce
air at pressure of 0.35 MPa and at 293 K into the pipeline to lower the pressure
drop by 25%, whilst maintaining the same flow rate of suspension. Assuming
that the plug model is applicable, calculate the required superficial velocity
of air.

4.4   It is required to transport gravel particles (8-mm size, density 2650 kg/m³) as a   (a)
suspension in a pseudoplastic polymer solution ($m = 0.25\,Pa\,s^n$, $n = 0.65$, density 1000 kg/m³) in a 42-mm diameter pipe over a distance of 1 km. The volumetric flow rate of the mixture is 7.5 m³/h when the volumetric concentration of
gravel in the discharged mixture is 22% (by volume). If the particles are conveyed in the form of a moving bed in the lower portion of the pipe, estimate the
pressure drop across this pipeline and the pumping power if the pump efficiency
is 45%. What is the rate of conveyance of gravel in kg/h?

4.5 A 13% (by volume) phosphate slurry ($m = 3.7\,\mathrm{Pa\,s}^n$, $n = 0.18$, density 1230 kg/ m^3) is to be pumped through a 50-mm diameter horizontal pipe at mean slurry velocities ranging from 0.2 to 2 m/s. It is proposed to pump this slurry in the form of a two-phase air-slurry mixture. The following data have been obtained: (b)

Superficial slurry velocity (m/s)	Superficial gas velocity (m/s)	$\dfrac{-\Delta p_{TP}/L}{-\Delta p_L/L}$
0.24	0.2	0.68
	0.5	0.48
	1	0.395
	2	0.40
	3	0.41
	4	0.44
	5	0.52
0.98	0.2	0.92
	0.5	0.90
	1.0	0.96
	1.4	1.08
	2	1.20
	3	1.44
	4	1.62
	5	1.85
1.5	0.5	1.22
	1.0	1.38
	1.5	1.50
	2	1.72
	2.4	1.86
1.95	0.4	1.38
	0.8	1.53
	1.0	1.60
	1.5	1.70

Are these data consistent with the predictions of the simple plug flow model? Compare these values of the two-phase pressure drop with those calculated using equations (4.22), (4.24) and (4.26).

4.6 A sewage sludge ($\rho = 1350\,\mathrm{kg/m}^3$; $m = 5.6\,\mathrm{Pa\,s}^n$; $n = 0.35$) is flowing together with air in a 50-mm diameter pipe. The following pressure drop – flow rate data have been reported by Jinming and Jingxuan (*Civil. Eng. Environ. Syst.* **23** (2006) 1). The gas pressure in the test section was ~0.08 MPa and the pipe line operates at 293 K. Evaluate the average liquid holdup in the pipe over this range of conditions and check whether the experimental values of the two-phase pressure drop are consistent with the predictions of methods presented in Chapter 4. (b)

| Volumetric flow rate (m³/s) | | Two-phase pressure drop |
Liquid	Gas	(Pa/m) $\times 10^4$
0.004	0.004	0.50
	0.006	0.54
	0.008	0.58
	0.012	0.65
	0.016	0.72
0.008	0.008	1.54
	0.012	1.70
	0.020	2.05
	0.024	2.20
0.01	0.010	2.08
	0.015	2.28
	0.020	2.53
	0.025	2.78
	0.030	2.93

5.1 Calculate the free falling velocity of a plastic sphere (d = 3.18 mm, density   (a)
1050 kg/m³) in a polymer solution which conforms to the power-law model with
m = 0.082 Pa s^n, n = 0.88 and has a density of 1000 kg/m³. Also, calculate the viscosity of a hypothetical Newtonian fluid of the same density in which this sphere
would have the same falling velocity. To what shear rate does this viscosity correspond for?

5.2 The following data for terminal velocities have been obtained for the settling rate   (a)
of spherical particles of different densities in a power-law type polymer solution
(m = 0.49 Pa s^n, n = 0.83, ρ = 1000 kg/m³).

Sphere diameter (mm)	Sphere density (kg/m³)	Settling velocity (mm/s)
1.59	4010	6.6
2.00	4010	10.0
3.175	4010	30
2.38	7790	45

Estimate the mean value of the drag correction factor for this power-law fluid.
How does this value compare with that listed in Table 5.1?

5.3 For the sedimentation of a sphere in a power-law fluid in the Stokes' law regime,   (a)
what error in sphere diameter will lead to an error of 1% in the terminal falling
velocity? Does the permissible error in diameter depend upon the value of the
power-law index? If yes, calculate its value over the range $1 \geq n \geq 0.1$. What is
the corresponding value for a Newtonian liquid?

5.4 Estimate the terminal falling velocity of a 5-mm steel ball (density 7790 kg/m³) in   (b)
a power-law fluid (m = 0.3 Pa s^n, n = 0.6 and density 1010 kg/m³).

5.5 The rheological behaviour of a china clay suspension of density $1200\,\mathrm{kg/m^3}$ (b)
is well approximated by the Herschel–Bulkley fluid model with consistency
coefficient of $11.7\,\mathrm{Pa\,s^n}$, flow behaviour index of 0.4 and yield stress of $4.6\,\mathrm{Pa}$.
Estimate the terminal falling velocity of a steel ball, 5-mm diameter and density
$7800\,\mathrm{kg/m^3}$. What is the smallest steel ball which will just settle under its own
weight in this suspension?

5.6 A 7.5-mm diameter PVC ball (density $1400\,\mathrm{kg/m^3}$) is settling in a power-law (b)
fluid ($m = 3\,\mathrm{Pa\,s^n}$, $n = 0.6$, density $1000\,\mathrm{kg/m^3}$) in a 30-mm diameter cylindrical tube. Estimate the terminal falling velocity of the ball.

5.7 Estimate the hindered settling velocity of a 30% (by volume) deflocculated sus- (a)
pension of $50\,\mu\mathrm{m}$ (equivalent spherical diameter) china clay particles in a polymer solution following the power-law fluid model with $n = 0.7$ and $m = 3\,\mathrm{Pa\,s^n}$,
in a 30-mm diameter tube. The densities of china clay particles and the polymer
solution are 2400 and $1000\,\mathrm{kg/m^3}$, respectively.

5.8 In a laboratory size treatment plant, it is required to pump the sewage sludge (a)
through a bed of porcelain spheres packed in a 50-mm diameter column. The
rheological behaviour of the sludge (density $1008\,\mathrm{kg/m^3}$) can be approximated
by a power-law model with $m = 3.8\,\mathrm{Pa\,s^n}$ and $n = 0.4$. Calculate the diameter of the spherical packing (voidage 0.4) which will be required to obtain a
pressure gradient of $8\,\mathrm{MPa/m}$ at a flow rate of $3.6\,\mathrm{m^3/h}$. What will be the flow
rates for the same pressure gradient if the nearest available packing sizes are 25%
too large and 25% too small? Assume the voidage remains at the same level.

5.9 Estimate the size of the largest steel ball (density $7800\,\mathrm{kg/m^3}$) which would (b)
remain embedded without settling in a viscoplastic suspension with a density of
$1040\,\mathrm{kg/m^3}$ and yield stress of $20\,\mathrm{Pa}$?

5.10 A polymeric melt exhibiting power-law behaviour ($m = 10^4\,\mathrm{Pa\,s^n}$, $n = 0.32$,
density $960\,\mathrm{kg/m^3}$) is to be filtered by using a sand-pack composed of $50\,\mu\mathrm{m}$
sand particles; the bed voidage is 37%. The pressure drop across a 100-mm
deep bed must be in the range 540–$1130\,\mathrm{MPa}$. Estimate the range of volumetric
flow rates which can be processed in a column of 50-mm diameter.

Over the relevant range of shear rates, the flow behaviour of this melt can
also be well approximated by the Bingham plastic model. What would be the
appropriate values of the plastic viscosity and the yield stress?

5.11 Estimate the minimum fluidizing velocity for a bed consisting of 3.57-mm glass (a)
spheres (density $2500\,\mathrm{kg/m^3}$) in a 101-mm diameter column using a power-law
polymer solution ($m = 0.35\,\mathrm{Pa\,s^n}$, $n = 0.6$ and density $1000\,\mathrm{kg/m^3}$) if the bed
voidage at the incipient fluidized condition is 37.5%. If the value of the fixed
bed voidage is in error by 10%, what will be the corresponding uncertainty in
the value of the minimum fluidizing velocity?

5.12 Estimate the terminal falling velocity of each of the following particles in a sta- (b)
tionary power-law liquid ($m = 0.02\,\mathrm{Pa\,s^n}$; $n = 0.53$; $\rho = 1020\,\mathrm{kg/m^3}$).

(i) Steel sphere of diameter 3.5 mm and density $7780\,\mathrm{kg/m^3}$.
(ii) Steel cylinders of the same volume as the sphere in part (i), but of length-
to-diameter ratios of 0.75, 1.5 and 5. Assume the cylinders to settle with
their axis oriented normal to the direction of gravity.
(iii) Gravel chips of the same volume as the sphere in part (i) (density of
$2700\,\mathrm{kg/m^3}$) having sphericity of 0.8 and the equal projected area circle
diameter $d_n = 3.8\,\mathrm{mm}$.

5.13   A house hold cleaning liquid contains 2% (by volume) abrasive phosphate par-   (b)
ticles ($d = 60$–$75 \, \mu m$; $\rho_s = 2700 \, kg/m^3$) suspended in a shear-thinning base ($m = 0.035 \, Pa \, s^n$; $n = 0.4$; $\rho = 1030 \, kg/m^3$). Calculate the rate of sedimentation
when the product is stored in 50-mm diameter plastic bottles. What should be
the value of the power-law consistency, m, if it is desired to reduce the rate of
sedimentation by 50%? Assume that the value of the power-law index is relatively insensitive to the minor changes in the composition of the suspending base.

6.1   Calculate the thermal conductivity of a 35% (by volume) suspensions of alumina   (a)
(thermal conductivity $= 30 \, W/mK$) and thorium oxide (thermal conductivity $= 14.2 \, W/mK$) in water and in carbon tetra chloride at 293 K.

6.2   A power-law non-Newtonian solution of a polymer is to be heated from 288   (b)
to 303 K in a concentric tube heat exchanger. The solution will flow at a mass
flow rate of 210 kg/h through the inner copper tube of 31.75-mm inside diameter. Saturated steam at a pressure of 0.46 bar and a temperature of 353 K is to
be condensed in the annulus. If the heater is preceded by a sufficiently long
unheated section for the velocity profile to be fully established prior to entering
the heater, determine the required length of the heat exchanger. Physical properties of the solution at the mean temperature of 295.5 K are density $= 850 \, kg/m^3$,
heat capacity $= 2100 \, J/kgK$, thermal conductivity $= 0.69 \, W/mK$, flow behaviour index $= 0.6$.

Temperature (K)	288	303	318	333	353	368
Consistency coefficient, m(Pa s^n)	10	8.1	6.3	4.2	2.3	1.3

Initially assume a constant value of the consistency index and subsequently
account for its temperature dependence using equation (6.36). Also, ascertain
the importance of free convection effects in this case, using equations (6.37)
and (6.38). The coefficient of thermal expansion, β, is $3 \times 10^{-4} \, K^{-1}$.

6.3   A coal-in-oil slurry which behaves as a power-law fluid is to be heated in a dou-   (c)
ble-pipe heat exchanger with steam condensing on the annulus side. The inlet
and outlet bulk temperatures of the slurry are 291 and 308 K, respectively. The
heating section (inner copper tube of 40-mm inside diameter) is 3-m long and is
preceded by a section sufficiently long for the velocity profile to be fully established. The flow rate of the slurry is 400 kg/h and its thermo-physical properties
are density $= 900 \, kg/m^3$, heat capacity $= 2800 \, J/kg \, K$, thermal conductivity $= 0.75 \, W/mK$. In the temperature interval $293 \le T \le 368 \, K$, the flow behaviour
index is nearly constant and is equal to 0.52.

Temperature (K)	299.5	318	333	353	368
Consistency coefficient, m(Pa s^n)	8.54	6.3	4.2	2.3	1.3

Calculate the temperature at which the steam condenses on the tube wall.
Neglect the thermal resistance of the inner copper tube wall. Do not neglect the
effects of free convection.

6.4   A power-law solution of a polymer is being heated in a 1.8-m long tube   (c)
heater from 291 to 303 K at the rate of 125 kg/h. The tube is wrapped with an

electrical heating coil to maintain a constant wall flux of $5\,kW/m^2$. Determine the required diameter of the tube. The physical properties of the solution are density $= 1000\,kg/m^3$, heat capacity $= 4180\,J/kg\,K$, thermal conductivity $= 0.56\,W/mK$, flow behaviour index $n = 0.33$ and consistency coefficient $m = (26 - 0.0756T)Pa\,s^n$, in the range $288 \leq T \leq 342\,K$.

Also, evaluate the mean heat transfer coefficient if a mean value of the consistency coefficient is used. How significant is free convection in this case? Also, determine the temperature of the tube wall at its midpoint and at the exit.

6.5 A power-law solution of a polymer (density $1000\,kg/m^3$) is flowing through a 3-m long, 25-mm inside diameter tube at a mean velocity of 1 m/s. Saturated steam at a pressure of 0.46 bar and a temperature of 353 K is to be condensed in the annulus. If the polymer solution enters the heater at 318 K, at what temperature will it leave? Neglect the heat loss to the surroundings. The thermo-physical properties of the solution are heat capacity $= 4180\,J/kg\,K$, thermal conductivity $= 0.59\,W/mK$, flow behaviour index $n = 0.3$.   (b)

Temperature (K)	303	313	323	333
Consistency coefficient, $m(Pa\,s^n)$	0.45	0.27	0.103	0.081

How many such tubes in parallel would be needed to heat up 17 ton/h of this solution? What will be the exit fluid temperature when the flow rate is 20% above, and 20% below, the value considered above?

6.6 A power-law fluid is heated by passing it under conditions of laminar flow through a long tube whose wall temperature varies in the direction of flow. For constant thermophysical properties, show that the Nusselt number in the region of fully developed (hydrodynamical and thermal) flow is given by: (c)

$$Nu = \frac{hd}{k} = \frac{6}{\phi(n)}\left(\frac{n+1}{3n+1}\right)$$

where

$$\phi(n) = \left(\frac{21}{20} - \frac{5n}{3n+1} + \frac{9n}{5n+1} - \frac{4n}{6n+1}\right)$$

Assume that the temperature difference between the fluid and the tube wall is given by a third-degree polynomial:

$$\theta = T - T_w = a_0 y + b_0 y^2 + c_0 y^3$$

where y is the radial distance from the pipe wall and a_0, b_0, c_0 are constant coefficients to be evaluated by applying suitable boundary conditions.

6.7 A power-law fluid ($\rho = 1040\,kg/m^3$; $C_p = 2090\,J/kgK$; $k = 1.21\,W/mK$) is being heated in a 0.0254-m diameter, 1.52-m long heated tube at the rate of 0.075 kg/ s. The tube wall temperature is maintained at 93.3°C by condensing steam on the outside. Estimate the fluid outlet temperature for the feed temperature of 37.8°C. While the power-law index is approximately constant at 0.35 in the   (b) temperature interval $37.8 \leq T \leq 93.3°C$, the consistency coefficient, m, of the fluid varies as

$$m = 1.275 \times 10^4 \exp\left[-0.01452(T + 273)\right]$$

where m is in Pa s^n.

Is free convection significant in this example?

7.1 For the laminar boundary layer flows of incompressible Newtonian fluids over a wide plate, Schlichting (*Boundary Layer Theory*, 6th edition, Mc Graw Hill, New York, 1965) showed that the following two equations for the velocity distributions give values of the shear stress and friction factor which are comparable with those obtained using equation (7.10):   (b)

(i) $\dfrac{V}{V_0} = 2\left(\dfrac{y}{\delta}\right) - 2\left(\dfrac{y}{\delta}\right)^3 + \left(\dfrac{y}{\delta}\right)^4$

(ii) $\dfrac{V}{V_0} = \sin\dfrac{\pi}{2}\left(\dfrac{y}{\delta}\right)$

Sketch these velocity profiles and compare them with the predictions of equation (7.10). Do velocity profiles (i) and (ii) above satisfy the required boundary conditions?

Obtain expressions for the local and mean values of the wall shear stress and friction factor (or drag coefficient) for the laminar boundary layer flow of an incompressible power-law fluid over a flat plate? Compare these results with the predictions presented in Table 7.1 for different values of the power-law index.

7.2 A polymer solution (density 1000 kg/m3) is flowing on both sides of a plate 250-mm wide and 500-mm long; the free stream velocity is 1.75 m/s. Over the narrow range of shear rates encountered, the rheology of the polymer solution can be adequately approximated by both the power-law ($m = 0.33$ Pa s^n and $n = 0.6$) and the Bingham plastic model (yield stress = 1.75 Pa and plastic viscosity = 10 m Pa s). Using each of these models, evaluate and compare the values of the shear stress and the boundary layer thickness 100 mm downstream from the leading edge, and the total frictional force exerted on the two sides of the plate.   (b)

At what distance from the leading edge will the boundary layer thickness be half of the value calculated above?

7.3 A dilute polymer solution at 293 K flows over a heated plane surface (250-mm wide × 500-mm long) maintained at 301 K. The thermo-physical properties (density, heat capacity and thermal conductivity) of the polymer solution are close to those of water at the same temperature. The rheological behaviour of this solution can be approximated by a power-law model with $n = 0.43$ and $m = 0.3 - 0.00033\,T$, where m is in Pa s^n and T is in K.   (b)

Evaluate:

(i) the momentum and thermal boundary layer thicknesses at distances of 50, 100 and 200 mm from the leading edge, when the free stream velocity is 1.6 m/s.

(ii) the rate of heat transfer from one side of the plate.

(iii) the frictional drag experienced by the plate.

(iv) the fluid velocity required to increase the rate of heat transfer by 25% while maintaining the fluid and the surface temperatures at the same values.

7.4 A polymer solution at 298 K flows at 1.1 m/s over a hollow copper sphere of 25-mm diameter, maintained at a constant temperature of 318 K (by steam condensing inside the sphere). Estimate the rate of heat loss from the sphere. The thermo-physical properties of the polymer solution are approximately those of water, the power-law constants in the temperature interval $298 \leq T \leq 328$ K are given below: flow behaviour index, $n = 0.40$ and consistency, $m = 30 - 0.05\,T$ (Pa s^n) where T is in K. (b)

What would be the rate of heat loss from a cylinder 25 mm in diameter and 100-mm long oriented with its axis normal to flow?

Also, estimate the rate of heat loss by free convection from the sphere and the cylinder to a stagnant polymer solution under otherwise identical conditions.

7.5 A 250-mm square plate heated to a uniform temperature of 323 K is immersed vertically in a quiescent slurry of 20% (volume) TiO_2 in water at 293 K. Estimate the rate of heat loss by free convection from the plate when the rheology of the slurry can be adequately described by the power-law model: $n = 0.28$, $m = 10 - 0.2(T - 20)$ (Pa s^n) where T is in °C. (b)

The physical properties of TiO_2 powder are density $= 4260$ kg/m^3, heat capacity $= 943$ J/kg K and thermal conductivity $= 5.54$ W/mK. The coefficient of thermal expansion of the slurry, $\beta = 2 \times 10^{-4}$ K^{-1}.

7.6 A polymer solution (density 1022 kg/m^3) flows over the surface of a flat plate at a free stream velocity of 2.25 m/s. Estimate the laminar boundary layer thickness and surface shear stress at a point 300 mm downstream from the leading edge of the plate. Determine the total drag force on the plate from the leading edge to this point. What is the effect of doubling the free stream velocity? (a)

The rheological behaviour of the polymer solution is well approximated by the power-law fluid model with $n = 0.5$ and $m = 1.6$ Pa s^n.

7.7 A china clay suspension (density 1200 kg/m^3, $n = 0.42$, $m = 2.3$ Pa s^n) flows over a plane surface at a mean velocity of 2.75 m/s. The plate is 600-mm wide normal to the direction of flow. What is the mass flow rate within the boundary layer at a distance of 1 m from the leading edge of the plate? (a)

Also, calculate the frictional drag on the plate up to 1 m from the leading edge. What is the limiting distance from its leading edge at which the flow will be laminar within the boundary layer?

8.1 A standard Rushton turbine at an operating speed of 1 Hz is used to agitate a power-law liquid in a cylindrical mixing vessel. Using a similar agitator rotating at 0.25 Hz in a geometrically similar vessel, with all the linear dimensions larger by a factor of 2, what will be the ratio of the power inputs per unit volume of fluid in the two cases? Assume the mixing to occur in the laminar region. What will be the ratio of total power inputs in two cases? Plot these ratios as a function of the flow behaviour index, n. (a)

8.2 How would the results in Problem 8.1 differ if the main flow in the vessel were fully turbulent? Does the flow behaviour index have an influence on these values? (a)

8.3 A fermentation broth (density 890 kg/m^3) behaves as a power-law fluid with $n = 0.35$ and $m = 7.8$ Pa s^n. It must be stirred in a cylindrical vessel by an agitator 150 mm in diameter of geometrical arrangement corresponding to configuration A-A in Table 8.2. The rotational speed of the impeller is to be in the range 0.5–5 Hz. Plot both the power input per unit volume of liquid and the total power input as functions of the rotational speed in the range of interest. (a)

Due to fluctuations in the process conditions and the composition of feed, the flow behaviour index remains constant, at ~0.35, but the consistency index varies by ±25%. What will be the corresponding variations in the power input?

8.4 The initial cost of a mixer including its impeller, gear box and motor is closely related to the torque, T, requirement rather than its power. Deduce the relationship between torque and size of the impeller for the same mixing time as a function of geometrical scale, for turbulent conditions in the vessel. Does your answer depend upon whether the fluid is Newtonian or inelastic shear-thinning in behaviour? What will be the ratio of torques for a scale up factor of 2? (b)

8.5 A viscous material (power-law rheology) is to be processed in a mixing vessel under laminar conditions. A sample of the material is tested using a laboratory scale mixer. If the mixing time is to be the same in the small- and large-scale equipments, estimate the torque ratio for a scale up factor of 5 for the range $1 \geq n \geq 0.2$. (a)

Additionally, deduce the ratio of mixing times if it is desired to keep the torque/unit volume of material the same on both scales.

8.6 Tests on a small-scale tank 0.3 m in diameter (impeller diameter of 0.1 m, rotational speed 250 rpm) have shown that the blending of two miscible liquids (aqueous solutions, density and viscosity approximately the same as water at 293 K) is completed after 60 s. The process is to be scaled-up to a tank of 2.5-m diameter using the criterion of constant (impeller) tip speed. (b)

(i)   What should be the rotational speed of the larger impeller?
(ii) What power would be required on the large scale?
(iii) What will be the mixing time in the large tank?

For this impeller–tank geometry, the Power number levels off to a value of 6 for $Re > 5 \times 10^4$.

8.7 An agitated tank, 3 m in diameter, is filled to a depth of 3 m with an aqueous solution whose physical properties correspond to those of water at 293 K. A 1-m diameter impeller is used to disperse gas and, for fully turbulent conditions, a power input of 800 W/m³ is required. (c)

(i)   What power will be required by the impeller?
(ii) What should be the rotational speed of the impeller?
(iii) If a 1/10 size small pilot scale tank is to be constructed to test the process, what should be the impeller speed to maintain the same level of power consumption, i.e., 800 W/m³?

Assume that at the low gas used, power number–Reynolds number relationship will not be affected and that under fully turbulent conditions, the power number is equal to 6.

8.8 A power-law fluid is to be warmed from 293 to 301 K in a vessel (2 m in diameter) fitted with an anchor agitator of 1.9 m diameter rotating at 60 rpm. The tank, which is filled to a 1.75-m depth, is fitted with a helical copper coil of diameter 1.3 m (25 mm od and 22 mm id copper tubing) giving a total heat transfer area of 3.8 m². The heating medium, water flowing at a rate of 40 kg/min, enters at 323 K and leaves at 313 K. The thermal conductivity, heat capacity and density of the fluid can be taken as the same as for water. The values of the power-law constants are $n = 0.45$ and $m = 25 - 0.05\,T$ (Pa s^n) in the range 293 K $\leq T \leq$ 323 K. (b)

Estimate the overall heat transfer coefficient and the time needed to heat a single batch of liquid.

8.9   A 75 liter vessel (460 mm in diameter) fitted with a six-bladed Rushton turbine   (b)
($D = 150$ mm) is used to homogenize a shear-thinning fluid ($m = 40$ Pa s^n; $n = 0.25$; $\rho = 1000$ kg/m^3). Bearing in mind the desired degree of homogenization and the shear-sensitivity of the product, after numerous tests, it is established that the optimum process performance is achieved at $N = 240$ rpm. The power consumption under these conditions is 25 W. It is proposed to scale up this process to a 3.75 m^3 vessel using the same geometrical configuration. Obtain the power input to the large unit if scale up criterion is (i) equal impeller tip speed, (ii) equal power input/volume of fluid, (iii) equal torque/unit volume of fluid, and (iv) equal mixing times in two scales.

8.10   It is proposed to design an industrial scale static mixer (Sulzer SMX type)   (b)
to disperse 10% of an additive into a base chemical ($\rho = 900$ kg/m^3; $\mu = 100$ mPa s) which is being processed at the rate of 30 m^3/h. Preliminary laboratory experiments on a 50-mm diameter mixer of the same kind suggest that the optimal performance (COV = 0.05) is obtained at the Reynolds number of 100. Calculate the diameter of the industrial unit to achieve the same level of performance.

Index

White, J.L., 36
Whitelaw, J.H., 177, 386
Whitlock, M., 15
Whitmore, R.L., 258, 259
Whitney, M.J., 258, 259
Whorlow, R.W., 29, 76, 96, 97
Wichterle, K., 391, 395, 402, 403, 404
Wiest, J.M., 34
Wilkens, R.J., 417
Wilkins, B., 213
Wilkinson, W.L., 139, 158, 192, 294, 295,
      335, 338, 339, 426, 428, 429
Williams, M.C., 91, 395
Wilson, K.C., 133, 137, 150, 154
Windhab, E.J., 38, 104
Winter, H.H., 338
Wisdom, D.J., 396, 420, 421
Wissler, E.H., 177, 181, 327
Withers, P.M., 244
Wojs, K., 151
Wollersheim, D.E., 372
Woodcock, L.V., 15
Wronski, S., 303
Wu, J., 351
Wu, Y.-X., 210, 222
Wunderlich, T., 104

Xie, C., 180
Xu, J.-Y., 210, 222
Xuereb, C., 378, 382, 388, 389, 390, 392, 396,
      419, 424

Yagi, H., 423
Yamamoto, K., 392
Yan, J., 94
Yanagimoto, M., 408, 409
Yap, C.Y., 396, 408, 411, 415
Yarusso, B.J., 14, 167
Yield-pseudo plastic behaviour, 6, 13, 115
Yield stress:
    apparent, 36, 91, 95
    temperature dependence, 158
    vane method, 23
Yim, J., 302
Yokoyama, T., 408, 409
Yoo, J.Y., 338
Yoo, S.S., 339
Yoshida, F., 423
Yoshimura, A.S., 86, 95
Young–Hoon, A., 213, 215, 222
Young's modulus, 24–5, 26
Yovanovich, M.M., 363
Yu, A.B., 299

Zaki, W.N., 272
Zeng, Y., 104
Zenit, R., 286
Zero shear viscosity, 7, 43, 44
Zhang, X.-M., 12
Zhou, J.Z.Q., 12, 300
Zitny, R., 395
Zuritz, C.A., 244, 245
Zwaneveld, A., 336